Mathematics 10

ADDISON-WESLEY SECONDARY MATHEMATICS

Authors

Brendan Kelly
*Co-ordinator of Pure
 and Applied Sciences
Halton Board of Education
Burlington, Ontario*

Bob Alexander
*Assistant Co-ordinator
 of Mathematics
Toronto Board of Education
Toronto, Ontario*

Paul Atkinson
*Vice Principal
Bluevale Collegiate Institute
Waterloo, Ontario*

ADDISON-WESLEY PUBLISHERS

Don Mills, Ontario • Menlo Park, California • Reading, Massachusetts
Amsterdam • London • Manila • Paris • Sydney • Singapore • Tokyo

Layout and Illustrations: Acorn Technical Art Inc.

Special Features: Glyphics Inc., Designers and Consultants

Photo Credits

Addison-Wesley photo library, viii, ix, 71, 110, 113, 115, 118, 122, 125, 127, 144, 149, 306, 355, 372, 387
David Alexander, 302
Canada Steamship Lines Inc., 180, 225
Canadian Forces Photo, 5, 377, 381
G.L. Christie, 243
CNE, 218
CP Photo, x
De Havilland Aircraft of Canada Ltd, 15, 79, 98, 103
Don Mills Civitan Hockey League, 126
Mr. and Mrs. Geoffrey E. Freeman, 77
Gouvernement du Québec, la Direction Générale du Tourisme, 249
French Consulate, 347
Kansas City Royals, 145, 323
Keystone Press Agency, 37
V. Last, 176
Miller Services Ltd, 177, 217
NASA, vii, 187
Blair Needham, 67, 214
NFB-Photothèque, 171
Noritsu, 173
Ontario Department of Transportation and Communications, 1, 99, 208
Ontario Hydro, 142, 211
Ontario Ladies Golfing Association, 320, 327
Ontario Ministry of Agriculture and Food, 155
Ontario Ministry of Industry and Tourism, 140
Peugeot Canada Ltd, 139
Sandia Laboratories, x
Sitmar Cruises, 267
Toronto Institute of Medical Technology, ix
Toronto Star, 133, 315
Travel Alberta, 330-331, 370
Vancouver Whitecaps, 254
Volkswagen Canada Inc., 213

Copyright © 1982 Addison-Wesley (Canada) Ltd. All rights reserved. No part of this publication may be reproduced, stored in a retrieval system, or transmitted, in any form or by any means, electronic, mechanical, photocopying, recording, or otherwise, without the prior written permission of the publisher.

Written, printed, and bound in Canada

B C D E F - ASP - 87 86 85   ISBN 0-201-18602-0

Contents

1 The Real Numbers

1-1	The Natural Numbers	2
1-2	The Integers and Absolute Value	6
	Mathematics Around Us: The Bermuda Triangle	10
	The Mathematical Mind: Goldbach's Conjecture	11
1-3	The Rational Numbers	12
1-4	The Irrational Numbers	16
	Computer Power: Scientific Notation	19
1-5	The Pythagorean Theorem	20
1-6	Multiplication of Radicals	24
1-7	Addition and Subtraction of Radicals	27
1-8	Combined Operations with Radicals	30
1-9	Division of Radicals	32
	Review Exercises	35

2 Algebraic Expressions

2-1	Variables, Coefficients, Terms, and Exponents	38
2-2	Integral Exponents	41
2-3	The Exponent Laws	45
	The Mathematical Mind: A Famous Unsolved Problem	48
2-4	Operations With Monomials: Addition and Subtraction	50
2-5	Operations With Monomials: Multiplication and Division	52
2-6	Sums and Differences of Polynomials	55
2-7	Products of Monomials and Polynomials	56
2-8	Products of Binomials and Trinomials	59
2-9	Common Factors	62
	Computer Power: The Euclidean Algorithm	66
	Mathematics Around Us: Meeting the World's Water Needs	67
2-10	Factoring Trinomials of the Form $x^2 + bx + c$	68
2-11	Factoring Trinomials of the Form $ax^2 + bx + c$	72
2-12	Factoring a Difference of Squares	75
2-13	Using Factoring to Evaluate Polynomials	78
2-14	Equivalent Rational Expressions	81
2-15	Multiplying and Dividing Rational Expressions	84
2-16	Adding and Subtracting Rational Expressions With Monomial Denominators	88
2-17	Adding and Subtracting Rational Expressions	92
2-18	Applications of Rational Expressions	97
	Review Exercises	100

| | | Cumulative Review | 103 |
| | | Problem-Solving Strategy: Use a Model. | 106 |

3 Equations

	3-1	Solving Equations in One Variable	108
	3-2	Solving Equations in Two Variables	112
	3-3	Solving Pairs of Linear Equations by Comparison	116
		The Mathematical Mind: The Rhind Papyrus	119
	3-4	Solving Pairs of Linear Equations by Substitution	120
	3-5	Solving Pairs of Linear Equations by Addition or Subtraction	123
		Mathematics Around Us: A Matter of Interest	127
	3-6	Translating into the Language of Algebra	129
	3-7	Writing Equations	132
	3-8	Solving Problems Using Equations: Part I	134
	3-9	Solving Problems Using Equations: Part II	138
	3-10	Solving Quadratic Equations	143
		Problem-Solving Strategy: Consider Different Cases.	146
		Review Exercises	148
		Computer Power: A Formula for Linear Systems	150

4 Graphing Equations and Inequalities

	4-1	Graphing a Linear Equation	152
	4-2	Graphing a Pair of Linear Equations	155
		The Mathematical Mind: Diophantine Equations	158
	4-3	Graphing Linear Inequalities	160
	4-4	Maximum and Minimum Values of Linear Expressions	165
		Problem-Solving Strategy: Work Backwards to the Solution.	168
		Mathematics Around Us: The World's Tallest Building	169
	4-5	Linear Programming	170
		Review Exercises	175

5 Coordinate Geometry and Relations

	5-1	Distance in the Plane	178
	5-2	Midpoint of a Line Segment	181
	5-3	Slope of a Line Segment	183
	5-4	Slopes of Perpendicular Lines	187
	5-5	Using Slope to Graph a Linear Equation	190
		Mathematics Around Us: Postal Rates Take a Licking.	194
	5-6	Finding the Equation of a Line	196
	5-7	Direct Variation	200
	5-8	Partial Variation	204
	5-9	Inverse Variation	208
		Problem-Solving Strategy: Reduce to a Simpler Problem.	212

5 - 10	Quadratic Relations	214	
5 - 11	Other Non-Linear Relations	217	
	Review Exercises	221	
	Cumulative Review	223	

Transformations 6

6 - 1	Transformations as Mappings	226
6 - 2	Translations	229
6 - 3	Rotations	233
6 - 4	Reflections	238
6 - 5	Successive Transformations	243
6 - 6	Isometries and Congruence	249
6 - 7	Dilatations	254
	Mathematics Around Us: The Pantograph	259
6 - 8	Applications of Transformations	260
	Computer Power: Finding Minimum Distances	263
	Review Exercises	264
	The Mathematical Mind: A Genius From India	266

Geometry 7

7 - 1	What Are Deductions?	268
7 - 2	Congruent Triangles: SAS	273
7 - 3	Congruent Triangles: SSS	276
7 - 4	Congruent Triangles: ASA	279
	The Mathematical Mind: A Best Seller From Way Back	282
7 - 5	Parallel Lines	284
7 - 6	Angles in a Triangle	287
7 - 7	Congruent Triangles: AAS	292
	Problem-Solving Strategy: Use Indirect Proof.	295
7 - 8	Some Theorems and Their Converses	297
	Mathematics Around Us: How Fast Do Glaciers Move?	302
7 - 9	The Pythagorean Theorem and Its Converse	304
7 - 10	Proofs Using Transformations	309
	Review Exercises	313

Statistics and Probability 8

8 - 1	Interpreting Graphs	316
8 - 2	Measures of Central Tendency: Simple Data	320
8 - 3	Measures of Central Tendency: Grouped Data	324
8 - 4	Measures of Dispersion	327
	Mathematics Around Us: Estimating Wildlife Populations	330
8 - 5	Sampling and Predicting	332
8 - 6	Probability	335
8 - 7	The Probability of Two or More Events	339

	8-8	Monte Carlo Methods	342
	8-9	Monte Carlo Methods with Random Numbers	345
		Computer Power: Monte Carlo Methods With Computers	348
		Review Exercises	350
		Cumulative Review	351
		Mathematics Around Us: Gwennap Pit	354

9 Trigonometry

	9-1	Angle of Inclination	356
	9-2	The Tangent Ratio in Right Triangles	360
	9-3	The Sine and Cosine Ratios in Right Triangles	363
	9-4	Solving Right Triangles	367
		Mathematics Around Us: Gondola Lifts in Banff and Jasper National Parks	370
		Computer Power: Investigating Repeating Decimals	371
	9-5	Applications of the Trigonometric Ratios	372
		Review Exercises	374

10 Vectors

	10-1	Vectors as Directed Line Segments	378
	10-2	Equal Vectors	381
	10-3	Vectors as Ordered Pairs	383
		The Mathematical Mind: The Four Squares Problem	386
		Mathematics Around Us: Designing Carton Sizes	387
	10-4	Addition of Vectors: The Triangle Law	388
	10-5	Addition of Vectors: The Parallelogram Law	391
	10-6	Scalar Products of Vectors	393
	10-7	Vector Proofs in Geometry	396
		Review Exercises	399
		Table of Square Roots	402
		Table of Trigonometric Ratios	403
		Glossary	404
		Answers	408
		Index	477

Prologue

Mathematics, a product of the human mind,...

How can the solar system be explored?

In 1977, two unmanned spacecraft, Voyager I and Voyager II, were launched to travel to the outer planets. Mathematics was used in all aspects of these missions. Astronomers calculated that Jupiter, Saturn, Uranus, and Neptune would all be in favorable positions in their orbits in the 1980's. Engineers used computers to find the exact positions of the spacecraft at all times. Mathematics was also used to schedule the activities of the thousands of people who were involved in the project.

...has been used to solve significant problems in the past,...

What is the shape of the Universe?

For centuries people believed that the world was flat. Eventually it was discovered that Earth was round and that anyone travelling far enough in any direction would, in time, return to the starting point. In 1917, Albert Einstein made known his General Theory of Relativity which suggests that, like Earth, the Universe is finite. A space traveller covering a distance of 10^{11} light-years in any given direction would eventually return to the original position.

and of the present.

The only way to find the average life span of persons living today would be to wait until they all died and then average the ages at which they died. However, we can obtain a very close estimate of the average life span using a method called *sampling*. With this method, data based on a small fraction of the population is used to obtain a good estimate of the human life span.

What is the average life span of human beings?

Mathematics and mathematicians...

When you borrow money for a major purchase, you agree to repay a certain amount each month. Part of this monthly payment is interest and part goes to reduce the amount still owed. Computers are used to calculate the monthly payment and also the outstanding amount on the loan at any time.

How are loan repayments determined?

...will continue to play...

In this century, economists and applied mathematicians have attempted to create mathematical models to help in decision-making, sales predictions, and pricing strategies. New fields of mathematics, such as game theory and linear programming, have been created and applied to the problems involved. Since there are many variables, computers are used to solve many of these problems.

What prices will bring maximum profit?

...an increasingly more important role...

Genetics is the study of inherited characteristics. This important field of science relies heavily on a branch of mathematics called probability theory. Using probability theory, geneticists can predict the likelihood that a particular person will inherit a hereditary disease. They can also predict the likelihood that a contagious disease contracted by a given number of people will generate an epidemic.

What is the chance of inheriting a hereditary disease?

...in solving the complex problems...

To study the effect of acid rain on the fish in Canadian lakes, it is important to know the number of fish in a given lake. Scientists have developed methods of estimating this number, and even the number of different species of fish. They can also tell how reliable their estimates are.

How does acid rain affect the fish in a lake?

...of the future.

As a source of power, the sun has many advantages—it is readily available, non-polluting, and abundant. The reflecting surface of a solar furnace has a parabolic cross section. This permits the parallel rays from the sun to be reflected through the same point, or focus, where the temperature can reach a very high figure. If a furnace is situated at this point, sufficient heat and power can be generated to supply the needs of thousands of homes.

What is the best shape for the reflecting surface of a solar furnace?

1 The Real Numbers

Two speedboats make a test run round a closed circuit on Lake Winnipeg starting at the same place and time. Boat *A* makes a lap every 80 s, boat *B* makes a lap every 90 s. When will the two speedboats again be level with one another? (See *Example 4* in Section 1-1.)

1-1 The Natural Numbers

Mathematics began with the set of **natural numbers** $\{1, 2, 3, \ldots\}$. Though these have been studied for thousands of years, some of their properties are still not fully understood. For instance, there are still some unresolved problems involving prime numbers.

A **prime number** is a number that has only two factors, itself and 1. Except for the number 1, numbers that are not prime are called **composite numbers**. Composite numbers can be expressed as a product of primes.

The number 1, having only one factor, is neither prime nor composite.

Example 1. Express each of these numbers as a product of primes.
a) 65   b) 120   c) 113

Solution.
a) $65 = 5 \times 13$
Since 5 and 13 are primes, 65 cannot be factored further. 5×13 is called the **prime factorization** of 65.

b) $120 = 12 \times 10$
$= 4 \times 3 \times 2 \times 5$
$= 2^3 \times 3 \times 5$

c) If 113 can be expressed as a product of prime factors, at least one of the factors must be less than 11. Since, 2, 3, 5, and 7 are the only primes less than 11 and none of them is a factor of 113, then 113 is a prime.

Two factors greater than or equal to 11 would yield a product greater than 113.

In factoring 120 (*Example 1b*), the first step could have been written in other ways, such as 3×40, or 6×20. No matter how 120 is factored into primes, three factors of 2, one factor of 3, and one factor of 5 are always obtained. This is the prime factorization property of composite numbers, which may be stated:

> Every composite number has one, and only one, prime factorization.

Example 1c suggests that determining whether or not a number is prime requires only that the prime factors equal to or less than its square root be checked. Why?

Two different composite numbers may have a common factor. For example, 12 is a common factor of 24 and 60 because
$$24 = 12 \times 2 \quad \text{and} \quad 60 = 12 \times 5.$$
The **greatest common factor** (g.c.f.) of two numbers is the greatest factor common to each.

Example 2. Find the g.c.f. of 48 and 60.

Solution. The set of factors of 48 is {1, 2, 3, 4, 6, 8, 12, 16, 24, 48}.
The set of factors of 60 is {1, 2, 3, 4, 5, 6, 10, 12, 15, 20, 30, 60}.
The numbers in color are the common factors of 48 and 60. The g.c.f. of 48 and 60 is 12.

If a number a is a factor of a number b, then b is a multiple of a. For example:
The set of multiples of 2 is {2, 4, 6, 8, 10, 12, ...}.
The set of multiples of 3 is {3, 6, 9, 12, 15, 18, ...}.
Note that 6 and 12 are multiples of 2 and multiples of 3. That is, 6 and 12 are common multiples of 2 and 3; 6 is the *least* common multiple. The **least common multiple** (l.c.m.) of two numbers is the least number that is a multiple of each.

Example 3. Find the l.c.m. of 48 and 60.

Solution. The set of multiples of 48 is {48, 96, 144, 192, 240, 288, ...}.
The set of multiples of 60 is {60, 120, 180, 240, 300, 360, ...}.
The set of common multiples of 48 and 60 is {240, 480, 720, ...}.
The l.c.m. is 240.

Example 4. Two speedboats make test runs round a closed circuit on Lake Winnipeg starting at the same place and time. Boat A makes a lap every 80 s, boat B makes a lap every 90 s. When will the two speedboats again be level with one another?

Solution. Boat A passes the starting point every 80 s.
Boat B passes the starting point every 90 s.
They will again be level at intervals that are common multiples of 80 and 90. The shortest interval is the l.c.m. of 80 s and 90 s, that is, 720 s. They will next be level with one another 12 min after the start.

Exercises 1-1

A 1. Which of the following are prime numbers?
   a) 34   b) 19   c) 71   d) 85   e) 97
   f) 109 g) 129 h) 152 i) 177 j) 193

2. Express as a product of prime factors:
   a) 18 b) 28 c) 42 d) 56 e) 64
   f) 76 g) 91 h) 110 i) 143 j) 187

3. Find all the factors of the following numbers:
   a) 12 b) 16 c) 18 d) 25 e) 32
   f) 36 g) 60 h) 100 i) 144 j) 200

4. Find the g.c.f. of each of these pairs of numbers:
   a) 24, 36 b) 45, 75 c) 42, 70 d) 35, 60
   e) 36, 60 f) 52, 65 g) 96, 144 h) 64, 112
   i) 56, 104 j) 54, 90 k) 105, 175 l) 126, 294

5. Find the l.c.m. of each of these pairs of numbers:
   a) 24, 36 b) 16, 24 c) 20, 35 d) 32, 40
   e) 18, 30 f) 45, 75 g) 60, 84 h) 72, 120

B 6. Are all the factors of a composite number less than its square root? Illustrate your answer with three examples.

7. Which natural numbers have an odd number of factors? Why?

8. Three primes have a sum of 146. The greatest is equal to the sum of the other two. What are the primes?

9. Write each of the following as the sum of consecutive natural numbers in as many ways as possible:
   a) 15 b) 30 c) 105

10. Some primes can be written in the form $n^2 + 1$, where n is a natural number. Example: $17 = 4^2 + 1$.
    a) Find three more primes of the form $n^2 + 1$.
    b) If $n^2 + 1$ is a prime greater than 2, explain why n must be even.

11. Some primes can be written in the form $2^n - 1$, where n is a natural number. Example: $7 = 2^3 - 1$.
    a) Find three more primes of the form $2^n - 1$.
    b) If $2^n - 1$ is a prime greater than 3, explain why n must be odd.

12. Find three primes that have a product of 646.

13. The lights on one airport tower flash every 16 s. The lights on another tower flash every 20 s. What is the time interval between simultaneous flashes?

1-1 The Natural Numbers

14. A textbook, having 14 chapters of equal length, is made by sewing together several 32-page booklets. What is the least number of pages that the book may have?

15. When a band marches in rows of 2, 3, 4, 5, or 6, there is always one member left over.
    a) What is the least number of members in the band?
    b) What is the least number of members in the band if there are no members left over when marching in rows of 7?

16. Two gears, A and B, are assembled with the timing arrows in contact. How many rotations does each gear make between successive contacts of the arrows if the numbers of teeth on the gears are
    a) $A:18, B:12$? b) $A:25, B:20$?
    c) $A:28, B:21$? d) $A:25, B:16$?
    e) $A:32, B:28$? f) $A:35, B:28$?

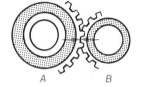

17. Two stock cars race on a 5 km circuit and pass a marker simultaneously. How long will it take the faster car to gain one lap on the slower if their average speeds are
    a) 120 km/h and 100 km/h? b) 150 km/h and 125 km/h?

18. If a certain positive number is doubled, the result is a perfect square. If it is tripled, the result is a perfect cube. Find the number.

19. The number 64 is both a perfect square (since $64 = 8^2$) and a perfect cube (since $64 = 4^3$). Find other integers which are both perfect squares and perfect cubes.

C 20. Copy and complete the table:

a, b	g.c.f.	l.c.m.	g.c.f. × l.c.m.	ab
9, 12	3	36	108	108
18, 45				
20, 35				
40, 48				

What is the relation between the g.c.f., the l.c.m., and the product of a and b?

21. a) Find the g.c.f. for each pair of numbers:
       i) 50, 51   ii) 189, 190   iii) 278, 279
    b) What conclusion do you draw from your answer to (a)?

1-2 The Integers and Absolute Value

If numbers are used to represent opposite quantities, such as gains and losses, or temperatures above and below freezing, it is convenient to extend the set of natural numbers to include zero and negative numbers. The result is the set of integers, I.

$$I = \{\ldots, -3, -2, -1, 0, 1, 2, 3, \ldots\}$$

Just as there is an arithmetic for natural numbers, there is also an arithmetic for integers.

Example 1. Simplify: a) $(+17) + (-10) - (-3) - (+9)$;
b) $(-5)[(+3)(-2) - (-8)(-1)]$.

Solution. a) $(+17) + (-10) - (-3) - (+9)$
$= (+17) + (-10) + (+3) + (-9)$
$= 17 - 10 + 3 - 9$
$= 1$

b) $(-5)[(+3)(-2) - (-8)(-1)]$
$= (-5)[(-6) - (+8)]$
$= (-5)[-14]$
$= 70$

Example 2. If $x = -3$, $y = -2$, and $z = 5$, find the value of the expression $5x^2y - 3xz^2$.

Solution. $5x^2y - 3xz^2 = 5(-3)^2(-2) - 3(-3)(5)^2$
$= 5(9)(-2) + 9(25)$
$= -90 + 225$
$= 135$

On the number line, the integers 3 and -3 are each located 3 units from 0. Each is said to have absolute value 3.

$$|3| = 3 \quad \text{and} \quad |-3| = 3$$

Read: "absolute value of -3".

In general, the **absolute value** of any number is its distance from 0 on the number line, and it is always positive. The absolute value of zero is zero.

Example 3. Simplify: a) $|-17|$; b) $|5| - |-12|$;
c) $3[5 - |-8| + 2|4 - 10| - 15]$.

Solution. a) $|-17| = 17$ b) $|5| - |-12| = 5 - 12$
$= -7$

c) $\quad 3[5 - |-8| + 2|4 - 10| - 15]$
$= 3[5 - 8 + 2(6) - 15]$
$= 3[-6]$
$= -18$

Example 4. The area, A, of a triangle with vertices $(0, 0)$, (a, b), and (c, d) is given by the formula: $A = \frac{1}{2}|ad - bc|$.

Find the area of a triangle with vertices $(0, 0)$, $(-6, 4)$, $(-2, 6)$.

Solution. Substitute $(-6, 4)$ for (a, b) and $(-2, 6)$ for (c, d) in the above formula:

$A = \frac{1}{2}|(-6)(6) - (4)(-2)|$

$= \frac{1}{2}|-36 + 8|$

$= \frac{1}{2}|-28|$, or 14

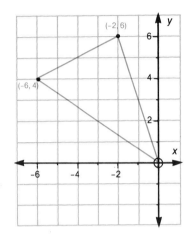

The area of the triangle is 14 units^2.

Example 5. An oil company plans to put a service station on a triangular lot formed by three roads. The lot is drawn to scale on a grid. If each square on the grid represents an area of 100 m^2, what is the area of the lot?

Solution. Using the formula in *Example 4*,

area of $\triangle OAB = \frac{1}{2}|(-2)(-1) - (6)(6)|$

$= \frac{1}{2}|-34|$

$= 17$ grid squares

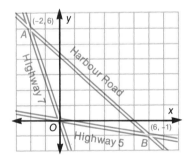

Since each grid square represents 100 m^2, the area of the lot is 1700 m^2.

Exercises 1-2

A 1. Simplify:
   a) $8 - 13$
   b) $19 - (-26)$
   c) $-9 - 17$
   d) $15 - 8 - (-2)$
   e) $-6 + (-4) - (-19)$
   f) $14 - (-9) + 5$

2. Simplify:
   a) $-23 - 8 - (-11)$
   b) $(-5)(4)(-2)$
   c) $24 \div (-3) \div 2$
   d) $(-15)(-3) \div (-5)$
   e) $(-2)(-3)(-4)(-5)$
   f) $-18 - 9 - (-7)$

3. Simplify:
   a) $16 - 2(-5)$
   b) $3 + 7(-5)$
   c) $-23 - 2(-4)$
   d) $(-9)(-12) - 24(3)$
   e) $-27 + 6(-9) + 5$
   f) $(-7)(3) + (-10) - 12$

4. Simplify:
   a) $12 \div (-3) + 6(-5)$
   b) $-9 + 6(2) - 7$
   c) $-25 \div 5(-2) - 6$
   d) $54 - 6(-3) + 15$
   e) $24 - 3(5) - 4$
   f) $45 \div 3 + 3 - 11 \times 2$

5. Simplify:
   a) $-3[5(-2) - 6 + 9]$
   b) $6 + 4[12 \div 2 - (-3)(-1)]$
   c) $2[4 + 18 \div (-3) - 2 \times 5]$
   d) $[3 - (4 \div 2 \times 10)][2 - 7]$

6. Simplify:
   a) $2[(-4)(9) - (-12)] \div (7 - 10)$
   b) $[-5 - 12 \times 3] - 4(-2)$
   c) $5[-12 \div 3 + (4 - 8) \div (6 - 8)]$
   d) $8[3 + 5 - 2(6 - 4) - 1]$
   e) $-7[2(1 - 4) - 5(3 - 7) - 11]$

7. Simplify:
   a) $|-29|$
   b) $-|-11|$
   c) $|107|$
   d) $|14| - |-9|$
   e) $|-6| - |-21|$
   f) $|-72| + |-17|$

8. Simplify:
   a) $2|-8| - |-19|$
   b) $|-32| - 4|-8|$
   c) $-|-6| + |-14|$
   d) $9 + |-6| - |-2|$
   e) $3|-7| - 7|-11|$
   f) $-5|-2| + 2|5|$

B 9. Simplify:
   a) $|-20 - (-37)|$
   b) $-|-15 - 9|$
   c) $|4 - 9 - 16|$
   d) $2|-19 - (-11)|$
   e) $3|4 - 10| + 5|2 - 6|$
   f) $5|6 - 2| - 7|-5|$

10. Simplify:
    a) $|12 - 19| - |19 - 12|$
    b) $-4|7 - 13| - 2|9 - 14|$
    c) $11|2 - 9| - 4|-3|$
    d) $3|-5| + |2(-7) - 4|$
    e) $2[5 - |-9| - 3|2 - 4|]$
    f) $4[-|6 - 15| - 2|4 - 9|]$

11. If $x = -2$, $y = 3$, and $z = -1$, evaluate:
    a) $3x^2yz$
    b) $5x - 2y + 7z$
    c) $2y(-4x) - 3y(-2z)$
    d) $4xyz - 3xy + 7yz$
    e) $xy^2 - xz^2 - xyz$
    f) $7xyz^2 + 4y^2 - xz^2$
    g) $3y^2z - 2x^2y - xz$
    h) $8xyz - 5x^2y^2 + 7xz^3$
    i) $\dfrac{5x + 2y}{xz} + \dfrac{xy}{z}$
    j) $\dfrac{6y - 5(y - z)}{2x + y}$

12. Find the areas of the triangles with one vertex at the origin and the other vertices having coordinates
    a) (10, 5) and (2, 8);
    b) (−6, −3) and (2, 5);
    c) (−12, −4) and (10, −3);
    d) (−3, 9) and (−10, −3).

13. Find the areas of the parallelograms with vertices
    a) (0, 0), (7, 2), (10, 7), (3, 5);
    b) (0, 0), (−3, 4), (5, 10), (8, 6).

14. Mr. Drake plans to sell a triangular section of his farm for a housing-development project. On a map, the grid references for the three vertices of the section he is selling are (0, 0), (8, 3), and (2, 12), where each number stands for a distance in hectometres (hm). What is the area of the section in hectares (ha)?

    1 hm = 100 m
    1 ha = 1 hm²

15. Mrs. Drake's farm is in the shape of a parallelogram and is enclosed by a wire fence. On a map, the grid references for the four corner posts of the farm are (0, 0), (2, 3), (8, 1), and (6, −2), where each number stands for a distance in kilometres. What is the area of the farm?

C 16. Find the areas of the triangles with vertices
    a) (6, 0), (0, 4), (4, 6);
    b) (6, −3), (3, 5), (−5, −1).

17. Find all the values of x which satisfy each equation:
    a) $|x + 1| = 7$
    b) $|2x - 1| = 9$
    c) $|5x - 7| = -3$

Mathematics Around Us

The Bermuda Triangle

Over the years, many ships and airplanes have inexplicably disappeared in a region of the Atlantic Ocean. This region has come to be known as the *Bermuda Triangle*. Several books have been written and much research undertaken about this region and the disappearances. The books and the research have done little to solve the mystery of the Bermuda Triangle. To date, no traces of the vanished ships and airplanes have been found.

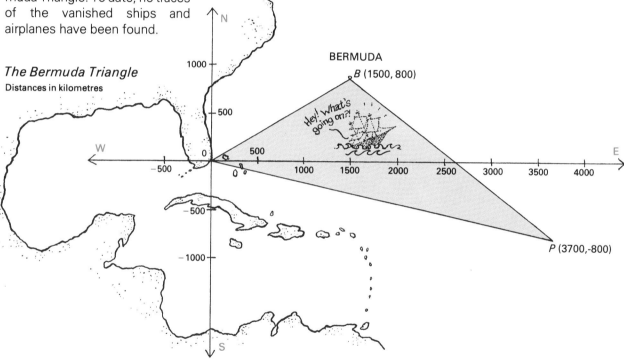

The Bermuda Triangle
Distances in kilometres

One side of the Bermuda Triangle is a line segment from Bermuda to the southeast coast of Florida. The other sides extend in a southeasterly direction from Bermuda and Florida to a point in mid-Atlantic. The map shows the approximate location of the Triangle.

Question

- What is the approximate area of the Bermuda Triangle?

THE MATHEMATICAL MIND

Goldbach's Conjecture

1
In 1742, the German mathematician C. Goldbach suggested that every even number greater than 2 is the sum of two primes.

```
4 = 2 + 2
6 = 3 + 3
8 = 5 + 3
10 = 7 + 3
12 = 7 + 5
14 = 11 + 3
```

2
No one has been able to prove that this is true for all even numbers...

```
76 = 59 + 17
78 = 71 + 7
80 = 37 + 43
82 = 71 + 11
```

3
...yet no even number has been found which is not the sum of two primes.

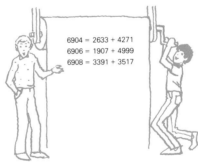

```
6904 = 2633 + 4271
6906 = 1907 + 4999
6908 = 3391 + 3517
```

4
As you can see, the conjecture is true for all even numbers (except 2) up to 100 000.

```
99 998 = 99 991 + 7
100 000 = 39 113 + 60 887
```

1. Express as the sum of two primes:
   a) 16   b) 24
   c) 64   d) 72
   e) 88   f) 100
   g) 144   h) 200

2. Is every odd number greater than 3 the sum of two primes? Explain.

3. a) Is it possible for a prime to be the sum of two primes?
   b) Is every prime the sum of two primes?

4. How many even numbers less than 50 can be expressed as the sum of two primes in:
   a) 2 ways?   b) 3 ways?
   c) 4 ways?   d) 5 ways?

1-3 The Rational Numbers

The sum, difference, and product of two integers is an integer. But the quotient of two integers is not necessarily an integer. The set of all quotients, $\frac{m}{n}$, where m and n are integers and $n \neq 0$, is the **set of rational numbers**, Q.

$$Q = \{\tfrac{m}{n} \mid m, n \text{ are integers and } n \neq 0\}$$

The rules of arithmetic with integers apply also to rational numbers.

Example 1. Simplify: $\dfrac{2[6 - 3(-4)]}{(+4)(-3)} - \dfrac{(-3)}{4}$

Solution.
$$\dfrac{2[6 - 3(-4)]}{(+4)(-3)} - \dfrac{(-3)}{4} = \dfrac{2(6+12)}{-12} + \dfrac{3}{4}$$
$$= -\dfrac{36}{12} + \dfrac{9}{12}$$
$$= -\dfrac{27}{12}, \text{ or } -\dfrac{9}{4}$$

Example 2. If $x = \dfrac{-2}{3}$ and $y = \dfrac{2}{-5}$, find the value of the expression $\dfrac{3x^2 - xy}{2y}$.

Solution.
$$\dfrac{3x^2 - xy}{2y} = \dfrac{3(\tfrac{-2}{3})^2 - (\tfrac{-2}{3})(\tfrac{2}{-5})}{2(\tfrac{2}{-5})}$$
$$= \dfrac{3(\tfrac{4}{9}) - \tfrac{4}{15}}{-\tfrac{4}{5}}$$
$$= (\tfrac{4}{3} - \tfrac{4}{15}) \times (-\tfrac{5}{4})$$
$$= (\tfrac{20 - 4}{15}) \times (-\tfrac{5}{4})$$
$$= \tfrac{16}{15} \times (-\tfrac{5}{4}), \text{ or } -\tfrac{4}{3}$$

Rational numbers may be expressed in various forms. The answer to *Example 1* may be written: $-\dfrac{9}{4}$, or $-2\dfrac{1}{4}$, or -2.25. And the

1-3 The Rational Numbers   13

answer to *Example 2* may be written: $-\frac{4}{3}$, or $-1\frac{1}{3}$, or $-1.\overline{3}$.

Any rational number in the form $\frac{m}{n}$ can be expressed in decimal form by dividing the numerator by the denominator. The decimal will either terminate or repeat. Conversely, any terminating or repeating decimal can be expressed in the form $\frac{m}{n}$.

Example 3. Express these numbers in the form $\frac{m}{n}$, where $n \neq 0$:

a) 5.23   b) $0.\overline{4}$ c) $3.5\overline{37}$

Solution. a) 5.23 means $5\frac{23}{100}$, or $\frac{523}{100}$.

b) $0.\overline{4}$ may be written 0.4444....
Let $x = 0.444...$ (i)
$10x = 4.444...$ (ii)
Subtract (i) from (ii).
$9x = 4$
$x = \frac{4}{9}$
Therefore, $0.\overline{4} = \frac{4}{9}$

To eliminate the repeating decimal, multiply by an appropriate power of 10 and subtract.

c) $3.5\overline{37}$ may be written 3.537 373 7....
Let $x = 3.537\ 373\ 7...$ (i)
$10x = 35.373\ 737.....$ (ii)
$1000x = 3537.373\ 737....$ (iii)
Subtract (ii) from (iii).
$990x = 3502$
$x = \frac{3502}{990}$
Therefore, $3.5\overline{37} = \frac{3502}{990}$

The solution to *Example 3c* was not reduced to lowest terms since it expresses $3.5\overline{37}$ in the form $\frac{m}{n}$, which is what was required.

Exercises 1-3

A 1. Simplify:

a) $-\frac{2}{3} \times \frac{7}{8}$ b) $\frac{5}{-8} \times \frac{-4}{15}$ c) $\frac{3}{5} \times \frac{-6}{-11}$

d) $\frac{11}{6} \times (-4)$ e) $-\frac{25}{4} \times \frac{5}{2}$ f) $\frac{-15}{8} \times \left(\frac{-13}{5}\right)$

g) $\frac{3}{7} \div \left(\frac{-9}{14}\right)$ h) $-\frac{13}{4} \div \frac{2}{-3}$ i) $-\frac{11}{2} \div \frac{7}{3}$

2. Simplify:
   a) $\frac{-5}{8} - \frac{3}{8}$
   b) $\frac{13}{-24} + \frac{-7}{24}$
   c) $\frac{-2}{3} + \frac{-4}{9}$
   d) $\frac{-7}{8} - \frac{-1}{4}$
   e) $\frac{-4}{9} + \frac{17}{-21}$
   f) $\frac{-3}{-4} - \frac{-2}{3}$

3. Simplify:
   a) $\frac{137}{24} - (-\frac{19}{8})$
   b) $-\frac{32}{15} + \frac{19}{6}$
   c) $-\frac{110}{9} - \frac{29}{6}$
   d) $\frac{14}{3} + (-\frac{31}{4})$
   e) $-\frac{14}{3} - (\frac{31}{4})$
   f) $\frac{17}{6} - \frac{22}{8}$

4. If $x = \frac{-3}{4}$ and $y = \frac{1}{-3}$, evaluate:
   a) $2xy - 6y^2$
   b) $3x - 2y + 5xy$
   c) $4x^2 - 3y^2$
   d) $6x^2y + 2y^2$
   e) $3x^2 - \frac{1}{4}xy$
   f) $24xy^2 - 18xy$

5. Express in decimal form:
   a) $\frac{-5}{8}$
   b) $\frac{4}{-9}$
   c) $3\frac{1}{7}$
   d) $-\frac{47}{12}$
   e) $5\frac{2}{5}$
   f) $\frac{-17}{11}$
   g) $\frac{-7}{13}$
   h) $-2\frac{5}{6}$

B 6. Express as a common fraction:
   a) 1.37
   b) $0.\overline{45}$
   c) $-6.\overline{7}$
   d) 2.875
   e) $-0.\overline{517}$
   f) -12.0125
   g) $4.1\overline{32}$
   h) $4.0\overline{12}$
   i) $1.\overline{1}$
   j) $-3.1\overline{425}$
   k) $0.5\overline{18}$
   l) $-0.312\overline{46}$
   m) $-0.\overline{9}$
   n) $0.4\overline{9}$
   o) $2.\overline{571\,428}$

7. Simplify:
   a) $\frac{5}{2} - \frac{11}{3} + \frac{5}{4}$
   b) $\frac{5}{2} - \frac{5}{4} \div \frac{4}{5}$
   c) $\frac{-6}{5} + \frac{10}{-2} \times (\frac{-3}{5})$
   d) $(\frac{3}{-4} - \frac{-3}{4}) \div 2$
   e) $-6(\frac{4}{5} - \frac{1}{2})$
   f) $\frac{3}{5}(-\frac{1}{2})(-\frac{6}{3}) + \frac{1}{5}$
   g) $\frac{3}{4} \times (\frac{1}{-2}) + \frac{5}{6} \times (\frac{-1}{3})$
   h) $\frac{3}{8} \times \frac{2}{3} - \frac{1}{2}(\frac{-5}{6}) + \frac{3}{5}(-\frac{3}{4})$
   i) $[\frac{5}{2} \div (\frac{-4}{5})] - [\frac{3}{-4} \times (\frac{-8}{9})]$

8. If $x = \frac{-3}{4}$ and $y = \frac{1}{-3}$, evaluate:

   a) $\frac{2x^2 + 4y}{15y^2}$
   b) $\frac{12xy + 4y}{5y}$
   c) $\frac{2x + 7y - xy}{7x^2}$

9. The cost, C, in dollars per hour, of operating a certain type of aircraft is given by the formula:
$$C = 950 + \frac{m}{250} + \frac{22\,500\,000}{m}$$
where m is the cruise altitude in metres. Find the hourly cost of operating the aircraft at 7500 m; at 9000 m.

10. The power, P, in kilowatts, delivered by a high voltage power line is given by the formula: $P = I(132 - \frac{1}{10}I)$, where I is the current in amperes. What power is available when the current is 440 A (amperes)?

11. The interest rate, r, on a loan is given by the formula
$$r = \frac{24c}{A(n-1)} \times 100\%$$
where c is the interest charged, A is the amount borrowed, and n is the number of monthly payments. Find the interest rate for
   a) $c = \$125$, $A = \$1000$, $n = 16$;
   b) $c = \$60$, $A = \$900$, $n = 11$.

12. If o and i are the distances of an object and its image respectively from a concave mirror of focal length f, the formula relating them is $\frac{1}{o} + \frac{1}{i} = \frac{1}{f}$.

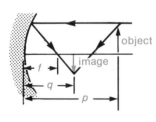
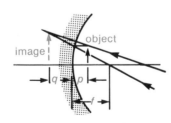

For a concave mirror with a focal length of 20 cm, find
   a) i when o is: i) 40 cm,   ii) 80 cm,   iii) 100 cm;
   b) o when i is: i) 60 cm,   ii) 80 cm,   iii) 20 cm;
   c) What is the meaning of your answer to b(iii)?

13. Express as a repeating decimal:
   a) $0.\overline{3} + 0.\overline{2}$
   b) $0.\overline{3} + 0.\overline{6}$
   c) $0.\overline{42} - 0.\overline{20}$
   d) $0.\overline{3} \times 0.25$
   e) $0.\overline{3} \times 0.\overline{6}$
   f) $0.\overline{6} \times 0.\overline{2}$

1-4 The Irrational Numbers

Recall from the previous section that the decimal representation of any rational number either terminates or repeats. Conversely, any terminating or repeating decimal can be expressed in the form $\frac{m}{n}$, where $m, n \in I$, and $n \neq 0$. It follows that any number, such as π, that is a non-repeating, non-terminating decimal is not rational. Such numbers are said to be **irrational**.

(*m* and *n* belong to the set of integers.)

> An irrational number is one that cannot be expressed in the form $\frac{m}{n}$, where $m, n \in I$, and $n \neq 0$. The decimal representation of an irrational number neither terminates nor repeats.

Example 1. Which of the following numbers appears to be irrational?
   a) $x = 0.123\,456\,789\,101\,112\ldots$
   b) $y = 3.131\,131\,131\,131\ldots$
   c) $\sqrt{2} = 1.414\,213\,562\,373\ldots$

Solution.
   a) Although there is a pattern in the decimal representation of x, there is no sequence of digits that repeats. Therefore, x appears to be irrational.
   b) y appears to be rational since the sequence 131 repeats.
   c) There is no repeating sequence of digits. Therefore, $\sqrt{2}$ appears to be irrational.

In *Example* 1c, it is not certain that $\sqrt{2}$ is irrational since a sequence of digits that repeats could occur farther out in the decimal expansion. However, it has been proved that $\sqrt{2}$ is irrational. In fact, **any number of the form $\sqrt{x}$, where $x > 0$ and is not the square of a rational number, is irrational**.

Example 2. Which of the following are irrationals?
$$\sqrt{3} \quad \sqrt{16} \quad \sqrt{20} \quad \sqrt{1.44} \quad \sqrt{\tfrac{4}{9}} \quad \sqrt{\tfrac{4}{5}}$$

Solution. $\sqrt{3}, \sqrt{20},$ and $\sqrt{\tfrac{4}{5}}$ are irrational since 3, 20, and $\tfrac{4}{5}$ are not perfect squares. The others are rational since the numbers are perfect squares:
$$\sqrt{16} = 4 \qquad \sqrt{1.44} = 1.2 \qquad \sqrt{\tfrac{4}{9}} = \tfrac{2}{3}$$

1-4 The Irrational Numbers

The set of natural numbers, N, the set of integers, I, and the set of rational numbers, Q, are all subsets of the set of **real numbers**, R. The relationship of the sets can be represented in a diagram. The region of R not included in Q represents the set of irrational numbers.

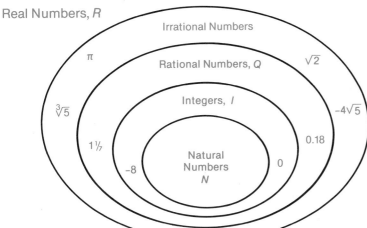

Exercises 1-4

A 1. Which numbers appear to be irrational?
   a) 2.147 474 747 474...
   b) −6.132 133 134...
   c) 72.041 296 47....
   d) 0.165 165 516 555...
   e) −2.236 067 977 49....
   f) −4.317 495
   g) 37.135 713 5.....
   h) −1.732 050 807 5.....
   i) 0.010 010 001 00....
   j) −127.721 127 721...

2. Which of the following are irrational?
   a) $\sqrt{21}$
   b) $\sqrt{16}$
   c) $\sqrt{2\frac{1}{4}}$
   d) $\sqrt{200}$
   e) $\sqrt{2.5}$
   f) $\sqrt{0.25}$
   g) $\sqrt{91}$
   h) $\sqrt{\frac{12}{75}}$
   i) $\sqrt{169}$
   j) $\sqrt{0.09}$
   k) $\sqrt{\frac{18}{98}}$
   l) $\sqrt{111}$

3. Which of the following are irrational?
   a) $5\sqrt{2}$
   b) $5 + \sqrt{2}$
   c) $5 - \sqrt{2}$
   d) $2\sqrt{36}$
   e) $7\sqrt{7}$
   f) $\sqrt{5} + \sqrt{2}$
   g) $\sqrt{7 + 9}$
   h) $\sqrt{7} + \sqrt{9}$
   i) $6\sqrt{21}$
   j) $\sqrt{17 + 12}$
   k) $2\sqrt{3} + 3\sqrt{2}$
   l) $5 + \sqrt{9}$

B 4. Name two rational and two irrational numbers between the numbers in each pair:
   a) 3.65, 3.69
   b) −1.476, −1.47
   c) $0.3\overline{97}$, 0.397 647 28....
   d) $-5.3\overline{76}$, $-5.3\overline{7}$
   e) $\frac{8}{9}$, $\frac{9}{10}$
   f) 2.236 067..., 2.236 071 23......
   g) $7.\overline{654\,12}$, 7.654 126 554 4.....
   h) $-12.\overline{407}$, −12.407 407 740 777...

5. Classify each of the following as a natural number, an integer, a rational, or an irrational number:
   a) $\frac{3}{5}$
   b) $0.2\overline{17}$
   c) −6
   d) 41 275
   e) 6.121 121 112...
   f) $-2\frac{1}{4}$
   g) $\sqrt{27}$
   h) $\sqrt{225}$
   i) −0.027 613 24....
   j) $-\frac{19}{7}$
   k) $-\sqrt{3}$
   l) 49

C 6. Evaluate the following for $x = \sqrt{2}$. Is the result a rational number?
   a) $3x^2 - 7x - 5$
   b) $x^3 - 2x + 6$
   c) $5x^3 + 4x^2 - 10x - 3$
   d) $x^4 - 2x + 12$

7. Evaluate the following for $x = -\sqrt{3}$. Is the result a rational number?
   a) $x^2 + 2x - 7$
   b) $5x^2 - 8x + 1$
   c) $x^3 + x^2 - 3x - 3$
   d) $2x^4 - 2x^3 - 6x + 5$

8. On a calculator with a $\boxed{\sqrt{}}$ key, $\sqrt{3}$ = 1.732 050 8. 1.732 050 8 is rational, but $\sqrt{3}$ is irrational. Can a number be both rational and irrational? Explain.

9. On a calculator with a $\boxed{\sqrt{}}$ key, $\sqrt{0.111\,111\,1}$ = 0.333 333 3.
   a) Write fractions in the form $\frac{m}{n}$ for 0.111 111 1 and 0.333 333 3. Use these fractions to explain why the above pattern appeared.
   b) Find another similar result.

Computer Power

Scientific Notation

Scientific notation is a convenient way to write numbers which are very large or very small.

For the number: 120 000 000 000 1.2×10^{11}

place the decimal point after the first non-zero digit. As its true position is 11 places to the right (the positive direction), the exponent is positive and the number becomes ─────────

For the number: 0.000 000 000 167 1.67×10^{-10}

place the decimal point after the first non-zero digit. As its true position is 10 places to the left (the negative direction), the exponent is negative and the number becomes ─────────

Calculators and microcomputers use scientific notation to express numbers which are too large or too small to be displayed on the screen. The symbol ↑ or ∧ on a computer keyboard denotes raising to a power. To obtain the value of 2^5 on a computer, instruct the computer to print 2 ↑ 5 by pressing the following keys:

     ?   2   ↑   5   RETURN

(This symbol means "print".)

(This key tells the computer to process the calculation.)

Exercises

1. Key in these calculations and record the answers displayed:

Calculate	by keying in
2^{29}	? 2 ↑ 2 9 RETURN
2^{30}	? 2 ↑ 3 0 RETURN
2^{64}	? 2 ↑ 6 4 RETURN
10^{19}	? 1 0 ↑ 1 9 RETURN

Calculate	by keying in
2^{-2}	? 2 ↑ − 2 RETURN
2^{-7}	? 2 ↑ − 7 RETURN
10^{-3}	? 1 0 ↑ − 3 RETURN
10^{-14}	? 1 0 ↑ − 1 4 RETURN

2. What is the least power of 2 which is greater than 10^{18}?

3. Light travels at approximately 300 000 km/s, and a light-year is the distance light travels in one year.
   a) Is a light-year greater than 10^{12} km?
   b) The nearest star, Proxima Centauri, is 4.31 light-years away. Is this distance greater than 4×10^{13} km?

4. Money invested at 24% doubles every 3 years. How long would it take an investment of $1000 to reach a trillion dollars?

1-5 The Pythagorean Theorem

In the 6th century B.C., the Greek mathematician, Pythagoras, discovered this important theorem which bears his name:

> For any right triangle with sides a, b, and c, where c is the hypotenuse, $c^2 = a^2 + b^2$.

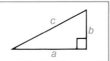

If the lengths of two sides of a right triangle are known, the Pythagorean theorem can be used to determine the length of the third side.

Example 1. Calculate the value of x rounded to one decimal place:

a)   b)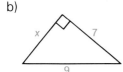

Solution.

a) $x^2 = 4^2 + 7^2$
$ = 16 + 49$
$ = 65$
$x = \sqrt{65}$
$ \doteq 8.1$

b) $9^2 = x^2 + 7^2$
$81 = x^2 + 49$
$81 - 49 = x^2$
$32 = x^2$
$x = \sqrt{32}$
$ \doteq 5.7$

Example 2. Express the length of side AB of $\triangle ABC$ in terms of x and y.

Solution.
$AC^2 = AB^2 + BC^2$
$x^2 = AB^2 + y^2$
$AB^2 = x^2 - y^2$
$AB = \sqrt{x^2 - y^2}$

Although the ancient Greeks knew that $\sqrt{2}$ was an irrational number, they were able to construct a line segment having a length of $\sqrt{2}$ units. They simply constructed an isosceles right triangle with the equal sides 1 unit in length. The length of the hypotenuse is $\sqrt{2}$ units.

By repeating this process, lengths of $\sqrt{3}$, $\sqrt{4}$, $\sqrt{5}$, $\sqrt{6}$,... can be constructed as shown below.

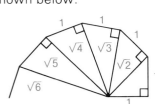

1-5 The Pythagorean Theorem

Example 3. Construct a line segment of length $\sqrt{11}$ units.

Solution. The procedure illustrated above could be continued until a length of $\sqrt{11}$ is obtained. A faster method is suggested by the diagram at the side. The length of AB is $\sqrt{11}$ units.

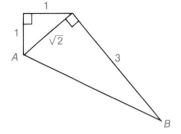

Example 4. A ground radar unit detects an aircraft at a height of 10 000 m, 5 km north and 12 km east of the unit. What is the actual distance of the aircraft from the radar unit?

Solution. Let R represent the position of the radar unit,
A represent the position of the aircraft,
P represent the point on the ground directly below the aircraft.
N and E are points on the north and east directions through R such that REPN is a rectangle.
The required distance is the length of the line segment RA. Since △REP is a right triangle,
$$\text{length of } RP = \sqrt{12^2 + 5^2}$$
$$= \sqrt{169}$$
$$= 13$$
Since △RPA is a right triangle,
$$\text{length of } RA = \sqrt{13^2 + 10^2}$$
$$= \sqrt{269}$$
$$\doteq 16.4$$
The aircraft is approximately 16.4 km from the radar unit.

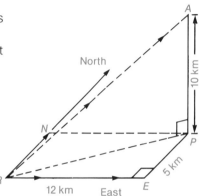

Exercises 1-5

A 1. Find x rounded to one decimal place:

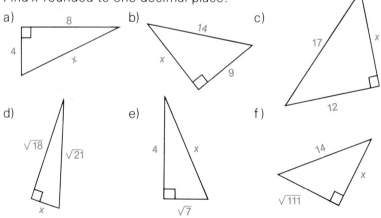

2. Find a:

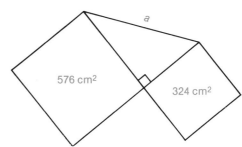

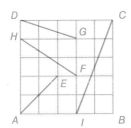

3. Square ABCD is divided into 25 small squares each of side length 1 cm. Find the lengths of these line segments shown in the diagram:
   a) AE   b) HF   c) DG   d) IC

4. Find the length, x, of this roof truss:

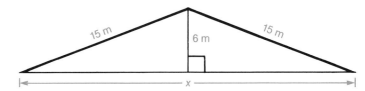

5. How much shorter is it to walk diagonally across a rectangular field than around the sides if the field is 250 m long by 110 m wide?

6. Construct a line segment of length
   a) $\sqrt{3}$ units;   b) $\sqrt{7}$ units;   c) $\sqrt{10}$ units;
   d) $\sqrt{27}$ units; e) $\sqrt{41}$ units; f) $\sqrt{21}$ units;

7. A pilot radios her position as 17 km south and 15 km west of an airport's control tower. The altimeter reads 8000 m. How far is the aircraft from the control tower?

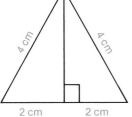

B 8. An equilateral triangle has sides of length 4 cm. Calculate:
   a) its height;   b) its area.

9. The area of △ABC (below, left) is 32 cm². What is the length of BD?

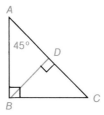
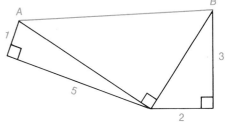

10. In the diagram above (right), find the length of AB.

1-5 The Pythagorean Theorem

11. An archaeologist measures a pyramid and finds that the base is square with a side length of 90 m. The slant height is 60 m. Calculate the height of the pyramid to the nearest metre.

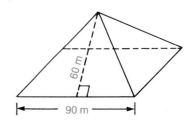

12. The tenth hole at the Silver Dunes Golf Club is a 90° dogleg. The score card shows the distances along the fairway to the circular green. What is the distance across the pond to the edge of the green?

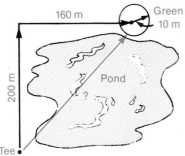

13. In the rectangular prism shown, PQ is a diagonal.
    a) Find the length of PQ if $PR = 5$, $RS = 3$, and $QS = 2$.
    b) Find a formula for the length, d, of the diagonal if the dimensions of the prism are a, b, and c.

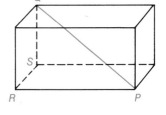

14. The line segments joining the vertices of a cube are of three different lengths.

edge   face diagonal   body diagonal

   a) If the edge length of the cube is 1, find the lengths of the other two segments joining the vertices.
   b) How many line segments of each length can be drawn joining the vertices?

C 15. a) How many line segments of different lengths can be drawn to join the midpoints of the edges of a cube?
   b) If the edge length of a cube is 1, find the length of the line segments identified in (a).
   c) How many line segments of each length can be drawn?

16. The top of a cylindrical storage tank, 55.3 m high and 28.4 m in diameter, is reached by means of a helical stairway that circles the tank exactly twice. Calculate the length of the stairway to the nearest metre.

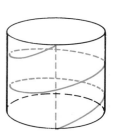

1-6 Multiplication of Radicals

Consider the expressions: $\sqrt{4} \times \sqrt{9}$ and $\sqrt{4 \times 9}$.

$\sqrt{4} \times \sqrt{9} = 2 \times 3$ $\qquad$ $\sqrt{4 \times 9} = \sqrt{36}$
$\qquad\qquad\quad = 6$ $\qquad\qquad\qquad\qquad = 6$

Therefore, $\sqrt{4} \times \sqrt{9} = \sqrt{4 \times 9}$

The same is true for irrationals. Consider the expressions: $\sqrt{3} \times \sqrt{5}$ and $\sqrt{3 \times 5}$. Square each expression.

$(\sqrt{3} \times \sqrt{5})^2 = (\sqrt{3})(\sqrt{3}) \times (\sqrt{5})(\sqrt{5})$
$\qquad\qquad\quad = 3 \times 5, \text{ or } 15$
$(\sqrt{3 \times 5})^2 = 3 \times 5$
$\qquad\qquad = 15$

Therefore, $\sqrt{3} \times \sqrt{5} = \sqrt{3 \times 5}$

The above results suggest the following property:

$$\sqrt{a} \times \sqrt{b} = \sqrt{ab}, a \geq 0, b \geq 0$$

Example 1. Simplify: a) $\sqrt{7} \times \sqrt{2}$ $\qquad$ b) $3\sqrt{2} \times \sqrt{5}$
$\qquad\qquad\qquad\quad$ c) $5\sqrt{3} \times 4\sqrt{2}$ $\qquad$ d) $4\sqrt{3} \times \sqrt{12}$

Solution. a) $\quad\sqrt{7} \times \sqrt{2}$ $\qquad\qquad$ b) $\quad 3\sqrt{2} \times \sqrt{5}$
$\qquad\qquad = \sqrt{7 \times 2}$ $\qquad\qquad\qquad = 3\sqrt{2 \times 5}$
$\qquad\qquad = \sqrt{14}$ $\qquad\qquad\qquad\quad = 3\sqrt{10}$

$\qquad\quad$ c) $\quad 5\sqrt{3} \times 4\sqrt{2}$ $\qquad\quad$ d) $\quad 4\sqrt{3} \times \sqrt{12}$
$\qquad\qquad = 20\sqrt{6}$ $\qquad\qquad\qquad = 4\sqrt{36}$
$\qquad\qquad\qquad\qquad\qquad\qquad\qquad = 4 \times 6$
$\qquad\qquad\qquad\qquad\qquad\qquad\qquad = 24$

Example 2. Express as a product:
$\qquad\qquad$ a) $\sqrt{21}$ $\qquad$ b) $\sqrt{30}$ $\qquad$ c) $\sqrt{20}$

Solution. a) $\sqrt{21} = \sqrt{7 \times 3}$
$\qquad\qquad\qquad = \sqrt{7} \times \sqrt{3}$
$\qquad\quad$ b) $\sqrt{30} = \sqrt{2 \times 3 \times 5}$
$\qquad\qquad\qquad = \sqrt{2} \times \sqrt{3} \times \sqrt{5}$
$\qquad\quad$ c) $\sqrt{20} = \sqrt{2 \times 2 \times 5}$
$\qquad\qquad\qquad = \sqrt{2} \times \sqrt{2} \times \sqrt{5}$

In *Example 2c*, since $\sqrt{2} \times \sqrt{2} = 2$, $\sqrt{20}$ may be written, $2\sqrt{5}$. An expression of the form $\sqrt{x}$ $(x > 0)$ is called an **entire radical**, and an expression of the form $a\sqrt{x}$, where a is any real number, is called a **mixed radical**.

1-6 Multiplication of Radicals

$$\sqrt{20} = 2\sqrt{5}$$

entire radical → ← mixed radical

These are also entire radicals: $\sqrt{3.5}$, $\sqrt{\frac{3}{2}}$.

These are also mixed radicals: $\frac{3}{2}\sqrt{6}$, $-2\sqrt{7}$.

Any number in the form $\sqrt{x}$, where x has a perfect square as a factor, can be expressed as a mixed radical.

Example 3. Simplify if possible: a) $\sqrt{18}$ b) $\sqrt{70}$ c) $\sqrt{48}$

Solution. a) 18 has 9 as a perfect-square factor
$$\sqrt{18} = \sqrt{9} \times \sqrt{2}$$
$$= 3\sqrt{2}$$

b) $\sqrt{70}$ cannot be simplified because no factor of 70 is a perfect square.

c) 48 has 16 as a perfect-square factor.
$$\sqrt{48} = \sqrt{16} \times \sqrt{3}$$
$$= 4\sqrt{3}$$

In *Example 3c*, both 4 and 16 are perfect-square factors of 48. If the smaller factor had been used, the result would have been $2\sqrt{12}$. This is not in simplest form because 12 is divisible by 4, a perfect square.

Example 4. Arrange in order from least to greatest:
$7\sqrt{2}$ $3\sqrt{7}$ $2\sqrt{15}$ $4\sqrt{6}$

Solution. Express each mixed radical as an entire radical.
$7\sqrt{2} = \sqrt{49} \times \sqrt{2}$, or $\sqrt{98}$
$3\sqrt{7} = \sqrt{9} \times \sqrt{7}$, or $\sqrt{63}$
$2\sqrt{15} = \sqrt{4} \times \sqrt{15}$, or $\sqrt{60}$
$4\sqrt{6} = \sqrt{16} \times \sqrt{6}$, or $\sqrt{96}$
Arranged from least to greatest:
$2\sqrt{15}$, $3\sqrt{7}$, $4\sqrt{6}$, $7\sqrt{2}$.

Exercises 1-6

A 1. Simplify:
   a) $\sqrt{7} \times \sqrt{8}$
   b) $\sqrt{11} \times \sqrt{14}$
   c) $\sqrt{8} \times (-\sqrt{18})$
   d) $2\sqrt{5} \times 3\sqrt{2}$
   e) $-4\sqrt{2} \times 3\sqrt{8}$
   f) $(-7\sqrt{3})(-5\sqrt{8})$

2. Simplify:
   a) $12\sqrt{3} \times (-3\sqrt{18})$
   b) $(-3\sqrt{5})(-5\sqrt{3})$
   c) $-7\sqrt{\frac{6}{35}} \times 2\sqrt{\frac{5}{9}}$
   d) $3\sqrt{\frac{6}{15}} \times 6\sqrt{\frac{10}{9}}$
   e) $-5\sqrt{0.3} \times 2\sqrt{0.7}$
   f) $4\sqrt{0.9} \times 11\sqrt{0.4}$

3. Express as a product of radicals:
   a) $\sqrt{24}$
   b) $\sqrt{18}$
   c) $\sqrt{45}$
   d) $\sqrt{28}$
   e) $\sqrt{72}$
   f) $\sqrt{60}$
   g) $\sqrt{39}$
   h) $\sqrt{65}$
   i) $\sqrt{96}$
   j) $\sqrt{120}$
   k) $\sqrt{126}$
   l) $\sqrt{105}$

4. Express in simplest form:
   a) $\sqrt{32}$
   b) $\sqrt{50}$
   c) $\sqrt{27}$
   d) $\sqrt{96}$
   e) $\sqrt{8}$
   f) $\sqrt{75}$
   g) $\sqrt{108}$
   h) $\sqrt{80}$

5. Express in simplest form:
   a) $\sqrt{147}$
   b) $\sqrt{54}$
   c) $\sqrt{76}$
   d) $\sqrt{180}$
   e) $3\sqrt{20}$
   f) $5\sqrt{18}$
   g) $6\sqrt{90}$
   h) $7\sqrt{242}$

6. Simplify:
   a) $2\sqrt{6} \times 3\sqrt{2}$
   b) $3\sqrt{5} \times 7\sqrt{10}$
   c) $-8\sqrt{6} \times 6\sqrt{8}$
   d) $5\sqrt{10} \times 4\sqrt{6}$
   e) $(-7\sqrt{12})(-2\sqrt{6})$
   f) $11\sqrt{3} \times 5\sqrt{6}$

7. Simplify:
   a) $\sqrt{24} \times \sqrt{18}$
   b) $2\sqrt{24} \times 5\sqrt{6}$
   c) $3\sqrt{20} \times 2\sqrt{5}$
   d) $2\sqrt{6} \times 7\sqrt{8} \times 5\sqrt{2}$
   e) $3\sqrt{7} \times 2\sqrt{6} \times 5\sqrt{2}$
   f) $4\sqrt{8} \times 3\sqrt{6} \times 7\sqrt{3}$

8. Arrange in order from least to greatest:
   a) $4\sqrt{3}$, $3\sqrt{5}$, $5\sqrt{2}$, $2\sqrt{10}$, $2\sqrt{13}$, $3\sqrt{6}$
   b) $-6\sqrt{2}$, $-3\sqrt{7}$, $-2\sqrt{17}$, $-4\sqrt{5}$, $-2\sqrt{21}$, $-5\sqrt{3}$
   c) $6\sqrt{0.1}$, $3\sqrt{0.7}$, $7\sqrt{0.05}$, $2\sqrt{0.8}$, $4\sqrt{0.5}$, $5\sqrt{0.3}$

B 9. Simplify:
   a) $\sqrt{24} \times \sqrt{54} \times \sqrt{18}$
   b) $\sqrt{20} \times \sqrt{32} \times \sqrt{18} \times \sqrt{125}$
   c) $\sqrt{27} \times \sqrt{12} \times \sqrt{45} \times \sqrt{80}$
   d) $5\sqrt{18} \times 3\sqrt{8} \times 6\sqrt{32}$
   e) $3\sqrt{20} \times 5\sqrt{8} \times 4\sqrt{180} \times 6\sqrt{72}$
   f) $0.8\sqrt{80} \times 0.125\sqrt{90} \times 0.5\sqrt{50}$

1-7 Addition and Subtraction of Radicals   **27**

10. Use the diagram below (left) to show that $\sqrt{8} = 2\sqrt{2}$.

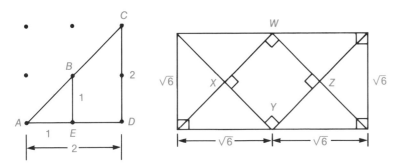

C 11. In the diagram above (right), what is the area of square *WXYZ*?

12. In the design on the cover of this book the side lengths of the black and green squares are 1 and 2 respectively.
   a) Find the side lengths of the remaining squares.
   b) Find the ratios of the side lengths of the squares indicated:
      i) blue : yellow   ii) brown : blue   iii) green : yellow
      iv) green : blue   v) black : yellow

1-7 Addition and Subtraction of Radicals

In the same way that $2x$ and $3x$ are called like terms, radicals such as $2\sqrt{3}$ and $3\sqrt{3}$ are called **like radicals**. Like radicals can be combined, that is, added or subtracted, using the distributive law. Radicals such as $2\sqrt{5}$ and $4\sqrt{7}$ are called **unlike radicals**; they cannot be combined.

Example 1. Simplify, if possible:
   a) $2\sqrt{3} + 3\sqrt{3}$
   b) $6\sqrt{2} - 4\sqrt{2} + \sqrt{2}$
   c) $4\sqrt{6} + 2\sqrt{10}$

Solution. a) $2\sqrt{3} + 3\sqrt{3} = (2 + 3)\sqrt{3}$
                                  $= 5\sqrt{3}$

b) $6\sqrt{2} - 4\sqrt{2} + \sqrt{2} = (6 - 4 + 1)\sqrt{2}$
                                    $= 3\sqrt{2}$

c) $4\sqrt{6} + 2\sqrt{10}$ cannot be combined because they are not like radicals. If required, the approximate numerical value can be found by substituting 2.4494 for $\sqrt{6}$ and 3.1622 for $\sqrt{10}$.

Radicals should be expressed in simplest form before attempting to combine them.

Example 2. Simplify:
a) $\sqrt{18} - \sqrt{2}$
b) $5\sqrt{5} - 2\sqrt{20} + \sqrt{125}$
c) $2\sqrt{98} + \sqrt{10} - 5\sqrt{8} - 3\sqrt{40}$

Solution.
a) $\sqrt{18} - \sqrt{2} = 3\sqrt{2} - \sqrt{2}$
$\phantom{\sqrt{18} - \sqrt{2}} = 2\sqrt{2}$

b) $ 5\sqrt{5} - 2\sqrt{20} + \sqrt{125}$
$= 5\sqrt{5} - 2 \times 2\sqrt{5} + 5\sqrt{5}$
$= 5\sqrt{5} - 4\sqrt{5} + 5\sqrt{5}$
$= 6\sqrt{5}$

c) $ 2\sqrt{98} + \sqrt{10} - 5\sqrt{8} - 3\sqrt{40}$
$= 2 \times 7\sqrt{2} + \sqrt{10} - 5 \times 2\sqrt{2} - 3 \times 2\sqrt{10}$
$= 14\sqrt{2} + \sqrt{10} - 10\sqrt{2} - 6\sqrt{10}$
$= 4\sqrt{2} - 5\sqrt{10}$

Although $\sqrt{a} \cdot \sqrt{b} = \sqrt{ab}$, it is not true that, in general, $\sqrt{a} + \sqrt{b} = \sqrt{a + b}$. Consider the expressions $\sqrt{4} + \sqrt{9}$ and $\sqrt{4 + 9}$.

$\sqrt{4} + \sqrt{9} = 2 + 3$ $\sqrt{4 + 9} = \sqrt{13}$
$\phantom{\sqrt{4} + \sqrt{9}} = 5$ $\doteq 3.6056$

Therefore, $\sqrt{4} + \sqrt{9} \neq \sqrt{4 + 9}$.

Similarly, in general, $\sqrt{a} - \sqrt{b} \neq \sqrt{a - b}$.

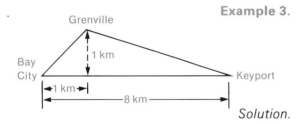

Example 3. Bay City is 8 km due west of Keyport and is linked to it by a straight stretch of railroad track. To travel from Bay City to Keyport by car one must go through Grenville, which is 1 km east and 1 km north of Bay City. How much farther is it by road than by train from Bay City to Keyport?

Solution. Distance from Bay City to Grenville: $\sqrt{2}$ km
Distance from Grenville to Keyport: $\sqrt{50}$ km
Distance from Bay City to Keyport by car:
$\sqrt{2} + \sqrt{50} = \sqrt{2} + 5\sqrt{2}$
$\phantom{\sqrt{2} + \sqrt{50}} = 6\sqrt{2}$ km
Distance from Bay City to Keyport by train: 8 km
Therefore, it is $6\sqrt{2} - 8$ km, or 0.49 km, farther by car than by train from Bay City to Keyport.

1-7 Addition and Subtraction of Radicals   **29**

Exercises 1-7

A   1. Simplify:
   a) $5\sqrt{7} - 3\sqrt{7}$
   b) $11\sqrt{6} + 5\sqrt{6}$
   c) $2\sqrt{13} - 8\sqrt{13}$
   d) $6\sqrt{19} - 31\sqrt{19}$
   e) $4\sqrt{3} + 29\sqrt{3}$
   f) $7\sqrt{15} - 2\sqrt{15}$

   2. Simplify:
   a) $4\sqrt{5} - 11\sqrt{5} + 3\sqrt{5}$
   b) $2\sqrt{10} + 7\sqrt{10} - 6\sqrt{10}$
   c) $5\sqrt{2} - 16\sqrt{2} + 29\sqrt{2}$
   d) $2\sqrt{6} - 6\sqrt{2} + 11\sqrt{6}$
   e) $4\sqrt{10} - 10\sqrt{10} + 3\sqrt{5}$
   f) $3\sqrt{5} - 9\sqrt{2} + 5\sqrt{5} - 2\sqrt{2}$

   3. Simplify:
   a) $\sqrt{40} + \sqrt{90}$
   b) $\sqrt{32} + \sqrt{8}$
   c) $\sqrt{12} - \sqrt{75}$
   d) $\sqrt{20} - \sqrt{45}$
   e) $\sqrt{50} - \sqrt{18}$
   f) $\sqrt{24} - \sqrt{96}$

   4. Simplify:
   a) $\sqrt{54} + \sqrt{150} - \sqrt{6}$
   b) $\sqrt{28} - \sqrt{63} + \sqrt{112}$
   c) $\sqrt{80} + \sqrt{45} - \sqrt{125}$
   d) $\sqrt{12} + \sqrt{27} + \sqrt{48}$
   e) $\sqrt{75} - \sqrt{3} + \sqrt{147}$
   f) $\sqrt{98} - \sqrt{72} - \sqrt{50}$

   5. Simplify:
   a) $2\sqrt{3} + 4\sqrt{12}$
   b) $5\sqrt{48} - 7\sqrt{3}$
   c) $3\sqrt{8} + 6\sqrt{18}$
   d) $4\sqrt{50} - 7\sqrt{32}$
   e) $2\sqrt{24} + 3\sqrt{54}$
   f) $6\sqrt{20} - 2\sqrt{45}$
   g) $3\sqrt{8} + 5\sqrt{18} - 6\sqrt{2}$
   h) $5\sqrt{28} - 3\sqrt{63} + 2\sqrt{112}$

   6. A straight stretch of railroad track connects Goshen to Humber, 16 km due west. The highway between the two towns passes through Ironton, 2 km east and 2 km north of Humber. How much farther is it to drive from Humber to Goshen than to take the train?

B 7. Simplify:
   a) $5\sqrt{12} - 2\sqrt{48} - 7\sqrt{75}$
   b) $3\sqrt{7} + 2\sqrt{11} - \sqrt{11} + 4\sqrt{7}$
   c) $\sqrt{48} - \sqrt{20} - \sqrt{27} - \sqrt{45}$
   d) $4\sqrt{18} - 2\sqrt{63} + \sqrt{175} + 5\sqrt{98}$
   e) $2\sqrt{12} + 3\sqrt{50} - 2\sqrt{75} - 6\sqrt{32}$
   f) $7\sqrt{24} + 3\sqrt{28} + 9\sqrt{54} + 6\sqrt{175}$
   g) $3\sqrt{20} - 2\sqrt{80} - 4\sqrt{48} - 5\sqrt{75}$

8. Simplify:
   a) $\frac{1}{2}\sqrt{8} + \frac{3}{5}\sqrt{50} - \frac{2}{3}\sqrt{18}$
   b) $2\sqrt{20} + \frac{3}{4}\sqrt{80} - \sqrt{125}$
   c) $7\sqrt{32} - \frac{1}{5}\sqrt{50} - \frac{2}{3}\sqrt{18} + \frac{3}{4}\sqrt{128}$
   d) $\frac{2}{5}\sqrt{125} - \frac{2}{3}\sqrt{243} - \frac{1}{3}\sqrt{45} + \frac{1}{2}\sqrt{48}$
   e) $\frac{2}{3}\sqrt{72} - \frac{2}{3}\sqrt{54} - \frac{1}{2}\sqrt{96} - \frac{5}{7}\sqrt{98}$

9. In $\triangle XYZ$, $\angle Y = 90°$, $XY = \sqrt{12}$, and $YZ = \sqrt{8}$.
   a) Find the length of XZ.
   b) Is it true that $\sqrt{12} + \sqrt{8} = \sqrt{20}$?

10. In $\triangle ABC$, $\angle B = 90°$ $AB = \sqrt{x}$, and $BC = \sqrt{y}$.
    a) Find an expression for the length of the hypotenuse, AC.
    b) Explain how the diagram shows that $\sqrt{x} + \sqrt{y} \ne \sqrt{x + y}$.

C 11. Is $\sqrt{x^2} = |x|$ for all real numbers, x? Explain.

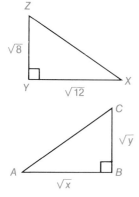

1-8 Combined Operations with Radicals

Products involving radicals can be expanded using the distributive law.

Example 1. Expand and simplify:
  a) $\sqrt{5}(4\sqrt{2} - \sqrt{5})$
  b) $3\sqrt{2}(5\sqrt{3} - \sqrt{8} + 4\sqrt{10})$

Solution. a) $\sqrt{5}(4\sqrt{2} - \sqrt{5}) = \sqrt{5} \times 4\sqrt{2} - \sqrt{5} \times \sqrt{5}$
$= 4\sqrt{10} - 5$

b) $3\sqrt{2}(5\sqrt{3} - \sqrt{8} + 4\sqrt{10})$
$= 3\sqrt{2} \times 5\sqrt{3} - 3\sqrt{2} \times \sqrt{8} + 3\sqrt{2} \times 4\sqrt{10}$
$= 15\sqrt{6} - 3\sqrt{16} + 12\sqrt{20}$
$= 15\sqrt{6} - 3 \times 4 + 12 \times 2\sqrt{5}$
$= 15\sqrt{6} - 12 + 24\sqrt{5}$

Example 2. Expand and simplify: $(5\sqrt{3} + 4\sqrt{7})(2\sqrt{3} - \sqrt{7})$

Solution. The diagram shows how the products are obtained.

$$(5\sqrt{3} + 4\sqrt{7})(2\sqrt{3} - \sqrt{7})$$
$$= 5 \times 2 \times (\sqrt{3})^2 - 5 \times \sqrt{3} \times \sqrt{7} +$$
$$\qquad\qquad 4 \times 2 \times \sqrt{7} \times \sqrt{3} - 4 \times (\sqrt{7})^2$$
$$= 30 - 5\sqrt{21} + 8\sqrt{21} - 28$$
$$= 2 + 3\sqrt{21}$$

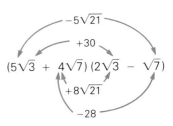

Exercises 1-8

A 1. Expand and simplify:
   a) $\sqrt{2}(\sqrt{5} - \sqrt{7})$
   b) $\sqrt{3}(\sqrt{11} + \sqrt{2})$
   c) $\sqrt{6}(\sqrt{13} - \sqrt{5})$
   d) $2\sqrt{3}(\sqrt{5} + 3\sqrt{7})$
   e) $\sqrt{5}(\sqrt{6} - \sqrt{10})$
   f) $4\sqrt{2}(3\sqrt{11} + 5\sqrt{13})$

2. Expand and simplify:
   a) $\sqrt{13}(\sqrt{3} + \sqrt{13})$
   b) $3\sqrt{3}(\sqrt{3} - 2\sqrt{6})$
   c) $3\sqrt{2}(2\sqrt{2} - 5\sqrt{8})$
   d) $2\sqrt{5}(3\sqrt{2} + 4\sqrt{3})$
   e) $6\sqrt{6}(3\sqrt{2} - 4\sqrt{3})$
   f) $2\sqrt{6}(3\sqrt{6} - 5\sqrt{8})$

3. Expand and simplify:
   a) $2\sqrt{3}(5\sqrt{7} - 2\sqrt{3} + 3\sqrt{6})$
   b) $4\sqrt{7}(2\sqrt{3} + 3\sqrt{6} - \sqrt{7})$
   c) $5\sqrt{2}(3\sqrt{18} + 7\sqrt{2} - 4\sqrt{8})$
   d) $12\sqrt{3}(2\sqrt{5} - 6\sqrt{6} - 3\sqrt{12})$

4. Expand and simplify:
   a) $-7\sqrt{5}(-2\sqrt{20} - 4\sqrt{30} + 3\sqrt{5})$
   b) $8\sqrt{6}(3\sqrt{2} - 4\sqrt{3} - 2\sqrt{6})$
   c) $2\sqrt{3}(5\sqrt{8} + 2\sqrt{3} - 4\sqrt{5} + 3\sqrt{6})$
   d) $4\sqrt{2}(2\sqrt{5} - 6\sqrt{3} - 2\sqrt{8} + 5\sqrt{32})$

5. Simplify:
   a) $(\sqrt{5} + \sqrt{2})(\sqrt{7} - \sqrt{3})$
   b) $(\sqrt{6} - \sqrt{3})(\sqrt{12} - \sqrt{6})$
   c) $(5 + 3\sqrt{2})(4 - \sqrt{2})$
   d) $(6 - 4\sqrt{2})(2 - 5\sqrt{2})$

6. Simplify:
   a) $(4\sqrt{3} + 2\sqrt{3})(2\sqrt{5} + 3\sqrt{2})$
   b) $(3\sqrt{6} - 5\sqrt{2})(2\sqrt{3} - 4\sqrt{5})$
   c) $(\sqrt{3} + \sqrt{2})(2\sqrt{3} + \sqrt{2})$
   d) $(11\sqrt{2} - 9\sqrt{7})(2\sqrt{6} + 2\sqrt{14})$

B 7. Simplify:
   a) $(2\sqrt{5} + 3\sqrt{2})^2$
   b) $(4\sqrt{2} - 5\sqrt{3})(4\sqrt{2} + 5\sqrt{3})$
   c) $(7 - 3\sqrt{2})^2$
   d) $(6\sqrt{3} - 2\sqrt{7})(6\sqrt{3} + 2\sqrt{7})$
   e) $(3\sqrt{5} - 2\sqrt{3})(3\sqrt{5} + 2\sqrt{3})$
   f) $(3\sqrt{6} + 2\sqrt{2})(2\sqrt{6} - 3\sqrt{2})$
   g) $(7\sqrt{3} - 5\sqrt{2})^2$
   h) $(3\sqrt{2} + 2\sqrt{5})(3\sqrt{2} - 2\sqrt{5})$
   i) $(6\sqrt{5} - 2\sqrt{10})^2$
   j) $(6\sqrt{8} - 3\sqrt{2})^2$

8. a) Square each of the following:
   $\sqrt{6} + \sqrt{10}, \quad \sqrt{13} + \sqrt{3}, \quad \sqrt{15} + \sqrt{1}, \quad \sqrt{5} + \sqrt{11}$
   b) Arrange the radical expressions in (a) in order from greatest to least.

9. Arrange in order from greatest to least:
   $\sqrt{18} - \sqrt{8}, \quad \sqrt{14} - \sqrt{12}, \quad \sqrt{20} - \sqrt{6}, \quad \sqrt{15} - \sqrt{11}$

1-9 Division of Radicals

Division is the inverse of multiplication.
   We know $56 \div 8 = 7$ because $7 \times 8 = 56$.
Similarly, $\sqrt{10} \div \sqrt{2} = \sqrt{5}$ because $\sqrt{5} \times \sqrt{2} = \sqrt{10}$.
This suggests the following property for radicals:

$$\frac{\sqrt{a}}{\sqrt{b}} = \sqrt{\frac{a}{b}}, \quad a \geq 0, b > 0$$

Example 1. Simplify: a) $\frac{2\sqrt{15}}{\sqrt{3}}$ b) $\frac{9\sqrt{24}}{2\sqrt{18}}$ c) $\frac{\sqrt{36}}{\sqrt{2}}$

Solution. a) $\frac{2\sqrt{15}}{\sqrt{3}} = 2\sqrt{\frac{15}{3}}$, or $2\sqrt{5}$

b) The radicals should be expressed in simplest form before dividing.
$$\frac{9\sqrt{24}}{2\sqrt{18}} = \frac{9 \times 2\sqrt{6}}{2 \times 3\sqrt{2}}$$
$$= \frac{3\sqrt{6}}{\sqrt{2}}, \text{ or } 3\sqrt{3}$$

c) $\frac{\sqrt{36}}{\sqrt{2}} = \sqrt{18}$ or $\frac{\sqrt{36}}{\sqrt{2}} = \frac{6}{\sqrt{2}}$
$= 3\sqrt{2}$

1 - 9 Division of Radicals   **33**

The answers to *Example 1c* found by two different methods must be equal. It can be shown that they are by multiplying numerator and denominator of $\frac{6}{\sqrt{2}}$ by $\sqrt{2}$.

$$\frac{6}{\sqrt{2}} = \frac{6}{\sqrt{2}} \times \frac{\sqrt{2}}{\sqrt{2}}$$

$$= \frac{6\sqrt{2}}{2}, \text{ or } 3\sqrt{2}$$

Equivalent to multiplication by 1.

This procedure is called **rationalizing the denominator**.

Example 2. Obtain equivalent expressions by rationalizing the denominator:

a) $\frac{1}{\sqrt{7}}$ b) $\frac{5}{\sqrt{20}}$ c) $\frac{4\sqrt{27}}{3\sqrt{72}}$

Solution. a) $\frac{1}{\sqrt{7}} = \frac{1}{\sqrt{7}} \times \frac{\sqrt{7}}{\sqrt{7}},$ or $\frac{\sqrt{7}}{7}$

b) $\frac{5}{\sqrt{20}} = \frac{5}{2\sqrt{5}} \times \frac{\sqrt{5}}{\sqrt{5}}$

$= \frac{5\sqrt{5}}{10},$ or $\frac{\sqrt{5}}{2}$

c) $\frac{4\sqrt{27}}{3\sqrt{72}} = \frac{4 \times 3\sqrt{3}}{3 \times 6\sqrt{2}}$

$= \frac{2\sqrt{3}}{3\sqrt{2}} \times \frac{\sqrt{2}}{\sqrt{2}}$

$= \frac{2\sqrt{6}}{6},$ or $\frac{\sqrt{6}}{3}$

Example 3. Simplify: a) $\frac{1}{\sqrt{12}} - \frac{1}{\sqrt{27}}$ b) $\frac{3}{\sqrt{12}} - \frac{5}{\sqrt{8}}$

Solution. a) $\frac{1}{\sqrt{12}} - \frac{1}{\sqrt{27}} = \frac{1}{2\sqrt{3}} - \frac{1}{3\sqrt{3}}$

$= \frac{\sqrt{3}}{6} - \frac{\sqrt{3}}{9}$

$= \frac{\sqrt{3}}{18}$

Multiplying numerator and denominator of each fraction by $\sqrt{3}$.

b) $\frac{3}{\sqrt{12}} - \frac{5}{\sqrt{8}} = \frac{3}{2\sqrt{3}} - \frac{5}{2\sqrt{2}}$

$= \frac{3\sqrt{3}}{6} - \frac{5\sqrt{2}}{4}$

$= \frac{6\sqrt{3} - 15\sqrt{2}}{12}$

$= \frac{2\sqrt{3} - 5\sqrt{2}}{4}$

Chapter 1

© 1979 United Feature Syndicate, Inc.

Exercises 1 - 9

A 1. Simplify:

  a) $\dfrac{\sqrt{24}}{\sqrt{3}}$ b) $\dfrac{\sqrt{56}}{\sqrt{8}}$ c) $\dfrac{\sqrt{72}}{\sqrt{6}}$ d) $\dfrac{3\sqrt{35}}{\sqrt{7}}$

  e) $\dfrac{6\sqrt{18}}{2\sqrt{6}}$ f) $\dfrac{5\sqrt{30}}{2\sqrt{15}}$ g) $\dfrac{4\sqrt{20}}{2\sqrt{5}}$ h) $\dfrac{3\sqrt{12}}{6\sqrt{3}}$

  2. Simplify:

  a) $\dfrac{12\sqrt{24}}{2\sqrt{12}}$ b) $\dfrac{4\sqrt{6}}{6\sqrt{\tfrac{2}{3}}}$ c) $\dfrac{3\sqrt{2}}{\sqrt{\tfrac{1}{2}}}$ d) $\dfrac{7\sqrt{\tfrac{1}{2}}}{\sqrt{\tfrac{1}{8}}}$

  3. Simplify:

  a) $\dfrac{3\sqrt{48}}{2\sqrt{27}}$ b) $\dfrac{6\sqrt{50}}{5\sqrt{18}}$ c) $\dfrac{4\sqrt{54}}{3\sqrt{12}}$ d) $\dfrac{3\sqrt{20}}{2\sqrt{10}}$

  e) $\dfrac{5\sqrt{24}}{2\sqrt{18}}$ f) $\dfrac{7\sqrt{32}}{5\sqrt{63}}$ g) $\dfrac{10\sqrt{27}}{3\sqrt{20}}$ h) $\dfrac{16\sqrt{24}}{4\sqrt{96}}$

  i) $\dfrac{4\sqrt{45}}{3\sqrt{54}}$ j) $\dfrac{3\sqrt{60}}{2\sqrt{27}}$ k) $\dfrac{2\sqrt{75}}{3\sqrt{20}}$ l) $\dfrac{9\sqrt{80}}{6\sqrt{18}}$

  4. Express in simplest form:

  a) $\dfrac{1}{\sqrt{6}}$ b) $\dfrac{3}{\sqrt{5}}$ c) $\dfrac{12}{\sqrt{17}}$ d) $\dfrac{10}{\sqrt{5}}$

  e) $\dfrac{6}{\sqrt{12}}$ f) $\dfrac{3}{2\sqrt{7}}$ g) $\dfrac{2\sqrt{6}}{5\sqrt{3}}$ h) $\dfrac{6\sqrt{10}}{3\sqrt{5}}$

  i) $\dfrac{3\sqrt{20}}{4\sqrt{12}}$ j) $\dfrac{8\sqrt{18}}{3\sqrt{75}}$ k) $\dfrac{10\sqrt{72}}{3\sqrt{54}}$ l) $\dfrac{6\sqrt{24}}{5\sqrt{27}}$

B 5. Simplify:

  a) $\dfrac{1}{\sqrt{5}} + \dfrac{1}{\sqrt{3}}$ b) $\dfrac{1}{\sqrt{2}} - \dfrac{1}{\sqrt{6}}$ c) $\dfrac{1}{\sqrt{3}} + \dfrac{1}{\sqrt{6}}$

  d) $\dfrac{2}{\sqrt{7}} - \dfrac{3}{\sqrt{5}}$ e) $\dfrac{4}{\sqrt{3}} + \dfrac{2}{\sqrt{10}}$ f) $\dfrac{3}{\sqrt{7}} - \dfrac{9}{\sqrt{6}}$

  6. Simplify:

  a) $\dfrac{3}{\sqrt{12}} + \dfrac{2}{\sqrt{18}}$ b) $\dfrac{5}{\sqrt{8}} - \dfrac{2}{\sqrt{6}}$ c) $\dfrac{7}{\sqrt{20}} - \dfrac{4}{\sqrt{12}}$

  d) $\dfrac{3\sqrt{2}}{\sqrt{12}} - \dfrac{5\sqrt{3}}{\sqrt{8}}$ e) $\dfrac{4\sqrt{3}}{\sqrt{18}} - \dfrac{2\sqrt{2}}{\sqrt{6}}$ f) $\dfrac{3\sqrt{5}}{\sqrt{20}} + \dfrac{4\sqrt{3}}{\sqrt{27}}$

Review Exercises

1. Express as the product of prime factors:
   a) 26   b) 44   c) 210   d) 182   e) 429

2. Find the g.c.f. of each pair of numbers:
   a) 15, 72   b) 39, 143   c) 20, 75   d) 14, 77

3. Find the l.c.m. of each pair of numbers:
   a) 20, 48   b) 30, 42   c) 40, 75   d) 36, 56

4. The lights on one airport tower flash every 35 s. The lights on another tower flash every 42 s. What is the interval between simultaneous flashes?

5. Simplify:
   a) $-2[3(-4) - 7 + 8]$
   b) $3 + 5[8 \div 2 - (-4)(-2)]$
   c) $[6 - (9 \div 3 \times 5)][7 - 6]$
   d) $4[-16 \div 4 + (3 - 7) \div (8 - 6)]$

6. If $x = -3$, $y = 2$, and $z = -1$, evaluate:
   a) $xz^2 - x^2z^2 - xyz$
   b) $4xy^2 - 4x^2y + 4xyz$

7. Simplify:
   a) $3|6 - 4 - 12|$ b) $5|7 - 2| - 3|2 - 8| + 2|3 - 7|$

8. Simplify:
   a) $\frac{5}{-8} - \frac{-3}{4}$
   b) $-\frac{11}{3} - (\frac{-7}{4})$
   c) $-\frac{13}{6} + \frac{17}{8}$

9. If $x = \frac{-1}{4}$ and $y = \frac{2}{-3}$, evaluate:
   a) $3x + 2y - 5xy$
   b) $3x^2 - \frac{1}{3}xy$

10. Express in decimal form:
    a) $\frac{2}{9}$ b) $\frac{-6}{11}$ c) $3\frac{3}{5}$ d) $-5\frac{1}{6}$

11. Express as a fraction:
    a) 3.23   b) $0.\overline{43}$ c) $1.\overline{27}$ d) $3.1\overline{13}$

12. Which of the following numbers are irrational?
    a) $\sqrt{48}$ b) $\sqrt{49}$ c) $\sqrt{1\frac{7}{9}}$ d) $\sqrt{3.6}$

13. Evaluate the following for $x = \sqrt{3}$. Say whether the result is rational or irrational.
    a) $x^2 + 3x - 9$ b) $2x^3 - x^2 - 6x - 1$

14. Find x, rounded to one decimal place:

    a) 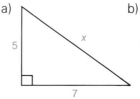 b) 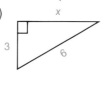 c)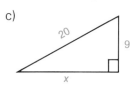

15. An Air Muskoka pilot radios his position as 12 km south and 5 km east of the airport tower. The altimeter reads 9000 m. How far is the aircraft from the tower?

16. Express in simplest form:
    a) $\sqrt{98}$ b) $\sqrt{112}$ c) $2\sqrt{45}$ d) $-3\sqrt{360}$

17. Simplify:
    a) $4\sqrt{6} \times 2\sqrt{3}$ b) $(-4\sqrt{12})(-3\sqrt{3})$
    c) $-5\sqrt{20} \times 2\sqrt{5}$ d) $-5\sqrt{6} \times 5\sqrt{12}$
    e) $(-3\sqrt{5})(-2\sqrt{15})$ f) $2\sqrt{8} \times 3\sqrt{6} \times \sqrt{12}$

18. Combine where possible:
    a) $3\sqrt{5} - 6\sqrt{5} + 8\sqrt{5}$ b) $4\sqrt{2} + 3\sqrt{2} - 2\sqrt{8}$
    c) $2\sqrt{8} + 3\sqrt{18} - \sqrt{50}$ d) $\frac{1}{2}\sqrt{27} + \frac{1}{3}\sqrt{108} - \frac{1}{4}\sqrt{12}$
    e) $3\sqrt{45} + \sqrt{180} - 2\sqrt{125} + \frac{1}{3}\sqrt{405} - \frac{1}{4}\sqrt{320}$

19. Expand and simplify:
    a) $\sqrt{2}(\sqrt{8} - \sqrt{3})$ b) $\sqrt{3}(\sqrt{2} - 3\sqrt{27})$ c) $\sqrt{5}(\sqrt{8} - \sqrt{3})$

20. Simplify:
    a) $(\sqrt{3} - \sqrt{2})(2\sqrt{3} + 3\sqrt{2})$ b) $(2\sqrt{2} - 3\sqrt{3})(2\sqrt{2} + 3\sqrt{3})$

21. Simplify:
    a) $\dfrac{\sqrt{24}}{\sqrt{2}}$ b) $\dfrac{3\sqrt{32}}{\sqrt{2}}$ c) $\dfrac{\sqrt{2}}{\sqrt{\frac{1}{2}}}$ d) $\dfrac{3\sqrt{12}}{4\sqrt{27}}$

22. Express in simplest form:
    a) $\dfrac{2\sqrt{3}}{\sqrt{2}}$ b) $\dfrac{3}{\sqrt{27}}$ c) $\dfrac{5}{\sqrt{50}}$ d) $\dfrac{3\sqrt{24}}{\sqrt{54}}$

2 Algebraic Expressions

In a training session, Susan coxes her boat crew to row with constant power both with and against the current. If the boat's speed upstream is u km/h and downstream is v km/h, and the distance each way is r km, write an expression for the average speed over the training course. (See *Example 2* in Section 2-18.)

2-1 Variables, Coefficients, Terms, and Exponents

In algebra, letters are often used to represent numbers. These letters are called **variables**. The table below reviews the meaning of variable and two other important words: **coefficient** and **term**.

Term	Coefficient	Variable	Meaning
$2x$	2	x	$2 \cdot x$
$-3y$	-3	y	$-3 \cdot y$
ab	1	a, b	$a \cdot b$
$-z^2$	-1	z	$-(z \cdot z)$
$(-z)^2$	1	z	$(-z)(-z)$
$-4y^3$	-4	y	$-4 \cdot y \cdot y \cdot y$
$7a^2b$	7	a, b	$7 \cdot a \cdot a \cdot b$
$6p^3q^2$	-6	p, q	$-6 \cdot p \cdot p \cdot p \cdot q \cdot q$
$\dfrac{x}{y^3}$	1	x, y	$\dfrac{x}{y \cdot y \cdot y}$

The numbers 3 and 2 in the term $-6p^3q^2$ are called **exponents**. The exponent 3 indicates that there are 3 factors of p in the term, and the exponent 2 indicates that there are 2 factors of q. It is important to remember the meaning of exponents when evaluating or operating with terms.

Example 1. Evaluate when $x = -3$ and $y = 2$:

a) $-x^2y^3$ b) $\dfrac{x}{y^4}$ c) $(-x)^3(-y^2)$

Solution. When $x = -3$ and $y = 2$:

a) $-x^2y^3 = -(-3)^2(2)^3$
$= -9 \times 8$
$= -72$

b) $\dfrac{x}{y^4} = \dfrac{-3}{2^4}$
$= -\dfrac{3}{16}$

c) $(-x)^3(-y^2) = (3)^3(-2^2)$
$= 27(-4)$
$= -108$

$-x = -(-3)$
$= 3$

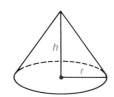

Example 2. The volume, V, of a cone is given by the formula $V = \dfrac{1}{3}\pi r^2 h$, where r is the radius of the base and h is the height. What is the volume of a conical pile of sand of height 4.6 m and radius 8.2 m?

2-1 Variables, Coefficients, Terms, and Exponents

Solution. When $r = 8.2$ and $h = 4.6$,
$$V = \frac{1}{3}\pi(8.2)^2 (4.6)$$
$$\doteq \frac{1}{3}(3.14)(8.2)^2(4.6)$$
$$\doteq 323.74$$

The pile contains about 324 m³ of sand.

π is approximately 3.14.

Exercises 2-1

A 1. Name the variables, coefficients and exponents:
   $-ab^3$, $4x^2y$, $-5m^3n^2$, $7x^4y^2z$, $\frac{1}{3}p^6q^2$, $6x$, 8

2. If $x = -2$ and $y = 3$, evaluate:
   a) xy
   b) x^2y
   c) xy^2
   d) $2x^2y^2$
   e) $\frac{x^3}{y^2}$
   f) $\frac{-3x^2}{y^3}$
   g) $-\frac{x^3y^2}{4}$
   h) $\left(\frac{-xy}{2}\right)^2$

B 3. The meaning of a term is given; write the term:
   a) $w + w + w$
   b) $s^2 + s^2 + s^2$
   c) $x \cdot x \cdot x$
   d) $-y - y - y - y$
   e) $-3a \cdot a \cdot b$
   f) $-3p \cdot p \cdot p \div q$

4. If $p = 2$ and $q = -4$, evaluate:
   a) p^4
   b) p^2q^2
   c) $\frac{p^5}{q^2}$
   d) $\left(\frac{p^2}{q}\right)^2$
   e) $\frac{12q^2}{p^3}$
   f) $\left(\frac{-2q}{p}\right)^3$

5. The meaning of a term is given; write the term:
   a) $\frac{1}{5}x + \frac{1}{5}x + \frac{1}{5}x + \frac{1}{5}x$
   b) $\frac{x}{2y} + \frac{x}{2y} + \frac{x}{2y}$
   c) $m \cdot m \cdot m \cdot n \cdot n$
   d) $\frac{p}{7q} \cdot \frac{p}{7q} \cdot \frac{p}{7q}$

6. The area of a rectangle is the product of its length and width.
   a) Write a formula in terms of length, l, and width, w, for
      i) the area, A;   ii) the perimeter, P.
   b) Find the area of a rectangle 15 m by 12 m.

7. The circumference of a circle is π times its diameter.
   a) Write a formula for the circumference, C, in terms of π and the radius, r.
   b) Find the circumference of a circle with radius 25 cm.

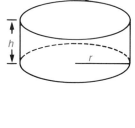

8. The area, A, of a circle is π times the square of its radius, r.
   a) Write a formula for the area in terms of π and r.
   b) Find the area of a circle with radius 15 cm.

9. The volume, V, of a cylinder of radius r and height h is given by the formula: $V = \pi r^2 h$. What is the volume of a cylindrical tank of radius 50 cm and height 3 m?

10. Write a formula for the area of the curved surface of a cylinder of length l and radius r.

11. In house construction, the safe load, m, in kilograms, that can be supported by a horizontal joist of length l metres is given by the formula:
$$m = \frac{4th^2}{l},$$

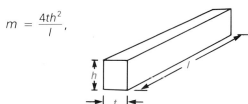

where t is the thickness and h the height in centimetres. What is the safe load for a beam 4 m long
   a) when $t = 5$ cm and $h = 10$ cm?
   b) when $t = 10$ cm and $h = 5$ cm?

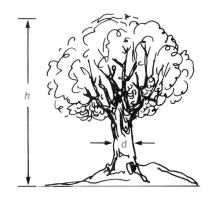

12. An estimate of the volume, V, in cubic metres, of wood in a tree is given by the formula:
$$V \doteq 0.000\,05 d^2 h,$$
where d is the average diameter of its trunk, in centimetres, and h is its height, in metres. Estimate the volume of wood in a tree when:
   a) $d = 40$ cm, $h = 10$ m;   b) $d = 75$ cm, $h = 30$ m.

13. The pressure, P, in kilopascals, exerted on the floor by the heel of a shoe is given by the formula:
$$P = \frac{100m}{x^2},$$
where m is the wearer's mass, in kilograms, and x is the width of the heel, in centimetres. Find the pressure exerted by
   a) a 90 kg man wearing shoes with heels 7 cm wide;
   b) a 60 kg woman wearing shoes with heels 2 cm wide.

C 14. A term contains the variables x and y. When $x = 3$ and $y = 2$, the value of the term is $\frac{9}{2}$. When $x = 4$ and $y = 5$, the value of the term is $\frac{16}{5}$. Find the term.

2-2 Integral Exponents

An expression of the form a^n is called a **power**.

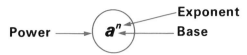

The definition of a power depends on whether the exponent is a positive integer, a negative integer, or zero.

Positive Integral Exponent

$$a^n = \underbrace{a \cdot a \cdot a \cdot \ldots a}_{n \text{ factors}}$$

Negative Integral Exponent

a^{-n} is defined to be the reciprocal of a^n.

$$a^{-n} = \frac{1}{a^n} \quad (a \neq 0)$$

Zero Exponent

a^0 is defined to be equal to 1.

$$a^0 = 1 \quad (a \neq 0)$$

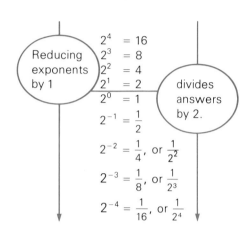

These definitions can be used to evaluate a power with any integral exponent.

Example 1. Simplify: a) 4^3 b) 3^{-2} c) $(-2)^{-3}$ d) $\left(\frac{1}{2}\right)^0$

Solution. a) $4^3 = (4)(4)(4)$
   $= 64$

b) $3^{-2} = \frac{1}{3^2}$
   $= \frac{1}{9}$

c) $(-2)^{-3} = \frac{1}{(-2)^3}$
   $= -\frac{1}{8}$

d) $\left(\frac{1}{2}\right)^0 = 1$

Evaluate powers first before doing any operations unless brackets indicate otherwise.

Example 2. If $m = -5$ and $n = 2$, evaluate.

a) $m^2 - n^{-1}$ b) $(-mn^2)^{-2}$

c) $\left(\frac{m}{n}\right)^{-3}$ d) $(m^{-1} + n^0)^{-1}$

Solution.
a) $m^2 - n^{-1} = (-5)^2 - 2^{-1}$
$= 25 - \frac{1}{2}$, or $24\frac{1}{2}$

b) $(-mn^2)^{-2} = (5 \times 4)^{-2}$
$= \frac{1}{20^2}$
$= \frac{1}{400}$

c) $\left(\frac{m}{n}\right)^{-3} = \left(-\frac{5}{2}\right)^{-3}$
$= \frac{1}{\left(-\frac{5}{2}\right)^3}$
$= \frac{1}{-\frac{125}{8}}$
$= -\frac{8}{125}$

d) $(m^{-1} + n^0)^{-1} = [(-5)^{-1} + 2^0]^{-1}$
$= \left[-\frac{1}{5} + 1\right]^{-1}$
$= \left[\frac{4}{5}\right]^{-1}$
$= \frac{5}{4}$

Expressions involving repeated factors can be written as powers with positive or negative integral exponents.

Example 3. Write as a power:

a) $(-r)(-r)(-r)(-r)(-r)$ b) $\left(\frac{1}{r}\right)\left(\frac{1}{r}\right)\left(\frac{1}{r}\right)$

Solution.
a) $(-r)(-r)(-r)(-r)(-r) = (-r)^5$

b) $\left(\frac{1}{r}\right)\left(\frac{1}{r}\right)\left(\frac{1}{r}\right) = \frac{1}{r \cdot r \cdot r}$
$= \frac{1}{r^3}$, or r^{-3}

Example 4. Write as a product of powers:

a) $\frac{x}{y} \cdot \frac{x}{y} \cdot \frac{x}{y} \cdot \frac{x}{y}$ b) $\frac{1}{x} \cdot \frac{1}{y} \cdot \frac{1}{x} \cdot \frac{1}{x} \cdot \frac{1}{y}$

Solution.
a) $\frac{x}{y} \cdot \frac{x}{y} \cdot \frac{x}{y} \cdot \frac{x}{y} = \frac{x \cdot x \cdot x \cdot x}{y \cdot y \cdot y \cdot y}$
$= \frac{x^4}{y^4}$, or $x^4 y^{-4}$

b) $\frac{1}{x} \cdot \frac{1}{y} \cdot \frac{1}{x} \cdot \frac{1}{x} \cdot \frac{1}{y} = \frac{1}{x \cdot y \cdot x \cdot x \cdot y}$
$= \frac{1}{x^3 y^2}$, or $x^{-3} y^{-2}$

Exercises 2 - 2

A 1. Simplify:
   a) 2^4
   b) 5^{-2}
   c) 3^{-1}
   d) $\left(\frac{1}{4}\right)^{-1}$
   e) $\left(\frac{2}{3}\right)^{-1}$
   f) $\left(\frac{3}{4}\right)^{-2}$
   g) $(0.5)^{-1}$
   h) $(1.5)^0$
   i) $(1.5)^{-2}$

  2. Simplify:
   a) 10^0
   b) $(-3)^{-2}$
   c) $\left(-\frac{1}{2}\right)^3$
   d) $\left(-\frac{2}{3}\right)^{-1}$
   e) $\left(-\frac{3}{5}\right)^{-2}$
   f) $(-1)^{-4}$
   g) $(0.1)^{-4}$
   h) $\frac{1}{2^{-3}}$
   i) $\frac{1}{10^{-1}}$

  3. Write as a power:
   a) $(-7)(-7)(-7)$
   b) $\frac{1}{b \times b \times b \times b \times b}$
   c) $\left(\frac{1}{x}\right)\left(\frac{1}{x}\right)\left(\frac{1}{x}\right)\left(\frac{1}{x}\right)$
   d) $\frac{1}{1.5 \times 1.5}$

  4. Write as a product of powers:
   a) $x \cdot x \cdot y \cdot y \cdot y$
   b) $x \cdot y \cdot x \cdot y \cdot x$
   c) $\frac{x \cdot x \cdot x}{z \cdot z}$
   d) $\left(\frac{a}{b}\right)\left(\frac{a}{b}\right)\left(\frac{a}{b}\right)\left(\frac{a}{b}\right)$

B 5. Simplify:
   a) $2^3 + 2^{-3}$
   b) $2^{-1} + 2^{-2}$
   c) $10^{-1} - 10^{-2}$
   d) $(3^2 - 2^2)^{-2}$
   e) $\frac{1}{(4^2 - 2^3)^{-1}}$
   f) $(10^{-2})(-3)^3$
   g) $(-1)^{-2} + (-2)^{-1}$
   h) $2^{-3} \div 3^{-2}$
   i) $(5^{-2})(2^{-3})$

  6. Evaluate for: i) $a = 2$, ii) $a = -2$:
   a) a^3
   b) $(-a)^3$
   c) $-a^3$
   d) a^{-3}
   e) $-(-a)^3$
   f) $-a^{-3}$

  7. Evaluate $x - x^{-1}$ for these values of x:
   a) 2
   b) -1
   c) -2
   d) 10
   e) $\frac{1}{2}$
   f) 0.1

  8. Evaluate for $x = 2$ and $y = -5$:
   a) $x^{-1} + y^0$
   b) $x^2 - y^{-2}$
   c) $x^{-1}y^2$
   d) $(-x^2y)^2$
   e) $(4x^{-2}y)^{-1}$
   f) $(2x^{-1} - y^2)^2$

9. Evaluate for $x = 2$, $y = -1$:
   a) $x^{-4}y$
   b) $(xy)^{-2}$
   c) $\left(\dfrac{1}{xy}\right)^{-1}$
   d) $(2x^2y^3)^{-1}$
   e) $y^{-2} - x^{-2}$
   f) $4x^{-2}y^{-7}$

10. Evaluate for $x = 2$, $y = -3$:
    a) $x^{-4}y$
    b) $(xy)^{-2}$
    c) $(-x^2y)^3$
    d) $2x^{-1} + 3y^{-1}$
    e) $3x^2y^{-3}$
    f) $\left(\dfrac{x}{y}\right)^{-2}$

11. Express the first number as a power of the second:
    a) 64, 4
    b) $\dfrac{1}{9}$, 3
    c) $-\dfrac{1}{32}$, -2
    d) 0.001, 10
    e) $\dfrac{1}{25}$, -5
    f) -0.125, -2
    g) 1, 5
    h) 2, $\dfrac{1}{2}$
    i) $\dfrac{3}{4}$, $\dfrac{4}{3}$

12. Express as powers with positive exponents:
    a) 100
    b) -1000
    c) $\dfrac{27}{64}$
    d) 0.16
    e) -32
    f) -0.001
    g) 7^{-3}
    h) $\dfrac{1}{2^{-6}}$
    i) $\left(\dfrac{3}{4}\right)^{-3}$

13. Express as powers with negative exponents other than -1:
    a) $\dfrac{1}{9}$
    b) $\dfrac{1}{36}$
    c) 0.0001
    d) $\dfrac{8}{27}$
    e) 0.04
    f) $-\dfrac{1}{8}$
    g) $\dfrac{1}{5^2}$
    h) 4^3
    i) $\left(\dfrac{2}{3}\right)^3$

C 14. Which is greater:
    a) $(1.5)^{-1}$ or $(5.1)^{-1}$?
    b) $\left(\dfrac{1}{4}\right)^3$ or $\left(\dfrac{1}{2}\right)^2$?
    c) 2^{-3} or 3^{-2}?
    d) $\left(\dfrac{1}{2}\right)^{-5}$ or 5^2?
    e) $(-5)^{-3}$ or $(-2)^{-5}$?
    f) $\left(\dfrac{7}{2}\right)^{-3}$ or $\left(\dfrac{2}{7}\right)^{-3}$?

15. Solve by inspection:
    a) $7^x = 1$
    b) $5^x = \dfrac{1}{5}$
    c) $3^x = \dfrac{1}{27}$
    d) $10^x = 0.0001$
    e) $(-2)^x = \dfrac{1}{16}$
    f) $x^{-1} = \dfrac{3}{2}$
    g) $x^{-3} = 8$
    h) $x^{-2} = 0.04$
    i) $\dfrac{1}{x} = (0.2)^{-2}$

2-3 The Exponent Laws

The definitions of integral exponents lead to some basic laws for working with exponents. Study the following examples and the laws they illustrate.

Example

$x^3 \cdot x^2 = x \cdot x \cdot x \cdot x \cdot x$
$= x^5$, or x^{3+2}

$x^5 \div x^3 = \dfrac{x \cdot x \cdot x \cdot x \cdot x}{x \cdot x \cdot x}$
$= x^2$, or x^{5-3}

$(x^3)^2 = (x \cdot x \cdot x)(x \cdot x \cdot x)$
$= x^6$, or $x^{3 \times 2}$

$(xy)^3 = xy \cdot xy \cdot xy$
$= x \cdot x \cdot x \cdot y \cdot y \cdot y$
$= x^3 y^3$

$\left(\dfrac{x}{y}\right)^2 = \dfrac{x}{y} \cdot \dfrac{x}{y}$
$= \dfrac{x^2}{y^2}$

Exponent Law

1. $x^m \cdot x^n = x^{m+n}$

2. $x^m \div x^n = x^{m-n}$ $(x \neq 0)$

3. $(x^m)^n = x^{mn}$

4. $(xy)^n = x^n y^n$

5. $\left(\dfrac{x}{y}\right)^n = \dfrac{x^n}{y^n}$ $(y \neq 0)$

The exponent laws may be used to simplify products and quotients involving powers.

Example 1. Simplify:

    a) $(x^3 y^2)(x^2 y^4)$ b) $\dfrac{a^5 b^3}{a^2 b^2}$ c) $\left(\dfrac{x^2}{z^3}\right)^2$

Solution. a) $(x^3 y^2)(x^2 y^4) = x^3 \cdot y^2 \cdot x^2 \cdot y^4$
$= x^3 \cdot x^2 \cdot y^2 \cdot y^4$
$= x^5 y^6$

    b) $\dfrac{a^5 b^3}{a^2 b^2} = \dfrac{a^5}{a^2} \cdot \dfrac{b^3}{b^2}$
$= a^3 b$

    c) $\left(\dfrac{x^2}{z^3}\right)^2 = \dfrac{x^2}{z^3} \cdot \dfrac{x^2}{z^3}$
$= \dfrac{x^4}{z^6}$

The expressions in *Example 1* can be simplified using either the laws of exponents or the definition of a positive integral exponent. This is also true for expressions containing negative integral exponents.

Example 2. Simplify: a) $x^{-3} \cdot x^5$ b) $m^2 \div m^{-3}$

Solution. a) $x^{-3} \cdot x^5 = x^{-3+5}$ 　　　　$x^{-3} \cdot x^5 = \frac{1}{x^3} \cdot x^5$

$\qquad\qquad\qquad = x^2$ 　　or　　　$= \frac{x^5}{x^3}$

$\qquad\qquad\qquad\qquad\qquad\qquad\qquad\qquad = x^2$

b) $m^2 \div m^{-3} = m^{2-(-3)}$ 　　　$m^2 \div m^{-3} = \frac{m^2}{\frac{1}{m^3}}$

$\qquad\qquad\qquad = m^5$ 　　or

$\qquad\qquad\qquad\qquad\qquad\qquad\qquad\qquad = m^2 \cdot m^3$

$\qquad\qquad\qquad\qquad\qquad\qquad\qquad\qquad = m^5$

Example 2 illustrates why a power with a negative integral exponent is defined as a reciprocal. With this definition, the laws of exponents apply to all powers with integral exponents.

Example 3. Simplify: a) $\frac{x^{-2}y}{x^{-4}y^{-3}}$ b) $(a^{-3}b^2)^3(a^2b^{-3})^2$

Solution. a) $\frac{x^{-2}y}{x^{-4}y^{-3}} = \frac{x^{-2}}{x^{-4}} \cdot \frac{y}{y^{-3}}$

$\qquad\qquad\qquad = x^{-2-(-4)}y^{1-(-3)}$

$\qquad\qquad\qquad = x^2y^4$

b) $(a^{-3}b^2)^3(a^2b^{-3})^2 = a^{-9}b^6a^4b^{-6}$

$\qquad\qquad\qquad\qquad\qquad = a^{-9+4}b^{6-6}$

$\qquad\qquad\qquad\qquad\qquad = a^{-5}b^0$

$\qquad\qquad\qquad\qquad\qquad = a^{-5}$

Exercises 2 - 3

A 1. Simplify:
　　a) $x^3 \cdot x^4$ 　　b) $a^2 \cdot a^5$ 　　c) $b^3 \cdot b^5$
　　d) $m^2 \cdot m^3$ 　　e) $c^2 \cdot c^3 \cdot c$ 　　f) $y^3 \cdot y^2 \cdot y^4$

2. Simplify:
　　a) $x^4 \div x^2$ 　　b) $y^7 \div y^3$ 　　c) $n^6 \div n^5$
　　d) $a^8 \div a^5$ 　　e) $x^4 \div x$ 　　f) $y^9 \div y^7$

3. Simplify:
　　a) $(x^3)^3$ 　　b) $(y^2)^3$ 　　c) $(n^3)^4$
　　d) $(a^2)^2$ 　　e) $(c^5)^2$ 　　f) $(a^2b^2)^3$
　　g) $(xy^3)^2$ 　　h) $(x^4y^2)^2$ 　　i) $(mn^2)^3$

4. Simplify:
　　a) $x^2 \div x^5$ 　　b) $c^3 \div c^4$ 　　c) $y^2 \div y^7$
　　d) $a^2 \div a^6$ 　　e) $b^5 \div b^8$ 　　f) $x^3 \div x^9$

2-3 The Exponent Laws

B 5. Simplify:
   a) $x^{-2} \cdot x^3$
   b) $m^{-3} \cdot m^{-1}$
   c) $a^6 \cdot a^{-2}$
   d) $y^4 \cdot y^{-4}$
   e) $x^5 \cdot x^{-1} \cdot x^{-3}$
   f) $b^4 \cdot b^{-3} \cdot b^2$

6. Simplify:
   a) $x^{-4} \div x^2$
   b) $y^3 \div y^{-2}$
   c) $m^5 \div m^{-5}$
   d) $a^{-3} \div a^{-3}$
   e) $c^{-1} \div c^{-2}$
   f) $x^{-2} \div x^{-4}$

7. Simplify:
   a) $(x^{-2})^3$
   b) $(y^{-1})^{-2}$
   c) $(m^{-3})^2$
   d) $(c^3)^{-3}$
   e) $(a^4)^{-1}$
   f) $(x^{-1}y^2)^{-1}$
   g) $(x^2y^{-3})^2$
   h) $(a^{-1}b^{-2})^{-1}$
   i) $(a^{-2}b^2)^{-2}$

8. Simplify:
   a) $(a^{-2}b^4)(a^2b^{-5})$
   b) $(x^{-2})^3 \div (x^3)^{-2}$
   c) $x^2y^{-2} \div y^{-1}$
   d) $(x^{-1}y^2)^{-3}(x^2)^{-1}$
   e) $(c^{-3}d)^{-1} \div (c^2d)^{-2}$
   f) $(m^3n^{-2})(m^{-1}n^2)^{-1}$

9. Evaluate for $x = -1$ and $y = 2$:
   a) $(x^3y^2)(x^2y^3)$
   b) $\dfrac{x^4y^5}{xy^3}$
   c) $(x^3y^2)^3$
   d) $(x^{-1}y^{-2})(x^{-2}y^{-3})$
   e) $\dfrac{x^{-3}y^{-2}}{x^2y^{-6}}$
   f) $(x^{-4}y^{-3})^{-2}$

10. Explain why the following are incorrect:
    a) $2^2 \times 2^3 = 2^6$
    b) $3^4 \times 3^2 = 9^6$
    c) $x^{12} \div x^6 = x^2$
    d) $(2x^2)^3 = 8x^5$
    e) $(2y^3)^2 = 2y^6$
    f) $6^8 \div 3^4 = 2^4$

11. a) Make a table of powers of 5 from 5^{-8} to 5^{10}.
    b) Use the table to evaluate the following without actually doing the arithmetic:
       i) 125×625
       ii) 625^2
       iii) $0.0016 \times 78\,125$
       iv) $125 \div 0.000\,32$
       v) $78\,125^{-1}$
       vi) $390\,625 \div 15\,625$

C 12. Simplify:
   a) $45x^{-3} \div 5x^5 \times 3x^{-7}$
   b) $4a^{-7} \times 12a^{-5} \div (-8a^{-4})$
   c) $(2x^2y^2)^{-1} \div (2x^2y)^{-2}$
   d) $\left(\dfrac{p^2}{q}\right)^{-3} \div \left(\dfrac{q}{p^6}\right)$
   e) $\left(\dfrac{-2x^2}{3y}\right)^{-3} \div \dfrac{1}{x}$
   f) $\left(\dfrac{a^{-2}b}{c^3}\right)^{-3} \div \left(\dfrac{a^5b^{-2}}{c^3}\right)^{-1}$

THE MATHEMATICAL MIND

A Famous Unsolved Problem

*Pierre de Fermat
1601–1665*

One of the most baffling problems in mathematics involves the natural numbers. This problem originated with the great French mathematician, Pierre de Fermat, who had the habit of writing his results, without the proofs, in the margins of his books. Therefore, his successors had to discover the proofs for themselves. He must have had proofs because, with only one exception, all his results have been proven true by others. The one exception is known as Fermat's Last Theorem.

Fermat's Last Theorem states that, if $n > 2$, there are no natural numbers that satisfy the equation:

$x^n + y^n = z^n$.

Fermat wrote this statement in the margin of a book, and added: "I have discovered a truly marvelous demonstration of this proposition which this margin is too narrow to contain."

A Short History of Fermat's Last Theorem

1
Some mathematicians have proved the theorem for specific values of n:

*Leonhard Euler
1707–1783
Proved for $n = 3$*

*A.M. Legendre
1752–1833
Proved for $n = 5$*

*E. Kummer
1810–1893
Proved for $n < 100$*

2

In their attempts to prove Fermat's Last Theorem for all values of n, these distinguished mathematicians developed proofs which turned out to contain errors:

G. Lamé 1840

E. Kummer 1843

F. Lindemann 1907

Lindemann worked for 7 years on his 63-page proof, but he made a mistake at the outset.

3

Mathematics departments of many universities charge a fee to evaluate proofs sent by amateur mathematicians.

> Dear Sir,
> We regret to inform you there is a mistake in your proof of Fermat's Last Theorem on page ___, line ___.
>
> Yours truly

4

In the 1970's, S.S. Wagstaff of the University of Illinois showed, using a computer, that Fermat's Last Theorem is true for all prime exponents less than 125 000. This means that if there are any natural numbers that satisfy;

$$x^n + y^n = z^n \ (n > 2),$$

the numbers involved in the equation would be so large that neither people nor computers would be able to work with them.

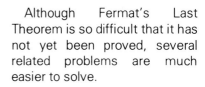

Although Fermat's Last Theorem is so difficult that it has not yet been proved, several related problems are much easier to solve.

1. Give examples of natural numbers that satisfy: $x^2 + y^2 = z^2$.

2. Give examples of natural numbers that satisfy:
   a) $x^2 + y^2 + z^2 = w^2$;
   b) $x^3 + y^3 + z^3 = w^3$;
   c) $x^2 + y^2 = z^2 + w^2$;
   d) $x^3 + y^3 = z^3 + w^3$.

3. Assuming that there are no natural numbers that satisfy the equation: $x^3 + y^3 = z^3$, explain why there could not be any natural numbers that satisfy the equation: $x^6 + y^6 = z^6$.

2-4 Operations With Monomials: Addition and Subtraction

Algebraic terms such as $5x^2$ and $-9x^3y$ are called **monomials**. A monomial is the product of a coefficient and one or more variables.

Monomials that have the same variables raised to the same exponent are called **like terms,** and they can be added or subtracted.

Example 1. Simplify: a) $6a - 4b - 2c - 5a + 5b + 3c$
b) $5x - 3x^2 + 4 - 7x^2 - 6x - 1$

Solution. a) $\quad 6a - 4b - 2c - 5a + 5b + 3c$
$= 6a - 5a - 4b + 5b - 2c + 3c$
$= a + b + c$

b) $\quad 5x - 3x^2 + 4 - 7x^2 - 6x - 1$
$= -3x^2 - 7x^2 + 5x - 6x + 4 - 1$
$= -10x^2 - x + 3$

(Group like terms.)

Example 2. Simplify: a) $6x^2y - 3xy^2 + 2y^2x - 5yx^2 + 2x^2y^2$
b) $3 - 4x + 7yx + 2xy^2 - 6xy - 8 - 2x^2y$

Solution. a) $\quad 6x^2y - 3xy^2 + 2y^2x - 5yx^2 + 2x^2y^2$
$= 2x^2y^2 + 6x^2y - 5yx^2 - 3xy^2 + 2y^2x$
$= 2x^2y^2 + x^2y - xy^2$

b) $\quad 3 - 4x + 7yx + 2xy^2 - 6xy - 8 - 2x^2y$
$= 2xy^2 - 2x^2y + 7yx - 6xy - 4x - 8 + 3$
$= 2xy^2 - 2x^2y + yx - 4x - 5$

In the last example, note that xy^2 and y^2x are like terms but xy^2 and x^2y are not. Why?

Exercises 2-4

A 1. Simplify:
a) $3x - 5y - x + 2y$
b) $7a + 6b - 11a + 9b$
c) $8c + 11d - 13c - 4d$
d) $2x - 4y - 9x - 3y$
e) $19m - 4n + 11m - 17n + 9n$
f) $5a + 2b - 8a - 7b + 3a$

2. Simplify:
   a) $5r - 9t + 16s - 11t - 14r + 16t$
   b) $8x - 5y + 3z - 13x - 22y + 11z$
   c) $4a + 7b + 3c - 10a - 4b + 8c$
   d) $13m - 5n + 6p + 11m + 17n - 15p$
   e) $2h - 9k - 14l - 14k + 2k + 8l$
   f) $4x - 9y + 7z - 17x - 13z + 7y - 5x$

3. Simplify:
   a) $a^2 + 4a - 5 + 3a^2 - 6a + 1$
   b) $-8m^2 + 6m - 9 - 13m + 3m^2 + 2$
   c) $-12s^2 - 4s + 11 + 7s^2 - 10s - 6$
   d) $5x^2 - 2x + 8 - 7x - 3 + x^2$
   e) $12p^2 + 9p - 4 - 7p^2 - 13p + 17$
   f) $14g^2 - 5g + 10 - 8g^2 - 6g - 19$

B 4. Simplify:
   a) $5xy + 7xz - 19xy - 4xz + 15xy$
   b) $4ab + 7ac - 5bc + 6ab - 3ac - bc$
   c) $7pq - 3pr + 8qr - 12pq + 11pr - 5qr$
   d) $mn + 8mp - 3np + 5mn - 3mp - 8np$
   e) $4de + 7df - 3ef - 9ed - 4df - 12ef$
   f) $12xy - 2xz + 7yx - 18xz - 9xz + 20xy$

5. Simplify:
   a) $3ab - 5ac + 9ab + 7ac$
   b) $6x^2 - 3xy + xy - 4x^2 - 11xy$
   c) $8y^2 + 7xy - 3xy - 12y^2 - 5xy$
   d) $41xy + 71xy - 37yx + 5xy - 22xy$
   e) $5m^2 - 9mn - 12m^2 + 4nm$
   f) $-14c^2 + 3cd + 5c^2 - 17cd - 9c^2$

6. Simplify:
   a) $5x^2y - 3y^2x + 2x^2y + 3xy^2 - 4y^2x^2$
   b) $2z^3y^2 - 3y^2z^3 + y^3z^3 - 3yz^3$
   c) $m^2n^2 - 7m + 2n - 2mn - 3n^2m^2$
   d) $3p^3q^2 - 7q^2p^2 + 3p^2q^3 - 3q^2p^3$
   e) $a^3b^2 - a^2b^3 - ab^3 - ba^3 + b^3a^2$
   f) $a^2 + b^2 - c^2 - 2ab - 2c^2 - b^2 + 2$

2-5 Operations With Monomials: Multiplication and Division

Products and quotients of monomials can be found using the exponent laws.

Using the law for multiplying powers	Using the law for dividing powers
Add the exponents. $(2x^6y^2)(-5x^2y) = -10x^8y^3$ Multiply the coefficients.	Subtract the exponents. $\dfrac{20x^6y^2}{-5x^2y} = -4x^4y$ Divide the coefficients.

Example 1. Simplify: a) $(5x^3y^4)(2xy^2)$ b) $(5ab^2)(-2a^2b)^3$
c) $\dfrac{9x^5y^4}{6x^3y}$ d) $\dfrac{(6m^2n^2)(8m^4n^3)}{(-4mn^2)^2}$

Solution. a) $(5x^3y^4)(2xy^2) = 10x^4y^6$
b) $(5ab^2)(-2a^2b)^3 = (5ab^2)(-8a^6b^3)$
$= -40a^7b^5$
c) $\dfrac{9x^5y^4}{6x^3y} = \dfrac{3}{2}x^2y^3$
d) $\dfrac{(6m^2n^2)(8m^4n^3)}{(-4mn^2)^2} = \dfrac{48m^6n^5}{16m^2n^4}$, or $3m^4n$

Many formulas for volume involve monomials.

Sphere

$V = \dfrac{4}{3}\pi r^3$

Cone

$V = \dfrac{1}{3}\pi r^2 h$

Cylinder

$V = \pi r^2 h$

Rectangular Prism

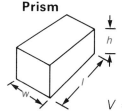

$V = lwh$

2-5 Operations With Monomials: Multiplication and Division   53

Example 2. Three golf balls fit snugly into a rectangular box. What fraction of the space in the box do they take up?

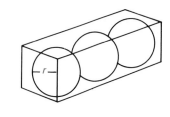

Solution. Each golf ball is a sphere. Let the radius be r.
Dimensions of the box (a rectangular prism):
$$2r \times 2r \times 6r$$
Volume of the box: $24r^3$

Volume of each golf ball: $\frac{4}{3}\pi r^3$

Volume of three golf balls: $3 \times \frac{4}{3}\pi r^3 = 4\pi r^3$

Fraction of the space occupied by the balls:
$$\frac{4\pi r^3}{24r^3} = \frac{\pi}{6}, \text{ or approximately } 0.52$$

The golf balls occupy slightly over half the space in the box.

Exercises 2 - 5

A 1. Simplify:
   a) $(7x^2)(5x^4)$ b) $(3a)(4a)$ c) $(-6m^3)(9m^2)$
   d) $(5n^2)(2n)$ e) $(-8y^5)(-7y^4)$ f) $(12p^5)(-3p^3)(-2p)$

2. Simplify:
   a) $(6x)(3y)$ b) $(8p^2)(4q)$ c) $(-5m^3)(-7n)$
   d) $(2ab)(7a)$ e) $(-9r)(5rs^3)$ f) $(4cd)(-11c)$

3. Simplify:
   a) $(-4ab)(2b^2)$ b) $(6m^2n^3)(-2m^3n^4)$
   c) $(7a^2b^2)(3ab^3)$ d) $(-8p^2q^5)(-3pq^2)$
   e) $(2xy)(3x^2y)(4x^3y)$ f) $(-3ab)(2ab^2)(a^2b)$

4. Simplify:
   a) $\dfrac{12x^5}{3x^2}$ b) $\dfrac{36m^6}{4m^2}$ c) $\dfrac{54a^8}{-6a^2}$
   d) $\dfrac{-24cd^3}{-6cd}$ e) $\dfrac{28m^2n^5}{-7m^2n^3}$ f) $\dfrac{-32a^3b^6}{-4a^3b^2}$

B 5. Simplify:
   a) $\dfrac{36r^6s^4}{8r^2s^2}$ b) $\dfrac{-45x^8y^5}{15x^4y^2}$
   c) $\dfrac{-35p^9q^6}{-14p^3q^2}$ d) $\dfrac{(3a^4b^2)(6ab^3)}{9a^3b^4}$
   e) $\dfrac{(4m^4n^2)(6m^8n^4)}{-3m^6n^6}$ f) $\dfrac{(-9x^3y^6)(8x^7y^4)}{(2x^2y^3)(-6x^3y)}$

6. Simplify:
   a) $(2x^5)(3x^2)^3$
   b) $(-3a^2b)(2a^4)^2$
   c) $(5s^2t^5)(3s^4)^2$
   d) $(-4pq^3)(-5p^3)^2$
   e) $(3m^2n^3)(-2mn^2)^3$
   f) $(5x^2y)^2(3x^2y^4)^3$

7. Simplify:
   a) $\dfrac{(2xy^2)^3(3x^5y^4)}{4x^2y^5}$
   b) $\dfrac{(-3p^2q^5)(-4pq^3)^2}{8p^4q^4}$
   c) $\dfrac{(-7a^2b^4)(4a^3b^5)}{(-2ab^2)^3}$
   d) $\dfrac{(-5x^4y^5)^2(2xy^2)^3}{(10x^3y^8)^2}$
   e) $\dfrac{(-4m^2n^4)^3(-3m^3n)^2}{(-6m^2n^3)^2}$
   f) $\dfrac{(3a^6b^2)^2(-2a^2b^4)^3}{(6ab)^2(-2a^4b)}$

8. Express these formulas in terms of the diameter, d:
   a) The area, A, of a circle: $A = \pi r^2$
   b) The surface area, A, of a sphere: $A = 4\pi r^2$
   c) The volume, V, of a sphere: $V = \dfrac{4}{3}\pi r^3$

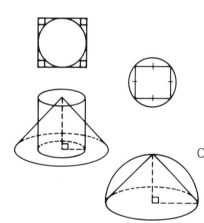

9. A circle is inscribed in a square. Find the ratio of their areas.

10. A square is inscribed in a circle. Find the ratio of their areas.

11. A cone has the same height as a cylinder but the radius of its base is twice as great. Find the ratio of their volumes.

C 12. The base of a hemisphere is also the base of an inscribed cone. Find the ratio of the volumes of the hemisphere and cone.

13. What is the height of a cylindrical silo with a capacity of $3x^2y^3$ units^3 and a base area of $2xy^2$ units^2? Calculate the height when $x = 7$ units and $y = 12$ units.

14. If $x = 5k$ and $y = -2k$, write the expression $x^2 + 2xy + y^2$ in terms of k, and simplify.

15. If $x:y = 3:2$, write the expression $x^2 - xy + y^2$
    a) in terms of x only;
    b) in terms of y only.

16. If $x = 2a^2b$ and $y = -ab^2$, write in terms of a and b:
    a) $3xy$
    b) $8xy^3$
    c) $-10x^3y^2$
    d) $(5xy)^2$
    e) $\sqrt{4x^2y^4}$

17. Tennis balls are packed in cans of 3. What fraction of the can is empty space?

2 - 6 Sums and Differences of Polynomials

Expressions formed by adding or subtracting monomials are called **polynomials**. Examples are:

$$y^2 \qquad 3x^3 \qquad 5x - 6y^2$$
$$3a^2b^3 - 5ab - 6b^3 + 7ab^2$$

Polynomials are added and subtracted in the same way as monomials. Like terms are combined. The term with the greatest exponent, or exponent sum, determines the **degree** of the polynomial.

Example 1. Simplify, and state the degree of the polynomial:
  a) $(2x - 5z + y) - (7x + 4y - 2z)$
  b) $(2x^3 - 4xy^2 + 5x^2y^2) + (3x^3 + 2x^2y - 6x^2y^2)$

Solution. a) $(2x - 5z + y) - (7x + 4y - 2z)$ *(Multiply the second expression by −1 using the distributive law.)*
  $= 2x - 5z + y - 7x - 4y + 2z$
  $= -5x - 3y - 3z$
  This is a first-degree polynomial.

  b) $(2x^3 - 4xy^2 + 5x^2y^2) + (3x^3 + 2x^2y - 6x^2y^2)$
  $= 2x^3 - 4xy^2 + 5x^2y^2 + 3x^3 + 2x^2y - 6x^2y^2$
  $= 5x^3 - x^2y^2 + 2x^2y - 4xy^2$
  The second term, x^2y^2, has the greatest exponent sum, 4. This is a fourth-degree polynomial.

To identify and collect like terms it is helpful to write all terms so that the variables in each appear in the same order, usually alphabetical.

Example 2. Simplify and state the degree of the polynomial:
$(6x^3y^2 - 2y^3x^2 - 3x^2y^2) - (5y^2x^2 - 2x^2y^3 + 4y)$

Solution. $(6x^3y^2 - 2y^3x^2 - 3x^2y^2) - (5y^2x^2 - 2x^2y^3 + 4y)$
$= (6x^3y^2 - 2x^2y^3 - 3x^2y^2) - (5x^2y^2 - 2x^2y^3 + 4y)$
$= 6x^3y^2 - 2x^2y^3 - 3x^2y^2 - 5x^2y^2 + 2x^2y^3 - 4y$
$= 6x^3y^2 - 8x^2y^2 - 4y$
The greatest exponent sum is 5. This is a fifth-degree polynomial.

Exercises 2 - 6

A 1. Simplify:
  a) $(6a + 9) + (4a - 5)$
  b) $(2x - 7y) + (8x - 3y)$
  c) $(3m^2 + 4) - (7 + 9m^2)$
  d) $(5p + 11) - (2p + 4)$
  e) $(17x^2 - 9x) - (6x^2 - 5x)$
  f) $(7k + 3l) - (12k - 8l)$

2. Simplify:
   a) $(3m^2 - 5m + 9) + (8m^2 + 2m - 7)$
   b) $(7a^3 - 2a^2 + 5a) + (-3a^3 + 6a^2 - 2a)$
   c) $(x^2 - 5x + 6y) - (4x^2 + 15x - 11y)$
   d) $(5t^2 - 13t + 17) + (9t^3 - 7t^2 + 3t - 26) - (16t^2 + 5t - 8)$
   e) $(2 + 8a^2 - a^3) + (2a^3 - 3a^2 - 8) - (12 - 7a^3 + 5a^2)$
   f) $(4x^2 - 7x + 3) - (x^2 - 5x + 9) - (8x^2 + 6x - 11)$

3. Simplify, and state the degree of the polynomial:
   a) $(5x^2 - 3x^2y + 2y^2x - y) - (2xy^2 - y - 5x^2 + 3yx^2)$
   b) $(p^3q^2 + 7q^2p^2 - 3p) + (2q - 6p^2q^2 - q^2p^3)$
   c) $(3m^2 - 3m + 7) - (2n^2 - 2n + 7)$
   d) $(2ab - 2ac - 2bc) + (2ca + 2cb - 2ba + 3)$
   e) $(xy + 3yx + 2x^2y - 3xy^2) - (2xy + 2yx^2)$

4. Add or subtract as indicated:
   a) $3x^2 - 6x + 7$
      $\underline{+ 7x^2 - 2x + 9}$

   b) $7z^3 - 6z^2 + z - 3$
      $\underline{- 2z^3 + 4z^2 - z + 4}$

   c) $5x^2y - 3xy + 2xy^2$
      $\underline{+ 3x^2y + 4xy - 3xy^2}$

   d) $6m^2n^2 - 4mn + 3mn^3$
      $\underline{- 2m^2n^2 + 2mn - 4mn^3}$

5. Simplify, and state the degree of the polynomial:
   a) $(3x^2 - 2y^2) + (y^2 - 2x^2) - (4x^2 + 2)$
   b) $(m^2 - n^2) - (n^2 - m^2) + (2n^2 + m^2)$
   c) $(4x^2y - 2yx) + (3yx^2 - 6xy^2) - (3x^2y^2 + 2y^2x^2 - xy)$
   d) $(a^2b^2 - b^2) - (3b^2a + 2a^2b) - (b^2 - b^2a^2 - b^2a)$

2-7 Products of Monomials and Polynomials

To multiply a polynomial by a monomial, use the distributive law. The result is another polynomial.

Example 1. Expand: a) $4x^2(2x - 5)$
  b) $-5a^3b(3ab - 2a^2 + b)$

Solution. a) $4x^2(2x - 5) = 8x^3 - 20x^2$
  b) $-5a^3b(3ab - 2a^2 + b) = -15a^4b^2 + 10a^5b - 5a^3b^2$

2-7 Products of Monomials and Polynomials

Example 2. Simplify: a) $4(2a - 3b + c) - 3(a + 5b - 2c)$
b) $2x(3x^2 - 5xy + y^2) - 5y^2(x - 2y)$

Solution.
a) $\quad 4(2a - 3b + c) - 3(a + 5b - 2c)$
$= 8a - 12b + 4c - 3a - 15b + 6c$
$= 5a - 27b + 10c$

b) $\quad 2x(3x^2 - 5xy + y^2) - 5y^2(x - 2y)$
$= 6x^3 - 10x^2y + 2xy^2 - 5xy^2 + 10y^3$
$= 6x^3 - 10x^2y - 3xy^2 + 10y^3$

Exercises 2 - 7

A 1. Simplify:
   a) $4(5x^2 + 10)$
   b) $7(2a - 5)$
   c) $-8(3k^2 - 2k)$
   d) $12(2b^2 - 3b + 9)$
   e) $-9(-5m^2 + 7m - 3)$
   f) $3(8p^2 - 5p + 7)$

2. Simplify:
   a) $3(x + 4) + 7$
   b) $-8(2a - 3) + 11a$
   c) $5(y + 2) - 7y$
   d) $4(7m - 5) - 13$
   e) $-6(3p^2 + 2p) + 5p^2$
   f) $7(5x - 3y) - 43x$

3. Simplify:
   a) $3(x + 2) + 2(x - 6)$
   b) $2(x + 9) - 3(x + 7)$
   c) $3(2a + 10b - 2c) - 6(a - 2b + 5c)$
   d) $3(2m - 4n + 3) - 5(-2m + 5n - 1)$
   e) $6(2x^2 - 5x) - 14(3x - x^2) + 3(x - x^2)$
   f) $-7(c - 3d + 5e) + 4(2c - 11d - 3e) - 5(3c - 7d + 2e)$

4. Simplify:
   a) $12x(5x - 4)$
   b) $3a(-7a + 2)$
   c) $6p(2p - q)$
   d) $-15n^2(6 - 9n)$
   e) $7m^3(3mn + 6)$
   f) $-8x^2(5x - 7y)$

B 5. Simplify:
   a) $3x^2(x + y) + 2x^2(3x + 5y)$
   b) $3a^3(2a - 5b) - 4a^3(2a + 3b)$
   c) $5p^2(4p - q) - 8p^2(2p - 7q)$
   d) $7m(2m - 5n + 3) + 2m(-3m + 9n - 4)$
   e) $4x^2(5x - 2y - 8) - 3x^2(4x + 8y - 2)$
   f) $6a^3(-3a + 7b - 4) - 8a^3(2a - 3b + 7)$

6. Simplify:
   a) $-2ab^2(ab - a^2 + b)$
   b) $3x^2y(xy^2 + xy - y)$
   c) $-5m^2n(3mn + mn^2 - n^2)$
   d) $5x(x - y) - 2y(x + y - 1) + y^2$
   e) $2b(b^2 - bc) - 2c(b - c) + (7bc - 4c^2)$
   f) $7x(x^2 - y^2) - 2xy - 2y(x^2 + y^2)$

7. Simplify, and state the degree of the polynomial:
   a) $3a^2b - 7a(a - 4) + (5ba^2 - 13a)$
   b) $6m(2mn - n^2) - 3(m^2n - 19mn^2) + 6n(-5mn - 7m^2)$
   c) $4xy(x + 2y) - 5x(2xy - 3y^2) - 3y(3x^2 + 7xy)$
   d) $5s(3s^2 - 7s + 2) + 2s(5s^2 + s - 6) - 9s(2s^2 + 6s - 4)$
   e) $4x(5x^2 - 2xy + y^2) - 9x(2x^2 + 6y^2 - xy) + 3x(x^2 - 5xy - y^2)$
   f) $5xy - 3y(2x - y) + 7x(8y - x) + 9(x^2 - 2y^2)$

8. Simplify:
   a) $(11a + 3)(9a)$
   b) $(3x - 5)(-2x)$
   c) $2.5x(16x + 12)$
   d) $(5a^2 + 2a - 3)(-4a)$
   e) $1.2n(3n - 7)$
   f) $5x(2x^2 - 6x + 3)$

9. Simplify:
   a) $5[(3p + 7) - 4q] + 2[(7p - 3) + q]$
   b) $8[2k - (3l + 5)] - 5[9k - (4l - 7)]$
   c) $3x[(x - 3) + y] - 2y[(y + 7) - x]$
   d) $2a[3a + 2(5a - 4)] + 3a[a - 4(2a + 7)]$
   e) $3x[x - 4x(y - 4)] - 2x^2(x - 2y + 3)$
   f) $5m[m - 2n(m + 3)] - 2m[3m - 4n(5m - 8)]$

10. If $x = a + b$ and $y = a - b$, write in terms of a and b, and simplify:
    a) $3x - 2y$
    b) $7x + 3y$
    c) $ax - by$
    d) $a^2y - b^2x$

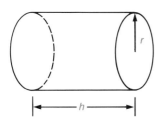

C 11. a) Write a formula for the surface area of a closed cylinder in terms of its radius and height.
   b) Find the surface area of a closed cylinder when
      i) the radius is 5 cm and the height is 24 cm;
      ii) the height is three times the radius;
      iii) the height is 5 units less than three times the radius;
      iv) the height is 7 units more than half the radius.

2-8 Products of Binomials and Trinomials

Polynomials with two terms are called **binomials**, and polynomials with three terms are called **trinomials**. Examples are:

Binomials: $2x - 5$ $\qquad$ $3a + 2b$ $\qquad$ $-4x^2y + 5xy$
Trinomials: $3x^2 - 4x + 2$ $\qquad$ $2m^2 - 3mn + n^2$

To find the product of two binomials, multiply each term of one binomial by both terms of the other binomial.

Example 1. Find the product: a) $(3x - 2)(x + 7)$
$\qquad\qquad\qquad\qquad\quad$ b) $(x + 5y)^2$

Solution. a) $\quad (3x - 2)(x + 7)$
$\qquad\quad = 3x(x + 7) - 2(x + 7)$
$\qquad\quad = (3x)(x) + (3x)(7) + (-2)(x) + (-2)(7)$
$\qquad\quad = 3x^2 + 21x - 2x - 14$
$\qquad\quad = 3x^2 + 19x - 14$

$\qquad$ b) $(x + 5y)^2 = (x + 5y)(x + 5y)$
$\qquad\qquad\qquad\quad = x^2 + 5xy + 5xy + 25y^2$
$\qquad\qquad\qquad\quad = x^2 + 10xy + 25y^2$

When squaring binomials, recognize and use these identities:

$$(a + b)^2 = a^2 + 2ab + b^2$$
$$(a - b)^2 = a^2 - 2ab + b^2$$

Example 2. Simplify: $(2x - 1)(3x + 4) - (2x - 3)^2$

Solution. $\quad (2x - 1)(3x + 4) - (2x - 3)^2$
$\qquad\quad = (6x^2 + 5x - 4) - (4x^2 - 12x + 9)$
$\qquad\quad = 6x^2 + 5x - 4 - 4x^2 + 12x - 9$
$\qquad\quad = 2x^2 + 17x - 13$

Example 3. What is the area of the shaded part of the rectangle?

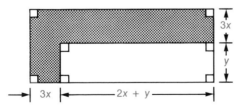

Solution.

Rectangle	Length	Width	Area	
Large	$5x + y$	$3x + y$	$(5x + y)(3x + y)$	...(i)
Small	$2x + y$	y	$(2x + y)y$	...(ii)

Shaded area: (i) − (ii)
$$= (5x + y)(3x + y) - (2x + y)y$$
$$= 15x^2 + 8xy + y^2 - 2xy - y^2$$
$$= 15x^2 + 6xy$$

The method of finding the product of two binomials can be extended to find the product of a binomial and a trinomial.

Example 4. Simplify: $(a - 2)(a^2 + 2a + 4) + 8(a + 1)(a - 1)^2$

Solution. $(a - 2)(a^2 + 2a + 4) + 8(a + 1)(a - 1)^2$
$$= (a^3 + 2a^2 + 4a - 2a^2 - 4a - 8) + 8(a^2 - 1)(a - 1)$$
$$= (a^3 - 8) + 8(a^3 - a^2 - a + 1)$$
$$= a^3 - 8 + 8a^3 - 8a^2 - 8a + 8$$
$$= 9a^3 - 8a^2 - 8a$$

$(a + 1)(a - 1)^2$
$= (a + 1)(a - 1)(a - 1)$
$= (a^2 - 1)(a - 1)$

Exercises 2 - 8

A 1. Find the product:
   a) $(4x + 1)(2x + 3)$
   b) $(5m - 2)(6m + 1)$
   c) $(2c - 3)(4c - 5)$
   d) $(3x - 7)(3x + 7)$
   e) $(4t + 5)(4t - 8)$
   f) $(8x - 2)(3x - 7)$

2. Square the binomial:
   a) $(g + h)$
   b) $(x + y)$
   c) $(m + n)$
   d) $(p - q)$
   e) $(k - l)$
   f) $(s - t)$

3. Find the product:
   a) $(3x + y)^2$
   b) $(x + 5y)^2$
   c) $(3a + 2)^2$
   d) $(5x - 9)^2$
   e) $(4m - 7)^2$
   f) $(-4y - 11)^2$

4. Find the product:
   a) $(2x + y)(3x + y)$
   b) $(3a + b)(5a - b)$
   c) $(2x - y)(3x + 4y)$
   d) $(x + 3y)(x + 4y)$
   e) $(5m - 2n)(7m - n)$
   f) $(6x - 2y)(3x - 7y)$

5. Find the product:
   a) $(3a + 5b)^2$
   b) $(2m - 7n)^2$
   c) $(6s - 2t)^2$
   d) $(4p - 3q)^2$
   e) $(8x - 3y)^2$
   f) $(5y^2 + 7z^2)^2$

6. Find the product:
   a) $(x - y)(x + y)$
   b) $(2m + 5n)(2m - 5n)$
   c) $(9a - 4b)(9a + 4b)$
   d) $(6x - 2y)(6x + 2y)$
   e) $(7p + 2q)(7p - 2q)$
   f) $(-12m + 9)(-12m - 9)$

2-8 Products of Binomials and Trinomials

B 7. Simplify:
  a) $(x - 3)(x + 2) + (2x - 5)(x - 1)$
  b) $(2x + 4)(3x - 2) + (5x - 2)(3x - 4)$
  c) $(2m + 3)(3m + 5) + (m - 7)(6m - 3)$
  d) $(c + 2d)(c + d) - (c - 2d)(c - d)$
  e) $(2a - 5b)(3a + b) - (6a - b)(4a + 7b)$
  f) $(3x - 2y)(5x - y) - (2x + 5y)(3x - y)$

8. Simplify:
  a) $(2x - 3)^2 + (x + 2)(3x - 5)$
  b) $(s - 2t)^2 - (s - 3t)(4s - 5t)$
  c) $(2x - y)(2x + y) - (x - 2y)^2$
  d) $(7m - 2n)(3m + 5n) - (4m - 11n)^2$
  e) $(5x - 2y)^2 - (5x + 2y)^2$
  f) $(2x^2 + 3)(x^2 - 5) - (3x^2 + 2)^2$

9. Find the product:
  a) $(3x - 1)(2x^2 + 3x - 4)$ b) $(m - 3)(4m^2 - 7m + 12)$
  c) $(2a - 5)(3a^2 + 8a - 9)$ d) $(3p + 2)(5p^2 - 6p + 2)$
  e) $(2x^2 - 7x - 4)(2x - 5)$ f) $(8y^2 + 3y - 7)(3y + 4)$

10. Find the product:
  a) $(3x + 4)(x - 5)(2x + 8)$
  b) $(4p - 3)(2p - 7)(5p - 6)$
  c) $(3x + 2)(2x - 5)(4x - 3)$
  d) $(2x - 5)(3x + 4)^2$
  e) $(5a - 3)^2(2a - 7)$ f) $(5m - 2)^3$

11. Find the perimeter and area of each shaded region:
  a)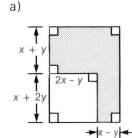
  b)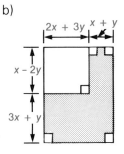
  c)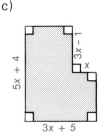

12. Simplify:
  a) $3(m - 5)(m + 2)$ b) $2(x + y)(3x - 5y)$
  c) $5(2p - 7)(3p - 4)$ d) $4x(x - 7y)(2x - 3y)$
  e) $3a(2a - 5b)(3a + b)$ f) $3x(2x + 4y)^2$

13. Simplify:
   a) $2(3x + 1)(x + 5) + 4(2x + 3)(3x + 5)$
   b) $5(2m - 4)(3m + 2) - 2(4m - 1)(3m - 4)$
   c) $3(2x + 5y)(7x - 8y) + 5(x + 3y)(2x + 4y)$
   d) $2(xy - 1)(xy + 1) - 5(xy - 2)(xy + 3)$
   e) $3(2a - 3b)(a + 2b) - 2(3a - b)^2$
   f) $4(xy - 2)^2 + 3(xy + 5)^2$

14. Simplify:
   a) $(2x - 5)(3x + 6) + (3x - 2)(4x + 9) - (5x - 3)(2x - 7)$
   b) $2(3s + 5)(2s - 2) - 5(3s^2 + 7s - 9) - (s + 6)^2$
   c) $(2x + 3)(3x^2 - 5x + 4) + (x - 4)(2x^2 + 3x - 7)$
   d) $(3m + 4)(m - 4n - 1) - (5m - 2)(3m - 6n - 8)$
   e) $(4y - 5)(3y + 2)^2 - (3y + 2)(4y - 5)^2$
   f) $4(3x - 2)(5x^2 + x + 6) - (2x + 6)(3x - 1)^2$

C 15. If $x = m + 3$, write these expressions in terms of m, and simplify:
   a) $x^2 + 5x + 2$ b) $3x^2 - 2x + 7$ c) $x^3 - 4x^2 + 3x - 5$

16. If $x = a + b$ and $y = a - b$, write these expressions in terms of a and b, and simplify:
   a) $3x^2 + y^2$ b) $x^2 - xy - 5y^2$ c) $4x^2 + 3xy + y^2$

2-9 Common Factors

The greatest common factor of two or more monomials is formed from the greatest common factor of their coefficients and the greatest common factor of their variables.

Example 1. Find the greatest common factor:
$3x^2y^3$, $-9x^3y^4$, $12x^3y^2$.

Solution. The greatest common factor of the coefficients is 3. The greatest common factor of the variables is x^2y^2. Therefore, the greatest common factor is $3x^2y^2$.

To factor a polynomial means to express it as a product. This can be done, using the distributive law, when all the terms share a common factor.

Example 2. Express as a product: $3x^2y^3 - 9x^3y^4 + 12x^3y^2$.

Solution. $3x^2y^3 - 9x^3y^4 + 12x^3y^2 = 3x^2y^2(y - 3xy^2 + 4x)$

Example 3. Factor, if possible: a) $12a^2b - 8ab^2$
b) $5x^2y^2 + 15x^3y^2 - 10x^4y$ c) $2m^2 + 3n$

Solution. a) $12a^2b - 8ab^2 = 4ab(3a - 2b)$
b) $5x^2y^2 + 15x^3y^2 - 10x^4y = 5x^2y(y + 3xy - 2x^2)$
c) $2m^2 + 3n$ cannot be factored using integers.

Some expressions have binomials or trinomials as common factors.

Example 4. Factor: a) $3a(4a + 5b) - 2b(4a + 5b)$
b) $(2 - 3x - x^2)x + 5(x^2 + 3x - 2)$

Solution. a) $3a(4a + 5b) - 2b(4a + 5b) = (4a + 5b)(3a - 2b)$
b) $\quad (2 - 3x - x^2)x + 5(x^2 + 3x - 2)$
$= (2 - 3x - x^2)x - 5(2 - 3x - x^2)$
$= (2 - 3x - x^2)(x - 5)$

common factor

Since
$(-1)(x^2 + 3x - 2)$
$= (2 - 3x - x^2)$

Example 5. The curved surface of a bar of circular cross section is to be nickel plated, the ends being left untreated. What fraction of the total surface area, S, of the bar will be plated?

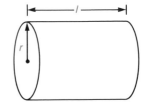

Solution. Let the radius of the bar be r and its length be l. Total surface area:
S = Area of ends + Area of curved surface
$= 2 \times \pi r^2 + 2\pi rl$

$$\frac{\text{Area of curved surface}}{\text{Total surface area}} = \frac{2\pi rl}{2\pi r^2 + 2\pi rl}$$
$$= \frac{2\pi rl}{2\pi r(r + l)}$$
$$= \frac{l}{r + l}$$

The fraction of the surface to be plated is $\dfrac{l}{r + l}$.

Exercises 2 - 9

A   1. Find the greatest common factor:
a) $21x^2, \ 28x^4, \ -14x$
b) $-3y^4, \ 8y^6, \ -6y^9$
c) $12a^3b, \ -16a^2b^2, \ -24a^2b^3$
d) $16m^2n^5, \ 24m^3n^4, \ 32m^5n^2$
e) $9x^2y^3, \ -12x^3y^2, \ 15x^2y^4$
f) $-54s^3t^2, \ 36s^5t, \ -72s^2t^2$

2. Express as products:
   a) $3x^2 + 6x$
   b) $8y^3 - 4y^2$
   c) $5p^3 - 15p^2$
   d) $24m^2n + 16mn^2$
   e) $12a^2b^2 + 18a^3b^2$
   f) $-28x^2y^3 - 35x^3y^2$

3. Express as products:
   a) $3w^2 - 7w^3 + 4w$
   b) $-2x^5 + 4x^2 - 6x + 2x^3$
   c) $8x^2 - 12x^4 + 16$
   d) $5ab^2 + 10ab - 15a^2b$
   e) $51x^2y + 39xy^2 - 72xy$
   f) $9m^4n^2 - 6m^3n^3 + 12m^2n^4$

4. Factor:
   a) $5y - 10$
   b) $8m + 24$
   c) $6 + 12x^2$
   d) $35a + 10a^2$
   e) $49b^2 - 7b^3$
   f) $35z^2 - 14z^6$

5. Factor:
   a) $3x^2 + 12x - 6$
   b) $3x^2 + 5x^3 + x$
   c) $a^3 + 9a^2 - 3a$
   d) $3x^2 + 6x^3 - 12x$
   e) $16y^2 - 32y + 24y^3$
   f) $8x^2y - 32xy^2 + 16x^2y^2$

6. Factor:
   a) $6b^7 - 3b + 12$
   b) $5y^3 + 6y^2 + 3y$
   c) $16x + 32x^2 + 48x^4$
   d) $12y^4 - 12y^2 + 24y^3$
   e) $9a^3 + 7a^2 + 18a$
   f) $10z^3 - 15z^2 + 30z$

7. Factor:
   a) $25xy + 15x^2$
   b) $14m^2n - 21mn^2$
   c) $9a^2b^3 - 12a^2b^2$
   d) $4x^2y - 16xy^2$
   e) $12p^2q + 18pq^2$
   f) $27m^3n^2 - 15m^2n^3$

8. Factor:
   a) $10a^3b^2 + 15a^2b^4 - 5a^2b^2$
   b) $12mn^2 - 8mn - 20m^2n$
   c) $20x^2y - 15x^2y^2 + 25x^3y^2$
   d) $7a^3b^3 + 14a^2b^2 - 21ab^2$
   e) $8x^4y^4 - 16x^3y^3 + 32x^2y^2$
   f) $18a^2bc - 6abc + 30abc^2 - 24ab^2c^2$

B 9. Express as a product:
   a) $3x(a + b) + 7(a + b)$
   b) $m(2x - y) - 5(2x - y)$
   c) $(x + 4)x^2 + (x + 4)y^2$
   d) $5x(a + 3b) - 9y(a + 3b)$
   e) $10y(x - 3) + 7(x - 3)$
   f) $7w(x + w) - 10(w + x)$

2-9 Common Factors **65**

10. Express as a product:
    a) $3x^2(x - 7) + 2x(x - 7) + 5(x - 7)$
    b) $2m(a - b) - 3n(b - a) - 7(a - b)$
    c) $5a^2(x^2 + y) - 7a(x^2 + y) + 8(x^2 + y)$
    d) $6a(b - a) + 4b(a - b) - 3(b - a)$
    e) $4m^2(2x + y) - 3m(2x + y) - 7(2x + y)$
    f) $2x^2(3a - 2b) + 5x(3a - 2b) - 9(2b - 3a)$

11. Find the area of the shaded region:
    a)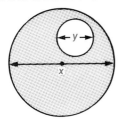
    b)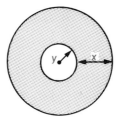

12. Four congruent circular arcs are drawn with centres at the vertices of a square of side x. Find an expression for the area of the figure that they form in terms of x.

    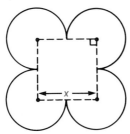

13. A cylindrical silo has a hemispherical top. Find expressions in terms of r and h to represent:
    a) the volume of the silo;
    b) the total surface area.

C 14. Factor:
    a) $x^2 + 3x + xy + 3y$
    b) $x^3 + x^2 + x + 1$
    c) $5am + a + 10bm + 2b$
    d) $3x^2 - 6xy + 5x - 10y$
    e) $5m^2 + 10mn - 3m - 6n$
    f) $2a^2 - 6ab - 3a + 9b$

COMPUTER POWER

When you finish entering the program, type R U N. Then type one number followed by a comma and the second number.

The Euclidean Algorithm

It is time-consuming to list the factors of two numbers, such as 270 and 84, in order to find their g.c.f. There is a more efficient method known as the **Euclidean algorithm**. It depends on the principle that if a is a factor of both b and $b + c$, it is also a factor of c.

Example. Find the g.c.f. of 270 and 84.

Solution. Divide 270 by 84:

$$270 = 3(84) + 18$$

The common factors of 270 and 84 are the same as the common factors of 84 and 18. Therefore, the g.c.f. of 270 and 84 is the same as the g.c.f. of 84 and 18. Divide 84 by 18:

$$84 = 4(18) + 12$$

The g.c.f. of 84 and 18 is the same as the g.c.f. of 18 and 12. The division process could be continued further but the g.c.f. of 18 and 12 is easily seen to be 6. Therefore, the g.c.f. of 270 and 84 is 6.

$$\begin{array}{r} 3 \\ 84{\overline{\smash{)}270}} \\ \underline{252} \\ 18 \end{array}$$

$$\begin{array}{r} 4 \\ 18{\overline{\smash{)}84}} \\ \underline{72} \\ 12 \end{array}$$

Because it involves repeated division, the Euclidean algorithm is ideally suited for computer programming. A simple BASIC program designed to find the g.c.f. of two positive integers follows.

```
10  INPUT "WHAT ARE THE TWO NUMBERS"; A, B
20  C = A − INT (A / B) * B
30  IF C = 0 THEN 50
40  A = B:B = C: GOTO 20
50  PRINT "THE G.C.F. IS "B
```

Exercises

Use the preceding program.

1. Find the g.c.f. of each pair of numbers:
    a) 391, 306   b) 1323, 884   c) 41 992, 30 508

2. Two numbers are said to be *relatively prime* if their g.c.f. is 1. Which of these pairs of numbers are relatively prime?
    a) 27, 16   b) 88, 56   c) 308, 273
    d) 9603, 5841   e) 12 507, 8266   f) 18 886, 10 887

3. Find the g.c.f. of these three numbers:
    a) 432, 288, 243   b) 3584, 3328, 2992

Mathematics Around Us

Meeting the World's Water Needs

The demand for fresh water for human consumption increases not only as the world's population increases but also as the standard of living improves. With our lawns, swimming pools, and cleaning methods, we are using water at a far greater rate than did our great-grandparents of only 100 years ago. Where is all the water to come from?

One of the ideas that has been put forward is to tow large icebergs from the polar regions to suitable points adjacent to large coastal cities. It would then be relatively simple technically to melt the ice and pump the water into the city's supply. As large, industrial cities in Canada require an estimated 800 L of fresh water per person per day, no possible source can be overlooked.

Questions

1. The population of Vancouver is about 1 500 000. How many cubic metres of fresh water does the city need each day?

2. The visible portion of an iceberg (11% of the whole iceberg) is 1 km long, 500 m wide, and 20 m high.
   a) What is the total volume of the iceberg?
   b) How many days' supply of water could be available to Vancouver from this iceberg assuming that 75% is recoverable?

2-10 Factoring Trinomials of the Form $x^2 + bx + c$

The result of multiplying one binomial by another binomial is often a trinomial.

$$(x + 6)(x + 2) = x^2 + 8x + 12$$

Product of +6 and +2
Sum of +6 and +2

To express a trinomial of the form $x^2 + bx + c$ as a product, two integers must be found that have a product of c and a sum of b.

Example 1. Factor: a) $x^2 + 11x + 24$ b) $m^2 - 10m + 21$
c) $a^2 + a - 12$

Factors of +24	
24,	1
12,	2
8,	3
6,	4

Solution. a) $x^2 + 11x + 24 = (x + 3)(x + 8)$

Product is +24. Sum is +11.

b) $m^2 - 10m + 21 = (m - 3)(m - 7)$

Product is +21. Sum is -10.

c) $a^2 + a - 12 = (a + 4)(a - 3)$

Product is -12. Sum is +1.

In some trinomials of the form $x^2 + bx + c$, b and c may be monomials in another variable. To factor, b must be the sum of two monomials and c the product.

Example 2. Factor: a) $x^2 + 12xy + 35y^2$
b) $p^2 - 3pq - 28q^2$
c) $m^2 - 10mn + 25n^2$

Solution. a) $x^2 + 12xy + 35y^2 = (x + 7y)(x + 5y)$

Product is $35y^2$. Sum is $12y$.

b) $p^2 - 3pq - 28q^2 = (p - 7q)(p + 4q)$

Product is $-28q^2$. Sum is $-3q$.

c) $m^2 - 10mn + 25n^2 = (m - 5n)(m - 5n)$

Product is $25n^2$. Sum is $-10n$.

2-10 Factoring Trinomials of the Form $x^2 + bx + c$

Example 3. For what integral values of k is $x^2 - kx + 15$ a product of two binomials?

Solution. $-k$ is the sum of two integers that have a product of 15. Since only integers are to be considered, there are just the four possibilities shown in the table.

Numbers with a product of 15	Sum $(-k)$	Value of k
3, 5	8	−8
−3, −5	−8	8
15, 1	16	−16
−15, −1	−16	16

If the terms of the trinomial have a common factor, it should be removed before the other factors are found.

Example 4. Factor:
    a) $5x^2 - 45x + 90$
    b) $3x^4 + 9x^3 - 12x^2$
    c) $-2a^4 + 2a^3b + 12a^2b^2$

Solution.
a) $5x^2 - 45x + 90 = 5(x^2 - 9x + 18)$
$$= 5(x - 6)(x - 3)$$

b) $3x^4 + 9x^3 - 12x^2 = 3x^2(x^2 + 3x - 4)$
$$= 3x^2(x + 4)(x - 1)$$

c) $-2a^4 + 2a^3b + 12a^2b^2 = -2a^2(a^2 - ab - 6b^2)$
$$= -2a^2(a - 3b)(a + 2b)$$

Exercises 2 - 10

A   **1.** Factor:
- a) $x^2 + 10x + 24$
- b) $m^2 + 5m + 6$
- c) $a^2 + 10a + 16$
- d) $p^2 + 8p + 16$
- e) $y^2 + 13y + 42$
- f) $d^2 + 11d + 18$

2. Factor
- a) $x^2 - 7x + 12$
- b) $c^2 - 17c + 72$
- c) $a^2 - 7a + 6$
- d) $b^2 - 12b + 32$
- e) $s^2 - 12s + 20$
- f) $x^2 - 11x + 28$

3. Factor:
   a) $y^2 + 4y - 5$
   b) $b^2 + 19b - 20$
   c) $p^2 + 15p - 54$
   d) $x^2 + 12x - 28$
   e) $a^2 + 2a - 24$
   f) $k^2 + 3k - 18$

4. Factor:
   a) $x^2 - 2x - 8$
   b) $n^2 - 5n - 24$
   c) $a^2 - a - 20$
   d) $d^2 - 4d - 45$
   e) $m^2 - 9m - 90$
   f) $y^2 - 2y - 48$

5. Factor:
   a) $x^2 + 14xy + 24y^2$
   b) $x^2 + 9xy + 18y^2$
   c) $p^2 + 15pq + 26q^2$
   d) $a^2 + 17ab + 72b^2$
   e) $k^2 + 11kl + 30l^2$
   f) $v^2 + 13vw + 36w^2$

6. Factor:
   a) $m^2 - 15mn + 50n^2$
   b) $a^2 - 11ab + 30b^2$
   c) $c^2 - 9cd + 20d^2$
   d) $t^2 - 13tv + 12v^2$
   e) $r^2 - 10rs + 9s^2$
   f) $x^2 - 16xy + 60y^2$

7. Factor:
   a) $a^2 + 8ab - 20b^2$
   b) $x^2 + 4xy - 96y^2$
   c) $b^2 + 7bc - 18c^2$
   d) $s^2 + 21st - 100t^2$
   e) $x^2 + 40xy - 41y^2$
   f) $y^2 + yz - 20z^2$

8. Factor:
   a) $m^2 - mn - 72n^2$
   b) $x^2 - 6xy - 16y^2$
   c) $p^2 - pq - 90q^2$
   d) $c^2 - 16cd - 80d^2$
   e) $m^2 - 7mn - 60n^2$
   f) $x^2 - 12xy - 45y^2$

9. Factor:
   a) $a^2 + 9a + 18$
   b) $m^2 - 5m - 36$
   c) $x^2y^2 - 15xy + 54$
   d) $p^2q^2 - 20pq + 51$
   e) $c^2d^2 - 9cd - 36$
   f) $x^2y^2 - 21xy - 72$
   g) $a^2b^2 + 19ab + 48$
   h) $m^2n^2 - 28mn - 60$

10. Find integral values of k so that each trinomial is the product of two binomials:
    a) $x^2 + kx + 18$
    b) $a^2 - ka - 12$
    c) $m^2 + km + 36$
    d) $y^2 - ky - 60$
    e) $n^2 - kn + 39$
    f) $b^2 + kb - 51$

2-10 Factoring Trinomials of the Form $x^2 + bx + c$ 71

B 11. Find five integral values of c so that each trinomial is the product of two binomials:
   a) $x^2 + 5x + c$
   b) $m^2 - 4m + c$
   c) $y^2 - 2y - c$
   d) $a^2 + 12a - c$
   e) $n^2 - 3n - c$
   f) $b^2 - 7b + c$

12. Factor:
   a) $4x^2 + 28x + 48$
   b) $9x^2 + 54x + 45$
   c) $2x^2 + 8x + 6$
   d) $5x^2 - 30x + 40$
   e) $15x^2 - 18xy + 3y^2$
   f) $60m^2 + 27mn + 3n^2$

13. Factor:
   a) $5a^2 + 15a - 20$
   b) $2a^2 - 8a - 24$
   c) $3p^2 + 15pq - 108q^2$
   d) $4y^2 - 16yz - 48z^2$
   e) $2m^2 + 2mn - 112n^2$
   f) $7x^2 - 84xy - 196y^2$

14. Factor:
   a) $3mn^2 + 18mn + 24m$
   b) $5am^2 - 40am + 35a$
   c) $15a^3 + 90a^2b + 135ab^2$
   d) $4x^2y - 24xy + 36y$
   e) $3p^2q^2 + 36pq^2 + 96q^2$
   f) $7c^2d - 35cd^2 + 42d^3$

15. Factor:
   a) $4x^2y - 20xy - 56y$
   b) $-6p^2 - 6pq + 180q^2$
   c) $7x^3y^3 - 63x^2y^2 + 140xy$
   d) $3x^2y + 24xy^2 - 60y^3$
   e) $4a^3 - 4a^2b - 48ab^2$
   f) $5x^3y^2 + 10x^2y^3 - 120xy^4$

16. Factor:
   a) $(2x)^2 + 5(2x) + 6$
   b) $(3a)^2 + 7(3a) + 12$
   c) $(7p)^2 + 13(7p) + 42$
   d) $(5m)^2 + 3(5m) - 18$
   e) $(4y)^2 + 5(4y)z - 14z^2$
   f) $(3x)^2 - 4(3x)y - 21y^2$

17. Factor:
   a) $(x + y)^2 + 9(x + y) - 10$
   b) $(p - 2q)^2 - 11(p - 2q) + 24$
   c) $(3y - 4)^2 - 2(3y - 4) - 63$
   d) $(x^2 + 4x)^2 + 8(x^2 + 4x) + 15$
   e) $(2m - n)^2 - (2m - n)p - 20p^2$
   f) $3(2x + 4)^2 + 12(2x + 4)y - 36y^2$

C 18. If $x = a + b$ and $y = a - b$, write in terms of a and b, and simplify:
   a) $x^2 + 2xy + y^2$
   b) $x^2 - 5xy + 6y^2$
   c) $x^2 + 4xy - 12y^2$
   d) $x^2y^2 - xy - 2$

2-11 Factoring Trinomials of the Form $ax^2 + bx + c$

A method of factoring a trinomial of the form $ax^2 + bx + c$ is indicated from a study of how one is formed from the product of two binomials.

$$(3x + 2)(x + 5) = 3x(x + 5) + 2(x + 5)$$
$$= 3x^2 + 15x + 2x + 10$$
$$= 3x^2 + 17x + 10$$

The integers 15 and 2 have a sum of 17 and a product of 30—the same as the product of 3 and 10.

A trinomial of the form $ax^2 + bx + c$ can be factored simply if two integers can be found with a sum b and a product ac.

Example 1. Factor: $3x^2 - 10x + 8$

Solution.

$3x^2 - 10x + 8$ What two integers have a sum of -10 and a product of 24?

The integers needed are -6 and -4, and the trinomial can be factored by first writing the second term as $-6x - 4x$.

$$3x^2 - 10x + 8 = 3x^2 - 6x - 4x + 8$$
$$= 3x(x - 2) - 4(x - 2)$$
$$= (x - 2)(3x - 4)$$

Example 2. Factor: $8w^2 + 14w - 15$

Solution.

$8w^2 + 14w - 15$ What two integers have a sum of 14 and a product of -120?

The integers needed are 20 and -6.
$$8w^2 + 14w - 15 = 8w^2 + 20w - 6w - 15$$
$$= 4w(2w + 5) - 3(2w + 5)$$
$$= (2w + 5)(4w - 3)$$

The order in which the **decomposition** of the second term is written is unimportant. The solution of *Example 1* could have proceeded as follows:

$$3x^2 - 10x + 8 = 3x^2 - 4x - 6x + 8$$
$$= x(3x - 4) - 2(3x - 4)$$
$$= (3x - 4)(x - 2)$$

As always, look for a common monomial factor before factoring any trinomial.

2-11 Factoring Trinomials of the Form $ax^2 + bx + c$ 73

Example 3. Factor: a) $10y^2 - 28y + 16$;
  b) $9x^2y + 3xy - 90y^2$.

Solution. a) $10y^2 - 28y + 16 = 2(5y^2 - 14y + 8)$
  $= 2(5y^2 - 10y - 4y + 8)$
  $= 2[5y(y-2) - 4(y-2)]$
  $= 2(y-2)(5y-4)$

  b) $9x^2y + 3xy - 90y = 3y(3x^2 + x - 30)$
  $= 3y(3x^2 + 10x - 9x - 30)$
  $= 3y[x(3x + 10) - 3(3x + 10)]$
  $= 3y(3x + 10)(x - 3)$

Exercises 2-11

A 1. Find two integers with the following properties:

	Product	Sum
a)	6	5
b)	18	9
c)	15	-8

	Product	Sum
d)	-15	-2
e)	-30	7
f)	-24	-2

2. Factor:
  a) $2x^2 + 11x + 12$
  b) $2k^2 + 3k + 1$
  c) $2x^2 + 5x + 2$
  d) $3x^2 + 13x + 4$
  e) $3s^2 + 4s + 1$
  f) $3t^2 + 19t + 28$

3. Factor:
  a) $6x^2 + 11x + 3$
  b) $8x^2 + 10x + 3$
  c) $6x^2 + 17x + 12$
  d) $2x^2 + 13x + 15$
  e) $4x^2 + 19x + 21$
  f) $6x^2 + 25x + 14$

4. Factor:
  a) $2m^2 - 11m + 12$
  b) $2x^2 - 11x + 15$
  c) $3x^2 - 14x + 8$
  d) $3y^2 - 22y + 7$
  e) $6x^2 - 19x + 10$
  f) $4t^2 - 23t + 15$

B 5. Factor:
  a) $8m^2 - 14m + 3$
  b) $10x^2 - 29x + 10$
  c) $14x^2 - 13x + 3$
  d) $6m^2 - 17m + 12$
  e) $3a^2 - 16a + 16$
  f) $32x^2 - 20x + 3$

6. Factor:
    a) $3x^2 + 7x - 6$
    b) $5y^2 + 19y - 4$
    c) $6x^2 + 7x - 20$
    d) $6k^2 + 5k - 4$
    e) $6x^2 + 17x - 14$
    f) $15x^2 + 29x - 14$

7. Factor:
    a) $2m^2 - m - 21$
    b) $12x^2 - 7x - 10$
    c) $10x^2 - 19x - 15$
    d) $6x^2 - 11x - 35$
    e) $6x^2 - x - 12$
    f) $24s^2 - 13s - 2$

8. Factor:
    a) $9x^2y^2 - 48x^2y + 48x^2$
    b) $6m^2n - 26mn + 24n$
    c) $20x^2y^3 - 85xy^3 - 75y^3$
    d) $4x^3y + 21x^2y^2 - 18xy^3$
    e) $20m^4n^2 - 10m^3n^2 - 210m^2n^2$
    f) $-18x^2 + 21x + 9$

C 9. Find two integers with the following properties:

	Product	Sum
a)	300	37
b)	72	17
c)	−120	2

	Product	Sum
d)	−220	−12
e)	264	−34
f)	−462	−1

10. Factor:
    a) $9k^2 - 24k + 16$
    b) $15t^2 - st - 2s^2$
    c) $12x^2 - 5xy - 2y^2$
    d) $4x^2 + 21xy - 18y^2$
    e) $20m^2 - 7mn - 3n^2$
    f) $21x^2 + 25xy - 4y^2$

11. Factor:
    a) $21x^2 + 17x - 30$
    b) $72x^2 + 11x - 6$
    c) $15x^2 - 28x - 32$
    d) $48x^2 - 22xy - 15y^2$
    e) $2m^2 + 9m - 35$
    f) $40y^2 + yz - 6z^2$
    g) $3 - 26xy + 35x^2y^2$
    h) $6 - 17mn - 45m^2n^2$

12. Factor:
    a) $18x^2y + 51xy - 135y$
    b) $20x^3y^2 - 94x^2y^2 + 84xy^2$
    c) $36m^3n - 21m^2n^2 - 36mn^3$
    d) $30a^2b^2c^2d^2 - 8a^2b^2cd - 70a^2b^2$
    e) $-30x + 62x^2 + 48x^3$
    f) $420x^2 + 230x^2y - 100x^2y^2$

2-12 Factoring a Difference of Squares

A polynomial that can be expressed in the form $x^2 - y^2$ is called a **difference of squares**. The factors are:

$$x^2 - y^2 = (x - y)(x + y)$$

Using the above identity, it is always possible to express a difference of squares as a product.

Example 1. Factor: a) $36x^2 - 49$ b) $16m^2 - 121n^2$

Solution. a) $36x^2 - 49 = (6x - 7)(6x + 7)$
b) $16m^2 - 121n^2 = (4m - 11n)(4m + 11n)$

Removing a common factor will sometimes result in terms that are perfect squares.

Example 2. Factor: a) $8m^2 - 2n^2$ b) $3a^3 - 12ab^2$

Solution. a) $8m^2 - 2n^2 = 2(4m^2 - n^2)$
$= 2(2m - n)(2m + n)$
b) $3a^3 - 12ab^2 = 3a(a^2 - 4b^2)$
$= 3a(a - 2b)(a + 2b)$

An additional step is required if one of the factors is itself a difference of squares.

Example 3. Factor: $81a^4 - 16$

Solution. $81a^4 - 16 = (9a^2 - 4)(9a^2 + 4)$
$= (3a - 2)(3a + 2)(9a^2 + 4)$

A difference of squares may have terms that are not monomials.

Example 4. Factor: a) $(x + 3)^2 - y^2$ b) $x^4 - (2x - 1)^2$

Solution. a) $(x + 3)^2 - y^2 = [(x + 3) - y][(x + 3) + y]$
$= (x - y + 3)(x + y + 3)$
b) $x^4 - (2x - 1)^2 = [x^2 - (2x - 1)][x^2 + (2x - 1)]$
$= (x^2 - 2x + 1)(x^2 + 2x - 1)$
$= (x - 1)^2(x^2 + 2x - 1)$

Differences of squares often occur in applications of the Pythagorean theorem.

Example 5. A ship leaves port A and travels due south to port B. From port B it travels to port C which is due east of A. It has then travelled a distance of 121 km. If B is 9 km closer to A than to C, how far east of port A is port C?

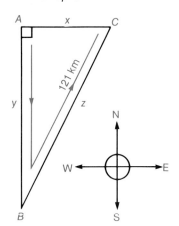

Solution. Let x, y, and z, represent respectively the distances AC, AB, and BC. Then from right triangle ABC,
$$x^2 = z^2 - y^2$$
$$= (z - y)(z + y)$$
Since $z + y = 121$ and $z - y = 9$,
$$x^2 = (121)(9)$$
$$= (11^2)(3^2)$$
$$x = (11)(3), \text{ or } 33$$
Port C is 33 km due east of port A.

Exercises 2 - 12

A 1. Factor:
   a) $x^2 - 49$ b) $4b^2 - 121$ c) $9m^2 - 64$
   d) $25y^2 - 144$ e) $49x^2 - 36$ f) $16y^2 - 81$

2. Factor:
   a) $100m^2 - 49$ b) $64b^2 - 1$ c) $121a^2 - 400$
   d) $25p^2 - 81$ e) $144m^2 - 49$ f) $36x^2 - 121$

3. Factor:
   a) $4s^2 - 9t^2$ b) $16x^2 - 49y^2$ c) $81a^2 - 64b^2$
   d) $p^2 - 36q^2$ e) $144y^2 - 81z^2$ f) $25m^2 - 169$

4. Factor:
   a) $16m^2 - 81n^2$ b) $64x^2 - 225y^2$ c) $49a^2 - 121b^2$
   d) $196x^2 - 25z^2$ e) $400a^2 - 324b^2$ f) $256p^2 - 625q^2$

5. Factor:
   a) $8m^2 - 72$ b) $6x^2 - 150$ c) $20x^2 - 5y^2$
   d) $12a^2 - 75$ e) $18p^2 - 98$ f) $80s^2 - 405$

6. Factor:
   a) $12x^3 - 27x$ b) $32m^3 - 98m$ c) $63a^2b - 28b$
   d) $125x^2y^2 - 180y^2$ e) $-98m + 32m^3$ f) $128x^2y - 50y^3$

7. Factor:
   a) $(x - y)^2 - z^2$ b) $(2a + b)^2 - 81$
   c) $(2c - 5)^2 - 121$ d) $x^2 - (y + z)^2$
   e) $81a^2 - (3a + b)^2$ f) $4(2x - y)^2 - 25z^2$

2-12 Factoring a Difference of Squares   77

B 8. Factor:
   a) $(x + 2)^2 - (x + 7)^2$
   b) $(5m - 2)^2 - (3m - 4)^2$
   c) $(2a + 3)^2 - (2a - 3)^2$
   d) $(3y + 8z)^2 - (3y - 8z)^2$
   e) $(3p - 7)^2 - (8p + 2)^2$
   f) $(2x - 1)^2 - (7x + 4)^2$

   9. Factor:
   a) $(x^2 - y^2)^2 - (x^2 + y^2)^2$
   b) $x^4 - (3x + 4)^2$
   c) $(x^2 - 4x)^2 - 144$
   d) $(x^2 + 5x)^2 - 36$
   e) $(a^2 - 10a)^2 - 576$
   f) $x^4 - (17x - 60)^2$

   10. A circular fountain, 150 cm in diameter, is surrounded by a circular flowerbed 325 cm wide. Find the area of the flowerbed.

   11. The formula for the volume, V, of a square-based pyramid is $V = \frac{1}{3}l^2 h$, where l is the length of a side of the base and h is its height.
   a) Express V in terms of s and h, where s is the slant height, that is, the height measured from the apex to the midpoint of a side of the base.
   b) Find the volume of the Great Pyramid of Khufu (Cheops), for which $h = 146$ m and $s = 186$ m.

   12. Explain how the two pieces of the diagram below (left) can be rearranged to show that $x^2 - y^2 = (x - y)(x + y)$.

   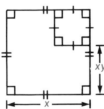

   13. A small square is cut from the corner of a large square (above, right). From the dimensions shown,
   a) find the area of the small square;
   b) find the area remaining when the small square is removed.

C 14. Factor, if possible:
   a) $8d^2 - 32e^2$
   b) $25m^2 - \frac{1}{4}n^2$
   c) $18x^2 y^2 - 50y^4$
   d) $10a^2 - 7b^2$
   e) $25s^2 + 49t^2$
   f) $p^2 - \frac{1}{9}q^2$
   g) $5x^4 - 80$
   h) $\frac{x^2}{16} - \frac{y^2}{49}$
   i) $7m^4 - 7n^4$

   15. Find all primes, p, such that $5p + 1$ is a perfect square.

2-13 Using Factoring to Evaluate Polynomials

Polynomials are evaluated by substituting a number for each variable. The computation is often simplified considerably by factoring the polynomial before substituting.

Example 1. Evaluate: $x^2 - 9xy + 18y^2$ for $x = 94$, $y = 16$.

Solution. $x^2 - 9xy + 18y^2 = (x - 3y)(x - 6y)$
When $x = 94$ and $y = 16$,
$(x - 3y)(x - 6y) = (94 - 48)(94 - 96)$
$= (46)(-2)$
$= -92$

In the above example, with no factoring the computation would be:
$x^2 - 9xy + 18y^2 = (94)^2 - 9(94)(16) + 18(16)^2$
$= 8836 - 13\,536 + 4608$
$= -92$

Even if a calculator is used for the arithmetic, factoring before substituting is the more efficient procedure.

Example 2. Evaluate: $a^4 - 16a^2b + 64b^2$ for $a = 15$, $b = 25$.

Solution. $a^4 - 16a^2b + 64b^2 = (a^2 - 8b)^2$
When $a = 15$ and $b = 25$,
$(a^2 - 8b)^2 = (15^2 - 8 \times 25)^2$
$= (225 - 200)^2$
$= 25^2$, or 625

Not all polynomials can be factored. However, any polynomial containing only one variable can be expressed in a **nested form** by successive grouping and factoring.

Example 3. Evaluate: $3x^4 - 5x^3 + 7x^2 - 4x + 2$ for $x = 3$.

Solution. $3x^4 - 5x^3 + 7x^2 - 4x + 2$
$= [3x^3 - 5x^2 + 7x - 4]x + 2$
$= [(3x^2 - 5x + 7)x - 4]x + 2$
$= [(\{3x - 5\}x + 7)x - 4]x + 2$
When $x = 3$, the value of this expression is
$[(\{9 - 5\}3 + 7)3 - 4]3 + 2 = [(19)3 - 4]3 + 2$
$= [53]3 + 2$
$= 161$

(Start with the innermost brackets and work out.)

An advantage of factoring in a nested form is that the computation can be made on a simple calculator by taking each term in order. Thus, for *Example 3*, the sequence of operations is:

2-13 Using Factoring to Evaluate Polynomials   **79**

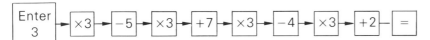

Example 4. A cylindrical concrete pipe, 12.65 m long, has an inside diameter of 48 cm and an outside diameter of 80 cm. Find, to the nearest cubic metre, the volume of concrete needed to make the pipe.

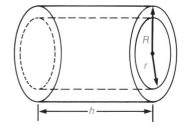

Solution. Let R, r, and h represent respectively the outside radius, the inside radius, and the length of the pipe.
Then, $\quad V = \pi R^2 h - \pi r^2 h$
$\quad\quad\quad = \pi h (R^2 - r^2)$
$\quad\quad\quad = \pi h (R - r)(R + r)$
When $h = 12.65$, $R = 0.40$, $r = 0.24$, and $\pi = 3.14$,
$\quad\quad V = (3.14)(12.65)(0.16)(0.64)$
$\quad\quad\quad = 4.07$
Approximately 4 m³ of concrete are needed.

Exercises 2 - 13

A 1. If $x = 78$ and $y = 15$, evaluate:
   a) $x^2 - 2xy - 15y^2$ b) $x^2 - 4xy - 12y^2$
   c) $x^2 - 6xy + 8y^2$ d) $x^2 - 25y^2$
   e) $x^2 - 14xy + 49y^2$ f) $x^2 - 9xy + 20y^2$

2. If $a = 27$ and $b = -13$, evaluate:
   a) $a^2 + 5ab + 6b^2$ b) $a^2 + 3ab - 4b^2$
   c) $a^2 + 17ab + 30b^2$ d) $a^2 - 16b^2$
   e) $2a^2 + 12ab + 18b^2$ f) $5a^2 - 5ab - 30b^2$

3. If $m = -29$ and $n = -17$, evaluate:
   a) $m^2 + 4mn - 12n^2$ b) $m^2 + 8mn + 15n^2$
   c) $m^2 - 3mn - 28n^2$ d) $m^2 - 9n^2$
   e) $3m^2 - 42mn + 147n^2$ f) $4m^2 + 8mn - 32n^2$

B 4. Evaluate:
   a) $53^2 - 47^2$ b) $77^2 - 73^2$
   c) $326^2 - 325^2$ d) $125^2 - 75^2$
   e) $128^2 - 112^2$ f) $46^2 - 24^2$

5. Evaluate:
   a) $93^2 + 2(93)(7) + 7^2$ b) $57^2 - 2(57)(17) + 17^2$
   c) $36^2 + 2(36)(39) + 39^2$ d) $84^2 - 6(84)(18) + 9(18)^2$
   e) $46^2 + 4(46)(25) - 12(25)^2$ f) $72^2 - 3(72)(9) - 40(9)^2$

6. Evaluate:
   a) 41 × 39
   b) 57 × 63
   c) 502 × 498
   d) 28 × 32
   e) 94 × 86
   f) 163 × 137

7. Evaluate:
   a) $5x^3 + 2x^2 - 7x + 8$ for $x = 4$
   b) $2m^3 - 5m^2 + 3m - 9$ for $m = 7$
   c) $6a^3 + a^2 - 4a + 12$ for $a = -5$
   d) $8p^3 - 7p^2 - 4p - 9$ for $p = 3$
   e) $2x^4 + 3x^3 - 5x^2 + 6x - 11$ for $x = -3$

8. Cardboard containers are made in a range of sizes, each size having a volume, V, given by the formula: $V = 2x^3 + 10x^2 - 48x$.
   a) Calculate the volume of a box when x is
      i) 8;   ii) 9;   iii) 11;   iv) 13.
   b) What are the dimensions of each size of box in terms of x?
   c) Determine the polynomial that represents the surface area of each size of box.

9. a) Evaluate $x^2 + 12x + 11$ for $x = 1, 2, 3, \ldots 10$.
   b) For what value of x is $x^2 + 12x + 11$ a perfect square?
   c) For what values of x are these polynomials perfect squares?
      i) $x^2 + 7x + 6$ ii) $x^2 + 8x - 9$ iii) $x^2 - 11x + 24$

C 10. A display of oranges is in the form of a square pyramid. The number of oranges, N, in the pyramid is given by the formula $N = \frac{1}{6}n + \frac{1}{2}n^2 + \frac{1}{3}n^3$ where n is the number of oranges along a bottom edge. How many oranges are in the display when n is
   a) 9?   b) 15?   c) 24?   d) 35?

Strip

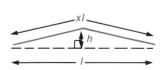

11. A metal strip of length l is secured at both ends to fixed points. When the strip is heated through a certain temperature range, the length of the strip increases by a factor x, causing the strip to buckle. Assume the heated strip forms two straight segments, as shown.
   a) Find an expression for the distance, h, through which the middle point of the strip moves.
   b) Find h for each strip listed in the table.

Metal	Length l	Expansion factor x
Steel	250 cm	1.0012
Brass	250 cm	1.0020
Aluminum	250 cm	1.0024

2-14 Equivalent Rational Expressions

In arithmetic,
we saw that the sum, difference, or product of any pair of integers is an integer, but that the quotient of two integers is not necessarily an integer. In order to include quotients of integers in arithmetic, the number system was extended to include **rational numbers**.

In algebra,
we saw that the sum, difference, or product of any two polynomials is a polynomial, but that the quotient of two polynomials is not necessarily a polynomial. In order to include quotients of polynomials in algebra, the set of algebraic expressions is extended to include **rational expressions**.

> Any algebraic expression that can be written as the quotient of two polynomials is called a **rational expression**.

These are rational expressions:

$$\frac{x+1}{x-3} \qquad \frac{3m^2 - n}{2m + 5} \qquad \frac{5a^3 - 2b^2 + 3}{2ab} \qquad 2x^2 + 4$$

These are not rational expressions:

$$\frac{2m^2 + n}{3\sqrt{n}} \qquad \frac{1 + 5\sqrt{a}}{b^2 + 3} \qquad 7\sqrt{x} + 5$$

Viewed as an algebraic "fraction", a rational expression can be reduced to lower terms by dividing numerator and denominator by a common factor.

Example 1. Reduce to lowest terms:

a) $\dfrac{18}{24}$ b) $\dfrac{-21y}{35x}$ c) $\dfrac{24m^3n}{56mn^2}$

Solution.

a) $\dfrac{18}{24} = \dfrac{\cancel{(6)}(3)}{\cancel{(6)}(4)} = \dfrac{3}{4}$

b) $\dfrac{-21y}{35x} = \dfrac{\cancel{(7)}(-3y)}{\cancel{(7)}(5x)} = \dfrac{-3y}{5x}$

c) $\dfrac{24m^3n}{56mn^2} = \dfrac{\cancel{(8mn)}(3m^2)}{\cancel{(8mn)}(7n)} = \dfrac{3m^2}{7n}$

Frequently, the factoring techniques of earlier sections must be used to reduce a rational expression to lowest terms.

Example 2. Simplify:
a) $\dfrac{3x^2 + 4x}{7x}$
b) $\dfrac{2m - 6}{3 - m}$
c) $\dfrac{x^2 - 3x - 18}{x^2 + x - 6}$
d) $\dfrac{81a^2 - b^2}{b^2 - 6ab - a^2}$

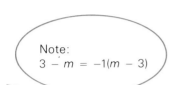

Note:
$3 - m = -1(m - 3)$

Solution.
a) $\dfrac{3x^2 + 4x}{7x} = \dfrac{\overset{1}{\cancel{x}}(3x + 4)}{\underset{1}{\cancel{7x}}}$, or $\dfrac{3x + 4}{7}$

b) $\dfrac{2m - 6}{3 - m} = \dfrac{2\overset{1}{\cancel{(m - 3)}}}{-1\underset{1}{\cancel{(m - 3)}}}$, or -2

c) $\dfrac{x^2 - 3x - 18}{x^2 + x - 6} = \dfrac{(x - 6)\overset{1}{\cancel{(x + 3)}}}{(x - 2)\underset{1}{\cancel{(x + 3)}}}$ or $\dfrac{x - 6}{x - 2}$

d) $\dfrac{81a^2 - b^2}{b^2 - 6ab - 27a^2} = \dfrac{\overset{-1}{\cancel{(9a - b)}}(9a + b)}{\underset{1}{\cancel{(b - 9a)}}(b + 3a)}$

$= -\dfrac{9a + b}{b + 3a}$

Notice in the above example that the word "simplify" is used to mean "reduce to lowest terms".

Exercises 2 - 14

A 1. Reduce to lowest terms:

a) $\dfrac{24x}{3}$ b) $\dfrac{36x}{-4y}$ c) $\dfrac{132a^4}{12a^2}$ d) $\dfrac{-28m^3}{7m}$

e) $\dfrac{5x^2}{15x}$ f) $\dfrac{-20s^3}{-35s^3}$ g) $\dfrac{102a^3}{17ab}$ h) $\dfrac{216m^2}{-18mn}$

2. Reduce to lowest terms:

a) $\dfrac{6m + 4}{2}$ b) $\dfrac{3a^2 + 3b}{3}$ c) $\dfrac{12x^2 - 8x}{4x}$

d) $\dfrac{12mn - 24m}{-6m}$ e) $\dfrac{12x^2y + 16xy}{4xy^2}$ f) $\dfrac{21ab - 28a}{7a}$

3. Reduce to lowest terms:

a) $\dfrac{3m}{3m - 6}$ b) $\dfrac{12x}{8x^2 - 20x}$

c) $\dfrac{5ab}{5a^2 + 10ab}$ d) $\dfrac{9xy}{6x^2 - 15xy}$

e) $\dfrac{6s^2t}{12s^2t + 9st^2}$ f) $\dfrac{10m^2n^2}{15m^2n - 25mn}$

2-14 Equivalent Rational Expressions

4. Reduce to lowest terms:

   a) $\dfrac{4x + 8}{2x + 4}$
   b) $\dfrac{a - 9b}{3a - 27b}$
   c) $\dfrac{2x - 10}{3x - 15}$
   d) $\dfrac{x - 5}{10 - 2x}$
   e) $\dfrac{3m - 12n}{20n - 5m}$
   f) $\dfrac{16 - 4a}{32 - 8a}$

5. Simplify:

   a) $\dfrac{2x^2 + 6x}{5x}$
   b) $\dfrac{2m^2 - 10m}{4m - 20}$
   c) $\dfrac{5x^2 + 7x}{3x}$
   d) $\dfrac{4x^2 - 12x}{x - 3}$
   e) $\dfrac{3x^2 - 6x}{14 - 7x}$
   f) $\dfrac{9 - 3x}{x - 3}$

6. Reduce to lowest terms:

   a) $\dfrac{a^2 + 7a + 10}{a^2 + 5a + 6}$
   b) $\dfrac{x^2 - 10x + 25}{x^2 - 4x - 5}$
   c) $\dfrac{x^2 - 16}{x^2 + 4x - 32}$
   d) $\dfrac{x^2 - 9}{x^2 + 6x + 9}$
   e) $\dfrac{r^2 + 7r + 12}{r^2 + 3r - 4}$
   f) $\dfrac{x^2 - 3x - 28}{x^2 - 9x + 14}$

B 7. Reduce to lowest terms:

   a) $\dfrac{4ab}{8ac}$
   b) $\dfrac{6a^2c}{8ab}$
   c) $\dfrac{-18x^2y}{3xy^2}$
   d) $\dfrac{45a^2bc^3}{60ab^2c}$
   e) $\dfrac{-8x^2yz}{-24xyz^2}$
   f) $\dfrac{15a^2bc^4}{-27a^2b^3c^2}$
   g) $\dfrac{-9m^2n}{-m^5}$
   h) $\dfrac{-25a^3b^2c}{40ab^7}$

8. Simplify:

   a) $\dfrac{5m - 5n}{3m - 3n}$
   b) $\dfrac{c - d}{d - c}$
   c) $\dfrac{2a - 2b}{3b - 3a}$
   d) $\dfrac{3xy - 18y^2}{12y^2 - 2xy}$
   e) $\dfrac{10xy - 15x^2y}{6x^2 - 4x}$
   f) $\dfrac{60a^2b^2 - 24ab}{16ab - 40a^2b^2}$

9. Simplify:

   a) $\dfrac{x^2 + 9xy + 18y^2}{2x^2 + 12xy}$
   b) $\dfrac{m^2 - 9mn + 20n^2}{3m^2 - 15mn}$
   c) $\dfrac{9a^2 - 16b^2}{6a^2 - 8ab}$
   d) $\dfrac{a^2 - ab - 6b^2}{a^2 + 2ab}$
   e) $\dfrac{9x^2 - 4y^2}{3x^2 - 2xy}$
   f) $\dfrac{m^2 + 2mn - 3n^2}{3m^2 + 9mn}$

10. Simplify:

   a) $\dfrac{c^2 - 5cd - 24d^2}{c^2 + 7cd + 12d^2}$
   b) $\dfrac{x^2 + xy - 30y^2}{x^2 + 11xy + 30y^2}$
   c) $\dfrac{a^2 + 10ab + 24b^2}{a^2 - 36b^2}$
   d) $\dfrac{4u^2 - 25v^2}{4u^2 + 20uv + 25v^2}$
   e) $\dfrac{m^2 - 2mn - 48n^2}{m^2 - 64n^2}$
   f) $\dfrac{x^2 - 36y^2}{x^2 - 3xy - 54y^2}$

11. Simplify:

a) $\dfrac{7x^2 - 21x}{7x^2 - 28x + 21}$ \hspace{1em} b) $\dfrac{5x^2 - 20}{x^2 + 14x + 24}$

c) $\dfrac{x^3 - 9x^2 + 20x}{x^3 - 25x}$ \hspace{1em} d) $\dfrac{32 - 2a^2}{4a^2 - 44a + 112}$

e) $\dfrac{3x^2 - 75}{6x^2 + 30x}$ \hspace{1em} f) $\dfrac{2x^3 - 28x^2 - 102x}{18x - 2x^3}$

C 12. Simplify if possible:

a) $\dfrac{4x + 2}{2}$ \hspace{1em} b) $\dfrac{2x + 6}{x + 3}$ \hspace{1em} c) $\dfrac{2x + 6x}{2}$

d) $\dfrac{2x - 1}{2x - 1}$ \hspace{1em} e) $\dfrac{x - 7}{7 - x}$ \hspace{1em} f) $\dfrac{x + 5}{x - 5}$

13. If $x = a + b$ and $y = a - b$, write the following expressions in terms of a and b and simplify:

a) $\dfrac{2x + 6y}{10x + 2y}$ \hspace{1em} b) $\dfrac{x^2 + 2xy + y^2}{x^2 - y^2}$ \hspace{1em} c) $\dfrac{6x^2 - 5xy + y^2}{6x^2 - xy - y^2}$

14. Simplify:

a) $\dfrac{x - 4}{2x^2 - 11x + 12}$ \hspace{1em} b) $\dfrac{2x - 7}{2x^2 - x - 21}$

c) $\dfrac{3y - 4}{3y^2 - 16y + 16}$ \hspace{1em} d) $\dfrac{2x^2 + 3xy + y^2}{3x^2 + 2xy - y^2}$

2-15 Multiplying and Dividing Rational Expressions

The same procedures are used to multiply and divide rational expressions as are used to multiply and divide rational numbers.

Example 1. Simplify: a) $\dfrac{x^2}{6} \times \dfrac{2y}{3x}$ \hspace{1em} b) $\dfrac{10t^2(r + 3)}{5(r - 3)} \times \dfrac{2(r - 3)}{rt}$

Solution. a) $\dfrac{x^2}{6} \times \dfrac{2y}{3x} = \dfrac{\cancel{x^2}^{\,x}}{\cancel{6}_{\,3}} \times \dfrac{\cancel{2y}^{\,1}}{\cancel{3x}_{\,1}}$, or $\dfrac{xy}{9}$

b) $\dfrac{10t^2(r + 3)}{5(r - 3)} \times \dfrac{2(r - 3)}{rt} = \dfrac{\cancel{10t^2}^{\,2t}(r + 3)}{\cancel{5(r - 3)}_{\,1}} \times \dfrac{2\cancel{(r - 3)}^{\,1}}{\cancel{rt}_{\,1}}$

$= \dfrac{4t(r + 3)}{r}$

2-15 Multiplying and Dividing Rational Expressions

Example 2. Simplify: a) $\dfrac{3a^3}{-5} \div \dfrac{(3a)^2}{10}$

b) $\dfrac{4(x-2y)}{x^2y} \div \dfrac{8(x-2y)}{xy^2(x+y)}$

Solution. a) $\dfrac{3a^3}{-5} \div \dfrac{(3a)^2}{10} = \dfrac{\cancel{3a^3}^{1a}}{\cancel{-5}_{3}} \times \dfrac{\cancel{10}^{-2}}{\cancel{9a^2}_{1}}$, or $-\dfrac{2a}{3}$ — *Multiply by the reciprocal.*

b) $\dfrac{4(x-2y)}{x^2y} \div \dfrac{8(x-2y)}{xy^2(x+y)} = \dfrac{\cancel{4(x-2y)}^{1}}{\cancel{x^2y}_{x\ \ 1}} \times \dfrac{\cancel{xy^2}^{1\,y}(x+y)}{\cancel{8(x-2y)}_{2\ \ 1}}$

$= \dfrac{y(x+y)}{2x}$

Sometimes expressions in the numerator or denominator must be factored before the whole expression can be simplified.

Example 3. Simplify: a) $\dfrac{x^2+7x+10}{x^2+x-6} \times \dfrac{x+3}{x+5}$

b) $\dfrac{3m-2n}{m^2-2mn+n^2} \div \dfrac{9m^2-4n^2}{3m^2n-3mn^2}$

Solution. a) $\dfrac{x^2+7x+10}{x^2+x-6} \times \dfrac{x+3}{x+5}$

$= \dfrac{(x+2)\cancel{(x+5)}^{1}}{\cancel{(x+3)}_{1}(x-2)} \times \dfrac{\cancel{(x+3)}^{1}}{\cancel{(x+5)}_{1}}$, or $\dfrac{x+2}{x-2}$

b) $\dfrac{3m-2n}{m^2-2mn+n^2} \div \dfrac{9m^2-4n^2}{3m^2n-3mn^2}$

$= \dfrac{\cancel{(3m-2n)}^{1}}{(m-n)\cancel{(m-n)}_{1}} \times \dfrac{3mn\cancel{(m-n)}^{1}}{\cancel{(3m-2n)}_{1}(3m+2n)}$

$= \dfrac{3mn}{(m-n)(3m+2n)}$

Exercises 2-15

A 1. Simplify:

a) $\dfrac{5}{8} \times \dfrac{2a}{3}$ b) $\dfrac{m^2}{4} \times \dfrac{2}{m}$ c) $\dfrac{-3x}{10} \times \dfrac{5x}{9}$

d) $\dfrac{2c^2}{15} \times \dfrac{5}{3c}$ e) $\dfrac{-6t}{35} \times \dfrac{14t}{3}$ f) $\dfrac{9r^2}{4} \times \dfrac{8}{3r}$

2. Simplify:
   a) $\dfrac{5}{8} \div \dfrac{3b}{4a}$
   b) $\dfrac{x^2}{14} \div \dfrac{x}{2}$
   c) $\dfrac{-6xy}{15} \div \dfrac{2x^2}{5}$
   d) $\dfrac{9a}{4} \div \dfrac{3b}{2}$
   e) $\dfrac{7m}{-3} \div \dfrac{5m}{-6}$
   f) $\dfrac{15a}{2} \div \dfrac{25c^2}{3a^2}$

3. Simplify:
   a) $\dfrac{8t}{21s^2} \times \dfrac{3s}{4}$
   b) $\dfrac{15x^2}{4x} \times \dfrac{6x^3}{5x^2}$
   c) $\dfrac{14e}{8} \times \dfrac{12f^2}{49e^3}$
   d) $\dfrac{-10x^3}{18x} \div \dfrac{-15x}{-27}$
   e) $\dfrac{8n}{-21} \div \dfrac{-4n}{7n}$
   f) $\dfrac{3x^2}{8y} \div \dfrac{9x}{28}$

4. Simplify:
   a) $\dfrac{3x^2}{2y} \times \dfrac{4y}{9x}$
   b) $\dfrac{4a}{3b} \times \dfrac{9b^2}{6a}$
   c) $\dfrac{2m}{9n} \div \dfrac{-4m}{3n^2}$
   d) $\dfrac{3x^3y^2}{6xy} \times \dfrac{4xy}{5x^2y^2}$
   e) $\dfrac{-8m^2n^5}{15mn^2} \div \dfrac{2m^4}{-25n^2}$
   f) $\dfrac{4c^2d}{8cd} \div \dfrac{3c^2d^3}{6cd^3}$

5. Simplify:
   a) $\dfrac{3a^2b}{12ab} \times \dfrac{8a^5b^4}{6ab^2}$
   b) $\dfrac{-5x^2y}{(2xy)^3} \times \dfrac{-12x^2y^2}{-6x^2y}$
   c) $\dfrac{2x^2y}{3xy} \times \dfrac{(6xy)^2}{4xy}$
   d) $\dfrac{12mn^2}{9mn} \div \dfrac{(3mn)^2}{6mn^2}$
   e) $\dfrac{2x}{3y} \times \dfrac{3y}{4z} \times \dfrac{4z}{5x}$
   f) $\dfrac{(2m)^2}{5n} \times \dfrac{10m}{8n} \div \dfrac{15m}{(4n)^2}$

6. Simplify:
   a) $\dfrac{2a}{a-3} \times \dfrac{7(a-3)}{4a}$
   b) $\dfrac{5(x-2)}{8x} \times \dfrac{2x}{15(x-2)}$
   c) $\dfrac{4(x-3)}{x+1} \div \dfrac{4}{x+1}$
   d) $\dfrac{12m^2}{5(m+4)} \times \dfrac{10(m+4)}{3m}$
   e) $\dfrac{3(s-2)}{4(s+5)} \div \dfrac{9(s-2)}{s+5}$
   f) $\dfrac{(5-a)3}{2a} \div \dfrac{3(a-5)}{4(a+1)}$

B 7. Simplify:
   a) $\dfrac{a+b}{3b} \times \dfrac{6b^2}{5(a+b)}$
   b) $\dfrac{3x^2y}{12x} \times \dfrac{4xy^3}{2xy}$
   c) $\dfrac{3y^3}{x^2-9} \times \dfrac{2x-6}{2y^2}$
   d) $\dfrac{3xy}{9x^2-12x} \times \dfrac{9x^2-16}{12y}$
   e) $\dfrac{10m^2n}{6m-9} \div \dfrac{25mn^2}{2m-3}$
   f) $\dfrac{4a^2-10}{a-3b} \div \dfrac{6a^2-15}{2a^2-18b^2}$

2-15 Multiplying and Dividing Rational Expressions

8. Simplify:

    a) $\dfrac{15x}{2x+6} \div \dfrac{10x}{3x+9}$
    b) $\dfrac{x^2-121}{x^2-4} \times \dfrac{x+2}{x-11}$
    c) $\dfrac{5x-10}{6x+6} \div \dfrac{2x-4}{x+1}$
    d) $\dfrac{y+2}{ay-by} \div \dfrac{y^2+2y}{ay^2-by^2}$
    e) $\dfrac{3xy}{x^2-4} \times \dfrac{(x-2)^2}{4y^2}$
    f) $\dfrac{(x+1)^2}{x^2-1} \times \dfrac{x^2-4}{(x+2)(x+1)}$

9. Simplify:

    a) $\dfrac{a^2-3a-10}{25-a^2} \div \dfrac{a+2}{a+5}$
    b) $\dfrac{x^2-2x-15}{x^2-9} \times \dfrac{x-3}{x-5}$
    c) $\dfrac{x^2+x-2}{x^2-x} \times \dfrac{x^2+x}{x^2-1}$
    d) $\dfrac{x^2-2x-15}{x^2-9} \times \dfrac{3-x}{x-5}$
    e) $\dfrac{a^2+2a-15}{a^2-8a+7} \times \dfrac{a^2-5a-14}{a^2+7a+10}$
    f) $\dfrac{x^2+5x+6}{x^2-5x+6} \div \dfrac{x^2-x-6}{x^2+x-6}$

C 10. Simplify:

    a) $\dfrac{x^2-16y^2}{6x^2y} \div \dfrac{x^2+xy-20y^2}{4x^3y^2}$
    b) $\dfrac{a^2+11ab+30b^2}{a^2-25b^2} \times \dfrac{3a^2-15ab}{6a^2+36ab}$
    c) $\dfrac{x^2+5xy+6y^2}{x^2+4xy-5y^2} \times \dfrac{x^2+3xy-10y^2}{x^2+xy-6y^2}$
    d) $\dfrac{m^2-9mn+14n^2}{m^2+7mn+12n^2} \div \dfrac{3m^2-21mn}{4m^3+16m^2n}$

11. Simplify:

    a) $\dfrac{3x^2+3x-6}{x^2y-7xy} \times \dfrac{x^2y-13xy+42y}{6x^2+12x}$
    b) $\dfrac{x^2+5xy+6y^2}{x^2+7xy+10y^2} \times \dfrac{x^2+6xy+5y^2}{x^2+2xy-3y^2}$
    c) $\dfrac{x+2y}{x-3y} \times \dfrac{x^2-9y^2}{x^2-4y^2} \div \dfrac{x+3y}{x-2y}$

12. If $x = a+b$ and $y = a-b$, write the following expressions in terms of a and b and simplify:

    a) $\dfrac{x^2-xy-12y^2}{x^2-2xy-3y^2} \times \dfrac{x^2+5xy+4y^2}{x^2-16y^2}$
    b) $\left(\dfrac{3x-21y}{6x+12y}\right)^2 \div \dfrac{x^2-49y^2}{2x^2+8xy+8y^2}$

2-16 Adding and Subtracting Rational Expressions With Monomial Denominators

In Section 2-14, rational expressions were reduced to lowest terms. To add and subtract rational expressions it is sometimes necessary to raise them to higher terms in order to obtain the lowest common denominator.

Example 1. Write an expression equivalent to $\dfrac{x+5}{x}$ with a denominator of:
a) $3x$; b) x^2; c) x^2y; d) $x(x-2)$.

Solution.
a) $\dfrac{x+5}{x} = \dfrac{x+5}{x} \times \dfrac{3}{3}$
$= \dfrac{3x+15}{3x}$

b) $\dfrac{x+5}{x} = \dfrac{x+5}{x} \times \dfrac{x}{x}$
$= \dfrac{x^2+5x}{x^2}$

c) $\dfrac{x+5}{x} = \dfrac{x+5}{x} \times \dfrac{xy}{xy}$
$= \dfrac{xy(x+5)}{x^2y}$
$= \dfrac{x^2y + 5xy}{x^2y}$

d) $\dfrac{x+5}{x} = \dfrac{x+5}{x} \times \dfrac{x-2}{x-2}$
$= \dfrac{(x+5)(x-2)}{x(x-2)}$
$= \dfrac{x^2 + 3x - 10}{x(x-2)}$

Rational expressions are added and subtracted in the same way that rational numbers are added and subtracted.

Example 2. Simplify:
a) $\dfrac{4}{3x} - \dfrac{7x}{6}$
b) $\dfrac{3}{8a} + \dfrac{5}{12a^2}$
c) $\dfrac{3y}{x} + \dfrac{7x}{y^2} - \dfrac{2x+1}{4y}$

Solution. a) The lowest common denominator is $6x$.
$\dfrac{4}{3x} - \dfrac{7x}{6} = \dfrac{4}{3x} \times \dfrac{2}{2} - \dfrac{7x}{6} \times \dfrac{x}{x}$
$= \dfrac{8}{6x} - \dfrac{7x^2}{6x}$
$= \dfrac{8 - 7x^2}{6x}$

2-16 Adding and Subtracting Rational Expressions With Monomial Denominators

b) The lowest common denominator is $24a^2$.

$$\frac{3}{8a} + \frac{5}{12a^2} = \frac{3}{8a} \times \frac{3a}{3a} + \frac{5}{12a^2} \times \frac{2}{2}$$

$$= \frac{9a}{24a^2} + \frac{10}{24a^2}$$

$$= \frac{9a + 10}{24a^2}$$

c) The lowest common denominator is $4xy^2$.

$$\frac{3y}{x} + \frac{7x}{y^2} - \frac{2x + 1}{4y}$$

$$= \frac{3y}{x} \times \frac{4y^2}{4y^2} + \frac{7x}{y^2} \times \frac{4x}{4x} - \frac{2x + 1}{4y} \times \frac{xy}{xy}$$

$$= \frac{12y^3}{4xy^2} + \frac{28x^2}{4xy^2} - \frac{xy(2x + 1)}{4xy^2}$$

$$= \frac{12y^3 + 28x^2 - 2x^2y - xy}{4xy^2}$$

Exercises 2-16

A 1. Write each expression with a denominator of $24mn$:

   a) $\frac{-2}{3mn}$ b) $\frac{5m}{8mn}$ c) $\frac{-5}{6m}$ d) $-\frac{13}{24}$

   e) $\frac{3}{-8}$ f) $\frac{7m}{12}$ g) $\frac{-7m}{8n}$ h) $\frac{11n}{-12m}$

2. Write an expression equivalent to:

   a) $\frac{2x}{3}$ with a denominator of: i) 6,   ii) $12x$,   iii) $3x^2$;

   b) $\frac{3m + 1}{4}$ with a denominator of: i) -12,   ii) $8x$,   iii) $16x^2$;

   c) $\frac{a - 7}{a}$ with a denominator of: i) $3a$,   ii) a^2,   iii) $5a^4$;

   d) $\frac{5y - 3}{2y}$ with a denominator of: i) $6y$,   ii) $4y^3$,   iii) $-2y^2$;

   e) $\frac{4x + 10}{2x^2}$ with a denominator of: i) $-4x^2$, ii) $20x^3$, iii) x^2.

3. For each pair of expressions write an equivalent pair with a common denominator:

   a) $\frac{x}{2}, \frac{x}{3}$ b) $\frac{2}{x}, \frac{x}{5}$ c) $\frac{3}{2a}, \frac{2}{a}$

   d) $\frac{5}{n}, \frac{2}{n^2}$ e) $\frac{1}{6x}, \frac{5}{8x}$ f) $\frac{x + 1}{5x^2}, \frac{x - 1}{4x}$

4. Simplify:
   a) $\frac{2}{x} - \frac{5}{x}$
   b) $\frac{4}{3x} + \frac{2}{3x}$
   c) $\frac{7}{4x} - \frac{5x}{4x}$
   d) $\frac{2}{3m^2} - \frac{9m}{3m^2}$
   e) $\frac{5a}{2a} + \frac{7a}{2a}$
   f) $\frac{16x}{5y^2} - \frac{11x}{5y^2}$

5. Simplify:
   a) $\frac{2a}{3} - \frac{4a}{5}$
   b) $\frac{2}{3a} - \frac{4}{5a}$
   c) $\frac{2a}{3} - \frac{4}{5a}$
   d) $\frac{2}{3a} - \frac{4a}{5}$
   e) $\frac{2}{3a} - \frac{4a}{5a^2}$
   f) $\frac{2a}{3a^2} - \frac{4}{5}$

6. Simplify:
   a) $\frac{2}{x} + \frac{5}{2x}$
   b) $\frac{4}{x} + \frac{27}{5x}$
   c) $\frac{7}{10x} + \frac{4}{15x}$
   d) $\frac{5}{2a} + \frac{3}{4a}$
   e) $\frac{7}{8m} + \frac{5}{6m}$
   f) $\frac{2}{9k} - \frac{5}{6k}$

7. Simplify:
   a) $\frac{7a}{10} - \frac{2a}{5} + \frac{3a}{10}$
   b) $\frac{5m}{6} - \frac{3m}{4} + \frac{m}{8}$
   c) $\frac{4x}{9} - \frac{2x}{3} + \frac{5x}{6}$
   d) $\frac{7c}{12} - \frac{5c}{9} - \frac{5c}{6}$
   e) $\frac{2e}{3} - \frac{5e}{6} + \frac{3e}{4}$
   f) $\frac{3m}{8} - \frac{2m}{3} + \frac{5m}{6}$

8. Simplify:
   a) $\frac{1}{2a} + \frac{1}{3a} + \frac{1}{4a}$
   b) $\frac{2}{3x} - \frac{3}{4x} - \frac{1}{2x}$
   c) $\frac{3}{8m} - \frac{2}{3m} + \frac{5}{6m}$
   d) $\frac{7}{6x} - \frac{5}{2x} - \frac{1}{3x}$
   e) $\frac{3}{4y} + \frac{2}{3y} - \frac{5}{6y}$
   f) $\frac{3}{8y} - \frac{5}{6y} + \frac{1}{4y}$

9. Simplify:
   a) $\frac{x+3}{x} + \frac{x-5}{x}$
   b) $\frac{5m-2}{m} - \frac{4m+7}{m}$
   c) $\frac{4a-2}{3a} + \frac{7a+11}{3a}$
   d) $\frac{6x+7}{5x^2} - \frac{2x-19}{5x^2}$
   e) $\frac{7m+4}{2m} - \frac{12m+11}{2m}$
   f) $\frac{2x-8}{4x} - \frac{10x-7}{4x}$

2-16 Adding and Subtracting Rational Expressions With Monomial Denominators

10. Simplify:
   a) $\dfrac{k-7}{4} - \dfrac{k+2}{5}$
   b) $\dfrac{c+5}{3} + \dfrac{c-8}{2}$
   c) $\dfrac{x+4}{3} - \dfrac{x+2}{4}$
   d) $\dfrac{m-5}{4} + \dfrac{m+3}{6}$
   e) $\dfrac{2a+3}{8} - \dfrac{5a-4}{6}$
   f) $\dfrac{4x-7}{6} + \dfrac{2x-7}{9}$

11. Simplify:
   a) $\dfrac{x+2}{3x} + \dfrac{2x-5}{2x}$
   b) $\dfrac{2n-7}{8n} - \dfrac{3n-4}{6n}$
   c) $\dfrac{5a-9}{6a} - \dfrac{3a+1}{9a}$
   d) $\dfrac{7x+3}{4x} - \dfrac{5x+2}{6x}$
   e) $\dfrac{2m-9}{4m} + \dfrac{7m+5}{8m}$
   f) $\dfrac{2a+3}{10a^2} - \dfrac{7a-4}{15a^2}$

B 12. Simplify:
   a) $\dfrac{5}{a} - \dfrac{5}{2a}$
   b) $\dfrac{2}{3m} - \dfrac{1}{2n}$
   c) $\dfrac{4}{x} + \dfrac{3}{xy}$
   d) $1 + \dfrac{a}{2b}$
   e) $\dfrac{2}{3a} + \dfrac{5}{b}$
   f) $\dfrac{3}{4m} - \dfrac{5}{6n}$

13. Simplify:
   a) $\dfrac{3}{x} - \dfrac{2}{y} + 1$
   b) $\dfrac{2}{a} - \dfrac{3}{b} + \dfrac{4}{c}$
   c) $\dfrac{1}{2x} + \dfrac{3}{4y} - \dfrac{5}{6z}$
   d) $\dfrac{1}{xy} - \dfrac{5}{x} - \dfrac{4}{y}$
   e) $\dfrac{2}{3a} + \dfrac{1}{4b} - \dfrac{5}{6ab}$
   f) $\dfrac{3}{8m} - \dfrac{2}{3n} + \dfrac{5}{6p}$

14. Simplify:
   a) $\dfrac{7x}{6y} + \dfrac{5y}{3x}$
   b) $\dfrac{5m}{3n} - \dfrac{4n}{3m}$
   c) $\dfrac{3a}{5b} - \dfrac{4b}{3a}$
   d) $\dfrac{3x}{2a} - \dfrac{4a}{5x}$
   e) $\dfrac{9p}{7q} + \dfrac{7q}{6p}$
   f) $\dfrac{3x}{5y} - \dfrac{2y}{3x}$

C 15. Consider the product: $\left(x + \dfrac{1}{x}\right)\left(x + \dfrac{2}{x}\right)$.
   a) Multiply the factors and then simplify the result.
   b) Simplify each factor and then multiply.
   c) Are the answers to (a) and (b) the same? Why?

16. Simplify:

a) $\left(a - \dfrac{1}{a}\right)\left(a - \dfrac{2}{a}\right)$

b) $\left(k + \dfrac{3}{k}\right)\left(k - \dfrac{5}{k}\right)$

c) $\left(2a - \dfrac{3}{a}\right)^2$

d) $\left(y - \dfrac{2}{y}\right)\left(y + \dfrac{2}{y}\right)$

17. a) Simplify: $\left(x + \dfrac{1}{x}\right)^2 - \left(x^2 + \dfrac{1}{x^2}\right)$.

b) If $x + \dfrac{1}{x} = 3$, find the value of $x^2 + \dfrac{1}{x^2}$.

c) If $x^2 + \dfrac{1}{x^2} = 8$, find the value of $x + \dfrac{1}{x}$.

18. Write as a sum or a difference:

a) $\dfrac{12 - a}{3}$

b) $\dfrac{2x - 5}{10x}$

c) $\dfrac{x^2 + xy}{xy}$

d) $\dfrac{7x^2 + x + 1}{x}$

19. If $\dfrac{x}{y} = \dfrac{3}{2}$, find the value of:

a) $\dfrac{y}{x}$;

b) $\dfrac{x + y}{y}$;

c) $\dfrac{x + y}{x}$;

d) $\dfrac{x + y}{x - y}$.

2-17 Adding and Subtracting Rational Expressions

The procedures of Section 2-16 are also used to add and subtract rational expressions with denominators that are not monomials.

Example 1. Simplify:

a) $\dfrac{m - 4}{m - 2} - \dfrac{m - 10}{m - 2}$

b) $\dfrac{5}{x} + \dfrac{6x}{x + 4}$

c) $a - \dfrac{3}{a + b}$

d) $\dfrac{x + 3}{x - 5} - \dfrac{x - 7}{x - 2}$

Solution. a) Since the terms have a common denominator, the numerators can be combined.

$$\dfrac{m - 4}{m - 2} - \dfrac{m - 10}{m - 2} = \dfrac{(m - 4) - (m - 10)}{m - 2}$$

$$= \dfrac{m - 4 - m + 10}{m - 2}$$

$$= \dfrac{6}{m - 2}$$

b) The lowest common denominator is $x(x + 4)$.

$$\frac{5}{x} + \frac{6x}{x + 4} = \frac{5(x + 4) + 6x(x)}{x(x + 4)}$$

$$= \frac{5x + 20 + 6x^2}{x(x + 4)}, \text{ or } \frac{6x^2 + 5x + 20}{x(x + 4)}$$

c) Since $a = \frac{a}{1}$, the lowest common denominator is $a + b$.

$$a - \frac{3}{a + b} = \frac{a(a + b) - 3}{a + b}$$

$$= \frac{a^2 + ab - 3}{a + b}$$

d) The lowest common denominator is $(x - 5)(x - 2)$.

$$\frac{x + 3}{x - 5} - \frac{x - 7}{x - 2} = \frac{(x + 3)(x - 2) - (x - 7)(x - 5)}{(x - 5)(x - 2)}$$

$$= \frac{(x^2 + x - 6) - (x^2 - 12x + 35)}{(x - 5)(x - 2)}$$

$$= \frac{x^2 + x - 6 - x^2 + 12x - 35}{(x - 5)(x - 2)}$$

$$= \frac{13x - 41}{(x - 5)(x - 2)}$$

Frequently, the denominators of rational expressions must be factored so that the lowest common denominator can be more readily determined.

Example 2. Simplify: a) $\frac{7}{2x + 4} - \frac{5}{3x + 6}$

b) $\frac{b}{a - 3} - \frac{b}{a + 3} + \frac{1}{a^2 - 9}$

Solution. a) $\frac{7}{2x + 4} - \frac{5}{3x + 6} = \frac{7}{2(x + 2)} - \frac{5}{3(x + 2)}$

$$= \frac{7 \times 3 - 5 \times 2}{6(x + 2)}$$ ← lowest common denominator

$$= \frac{21 - 10}{6(x + 2)}, \text{ or } \frac{11}{6(x + 2)}$$

b) $\frac{b}{a - 3} - \frac{b}{a + 3} + \frac{1}{a^2 - 9}$

$$= \frac{b(a + 3) - b(a - 3) + 1}{(a - 3)(a + 3)}$$

$a^2 - 9$
$= (a + 3)(a - 3)$

$$= \frac{ba + 3b - ba + 3b + 1}{(a - 3)(a + 3)}$$

$$= \frac{6b + 1}{(a - 3)(a + 3)}, \text{ or } \frac{6b + 1}{a^2 - 9}$$

When possible, the sum or difference of rational expressions should be reduced to lowest terms.

Example 3. Simplify: $\dfrac{x}{x^2 - 9x + 18} - \dfrac{x - 2}{x^2 - 10x + 24}$

Solution.
$$\dfrac{x}{x^2 - 9x + 18} - \dfrac{x - 2}{x^2 - 10x + 24}$$
$$= \dfrac{x}{(x - 6)(x - 3)} - \dfrac{x - 2}{(x - 6)(x - 4)}$$
$$= \dfrac{x(x - 4) - (x - 2)(x - 3)}{(x - 6)(x - 3)(x - 4)}$$
$$= \dfrac{x^2 - 4x - x^2 + 5x - 6}{(x - 6)(x - 3)(x - 4)}$$
$$= \dfrac{x - 6}{(x - 6)(x - 3)(x - 4)}, \text{ or } \dfrac{1}{(x - 3)(x - 4)}$$

(lowest common denominator)

Exercises 2-17

A 1. Simplify:

a) $\dfrac{3m - 5}{m + 3} + \dfrac{m + 4}{m + 3}$ b) $\dfrac{2s + 7}{s - 5} - \dfrac{6s - 4}{s - 5}$

c) $\dfrac{5k - 9}{k - 4} - \dfrac{2k + 3}{k - 4}$ d) $\dfrac{2x + 7}{x + 6} + \dfrac{2x + 11}{x + 6}$

e) $\dfrac{4m - 9}{2m + 1} - \dfrac{m - 4}{2m + 1}$ f) $\dfrac{3a - 8}{a^2 + 4} - \dfrac{7a + 3}{a^2 + 4}$

2. Simplify:

a) $\dfrac{4}{a - 3} - \dfrac{1}{a}$ b) $\dfrac{2}{y - 5} - \dfrac{6}{y}$

c) $\dfrac{7}{m} - \dfrac{3}{m - 4}$ d) $\dfrac{2c}{c - 1} - \dfrac{5}{c}$

e) $\dfrac{3x}{x + 2} - \dfrac{6}{x}$ f) $\dfrac{3}{x} + \dfrac{5}{x + 2}$

3. Simplify:

a) $\dfrac{3}{2a} - 4$ b) $\dfrac{7}{y + 1} - 2$ c) $4 - \dfrac{9}{n - 5}$

d) $x - \dfrac{2}{x + 4}$ e) $\dfrac{3}{s - 8} - 2s$ f) $\dfrac{2w}{w + 3} - 4w$

4. Simplify:

a) $\dfrac{4}{x - 1} - (x - 2)$ b) $\dfrac{4}{x - 1} - x - 2$

c) $x - 5 + \dfrac{2}{x - 3}$ 　　　d) $x + 3 + \dfrac{5}{x - 2}$

e) $\dfrac{2}{x - 4} - x - 8$ 　　　f) $4 - \dfrac{3}{x + 2} - x$

5. Simplify:

a) $\dfrac{2}{x + 5} + \dfrac{3}{x + 2}$ 　　　b) $\dfrac{4}{x - 3} - \dfrac{2}{x + 1}$

c) $\dfrac{x}{x + 1} - \dfrac{2}{x - 1}$ 　　　d) $\dfrac{3x}{x + 4} + \dfrac{2}{x + 4}$

e) $\dfrac{5x}{x - 1} - \dfrac{2x}{x + 3}$ 　　　f) $\dfrac{2x}{x + 5} + \dfrac{3x}{x - 3}$

6. Simplify:

a) $\dfrac{1}{x + 1} - \dfrac{1}{x - 1}$ 　　　b) $\dfrac{x - 3}{x - 2} + \dfrac{1}{x - 3}$

c) $\dfrac{3x - 1}{x + 7} - \dfrac{2x + 1}{x - 3}$ 　　　d) $\dfrac{x - 2}{x + 2} + \dfrac{x + 1}{x - 4}$

e) $\dfrac{x + 6}{x - 3} + \dfrac{x - 4}{x - 5}$ 　　　f) $\dfrac{x + 1}{x - 2} - \dfrac{x - 1}{x + 2}$

7. Simplify:

a) $\dfrac{6}{2x + 4} + \dfrac{9}{3x + 6}$ 　　　b) $\dfrac{3}{5x - 10} + \dfrac{7}{2x - 4}$

c) $\dfrac{5x}{3x + 9} - \dfrac{9x}{2x + 6}$ 　　　d) $\dfrac{5x}{10x - 15} - \dfrac{4x}{16x - 24}$

e) $\dfrac{2x + 5}{3x - 12} + \dfrac{2x}{x - 4}$ 　　　f) $\dfrac{3x}{2x + 8} - \dfrac{2x - 3}{3x + 12}$

8. Simplify:

a) $\dfrac{4x}{x^2 - 9x + 18} + \dfrac{2x - 1}{x - 6}$ 　　　b) $\dfrac{x - 7}{x^2 - 2x - 15} - \dfrac{3x}{x - 5}$

c) $\dfrac{2x}{x - 2} - \dfrac{3}{x^2 - 4}$ 　　　d) $\dfrac{4x + 1}{x + 3} + \dfrac{x - 6}{x^2 - 9}$

e) $\dfrac{3x}{x - 1} - \dfrac{2x}{x^2 + x - 2}$ 　　　f) $\dfrac{8x - 3}{x^2 - 7x + 12} - \dfrac{2x + 1}{x - 4}$

B 9. Simplify:

a) $\dfrac{2}{x} - \dfrac{3x}{x - 2}$ 　　　b) $\dfrac{7}{2(x + 3)} - \dfrac{4}{5(x + 3)}$

c) $\dfrac{3y}{2(y + 9)} + \dfrac{5y}{3(y + 9)}$ 　　　d) $\dfrac{5}{3(a - 7)} - \dfrac{2}{3(a + 1)}$

e) $\dfrac{3}{a + 1} + \dfrac{1}{a - 1}$ 　　　f) $\dfrac{3x}{x - 2} - \dfrac{4x}{x - 3}$

10. Simplify:

a) $\dfrac{3m}{2(m-1)} - \dfrac{5m}{2(m+1)}$ b) $\dfrac{x-1}{x(x+5)} + \dfrac{2x-3}{x(x-1)}$

c) $\dfrac{8a}{5(a-2)} + \dfrac{5a-1}{3(a+3)}$ d) $\dfrac{5k}{4(k-3)} - \dfrac{4k}{k-4}$

e) $\dfrac{3x-1}{2(x-2)} + \dfrac{5x+1}{x+7}$ f) $\dfrac{4m+3}{3(2m-1)} - \dfrac{m-5}{2(3m+7)}$

11. Simplify:

a) $\dfrac{x+3}{x^2+11x+24} - \dfrac{2x+10}{x^2+11x+30}$

b) $\dfrac{m-4}{m^2-8m+16} + \dfrac{3m+21}{m^2+12m+35}$

c) $\dfrac{3x+9}{x^2+5x+6} - \dfrac{2x-2}{x^2+x-2}$

d) $\dfrac{5m+25}{m^2+7m+10} - \dfrac{10m-20}{m^2-4}$

e) $\dfrac{4x^2-20x}{x^2+2x-35} + \dfrac{3x-6}{x^2-12x+20}$

f) $\dfrac{2x-6}{x^2-5x+6} - \dfrac{3x-12}{x^2-x-12}$

C 12. Simplify:

a) $\dfrac{3x^2+6xy}{3x} - \dfrac{4y^2-2xy}{2y}$

b) $\dfrac{x^2-5xy+6y^2}{x-3y} - \dfrac{x^2-xy-12y^2}{x-4y}$

c) $\dfrac{x^2-4xy-21y^2}{3x-21y} + \dfrac{x^2+2xy-24y^2}{2x+12y}$

d) $\dfrac{a-b}{a^2+2ab-3b^2} + \dfrac{a+b}{a^2-2ab-3b^2}$

13. If $a = \dfrac{1}{x}$ and $b = \dfrac{1}{y}$, write the following expressions in terms of x and y and simplify:

a) $a + b$ b) $a - b$ c) $\dfrac{a+b}{a-b}$

d) $\dfrac{1}{a+b} + \dfrac{1}{a-b}$ e) $\dfrac{a}{a-b} - \dfrac{b}{a+b}$

14. If $x = \dfrac{a+1}{a+2}$, write the following in terms of a and simplify:

a) $x + 1$ b) $\dfrac{x+1}{x}$ c) $\dfrac{x+1}{x+2}$

2-18 Applications of Rational Expressions

Many formulas in mathematics and science involve rational expressions. The expressions are almost always simpler than those worked with earlier in the chapter.

Example 1. The focal length, f, of a lens is related to the object and image distances, p and q respectively, by the formula $\frac{1}{f} = \frac{1}{p} + \frac{1}{q}$. Express f in terms of p and q.

Solution.
$$\frac{1}{f} = \frac{1}{p} + \frac{1}{q}$$
$$= \frac{q + p}{pq}$$
so that $f = \frac{pq}{q + p}$

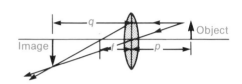

When developing a formula for a specific purpose, it is sometimes helpful to use numerical data before generalizing.

Example 2. In a training session, Susan coxes her boat crew to row with constant power both with and against the current.
a) If the boat's speed upstream is 14 km/h and downstream is 18 km/h, and the distance each way is 3 km, calculate the average speed over the training course.
b) Find an expression for the average speed if the distance each way is r km, and u km/h and v km/h are the upstream and downstream speeds respectively.

Solution. a) Time to row upstream (hours): $\frac{3}{14}$, or 0.214

Time to row downstream (hours): $\frac{3}{18}$, or 0.167

Total time (hours): 0.381

Average speed: $\frac{\text{Total distance}}{\text{Total time}} = \frac{6}{0.381}$, or 15.75

The boat's average speed is 15.75 km/h.

b) Time to row upstream (hours): $\frac{r}{u}$

Time to row downstream (hours): $\frac{r}{v}$

Total time (hours): $\frac{r}{u} + \frac{r}{v}$, or $\frac{r(v + u)}{uv}$

Average speed:

$$\text{Distance} \div \text{Time} = 2r \div \frac{r(v+u)}{uv}$$

$$= 2r \times \frac{uv}{r(v+u)}, \text{ or } \frac{2uv}{v+u}$$

The boat's average speed is $\frac{2uv}{v+u}$ km/h.

Example 3. An aircraft maintains an airspeed of V km/h between two airports d km apart on both the outbound and return trips. On the outbound trip there is a headwind of w km/h and on the return trip a tailwind of the same speed. Write an expression for the time taken for the round trip.

Solution. Time $= \frac{\text{distance}}{\text{groundspeed}}$,

where groundspeed = airspeed ± windspeed

Outbound time (hours): $\frac{d}{V-w}$

Return time (hours): $\frac{d}{V+w}$

Total time (hours): $\frac{d}{V-w} + \frac{d}{V+w}$

$$= \frac{d(V+w) + d(V-w)}{(V-w)(V+w)}$$

$$= \frac{dV + dw + dV - dw}{V^2 - w^2}$$

$$= \frac{2dV}{V^2 - w^2}$$

The time taken for the round trip is $\frac{2dV}{V^2 - w^2}$ hours.

Exercises 2 - 18

A 1. The lens formula is $\frac{1}{f} = \frac{1}{p} + \frac{1}{q}$. Calculate f when $p = 8$ and $q = 12$.

2. Guy cycles from his home to the store and back again.
   a) If the store is 12 km away and he averages 9 km/h going and 18 km/h returning because of the wind, calculate his average speed for the trip.
   b) If the store is d km away and he averages x km/h going and y km/h returning, find an expression for his average speed for the trip.

3. In *Example 3*, find an expression for the average speed of the aircraft for the round trip.

4. An aircraft maintains an airspeed of 325 km/h between two airports on both the outbound and return trips. On the outbound trip there is a tailwind of 75 km/h and on the return trip a headwind of the same speed. If the airports are 1000 km apart, how long does the aircraft take to fly the round trip?

5. The trip to a cottage is 150 km on a highway then 60 km on a gravel road. How long does the trip take if
   a) the average speed of 90 km/h on the highway is reduced by 40 km/h on the gravel road?
   b) the average speed of v km/h on the highway is reduced by x km/h on the gravel road?

6. The speed of a boat in still water is x km/h. It travels 200 km upriver and returns to its starting point. If the river current is y km/h, find the time for the round trip.

B 7. When two resistances, r and s, are connected in parallel, the total resistance, R, is given by the formula:

$$\frac{1}{R} = \frac{1}{r} + \frac{1}{s}.$$

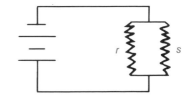

   a) Express R in terms of r and s.
   b) Express r in terms of R and s.
   c) Express s in terms of R and r.

8. What time is saved on a trip of 150 km if the average speed is
   a) increased from 80 km/h to 90 km/h?
   b) increased from v km/h to $(v + x)$ km/h?

9. A rectangular poster has an area of 4000 cm².
   a) Write an expression for its width when its length is l cm.
   b) Write an expression for the increase in its width when its length is decreased by x cm.
   c) Find the increase in width for $x = 15$ cm and
      i) $l = 90$ cm;   ii) $l = 120$ cm.

C 10. The density, d, in kilograms per cubic metre at $t°C$, of a space-age alloy is given by the formula:

$$d = \frac{3000}{1 + 0.000\ 125\ t}.$$

   a) Calculate the alloy's density at: i) 40°C,   ii) −20°C.
   b) At about what temperature will its density be 2992.5 kg/m³?

Review Exercises

1. Simplify:
   a) 2^0
   b) $(-2)^{-3}$
   c) $\left(-\dfrac{1}{3}\right)^2$
   d) $3^2 + 3^{-2}$
   e) $(2^3 - 3^2)^{-2}$
   f) $(3^{-2})(2^{-3})$

2. Evaluate for $x = 2$, $y = -1$:
   a) $x^{-3}y$
   b) $(xy)^{-3}$
   c) $\left(\dfrac{1}{x^2y}\right)^{-1}$
   d) $(3x^3y^2)^{-1}$
   e) $y^{-3} - x^{-2}$
   f) $4x^{-2}y^{-5}$

3. Simplify:
   a) $x^2 \cdot x^5$
   b) $c^3 \cdot c^4 \cdot c$
   c) $y^6 \div y^3$
   d) $a^8 \div a^3$
   e) $(x^2)^4$
   f) $(a^3b^2)^2$

4. Simplify:
   a) $x^{-3} \cdot x^2$
   b) $x^3 \cdot x^{-2} \cdot x^4$
   c) $x^{-4} \div x^2$
   d) $x^{-1} \div x^{-2}$
   e) $(x^{-3})^2$
   f) $(x^{-3}y^2)^{-1}(x^2)^{-2}$

5. Evaluate for $x = -1$, $y = 2$:
   a) $(x^2y^3)(x^3y^2)$
   b) $\dfrac{x^3y^5}{xy^3}$
   c) $(x^{-3}y^{-2})^{-2}$

6. Simplify:
   a) $4x^2 - xy + 3xy - 3x^2 + 7xy$
   b) $6a^2b - 2a^2b^2 - 11a^2b + 5a^2b^2 - 3ab$

7. Simplify:
   a) $(8xy)(-5y)$
   b) $(-3a^2b)(2ab^2)$
   c) $(3xy)(-2x^2y)(-4xy^2)$

8. Simplify:
   a) $\dfrac{35m^4n^3}{-5mn^2}$
   b) $\dfrac{(8x^4y^5)(3x^2y)}{6x^2y^2}$
   c) $\dfrac{(4a^2b^2)^2(-3a^4b^2)}{(2ab)^3(6a^2b)}$

9. Simplify:
   a) $(3x^2 + 17xy + 5y^2) - (12x^2 + 6y^2 - 3xy) + (-5xy + 3x^2 - 2y^2)$
   b) $(-7m^2 + 6mn + 3n^2) - (3m^2 + 2n^2 - 5mn) - (2m^2 + 3mn - 7n^2)$

10. Simplify:
    a) $2x(x + y) - 3x(2x - 3y)$
    b) $2a(3a - 5b) - a(2b + 3a)$
    c) $3xy(x - 2y) - 3x(2xy + 3y^2) - y(2x^2 + 5xy)$
    d) $5m(3mn - 2n^2) - 2(m^2n - 15mn^2) + 5n(-3mn - 5m^2)$

11. Find the perimeter and area of the shaded region of the rectangle shown.

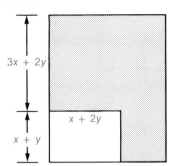

12. Factor:
    a) $8m^3 - 4m^2$
    b) $8y^2 - 12y^4 + 24$
    c) $28a^2 - 7a^3$
    d) $6a^2b^3c - 15a^2b^2c^2$
    e) $30x^2y - 20x^2y^2 + 10x^3y^2$
    f) $8mn^2 - 12mn - 16m^2n$

13. Factor:
    a) $m^2 + 8m + 16$
    b) $a^2 - 7a + 12$
    c) $y^2 - 2y - 8$
    d) $n^2 - 4n - 45$
    e) $s^2 - 15s + 54$
    f) $k^2 - 9k - 90$

14. Factor:
    a) $a^2 + 14ab + 24b^2$
    b) $m^2 + 9mn + 18n^2$
    c) $x^2 - xy - 20y^2$
    d) $c^2 + 21cd - 100d^2$

15. Factor:
    a) $3xy^2 + 18xy + 24x$
    b) $15m^3 + 90m^2n + 135mn^2$
    c) $7a^3b^3 - 63a^2b^2 + 140ab$
    d) $13s^3t^3 - 91s^2t^2 - 572st$

16. Factor:
    a) $b^2 - 36$
    b) $9k^2 - 1$
    c) $36x^2 - 49y^2$
    d) $4a^2 - 9b^2$
    e) $25m^2 - 81n^2$
    f) $1 - 64s^2$

17. Factor:
    a) $8a^2 - 72$
    b) $150 - 6n^2$
    c) $7x^4 - 7y^4$
    d) $27m^3 - 12m$
    e) $\dfrac{a^2}{36} - \dfrac{b^2}{49}$
    f) $125p^2q^2 - 180q^2$

18. A circular fountain, 250 cm in diameter, is surrounded by a circular flowerbed 375 cm wide. Find the area of the flowerbed.

19. If $x = 120$ and $y = 20$, evaluate:
    a) $x^2 - 2xy - 15y^2$
    b) $x^2 - 6xy + 8y^2$
    c) $x^2 - 10xy - 39y^2$
    d) $x^2 - 14xy - 95y^2$

20. Evaluate:
    a) $5y^3 + 2y^2 - 7y + 8$ for $y = 4$;
    b) $3m^4 - 5m^3 + 2m^2 - m - 19$ for $m = 6$.

21. Factor:
    a) $4x^2 - 7x + 3$
    b) $28x^2 - 9x - 4$

22. Reduce to lowest terms:
    a) $\dfrac{78a^3b}{6a^2}$
    b) $\dfrac{9m+6}{3}$
    c) $\dfrac{20y^2-8y}{4y}$
    d) $\dfrac{9ab-18a}{-3a}$
    e) $\dfrac{6xy}{3x^2+15xy}$
    f) $\dfrac{6m^2-15mn}{4m-10n}$

23. Simplify:
    a) $\dfrac{36-12x}{4x-12}$
    b) $\dfrac{a^2-25}{3a^2-15a}$
    c) $\dfrac{n^2-10n+24}{n^2-6n+8}$
    d) $\dfrac{a^2-9}{a^2+6a+9}$
    e) $\dfrac{b^2-3b-28}{b^2-9b+14}$
    f) $\dfrac{m^2-8m+15}{m^2-3m-10}$

24. Simplify:
    a) $\dfrac{9a^3}{4} \times \dfrac{8}{3a}$
    b) $\dfrac{-6mn}{15} \div \dfrac{2m^2}{5}$
    c) $\dfrac{4x^2y}{8xy} \div \dfrac{3x^2y^3}{6xy^3}$
    d) $\dfrac{15x^4y^2}{24xy^2} \times \dfrac{8x^2y}{5xy^2}$
    e) $\dfrac{2m^2n}{3mn} \times \dfrac{(6mn)^2}{4mn}$
    f) $\dfrac{2x}{3y} \times \dfrac{3z}{4y} \div \dfrac{4z}{5x}$

25. Simplify:
    a) $\dfrac{12a^2}{5(a+4)} \times \dfrac{10(a+4)}{3a}$
    b) $\dfrac{3(5-a)}{2a} \div \dfrac{3(a-5)}{8(a+1)}$
    c) $\dfrac{6mn^2}{2(2m-5)} \times \dfrac{3(2m-5)}{9m^2n}$
    d) $\dfrac{3x^2y(2x-y)}{15xy} \div \dfrac{2xy^2(2x-y)}{5x^2y^3}$

26. Simplify:
    a) $\dfrac{3b^3}{a^2-9} \times \dfrac{2a-b}{2b^2}$
    b) $\dfrac{m^2-2m+1}{m^2-1} \div \dfrac{m^2-4}{m^2+3m+2}$
    c) $\dfrac{x^2-3x-10}{25-x^2} \div \dfrac{x+2}{x+5}$
    d) $\dfrac{y^2+2y-15}{y^2-8y+7} \times \dfrac{y^2-5y-14}{y^2+7y+10}$

27. Simplify:
    a) $\dfrac{4}{3a}+\dfrac{5}{3a}$
    b) $\dfrac{3}{2x}-\dfrac{5}{3x}$
    c) $\dfrac{5}{m}+\dfrac{3}{4m}$
    d) $\dfrac{5x}{6}-\dfrac{3x}{4}+\dfrac{x}{8}$
    e) $\dfrac{2}{3a}-\dfrac{3}{4a}+\dfrac{1}{2a}$
    f) $\dfrac{7}{6n}-\dfrac{5}{2n}+\dfrac{1}{5n}$

28. Simplify:
    a) $\dfrac{a+3}{a}+\dfrac{a-5}{a}$
    b) $\dfrac{3x-2}{x}-\dfrac{2x+7}{x}$
    c) $\dfrac{y-5}{4}-\dfrac{y+2}{5}$
    d) $\dfrac{a+2}{3a}+\dfrac{2a-5}{2a}$
    e) $\dfrac{3x-7}{8x}-\dfrac{2x-4}{6x}$
    f) $\dfrac{3}{4x}-\dfrac{2}{3x}+1$

29. Simplify:

a) $\dfrac{5x-9}{x-4} - \dfrac{2x+3}{x-4}$ b) $\dfrac{2a}{a-1} - \dfrac{5}{a}$

c) $\dfrac{7}{x+1} - 2$ d) $\dfrac{m}{m+1} - \dfrac{2}{m-1}$

e) $\dfrac{y-3}{y-2} + \dfrac{1}{y-3}$ f) $\dfrac{3a-1}{a+7} - \dfrac{2a+1}{a-3}$

30. Simplify:

a) $\dfrac{5a}{10a-15} - \dfrac{4a}{16a-24}$ b) $\dfrac{2m+5}{3m-12} + \dfrac{2m}{m-4}$

c) $\dfrac{2k}{k-2} - \dfrac{3}{k^2-4}$ d) $\dfrac{8b-3}{b^2-7b+12} - \dfrac{2b+1}{b-4}$

e) $\dfrac{8x}{5(x-2)} + \dfrac{5x-1}{3(x+3)}$ f) $\dfrac{4x+3}{3(2x-1)} - \dfrac{x-5}{2(3x+7)}$

31. Barbara flew her airplane 500 km against the wind in the same time that it took her to fly it 600 km with the wind. If the speed of the wind was 20 km/h, what was the average speed of her airplane?

Cumulative Review (Chapters 1 - 2)

1. Find the g.c.f. and l.c.m. for each pair of numbers:
   a) 20, 45   b) 18, 60   c) 42, 390   d) 210, 308

2. Simplify:
   a) $[8 - (12 \div 4 \times 2)][10 - 24 \div 3]$
   b) $3[-21 \div 3 \times (-2) + 8(-6) \div (-3)]$

3. If $x = -2$, $y = 3$, and $z = -1$, evaluate:
   a) $x^2z - x^2y^2 - yz$ b) $2x^2z - 2x^2y + 2xz^2 + 2yz^2$

4. If $x = -\dfrac{1}{3}$ and $y = -\dfrac{3}{4}$, evaluate:
   a) $6x + 4y - 24xy$ b) $4x^2 - \dfrac{1}{3}xy + 4y^2$

5. Simplify:
   a) $\sqrt{320}$ b) $\sqrt{108}$ c) $-3\sqrt{180}$ d) $2\sqrt{216}$
   e) $3\sqrt{5} \times 4\sqrt{10}$ f) $(-2\sqrt{12})(3\sqrt{6})$
   g) $(-3\sqrt{15})(-2\sqrt{5})$ h) $3\sqrt{8} \times 2\sqrt{6} \times 4\sqrt{12}$

6. Find x, rounded to one decimal place:

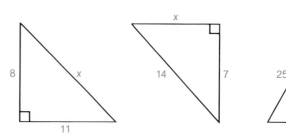

a) b) c)

7. Expand and simplify:
   a) $\sqrt{3}(\sqrt{8} - \sqrt{2})$
   b) $\sqrt{2}(\sqrt{3} - 2\sqrt{27})$
   c) $\sqrt{3}(\sqrt{5} - \sqrt{8})$
   d) $\sqrt{3}(\sqrt{108} - \sqrt{12})$

8. Combine where possible:
   a) $\frac{1}{3}\sqrt{27} + \frac{1}{2}\sqrt{108} - 3\sqrt{3}$
   b) $2\sqrt{32} - 4\sqrt{128} + \sqrt{98}$

9. Express in simplest form:
   a) $\frac{4\sqrt{32}}{\sqrt{2}}$
   b) $\frac{10}{\sqrt{50}}$
   c) $\frac{4\sqrt{54}}{3\sqrt{24}}$
   d) $\frac{\sqrt{3}}{\sqrt{\frac{1}{3}}}$

10. Simplify:
    a) $\left(-\frac{2}{5}\right)^{-2}$
    b) $\left(\frac{1}{4}\right)^{-3}$
    c) $3^{-1} + 3^{-2}$
    d) $\frac{1}{(3^2 - 2^2)^{-1}}$
    e) $(-1)^2 + (-2)^{-2}$
    f) $(-3)^0$

11. Simplify:
    a) $y^2 \cdot y^3 \cdot y$
    b) $x^5 \div x^8$
    c) $a^{-3} \cdot a^{-2}$
    d) $(x^{-2})^{-2}$
    e) $x^2 y^3 \div y^{-2}$
    f) $(x^{-2}y)^{-2}(x^2y^{-2})^{-1}$

12. Simplify:
    a) $(5x^2 - 12xy + 7y^2) - (10x^2 - 3y^2 - 4xy) + (-3xy + 2x^2 - y^2)$
    b) $3x(x + y) - 2x(5x - 2y) - 4x(3x + 5y)$

13. Simplify:
    a) $(-7x^2y)(6xy^2)$
    b) $(-4ab^3)(-3a^2b)$
    c) $(-x^2y)(3xy)^2(-4x^2y^2)$
    d) $\frac{(3a^2b^3)(-6a^3b^2)}{(2ab)^2(9ab^2)}$

14. Factor:
    a) $9x^3 - 18x^2$
    b) $14y^4 - 7y^3$
    c) $m^2 + 2m + mn + 2n$
    d) $a^2 + 8a + 16$
    e) $x^2 - x - 20$
    f) $y^2 - 4y - 21$
    g) $m^2 - 17m + 72$
    h) $c^2 - 2cd - 24d^2$
    i) $x^2 + 5xy - 24y^2$
    j) $x^2 - 20xy + 91y^2$

15. Factor:
    a) $2x^2y^2 + 12x^2y + 16x^2$
    b) $5m^3n + 20m^2n - 105mn$
    c) $8x^2 - 72$
    d) $180 - 5m^2$
    e) $8x^4 - 8y^4$

16. If $x = 90$ and $y = 10$, evaluate:
    a) $x^2 - xy - 12y^2$
    b) $x^2 - 2xy - 15y^2$
    c) $x^2 + 2xy - 35y^2$
    d) $x^2 - 8xy + 65y^2$

17. Simplify:
    a) $\dfrac{84x^3y^2}{6xy}$
    b) $\dfrac{x^2 - 36}{3x^2 - 18x}$
    c) $\dfrac{m^2 - m - 12}{m^2 - 2m - 15}$
    d) $\dfrac{a^2 - 16}{a^2 + 8a + 16}$
    e) $\dfrac{x^2 - 2x - 24}{x^2 - x - 20}$
    f) $\dfrac{y^2 + 12y + 36}{y^2 + y - 30}$

18. Simplify:
    a) $\dfrac{3x^3}{2y} \times \dfrac{16x^2y^2}{4x^4}$
    b) $\dfrac{2(5 - x)}{3x} \div \dfrac{2(x - 5)}{9(x + 1)}$
    c) $\dfrac{4c^3}{c^2 - 4} \times \dfrac{2c - 4}{3c}$
    d) $\dfrac{a^2 - 2a - 8}{16 - a^2} \div \dfrac{a^2 + a - 2}{a^2 + 3a - 4}$

19. Simplify:
    a) $\dfrac{5}{2x} - \dfrac{3}{5x}$
    b) $\dfrac{2}{3m} + \dfrac{5}{4m}$
    c) $\dfrac{3}{4a} - \dfrac{2}{3a} + \dfrac{5}{6a}$
    d) $\dfrac{x + 3}{5} - \dfrac{x - 4}{3}$

20. Simplify:
    a) $\dfrac{x}{x + 2} - \dfrac{3}{x - 2}$
    b) $\dfrac{3a}{8a - 12} + \dfrac{2a}{6a - 9}$
    c) $\dfrac{2x + 3}{2x - 8} - \dfrac{3x}{x - 4}$
    d) $\dfrac{3m}{2(m - 2)} + \dfrac{2m - 1}{3(m + 2)}$

106 Chapter 2

Use a Model

Problems that deal with 3-dimensional objects often present a problem in visualization. It is sometimes useful to draw a diagram to represent these objects, but in most cases an actual 3-dimensional model is more helpful.

Example. The diagrams show two different views of the same cube.

Sketch the missing letters on the cube as they would appear in this view:

Solution. Make a model of the cube and label it with the letters A, C, D, E, F as they appear in the views of the cube that are given.

Rotate the cube until the D is in the required position and copy the letters as they appear on the other two faces.

Exercises

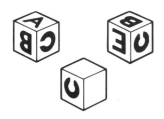

1. Two different views of the same cube are shown. On a copy of the third view showing two blank faces, sketch the letters that would appear on those faces.

2. Determine how to label the faces of two cubes with the digits 0 to 9 so that every day of the month can be represented using both cubes.

3. A right-hand rubber glove is turned inside out and worn on the left hand. What part of the left hand is the palm of the original glove touching?

3 Equations

The cost, C, in dollars, of printing n books is given by the formula $C = P + Rn$, where P is the cost of preparing the plates and R is the press-run cost per book. If the cost of printing 500 books is $3000 and the cost of printing 2000 books is $5250, find the cost of printing 6000 books. How many books can be printed for $10 000 ? (See *Example 2* of Section 3-3.)

3-1 Solving Equations in One Variable

This section reviews the techniques for solving equations in one variable.

To solve an equation, isolate the variable. This will require performing one or more operations on both sides of the equation.

Example 1. Solve: $3(x - 4) - 10 = 9 - 5(x + 3)$

Solution.

$$3(x - 4) - 10 = 9 - 5(x + 3)$$

↓ *Apply the distributive law.*

$$3x - 12 - 10 = 9 - 5x - 15$$

↓

$$3x - 22 = -5x - 6$$

↓ *Add 5x and 22 to both sides.*

$$3x + 5x = -6 + 22$$

↓

$$8x = 16$$

↓ *Divide both sides by 8.*

$$x = 2$$

Check. Substitute 2 for x in each side of the equation, and simplify each side independently.

$$\begin{aligned}
&3(x - 4) - 10 & & 9 - 5(x + 3) \\
&= 3(2 - 4) - 10 & & = 9 - 5(2 + 3) \\
&= 3(-2) - 10 & & = 9 - 25 \\
&= -16 & & = -16
\end{aligned}$$

Each side simplifies to the same number. $x = 2$ is the correct solution.

When an equation contains fractions, multiply both sides of the equation by a common denominator to obtain an equivalent equation without fractions.

Example 2. Solve: $\dfrac{5(y+6)}{4} + 3 = \dfrac{-(1-10y)}{3}$

Solution.

$$\dfrac{5(y+6)}{4} + 3 = \dfrac{-(1-10y)}{3}$$

↓

Multiply both sides by 12.

↓

$$15(y+6) + 36 = -4(1-10y)$$

↓

Apply the distributive law.

↓

$$15y + 90 + 36 = -4 + 40y$$

↓

$$15y + 126 = 40y - 4$$

↓

Subtract $15y$ and -4 from both sides.

↓

$$126 - (-4) = 40y - 15y$$

↓

$$130 = 25y$$

↓

$$5.2 = y$$

Check. Substitute 5.2 for y in each side of the equation.

$$\dfrac{5(y+6)}{4} + 3 \qquad\qquad \dfrac{-(1-10y)}{3}$$

$$= \dfrac{5(5.2+6)}{4} + 3 \qquad = \dfrac{-(1-10 \times 5.2)}{3}$$

$$= \dfrac{5(11.2)}{4} + 3 \qquad\quad = \dfrac{-(1-52)}{3}$$

$$= \dfrac{56}{4} + 3, \text{ or } 17 \qquad = \dfrac{51}{3}, \text{ or } 17$$

The solution is correct.

Equations are used to solve problems. The facts of a problem are translated into an equation, and the solution of the equation is used to answer the question asked by the problem.

Example 3. How much antifreeze must be added to 5 L of water to make a solution that is 40% antifreeze?

Solution. Let volume of antifreeze to be added be x L.

$$\frac{\text{Volume of antifreeze}}{\text{Total volume}} = \frac{40}{100}$$

$$\frac{x}{x+5} = \frac{40}{100}$$

$$100x = 40x + 200$$

$$60x = 200$$

$$x = \frac{10}{3}$$

About 3.3 L of antifreeze must be added.

Check. Volume of solution: $(5 + 3.3)$ L $= 8.3$ L

Percent of antifreeze: $\frac{3.3}{8.3} \times 100\% \doteq 40\%$

The solution is correct.

Exercises 3-1

A 1. Solve and check:
   a) $3x - 4 = 23$ b) $2m + 5 = -21$
   c) $-5a - 7 = -62$ d) $8y + 3 = -37$
   e) $2(3x - 5) = 32$ f) $5(3x + 4) = -10$
   g) $9x - (4x + 7) = 28$ h) $4x - (7x - 8) = -25$

2. Solve:
   a) $7m + 2 = 5m + 18$ b) $11x - 18 = 3 + 8x$
   c) $9x - 30 = -3(5x - 6)$ d) $13b - 12 = 49b + 24$
   e) $4(5 + x) = x + 5$ f) $6x + 10 = 15x + 64$
   g) $3(x - 5) = 2(5x - 11)$ h) $3(2r + 4) = 4(5r - 4)$

3. Solve:
   a) $x - 5 = 8 - 2(x + 2)$
   b) $4(x + 1) = 10 - (2x + 6)$
   c) $3 - (2 + 4x) = 4 + 2(3x + 1)$
   d) $12(a - 3) - 35 = 5(13 - a)$
   e) $3(y - 2) - 8 = 68 - 2(2y - 1)$
   f) $3x + 7(2 - x) = 14 - 9x - 3$
   g) $17x - 9(1 + x) = 4(3x - 1) + 7$
   h) $4(x - 9) + 52 = 2(3x + 17) + 2x$

4. Solve:

a) $\dfrac{7x}{6} = \dfrac{7}{2}$

b) $\dfrac{x}{5} + \dfrac{1}{2} = \dfrac{3}{10}$

c) $\dfrac{3m}{5} - \dfrac{1}{2} = \dfrac{7}{10}$

d) $\dfrac{2x}{9} + \dfrac{1}{3} = -\dfrac{1}{6}$

e) $\dfrac{1}{2}x - \dfrac{1}{3}x = \dfrac{7}{3}$

f) $\dfrac{2x}{5} + \dfrac{3}{4} = \dfrac{4x}{5} - \dfrac{1}{2}$

g) $\dfrac{4}{3}x + \dfrac{5}{6} = \dfrac{1}{2}x - \dfrac{5}{3}$

h) $\dfrac{x}{5} + \dfrac{x}{3} = \dfrac{2}{3} - \dfrac{2x}{5}$

B 5. Solve and check:

a) $2(5x - 11) + 7 = 3(x - 7) - 15$

b) $13x - (3x + 12) = 12x - (5x - 3)$

c) $2(x - 6) + 3x = 2(x + 2) - x$

d) $9(3 + y) - 16 = 8(y + 4) + 5$

e) $15 - 2(3 + 2x) = 4 + 3(2x - 5)$

f) $3x(x - 5) - 3x(x + 7) = 72$

g) $4(3x + 1) - 6(x - 3) = 4(2x - 7) + 34$

h) $2x(3x + 4) - 19 = x(6x - 2) + 31$

6. Solve:

a) $\dfrac{3x - 2}{6} = \dfrac{5}{3}$

b) $\dfrac{2x}{x - 3} = 5$

c) $\dfrac{3x - 1}{x + 1} = -2$

d) $\dfrac{5(x - 2)}{4} = \dfrac{6(x - 2)}{5}$

e) $\dfrac{-2 + x}{2x + 1} = 3$

f) $\dfrac{2x}{x + 3} = \dfrac{6x + 5}{3x - 1}$

g) $\dfrac{2x^2 - 2x + 8}{2x - 1} = x + 1$

h) $\dfrac{x - 2}{x + 3} = \dfrac{x + 4}{x - 1}$

i) $\dfrac{w - 5}{1 - w} = \dfrac{1 - w}{w + 5}$

7. How much water should be added to 5 L of antifreeze to make a solution that is 40% antifreeze?

8. A rectangular photograph, 25 cm by 20 cm, is centred in a frame with a length-to-width ratio of 6:5. What is the width of the border?

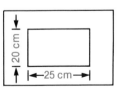

9. Two strips of equal width are cut from a rectangular piece of card, 15 cm by 12 cm, as shown. If the length-to-width ratio of the remaining rectangle is 4:3, find the width of the strips.

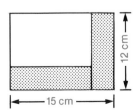

10. The coin box of a vending machine contains half as many quarters as dimes. If the total value of the coins is $22.50, how many dimes are there?

11. In a collection of nickels, dimes, and quarters, there are 10 more dimes than nickels and 5 more quarters than nickels. If there is a total of 45 coins, how many of each kind are there?

12. A man is five times as old as his son. The man's daughter is two years older than her brother. In ten years' time, the sum of their ages will be 81 years. How old is each now?

C 13. For which of these equations is $x = -3$
    i) a solution?   ii) the only solution?
    a) $3(2 - x) + x = 2(3 - x)$ b) $2 - 5(x + 1) = 3(9 - 5x)$
    c) $2(3x - 1) + x = 10 - 3(5 - 2x)$
    d) $5x + 4 - 3(x + 1) = (3 - 2x) + 2(2x - 1)$

14. If $x = 5$ is the solution of each equation, find the value of k:
    a) $2x - k = 3 - x$ b) $2 + 3x = 8 - (x - k)$
    c) $kx - 6 = 2x + k$ d) $2(x - 3) + k(1 + 2x) = k - x - 1$

3-2 Solving Equations in Two Variables

Equations may have more than one variable. An equation such as $2x + y = 8$ has two variables. To solve this equation means to find all the ordered pairs (x, y) that satisfy the equation. The first step is to express y in terms of x:

$$2x + y = 8$$
$$\downarrow$$
$$\boxed{\text{Subtract } 2x \text{ from both sides.}}$$
$$\downarrow$$
$$y = 8 - 2x$$

Using this expression, values of y can be found when values are given to x. For example:

x	-2	0	2	4	6	8
y	12	8	4	0	-4	-8

Each of the ordered pairs $(-2, 12)$, $(0, 8)$, $(2, 4)$, $(4, 0)$, $(6, -4)$, $(8, -8)$ is a solution of the equation. There are many others. The set of *all* solutions to the equation, $2x + y = 8$, called the **solution set**, is written as follows:

$\{(x,\ 8 - 2x)\ |\ x \in R\}$

- The set — of all ordered pairs of the form $(x, 8 - 2x)$
- where — x is any real number

Since every real number, x, yields a different ordered pair, the equation, $2x + y = 8$, has infinitely many solutions.

Example 1. Find four different ordered pairs (x, y) in the solution set of $3x - 2y = 4$.

Solution. Express y in terms of x: $3x - 2y = 4$
Subtract $3x$ from both sides. $-2y = 4 - 3x$
Divide both sides by -2. $y = -2 + \frac{3}{2}x$

x	-2	0	2	4
y	-5	-2	1	4

Any four values are given to x.

Example 2. Find the solution set of the equation $2x + 5y = -12$.

Solution. Express y in terms of x:
$$2x + 5y = -12$$
$$5y = -12 - 2x$$
$$y = \frac{-12 - 2x}{5}$$

The solution set is $\{(x, \frac{-12 - 2x}{5})\ |\ x \in R\}$.

Many formulas are equations in two variables. If the value of either variable is known, the value of the other can be found.

Example 3. The cost, C, in dollars, of renting a room for a party is given by the formula $C = 100 + 5n$, where n is the number of people in attendance.
  a) What is the cost if 70 people attend?
  b) If the cost was $650, how many people attended?

Solution. a) Substitute 70 for n: $C = 100 + 5(70)$
                                        $= 450$
If 70 people attend, the cost is $450.

b) Substitute 650 for C: $650 = 100 + 5n$
                             $550 = 5n$
                             $110 = n$
If the cost was $650, then 110 people attended.

Example 4. When the digits of a two-digit number are reversed, the new number is 54 greater than the original number. What was the original number?

Solution. Let the digits of the original number, in order, be x and y.

Original number: $10x + y$
New number: $10y + x$
Difference: $(10y + x) - (10x + y) = 54$
$$9y - 9x = 54$$
$$y - x = 6$$

Since x and y represent the digits of a number, the only acceptable solutions are $(1, 7)$, $(2, 8)$, and $(3, 9)$. The original number may be any one of 17, 28, or 39.

Exercises 3-2

A 1. If $y = 5 + 2x$, find

a) the value of y for the following values of x:

i) 3   ii) 10   iii) -7 iv) 2.5   v) $\frac{1}{2}$

b) the value of x for the following values of y:

i) 7   ii) 15   iii) 0   iv) -3 v) $\frac{1}{2}$

2. Find the solution set:

a) $x - y = 6$ b) $-x + y = 6$ c) $-x - y = 6$
d) $2x - y = 8$ e) $3x + y = 0$ f) $5x + y = 10$
g) $-2x + y = 6$ h) $4x - y = -8$ i) $4x - y = 12$

3. Solve:

a) $3x + 2y = 4$ b) $3x - 2y = -4$ c) $2x + 5y = 10$
d) $2x = 5y$ e) $-2x + 5y = 10$ f) $4x - 3y = 24$
g) $x + 3y = 9$ h) $3x - 5y = 30$ i) $2x + 6y = 0$

4. Solve:

a) $2x - y = 6$ b) $x + 3y = 3$ c) $8x + 4y = 16$
d) $3x - 2y = 12$ e) $3x + 6y = 12$ f) $5x - 2y = 20$
g) $4x + 3y = 24$ h) $6x - 4y = 0$ i) $10x - 8y = 40$

5. Find six ordered pairs of integers that satisfy each of the following:

a) The sum of the integers is 11.

b) The difference of the integers is 5.

c) Double the first integer plus the second is 10.
d) The first integer plus triple the second is 12.

6. When the digits of a two-digit number are reversed, the new number is 36 greater than the original number. What was the original number?

7. When the digits of a two-digit number are reversed, the sum of the new and original numbers is 88. What was the original number?

B 8. If $3x + 2y = n$, find
   a) the value of n for these values of x and y:
      i) (1, 1)   ii) (−1, 5)   iii) (2, −7)   iv) (−3, −1)
   b) three possible values of x and y for these values of n:
      i) 6   ii) 0   iii) −3   iv) 7

9. The cost, C, in cents, of making n photocopies is given by the formula $C = 70 + 6n$.
   a) What is the cost of making 10 copies?
   b) How many copies were made if the cost was
      i) $1.18?   ii) $3.10?   iii) $6.10?

10. The time it takes to pump C litres of water is given by the formula $t = \frac{1}{1500}C$, where t is the time in hours.
    a) How long will it take to fill a tank that has
       i) a capacity of 300 L?   of 5 kL?
       ii) dimensions: 4 m by 3 m by 2 m?
    b) If the pump operates continuously, how much water will be pumped between 18:30 one day and 08:30 the next?

11. Tickets for a school play cost $3 for adults and $2 for students. If Lorna sold $17 worth of tickets, how many of each did she sell?

12. Suzanne has $1.70 in quarters and dimes. Ivan has $1.70 in quarters and nickels. How many quarters, dimes, and nickels does each have?

C 13. Find, by inspection, an ordered pair in the solution set of both equations in each of the following:
   a) $x + y = 6$
      $2x + y = 8$
   b) $2x + 3y = 13$
      $x + 2y = 7$
   c) $x - y = 1$
      $5x + 2y = 5$
   d) $3x + y = -1$
      $2x + 3y = 4$

3-3 Solving Pairs of Linear Equations by Comparison

To solve a pair of equations such as $\begin{array}{l}3x + 2y = 1\\5x - y = -7\end{array}$ means to find all the ordered pairs (x, y), if any, which satisfy both equations. The solution is obtained by eliminating one of the variables. One method of doing this is to write both equations in terms of the same variable and compare the expressions.

Example 1. Solve: $\begin{array}{l}3x + 2y = 1 \dots ①\\5x - y = -7 \dots ②\end{array}$

Solution. For both equations, express y in terms of x:

$$3x + 2y = 1 \qquad\qquad 5x - y = -7$$
$$2y = 1 - 3x \qquad\qquad -y = -7 - 5x$$
$$y = \frac{1 - 3x}{2} \dots ③ \qquad\qquad y = 5x + 7 \dots ④$$

Since these two expressions for y must be equal:

$$\frac{1 - 3x}{2} = 5x + 7$$

(*y is eliminated by comparison.*)

Solve this equation for x:
$$1 - 3x = 10x + 14$$
$$-13x = 13$$
$$x = -1$$

Substitute -1 for x in either ③ or ④ to obtain the value of y:
$$y = 5(-1) + 7, \text{ or } 2$$

The solution of the pair of equations is $(-1, 2)$.

Check. When $x = -1$ and $y = 2$,

$$\begin{array}{ll} 3x + 2y & 5x - y \\ = 3(-1) + 2(2) & = 5(-1) - (2) \\ = -3 + 4, \text{ or } 1 & = -5 - 2, \text{ or } -7 \end{array}$$

In the above example, the solution could be found equally as well by expressing x in terms of y for both equations and comparing the expressions.

The solution of many real-life problems requires the solving of a pair of equations.

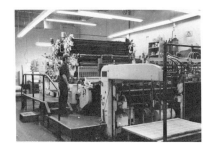

Example 2. The cost, C, in dollars, of printing n books is given by the formula $C = P + Rn$, where P is the cost of preparing the plates and R is the press-run cost per book. If the cost of printing 500 books is $3000 and the cost of printing 2000 books is $5250, find

a) the cost of the plates and the press-run cost per book;

b) the cost of printing 6000 books;

c) how many books can be printed for $10 000.

Solution. a) Substitute the two values given for C and n in the formula $C = P + Rn$ to obtain two equations:
$$3000 = P + 500R \ldots \text{①}$$
$$5250 = P + 2000R \ldots \text{②}$$
For both equations, express P in terms of R:
$$P = 3000 - 500R \ldots \text{③}$$
$$P = 5250 - 2000R \ldots \text{④}$$
Compare the expressions for P and solve for R:
$$3000 - 500R = 5250 - 2000R$$
$$1500R = 2250$$
$$R = 1.5$$
To find P, substitute 1.5 for R in ③:
$$P = 3000 - 500(1.5)$$
$$= 3000 - 750, \text{ or } 2250$$
The cost of preparing the plates is $2250 and the press-run cost per book is $1.50.

b) From (a), cost of printing n books:
$$C = 2250 + 1.5n$$
When $n = 6000$: $C = 2250 + 1.5(6000)$
$$= 2250 + 9000, \text{ or } 11\,250$$
The cost of printing 6000 books is $11 250.

c) Substitute 10 000 for C: $10\,000 = 2250 + 1.5n$
$$7750 = 1.5n$$
$$5167 \doteq n$$
5167 books can be printed for $10 000.

Exercises 3-3

A 1. Solve:

a) $2x + y = 7$
$5x + y = 16$

b) $3x - y = -8$
$x + y = 0$

c) $4x - y = -1$
$2x - y = 1$

d) $7x - 2y = 34$
$2x + y = 5$

e) $4x + y = 0$
$6x - 5y = -13$

f) $2x + 3y = -11$
$3x - y = 11$

g) $4x - 3y = 9$
$8x + 5y = -4$

h) $2x - 5y = 4$
$3x + 4y = -17$

i) $3x - 6y = 4$
$5x + 4y = 2$

2. For which pair of equations is $(-1, 1)$ a solution?

a) $5x + 6y = 1$
$6x + 2y = -3$

b) $3x + 4y = 1$
$5x - 3y = -8$

c) $3x - 4y = -6$
$3x + 3y = 1$

B 3. Solve:

a) $x + y = -6$
$2x + y = 12$

b) $6x + 5y = 9$
$4x - 3y = 25$

c) $4x - 3y = -7$
$-2x + 5y = 21$

d) $\frac{3}{4}x + y = -3$
$3x + \frac{1}{2}y = 9$

e) $x + y = 7$
$\frac{x}{3} + \frac{y}{4} = 2$

f) $3y + 2x = 2$
$4x + 9y = -1$

g) $\frac{x}{2} - \frac{y}{7} = 1$
$\frac{x}{4} - \frac{2y}{7} = -1$

h) $5x - \frac{1}{3}y = 3$
$10x + \frac{2}{3}y = 2$

i) $6x + 5y = \frac{7}{2}$
$y = 3x$

4. Find the values of k and m for $(2, -3)$ to be a solution of each pair of equations:

a) $4x + 2y = k$
$3x - y = m$

b) $2x - 5y = k$
$mx + 2y = 8$

c) $kx - 3y = 19$
$6x + my = 18$

5. Solve:

a) $\frac{1}{2}s + t = \frac{1}{4}$
$\frac{3}{10}s + \frac{3}{5}t = \frac{3}{20}$

b) $\frac{x + 2}{y} = \frac{2}{3}$
$\frac{x}{y - 3} = \frac{2}{3}$

c) $y = \frac{19 - 3x}{2}$
$x = \frac{5y + 5}{3}$

d) $m = \frac{3n - 7}{4}$
$7m - 3n = -1$

6. The cost, C, in dollars, of a wedding reception for n guests is given by the formula $C = A + Bn$, where A is the cost of renting the banquet hall and B is the cost per person. If the cost for 30 guests is $655 and for 50 guests is $825, find
  a) the cost to rent the hall and the cost per person;
  b) the cost of a reception for 175 guests;
  c) how many guests can be invited for a total cost of $2000.

7. The cost, C, in dollars, of translating n words from English to French is given by the formula $C = A + Wn$, where A is a fixed cost and W is the cost per word. If the cost for 200 words is $39 and for 700 words is $99, find
  a) the fixed cost and the cost per word;
  b) the cost of translating a 1500-word passage.

C 8. Solve:

a) $\frac{m - 2}{5} - \frac{n + 3}{6} = -1$
$\frac{m - 3}{3} + \frac{n - 5}{2} = 8$

b) $\frac{x + 3}{3} = 1 + \frac{3y - 1}{5}$
$\frac{2 - 3x}{4} = 5 + \frac{y + 3}{6}$

THE MATHEMATICAL MIND

The Rhind Papyrus

In 1858, Henry Rhind bought an ancient scroll, over 5 m long, in the village of Luxor, Egypt. It was written about 1650 B.C. by the scribe Ahmes, who called it "directions for knowing all dark things".

This manuscript, now known as the Rhind papyrus, is a collection of 85 problems in arithmetic and geometry. It is our principal source of information about ancient Egyptian mathematics.

A portion of the Rhind papyrus (British Museum)

Problem 26 of the Rhind papyrus is as follows:
A quantity and its $\frac{1}{4}$ added together become 15. What is the quantity?

Using modern notation, this problem calls for the solution of the equation: $x + \frac{1}{4}x = 15$. The Egyptians solved equations such as this using the "method of false position".

Step 1.
Choose any convenient value for x.
Assume that $x = 4$

Step 2.
Substitute this value for x in the left side of the equation.
$x + \frac{1}{4}x = 4 + \frac{1}{4}(4)$, or 5

Step 3.
Adjust assumed value to give correct value on the right side.

Since 5 must be multiplied by 3 to give 15, the assumed value must also be multiplied by 3. Therefore, the correct value of x is 4×3, or 12.

1. Solve these problems from the Rhind papyrus using the "method of false position".
   a) Problem 25: $x + \frac{1}{5}x = 21$
   b) Problem 24: $x + \frac{1}{7}x = 19$

2. Problem 63 concerns the rationing of food: "Directions for dividing 700 breads among four people, $\frac{2}{3}$ for one, $\frac{1}{2}$ for the second, $\frac{1}{3}$ for the third, $\frac{1}{4}$ for the fourth."

   Decide what this problem means, and then solve the problem using the "method of false position".

3. In several problems, procedures are given for finding the area of a circle. These are all equivalent to the relation: $A = (\frac{8}{9}d)^2$, where d is the diameter. By comparing this with the modern formula: $A = \pi r^2$, determine the value the Egyptians were actually using for π.

3-4 Solving Pairs of Linear Equations by Substitution

The solution of a pair of linear equations is found by eliminating one of the variables and solving the resulting equation for the other variable. Comparison is one method of elimination, substitution is another. The expression for one of the variables obtained from one equation is substituted in the other equation.

Example 1. Solve: $\begin{aligned} 2x - y &= 13 \quad \ldots \text{①} \\ 4x + 3y &= 1 \quad \ldots \text{②} \end{aligned}$

Solution. Choose equation ① and express y in terms of x.
$$2x - y = 13$$
$$y = 2x - 13 \quad \ldots \text{③}$$
Substitute this expression for y in equation ② and solve for x:
$$4x + 3(2x - 13) = 1$$
$$4x + 6x - 39 = 1$$
$$10x = 40$$
$$x = 4$$
Substitute 4 for x in ③ and solve for y:
$$y = 2(4) - 13$$
$$= -5$$
The solution of the pair of equations is $(4, -5)$.

y is eliminated by substitution.

Check. When $x = 4$ and $y = -5$,

$\begin{aligned} 2x - y & \\ = 2(4) - (-5) & \\ = 8 + 5, \text{ or } 13 & \end{aligned}$ $\qquad$ $\begin{aligned} 4x + 3y & \\ = 4(4) + 3(-5) & \\ = 16 - 15, \text{ or } 1 & \end{aligned}$

The solution is correct.

It makes no difference which variable is chosen for elimination. However, if one has a coefficient of ±1, its choice makes for easier work since fractions are avoided. In the next example, x is eliminated because in one of the equations it has a coefficient of 1.

Example 2. Solve: $\begin{aligned} x + 6y &= 9 \quad \ldots \text{①} \\ 3x - 2y &= -23 \quad \ldots \text{②} \end{aligned}$

Solution. Choose equation ① and express x in terms of y.
$$x + 6y = 9$$
$$x = 9 - 6y \quad \ldots \text{③}$$
Substitute $9 - 6y$ for x in ② and solve for y:
$$3(9 - 6y) - 2y = -23$$
$$27 - 18y - 2y = -23$$
$$-20y = -50$$
$$y = \frac{5}{2}$$

Substitute $\frac{5}{2}$ for y in ③ to find the value of x:
$$x = 9 - 6(\tfrac{5}{2})$$
$$= 9 - 15, \text{ or } -6$$

The solution of the pair of equations is $(-6, \frac{5}{2})$.

Check. When $x = -6$ and $y = \frac{5}{2}$,

$x + 6y$
$= -6 + 6(\tfrac{5}{2})$
$= -6 + 15, \text{ or } 9$

$3x - 2y$
$= 3(-6) - 2(\tfrac{5}{2})$
$= -18 - 5, \text{ or } -23$

The solution is correct.

Exercises 3-4

A 1. Solve:
 a) $x + y = 9$
    $2x + y = 11$
 b) $x + y = 1$
    $3x - y = 11$
 c) $x - y = 7$
    $2x + y = -10$
 d) $3x + y = 7$
    $5x + 2y = 13$
 e) $2x + 3y = 11$
    $5x - y = -15$
 f) $4x + y = -5$
    $2x + 3y = 5$
 g) $3x + 2y = 19$
    $2x - 3y = -9$
 h) $5y + 2x = -2$
    $5x - 2y = 24$
 i) $3x + 4y = 2$
    $-3y + 8x = 60$

2. For which pair of equations is $(-2, 5)$ a solution?
 a) $3x + y = 1$
    $2x + 3y = 11$
 b) $5x - 3y = -5$
    $3x + 2y = 4$
 c) $\frac{3}{2}x + \frac{2}{5}y = -1$
    $\frac{5}{4}x - \frac{3}{10}y = -4$

B 3. Solve:
 a) $3x - 4y = -15$
    $5x + y = -2$
 b) $2x + y = 2$
    $3x - 2y = 10$
 c) $3m - n = 5$
    $5m - 2n = 8$
 d) $4s - 3t = 9$
    $2s - t = 5$
 e) $5v + u = -17$
    $3u - 4v = 6$
 f) $x + 5y = -11$
    $4x - 3y = 25$
 g) $\frac{x}{2} + \frac{y}{2} = 7$
    $3x + 2y = 48$
 h) $\frac{a}{2} + \frac{b}{3} = 1$
    $\frac{a}{4} + \frac{2b}{3} = -1$
 i) $2x + 5y = -\frac{39}{2}$
    $9 = 4x - 2y$

4. Solve:
 a) $3x + 6y = 4$
    $x - 2y = 1$
 b) $7x + y = 13$
    $3x - 2y = 8$
 c) $4x + 6y = 1$
    $x + y = 4$
 d) $5x + 3y = 5$
    $2x + y = 8$
 e) $9x + 2y = 2$
    $1 - y = 4x$
 f) $8x + 4y = 1$
    $7x = -2y$
 g) $9x + 6y = 4$
    $8x + 3y = 9$
 h) $2x + 8y = 1$
    $x = 2y$
 i) $5x = 2y + 3$
    $10x + 3y = 13$

5. For what values of m and n is $(-1, 2)$ the solution of each pair of equations?
   a) $mx + 3y = 1$
      $2x + ny = -4$
   b) $mx + ny = 3$
      $4x + ny = -2$
   c) $mx + ny = 1$
      $mx + 5y = 7$
   d) $2mx + ny = 6$
      $3mx - 2ny = 2$

6. The value of a car, V, in dollars, n years after it is made is given by the formula $V = A - Dn$ ($n < 8$), where A is its value when new and D is the annual depreciation. If a car is worth $5400 after 3 years and $3000 after 5 years, find
   a) the value of the car when new and the annual depreciation;
   b) the value of the car after 4 years.

7. The weekly wage, W, in dollars, of a construction worker is given by the formula $W = R + Ht$, where R is the wage for the first 40 h, t is the number of hours overtime worked, and H is the overtime rate of pay in dollars per hour. If a worker receives $572 for working 44 h and $680 for working 50 h, find
   a) the wage for the first 40 h and the overtime rate;
   b) how much a worker receives for working 55 h;
   c) the overtime necessary to earn a total of $600.

8. The daily payroll for a construction crew is given by the formula $P = 80B + 40L$, where B is the number of bulldozer operators and L is the number of laborers. One day the payroll was $1520 and the total number of workers was 32.
   a) Find the number of bulldozer operators and the number of laborers working that day.
   b) How many of each kind of worker was used when
      i) the payroll was $1840 and the crew totalled 38?
      ii) the payroll was $1400 and the crew totalled 32?
      iii) the payroll was $1400 and there were three times as many laborers as bulldozer operators?

C 9. a) Combine the equations $\begin{array}{l} x = 2t - 1 \\ y = 2 - 3t \end{array}$ to form a single equation in which x and y are the only variables.
   b) Combine the equations $\begin{array}{l} x = s - 3 \\ y = 5s - 8 \end{array}$ to form a single equation in which x and y are the only variables.
   c) Solve the equations obtained from parts (a) and (b) as a pair.
   d) Use the answers to parts (a), (b), and (c) to solve the pair of equations $\begin{array}{l} 2t - 1 = s - 3 \\ 2 - 3t = 5s - 8 \end{array}$.

3-5 Solving Pairs of Linear Equations by Addition or Subtraction

A third method of eliminating a variable from a pair of linear equations is developed in this section. This method depends on two fundamental properties of equations illustrated below:

- Any ordered pair (x, y) that is the solution of two linear equations is also a solution of the equation formed by their sum or difference.

    Since $(2, 1)$ is the solution of $\begin{array}{l} 3x + 5y = 11 \\ x + 2y = 4 \end{array}$,

    it is also a solution of $4x + 7y = 15$, their sum,
    $$4(2) + 7(1) = 15$$

    and $2x + 3y = 7$, their difference.
    $$2(2) + 3(1) = 7$$

- The solution set of an equation is unchanged if both sides of the equation are multiplied (or divided) by the same non-zero number.
    $x + 2y = 4$ and $2x + 4y = 8$ have the same solution set.

Example 1. Solve: $\begin{array}{l} 3x - 5y = -9 \quad \dots \text{①} \\ 4x + 5y = 23 \quad \dots \text{②} \end{array}$

Solution. Since $-5y$ and $+5y$ occur in the equations, y can be eliminated by adding the two equations:

$$\begin{array}{r} 3x - 5y = -9 \\ 4x + 5y = 23 \\ \hline 7x = 14 \\ x = 2 \end{array}$$

Substitute 2 for x in ②: $4(2) + 5y = 23$
$$5y = 23 - 8, \text{ or } 15$$
$$y = 3$$

The solution of the pair of equations is $(2, 3)$.

Check. When $x = 2$ and $y = 3$,

$3x - 5y$ $\qquad$ $4x + 5y$
$= 3(2) - 5(3)$ $\qquad$ $= 4(2) + 5(3)$
$= 6 - 15$, or -9 $\qquad$ $= 8 + 15$, or 23

The solution $(2, 3)$ is correct.

Sometimes it is necessary to multiply one, or both, of the equations by a constant before one of the variables can be eliminated by addition or subtraction.

Example 2. Solve: $\begin{array}{l} 3x + 2y = 9 \quad \dots \text{①} \\ 4x - 7y = 41 \quad \dots \text{②} \end{array}$

Solution. Multiply ① by 7: $21x + 14y = 63$
Multiply ② by 2: $\underline{8x - 14y = 82}$
Add: $29x = 145$
$x = 5$
Substitute 5 for x in ①: $3(5) + 2y = 9$
$2y = 9 - 15$, or -6
$y = -3$
The solution of the pair of equations is $(5, -3)$.

Check. When $x = 5$ and $y = -3$,

$3x + 2y$
$= 3(5) + 2(-3)$
$= 15 - 6$
$= 9$

$4x - 7y$
$= 4(5) - 7(-3)$
$= 20 + 21$
$= 41$

The solution is correct.

If the value of the first variable found is not an integer, it may be easier to find the value of the other variable by elimination than by substitution.

Example 3. Solve: $2x + 3y = 8$ ..... ①
$5x - 4y = -6$ ... ②

Solution. Multiply ① by 5: $10x + 15y = 40$
Multiply ② by 2: $\underline{10x - 8y = -12}$
Subtract: $23y = 52$
$y = \frac{52}{23}$

Multiply ① by 4: $8x + 12y = 32$
Multiply ② by 3: $\underline{15x - 12y = -18}$
Add: $23x = 14$
$x = \frac{14}{23}$

The solution of the pair of equations is $(\frac{14}{23}, \frac{52}{23})$.

Check. When $x = \frac{14}{23}$ and $y = \frac{52}{23}$,

$2x + 3y$
$= 2(\frac{14}{23}) + 3(\frac{52}{23})$
$= \frac{28 + 156}{23}$
$= \frac{184}{23}$
$= 8$

$5x - 4y$
$= 5(\frac{14}{23}) - 4(\frac{52}{23})$
$= \frac{70 - 208}{23}$
$= -\frac{138}{23}$
$= -6$

The solution is correct.

3-5 Solving Pairs of Linear Equations by Addition or Subtraction

Exercises 3-5

A 1. Solve:
   a) $2x + 3y = 18$
      $2x - 3y = -6$
   b) $3x + 5y = 12$
      $7x + 5y = 8$
   c) $7x - 4y = 26$
      $3x + 4y = -6$
   d) $3x - 4y = 0$
      $5x - 4y = 8$
   e) $7x + 6y = 4$
      $5x - 6y = -28$
   f) $4x - 3y = 20$
      $6x - 3y = 24$
   g) $3x + 5y = 4$
      $3x + 2y = 7$
   h) $-5x + 2y = -1$
      $5x - 4y = -13$
   i) $4x + 5y = -19$
      $4x - 3y = 5$

2. Solve:
   a) $3x + y = 3$
      $2x + 3y = -5$
   b) $5x + 2y = 5$
      $3x - 4y = -23$
   c) $4x + 3y = 9$
      $2x - 7y = 13$
   d) $2x + 6y = 26$
      $5x - 3y = 11$
   e) $2x - 5y = -18$
      $8x - 13y = -58$
   f) $4x + y = -11$
      $3x - 5y = 9$
   g) $6x - 5y = -2$
      $2x + 3y = 18$
   h) $3x - 10y = 16$
      $4x + 2y = 6$
   i) $5x + 3y = 9$
      $x - 4y = 34$

B 3. Solve:
   a) $8x - 3y = 38$
      $4x - 5y = 26$
   b) $3x + 4y = 29$
      $2x - 5y = -19$
   c) $6a - 5b = \frac{4}{3}$
      $10a + 3b = 6$
   d) $3s + 4t = 18$
      $2s - 3t = -5$
   e) $2m - 5n = 29$
      $7m - 3n = 0$
   f) $5x + 8y = -2$
      $4x + 6y = -2$
   g) $6x - 2y = 21$
      $3y + 4x = 1$
   h) $3c + 7d = 3$
      $4d - 5c = 42$
   i) $9x + 5y = 7$
      $6x + 7y = -10$

4. Solve:
   a) $7x + 6y = 2$
      $x + 8y = -4$
   b) $8x - y = 16$
      $2x - 3y = 2$
   c) $5x = y$
      $-x + 3y = 3$
   d) $3x + 2y = 8$
      $x - 12y = -10$
   e) $3x - y = 5$
      $2x + 3y = 10$
   f) $4x + 3y = 3$
      $3x - 2y = -19$
   g) $\frac{x}{3} + \frac{y}{2} = \frac{1}{6}$
      $x - 6y = 8$
   h) $\frac{1}{2}x - \frac{2}{3}y = 6$
      $\frac{1}{4}x + \frac{1}{3}y = -1$
   i) $\frac{2x}{5} + 3y = -1$
      $4x - 10y = 15$

5. The annual cost, C, in dollars, of operating a car is given by the formula $C = F + Rn$, where F is a fixed cost, n is the number of kilometres travelled, and R is the cost per kilometre. If the annual cost of driving 12 000 km is $5410, and driving 15 000 km is $5800, find
   a) the fixed cost and the cost per kilometre;
   b) the cost of driving 20 000 km.

6. The cost, C, in dollars, for a school hockey team to play in a tournament is given by the formula $C = A + Pn$, where A is a fixed cost, n is the number of players, and P is the cost per player. If the cost of taking 17 players is $507.50, and taking 22 players is $620, what is the cost of taking 19 players?

7. Consider the pair of equations $\begin{array}{l} 2x - 3y = 10 \\ -4x + 6y = -20 \end{array}$.
   a) Check that the ordered pairs $(2, -2)$, $(5, 0)$, and $(8, 2)$ are all solutions.
   b) Find two more ordered pairs that are solutions.
   c) Attempt to find their solution by the method of addition or subtraction. Why does this pair appear to have many solutions?
   d) Write the solution set of this pair of equations.

8. Consider the pair of equations $\begin{array}{l} 2x - 3y = 10 \\ -4x + 6y = -30 \end{array}$.
   a) Attempt to find their solution by the method of addition or subtraction.
   b) Does the pair of equations have any solutions? Explain.

9. Which of the following pairs of equations has no solution?
   a) $\begin{array}{l} 3x + 5y = 9 \\ 7x - 3y = 6 \end{array}$
   b) $\begin{array}{l} 2x - 3y = 2 \\ 4x - 6y = 4 \end{array}$
   c) $\begin{array}{l} 5x - 3y = 1 \\ -15x + 9y = -5 \end{array}$

10. For what values of m and n is $(5, -3)$ the solution of each pair of equations?
    a) $\begin{array}{l} mx - y = 23 \\ nx + y = 12 \end{array}$
    b) $\begin{array}{l} mx + ny = 12 \\ mx - ny = 18 \end{array}$
    c) $\begin{array}{l} mx + ny = -11 \\ 2mx - 3ny = 8 \end{array}$
    d) $\begin{array}{l} 3mx + 4ny = 30 \\ 2mx - 5ny = 20 \end{array}$

11. Solve:
    a) $\begin{array}{l} \frac{1}{3}x + \frac{1}{4}y = 0 \\ x + y = -1 \end{array}$
    b) $\begin{array}{l} \frac{1}{2}x - \frac{1}{3}y = 1 \\ x + \frac{1}{4}y = 2 \end{array}$
    c) $\begin{array}{l} \frac{2}{3}x + \frac{1}{5}y = -2 \\ \frac{1}{3}x - \frac{1}{2}y = -7 \end{array}$
    d) $\begin{array}{l} \frac{x}{4} + \frac{y}{2} = 0 \\ x + y = 2 \end{array}$
    e) $\begin{array}{l} \frac{1}{3}x + \frac{1}{2}y = -\frac{1}{2} \\ \frac{1}{5}x - \frac{1}{3}y = \frac{27}{5} \end{array}$
    f) $\begin{array}{l} \frac{3}{4}x + \frac{y}{3} = \frac{11}{2} \\ \frac{2x}{5} - \frac{3y}{2} = -\frac{21}{10} \end{array}$

Mathematics Around Us

A Matter Of Interest

Simple Interest

Interest is the money paid for the use of money. If you have a savings account, the bank pays you interest for the use of your savings. If you borrow money from a bank, the bank charges you interest for the use of its money. The charge is expressed as a rate, usually a percent, called the *interest rate*.

The amount borrowed or placed on deposit is called the *principal*.

Questions

1. Copy and complete the table:

	Principal of a loan, P	Interest rate, r	Interest for 1 year, I	Amount owing at the end of 1 year, A
a)	$ 200	22%	22% of $200 = $44	$ 244
b)	$ 1000	18%		
c)	$ 3000	19.5%		
d)	$ 3200		$528	
e)	$17 500			$21 262.50

2. Using the definitions of P, r, I, and A in the table above, write a formula for:
   a) I in terms of P and r;
   b) A in terms of P and r.

3. A store's finance department charges 2% per month on overdue accounts. An account for $325.47 becomes due. If no payment is made, calculate:
   a) the interest charges at the end of the next month;
   b) the amount due at the end of the next month.

4. A bank pays 12.75% annual interest on Daily Interest Savings Accounts, interest being credited to the accounts at the end of each month. What interest will be credited at the end of the month to an account which has
   a) a balance of $100 through January?
   b) a balance of $100 on May 1 which is increased on May 9 by a deposit of $200?
   c) a balance of $600 on September 1 which is decreased on September 17 by a withdrawal of $150?

Compound Interest

When banks or trust companies lend money, they charge interest on interest, or *compound interest*. Depending on the terms of the loan, interest may be compounded annually, semi-annually, quarterly, or even monthly. The following example shows how interest is calculated when it is compounded annually.

Example. If no repayments are made on a loan of $1000, how much interest will be owing at the end of 3 years when the interest rate is 24% per annum compounded annually?

Solution.

Year	Amount Owing Start of Year	Year's Interest	Amount Owing End of Year
1	$1000	24% of $1000 = $240	$1240
2	$1240	24% of $1240 = $297.60	$1537.60
3	$1537.60	24% of $1537.60 = $369.02	$1906.62

The amount owing on the $1000 loan at the end of 3 years is $1906.62. That is, the interest is $906.62.

Questions

5. Laura has $800 in an account which gives 20% interest compounded annually. If she leaves it untouched, how much will she have in the account after 4 years? Complete the table.

Year	Amount Start of Year	Interest	Amount End of Year
1	$800	$160	$960
2	$960		
3			
4			

6. a) In Question 5, by what factor can the amount at the start of a year be multiplied to obtain the amount at the end of the year?
   b) Write an expression for the amount owing at the end of the 4th year; at the end of the nth year.

7. A savings account has interest compounded annually. Determine the amount in an account:
   a) after 3 years at 15% with a deposit of $200;
   b) after 5 years at 13.5% with a deposit of $450;
   c) after 4 years at 12.25% with a deposit of $850.

8. A store charges interest at the rate of 2% per month on overdue accounts. If $217.60 becomes due on one account and no repayments are made, calculate the amount owing when the account is 4 months overdue.

3-6 Translating into the Language of Algebra

Equations are used to solve problems. Before any equation can be written, the facts of the problem must be translated into the language of algebra. Examine closely these verbal expressions and their algebraic equivalents.

Verbal	Algebraic
A certain number	x
6 more than the number	$x + 6$
2 less than the number	$x - 2$
3 times the number	$3x$
5 less than 4 times the number	$4x - 5$
Keith's present age	a
His age 3 years ago	$a - 3$
His age 2 years from now	$a + 2$
The tens' digit of a two-digit number	x
The ones' digit of a two-digit number	y
The number	$10x + y$
The number with the digits reversed	$10y + x$
The sum of the digits	$x + y$
A certain number of dimes	d
A certain number of quarters	q
The value of the dimes (in cents)	$10d$
The value of the quarters (in cents)	$25q$
The total value of the coins (in cents)	$10d + 25q$
The total number of coins	$d + q$

When two numbers are related in a problem, a variable may be chosen to represent one number and the other number expressed in terms of that variable. Alternatively, the other number may be represented by a second variable and an equation written to express the relation between the two.

Verbal	Algebraic (1 variable)	Algebraic (2 variables)
The sum of two numbers is 72.	$x, 72 - x$	$x, y,$ where $x + y = 72$
One number exceeds twice another number by 3.	$y, 2y + 3$	$x, y,$ where $x - 2y = 3$
Two times one number plus three times another number is 24.		$x, y,$ where $2x + 3y = 24$
4 chocolate bars and 3 ice-cream cones cost $3.60.		$x, y,$ where $4x + 3y = 360$
The daily wages of two workers are in the ratio 4:3.	$4k, 3k$	$x, y,$ where $\dfrac{x}{y} = \dfrac{4}{3}$
Two consecutive integers	$x, x + 1$	$x, y,$ where $y - x = 1$
Three consecutive odd integers	$x, x + 2, x + 4$	

Exercises 3 - 6

Write algebraic equivalents for these verbal expressions:

A 1. a) A certain number b) 5 more than the number
     c) 4 less than the number d) Double the number

   2. a) A certain number b) A different number
      c) The sum of the two numbers
      d) Double the first number plus triple the second number

   3. a) A certain number b) A different number
      c) Five times the first number minus four times the second number
      d) The sum of one-half the first number and one-third of the second number

   4. a) A two-digit number
      b) The number with the digits reversed
      c) The sum of the digits of the number
      d) The sum of the original number and the number with the digits reversed.

5. a) A certain number of nickels
   b) A certain number of dimes
   c) The total value of the coins
   d) The total number of the coins

6. a) A certain number of $2 bills
   b) A certain number of $5 bills
   c) The total value of the bills
   d) The total number of the bills

7. a) A certain number of adult tickets
   b) A certain number of children's tickets
   c) The total number of tickets
   d) The total cost of the tickets at $4 per adult ticket and $2 per child's ticket

8. a) The length of a rectangle
   b) The width of a rectangle
   c) The perimeter of a rectangle
   d) The area of a rectangle

9. a) Colleen's present age   b) Her age 5 years ago
   c) Her age 2 years from now
   d) The number of years until she reaches age 25
   e) The number of years until she is double her present age

10. a) Darren's present age   b) Terry's present age
    c) The sum of their ages 3 years ago
    d) The sum of their ages 20 years from now

Represent each of the following algebraically:

11. Two numbers have a sum of 50.

12. One number exceeds another number by 15.

13. The sum of two consecutive numbers

14. The sum of three consecutive even numbers

15. The sum of three consecutive odd numbers

16. Two times one number exceeds five times another number by 10.

17. One-quarter of one number is 3 more than one-half of a second number.

18. The sum of five times one number and four times another number is equal to 40.

19. Divide 10 into two parts.

20. Jackie is 3 years older than Diana.

21. Three years ago Craig was twice as old as Stephen.

22. Four pencils and three pens cost $1.50.

23. Six adult tickets and four children's tickets cost $18.

24. A number of nickels and dimes has a total value of $2.

25. The perimeter of a rectangle with a length five times the width

26. The perimeter of a rectangle with a length 10 cm greater than the width

27. The perimeter of a rectangle where six times the width exceeds four times the length by 10 cm

28. Two numbers such that the square of the first number is double the second number

29. The sum of the squares of two consecutive numbers

30. When one number is divided by another number the quotient is 7 and the remainder is 5.

3-7 Writing Equations

Equations can be written only after the facts of a problem have been translated into the language of algebra. If the translation involves one variable, one equation is required. If the translation involves two variables, two equations are required. Many problems can be solved using one variable or two variables.

Example 1. The difference of two numbers is 22. Twice the smaller number exceeds the larger number by 17. Write the equation, or equations, needed to find the numbers.

Solution.

	One variable	or	Two variables
The smaller number:	x		x
The larger number:	$x + 22$		y
Equations:	$2x - (x + 22) = 17$		$y - x = 22$
			$2x - y = 17$

3-7 Writing Equations

For some problems, it is easier to use two variables.

Example 2. Four chocolate bars and three ice-cream cones cost $3.60. Two chocolate bars and one ice-cream cone cost $1.50. Find the cost of a chocolate bar and the cost of an ice-cream cone.

Solution. Let the cost of the chocolate bar be x cents.
Let the cost of the ice-cream cone be y cents.
Equations: $4x + 3y = 360 \dots ①$
  $2x + y = 150 \dots ②$

(4 chocolate bars and 3 ice-cream cones cost 360¢.)

Exercises 3-7

Write an equation, or equations, for each exercise. Do not solve the equations at this time.

A 1. Find two numbers that have a sum of 25 and a difference of 7.

2. Two numbers have a sum of 10. The sum of the first number plus three times the second number is 24. Find the numbers.

3. Two numbers have a sum of 7. Three times the larger number exceeds the smaller number by 15. Find the numbers.

4. Two numbers differ by 5. Four times the smaller number is 5 less than three times the larger number. Find the numbers.

5. Divide 10 into two numbers such that when the larger is doubled and the smaller is tripled, the sum is 22.

6. Divide 15 into two numbers such that when the larger is divided by 3 and the smaller is divided by 2, the quotients are equal.

7. Tom weighs 15 kg more than Bill. Together they weigh 95 kg. How much does each weigh?

8. Wendy and Sandra have a total mass of 100 kg. Five times Wendy's mass is the same as four times Sandra's mass. How heavy is each of them?

B 9. Alfred is twice as old as his brother. Four years ago he was four times as old. How old is Alfred now?

10. Alice is 6 years older than Lois. In two years Alice will be twice as old as Lois. How old are they now?

11. Karyn has $1.15 in dimes and quarters. There are 7 coins altogether. How many of each does she have?

12. Ken has $1.95 in nickels and dimes. There are 3 more nickels than dimes. How many of each does he have?

13. Find two consecutive integers such that the sum of twice the first integer and three times the second integer is 38.

14. Find two consecutive integers such that five times the first integer is 6 more than four times the second integer.

15. Find two consecutive even integers such that the sum of one-third of the smaller integer and one-quarter of the larger integer is 11.

16. Find three consecutive odd integers such that triple the first integer exceeds double the third integer by 9.

17. Find the length and width of a rectangle that has
    a) a perimeter of 48 cm and a length that is twice the width;
    b) a perimeter of 32 cm and a length that is 2 cm less than twice the width.

18. How much alcohol must be added to 1 L of a 20% solution of alcohol to increase its strength to 50%?

19. If 1 is added to the numerator of a fraction, the result is equivalent to $\frac{3}{4}$. If 1 is added to the denominator of the same fraction the result is equivalent to $\frac{2}{3}$. Find the fraction.

20. The denominator of a fraction exceeds the numerator by 3. If 5 is added to both numerator and denominator, the result is equivalent to $\frac{3}{4}$. Find the fraction.

3 - 8 Solving Problems Using Equations: Part I

When equations are used to solve problems, remember that one variable requires one equation, two variables require two equations.

Example 1. Economy Car Rentals charge for rental cars on the basis of the length of time the car is rented and the distance travelled. When Angela rented a car for 3 days and drove 160 km, the charge was $124. When she rented the same car for 5 days and drove 400 km, the charge was $240. How much does Economy Car Rentals charge per day and per kilometre?

3-8 Solving Problems Using Equations: Part I

Solution. Let x represent the daily rental charge in dollars.
Let y represent the charge per kilometre in dollars.
Then,
$$3x + 160y = 124 \ldots \text{①}$$
$$5x + 400y = 240 \ldots \text{②}$$
Multiply ① by 5: $\quad 15x + 800y = 620$
Multiply ② by 3: $\quad 15x + 1200y = 720$
Subtract: $\quad\quad\quad\quad\quad 400y = 100$
$$y = \frac{1}{4}$$
Substitute $\frac{1}{4}$ for y in ②: $5x + 400\left(\frac{1}{4}\right) = 240$
$$5x = 140$$
$$x = 28$$
The charge per day is $28. The charge per kilometre is $0.25.

Check.
3 days at $28: $ 84
160 km at
$0.25/km: $ 40
Total: $124

5 days at $28: $140
400 km at
$0.25/km: $100
Total: $240

The solution is correct.

Although a problem may ask for only one unknown to be found, it is sometimes better to write two equations using two variables and solve for just one of them.

Example 2. The monthly incomes of two students are in the ratio 5:4 and their expenses are in the ratio of 3:2. If both students save $100 per month, find their monthly incomes.

Solution. Let $5x$ and $4x$ represent their monthly incomes.
Let $3y$ and $2y$ represent their monthly expenses.
Then,
$$5x - 3y = 100 \ldots \text{①}$$
$$4x - 2y = 100 \ldots \text{②}$$
Since only the value of x is needed to solve the problem, eliminate y.
Multiply ① by 2: $\quad 10x - 6y = 200$
Multiply ② by 3: $\quad 12x - 6y = 300$
Subtract: $\quad\quad\quad\quad 2x \quad\quad = 100$
$$x = 50$$
The students' monthly incomes are 5(50), or $250, and 4(50), $200.

Check. Students' expenses are:
$250-$100, or $150, and $200-$100, or $100.
Ratio of expenses: 150:100, or 3:2
The solution is correct.

Some problems can be solved using one variable or two variables.

Example 3. Nirmala has $4.80 in nickels and quarters. She has 6 more nickels than quarters. How many of each kind does she have?

Solution.

	Two variables	One variable
Number of nickels:	x	x
Number of quarters:	y	$x - 6$
Value of nickels (cents):	$5x$	$5x$
Value of quarters (cents):	$25y$	$25(x - 6)$
Equations:	$x - y = 6$ .... ①	$5x + 25(x - 6) = 480$
	$5x + 25y = 480$ .... ②	$30x = 630$
	From ①: $y = x - 6$	$x = 21$
	Substitute $x - 6$ for y in ②:	She has 21 nickels
	$5x + 25(x - 6) = 480$	and 15 quarters.
	$30x = 630$	
	$x = 21$	
	From ①: $y = 21 - 6$, or 15	
	She has 21 nickels and 15 quarters.	

Compare both methods. Do you see that after a certain point they are essentially the same?

Exercises 3 - 8

A 1. The sum of two numbers is 46 and their difference is 8. Find the numbers.

2. The sum of two numbers is 56. The larger exceeds twice the smaller by 2. Find the numbers.

3. Find two numbers such that twice the first exceeds three times the second by 1, and three times the first exceeds twice the second by 14.

4. The perimeter of a rectangle is 64 cm. Twice the width is 4 cm more than the length. Find the dimensions of the rectangle.

5. Find two numbers with a sum of 76 and a difference of 42.

6. What two numbers have a difference of 5, and the sum of three times the smaller number and twice the larger is 25?

7. Find two consecutive integers such that the sum of three times the smaller and five times the larger is 93.

8. Find two consecutive even integers such that the sum of one-third of the smaller integer and one-quarter of the larger is 32.

9. Divide 12 into two numbers such that when one number is tripled and the other number is multiplied by 5, the sum is 36.

10. Divide 20 into two numbers such that when the larger number is divided by 3 and the smaller number by 4, the sum of the quotients is 6.

11. Divide 48 into two numbers such that when the larger number is divided by 5 and the smaller number by 3, the quotients are equal.

12. Arlene and Shaun have a total mass of 105 kg. Four times Arlene's mass is the same as three times Shaun's mass. How heavy is each of them?

13. Trinika weighs 7 kg less than Amanda. Together they weigh 101 kg. How much does each weigh?

B 14. Mr. Kwok paid $16.00 for theatre tickets for himself and 8 children. On another occasion, he paid $22.00 for himself, 3 adults, and 4 children. What are the prices of the theatre tickets for adults and children?

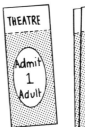

15. A sports club charges an initiation fee and a monthly fee. At the end of 5 months a member had paid a total of $170 and at the end of 10 months she had paid $295. What is the initiation fee?

16. The cost of 4 L of oil and 50 L of gasoline is $28.50. The cost of 3 L of oil and 35 L of gasoline is $20.30. Find the cost per litre of oil and gasoline.

17. A scientist worked for 6 days on a project and her assistant worked for 5 days earning a total of $780. They earned $500 on the next project which required 4 days' work from the scientist and 3 days' work from the assistant. How much does each earn per day?

18. Chris and Jerry's earnings in July were in the ratio 6:5. Their earnings in August were in the ratio 4:5. If Chris earned $790 in the 2 months and Jerry earned $10 more, how much did each earn each month?

19. A number of dimes and quarters have a total value of $19.75. If there are 5 fewer quarters than twice the number of dimes, find the number of dimes.

20. For a school play, Janis sold 6 adult tickets and 15 student tickets, and collected $48. Tom sold 8 adult tickets and 7 student tickets, and collected $38. Find the cost of adult and student tickets.

C 21. If 1 is added to the numerator of a fraction and 2 subtracted from the denominator, the result is equivalent to 2. If 1 is subtracted from the numerator of the same fraction and 2 added to the denominator, the result is equivalent to $\frac{1}{3}$. What is the fraction?

22. Find two numbers such that three times the second exceeds twice the first by 1, and the ratio of their sum to their difference is 9 : 2.

3 - 9 Solving Problems Using Equations: Part II

Solving problems using equations requires a combination of algebraic skill and a strategy for attacking the problem. One such strategy is to make a table.

Example 1. Candy A sells for $2.50/kg. Candy B sells for $5.00/kg. What quantities of each kind of candy should be used to make up a 100 kg mixture to sell for $4.00/kg?

Solution.

Candy	Mass kg	Price $/kg	Value $
A	x	2.50	2.50x
B	y	5.00	5.00y
Mixture	100	4.00	400

The second and fourth columns of the table give the equations:
$$x + y = 100 \quad \text{①}$$
$$2.5x + 5y = 400 \quad \text{②}$$
Multiply ① by 5: $\quad 5x + 5y = 500 \quad \text{③}$
Subtract ② from ③: $\quad 2.5x = 100$
$$x = 40$$
Substitute 40 for x in ①: $\quad 40 + y = 100$
$$y = 60$$
The mixture should contain 40 kg of candy A and 60 kg of candy B.

Check. 40 kg at $2.50/kg: $100
60 kg of $5.00/kg: $300
100 kg of mixture: $400
The solution is correct.

Example 2. A car averages 12.5 L/100 km in city driving and 7.5 L/100 km on the highway. In a week of mixed driving, the car used 35 L of fuel and travelled 400 km. Determine the distance travelled in highway driving.

3 - 9 Solving Problems Using Equations: Part II

Solution.

Type of driving	Distance km	Fuel consumption L/100 km	Fuel consumed L
City	x	12.5	$\frac{12.5x}{100}$
Highway	y	7.5	$\frac{7.5y}{100}$
Mixture	400		35

The second and fourth columns of the table give the equations:
$$x + y = 400 \ldots \text{①}$$
$$\frac{12.5x}{100} + \frac{7.5y}{100} = 35 \ldots \text{②}$$

Multiply ② by 100: $12.5x + 7.5y = 3500 \ldots$ ③
Multiply ① by 12.5: $12.5x + 12.5y = 5000 \ldots$ ④
Subtract ③ from ④: $5y = 1500$
$y = 300$

The car travelled 300 km in highway driving.

Check. Fuel consumed on highway: $\frac{300}{100} \times 7.5$ L, or 22.5 L.

Fuel consumed in city: $\frac{100}{100} \times 12.5$ L, or 12.5 L.

Total fuel consumed: 35 L
The solution is correct.

Example 3. The sum of the digits of a two-digit number is 12. The number formed by reversing the digits is 54 more than the original number. What is the original number?

Solution. Let the tens' digit be x and the ones' digit be y.
Original number: $10x + y$
Number with digits reversed: $10y + x$
Since their difference is 54:
$$(10y + x) - (10x + y) = 54$$
$$9y - 9x = 54$$
$$y - x = 6 \ldots \text{①}$$
Since sum of digits is 12: $x + y = 12 \ldots$ ②
Add ① and ②: $2y = 18$
$y = 9$
Substitute 9 for y in either ① or ②: $x = 3$
The original number is 39.

Check. $93 - 39 = 54$. The solution is correct.

Exercises 3-9

A 1. Standard quality coffee sells for $13.00/kg and prime quality coffee sells for $18.00/kg. What quantities of each should be used to produce 40 kg of a blend to sell for $16.00/kg.

2. A car averages 7.5 L/100 km in city driving and 5.0 L/100 km on the highway. In 300 km of highway and city driving it is found to have used 17.5 L of fuel.
    a) How far was the car driven on the highway?
    b) How much fuel did it use in the highway driving?

3. The sum of the digits of a two-digit number is 13. The number formed by reversing the digits is 27 more than the original number. Find the original number.

4. A boat at constant power travels 60 km upriver in 3 h and returns in 2 h. What is the speed of the boat relative to the water and what is the speed of the current?

5. The sum of the digits of a two-digit number is 11. The difference of the digits is 5. Find the number.

6. The difference of the digits of a two-digit number is 6. The sum of the number and the number formed by reversing the digits is 88. Find the number.

7. The ones' digit of a two-digit number is 5 more than the tens' digit. The new number formed by reversing the digits is eight times the sum of the digits. Find the number.

8. The sum of the digits of a two-digit number is 6. The new number formed by reversing the digits is equal to three times its ones' digit. Find the original number.

9. A two-digit number is equal to four times the sum of its digits. If it is increased by 3, the resulting number is equal to three times the sum of its digits. Find the original number.

10. A two-digit number is equal to seven times the sum of its digits.
    a) Show that the tens' digit must be double the ones' digit.
    b) Show that the number formed by reversing the digits is equal to four times the sum of its digits.

11. The coin box of a vending machine contains twice as many dimes as quarters. If the total value of the coins is $18.00, how many dimes are there?

12. A father is three times as old as his son. Six years ago he was five times as old as his son. How old are father and son now?

13. Three sisters have a total age of 68 years. The oldest is 7 years older than the youngest and 3 years older than the second sister. Find their ages.

14. A man is five times as old as his son. The man's daughter is 2 years older than her brother. In 10 years the sum of their three ages will be 81 years. How old are they now?

B 15. Last year, Mr. Bryson's age was half his mother's. Next year the sum of their ages will be 100 years. Find their present ages.

16. Kevin says to the owner of store A: "If you lend me as much money as I have with me, I'll spend $10 in your store." The owner agrees. Kevin then goes on to stores B and C in turn and makes the same deal. He then has no money left. How much did he start with?

17. A set of barbells consists of red and blue iron discs. Discs with the same color have the same mass. Using a balance, Mack found that 7 blue discs balance 1 red disc and a 5 kg mass, while 2 blue discs and a red disc balance a 4 kg mass. Find the masses of the red and blue discs.

18. Melanie has a pocketful of nickels, dimes, and quarters. She has 3 more dimes than nickels and 5 more quarters than nickels. If she has 29 coins in all, how many of each kind does she have?

19. Brent has a pocketful of nickels, dimes, and quarters. He has 3 fewer nickels than dimes and 3 more quarters than dimes. If the coins are worth a total of $4.20, how many of each kind does he have?

20. Find four consecutive even numbers such that if the first is increased by 2, the second decreased by 2, the third multiplied by 2, and the fourth divided by 2, the sum of the four resulting numbers is 58.

21. Find four consecutive odd numbers such that if the first is increased by 2, the second decreased by 3, the third multiplied by 4, and the fourth divided by 5, the sum of the four resulting numbers is 136.

22. A two-digit number is equal to six times the sum of its digits. The new number formed by reversing the digits is equal to five times the sum of its digits. Find the number.

23. The ones' digit of a two-digit number is 3 more than the tens' digit. The new number formed by reversing the digits is 27 more than the original number. Find the original number.

24. Lorraine buys 6 cheap golf balls and 4 expensive ones for $12.50. Bob buys 4 cheap and 3 expensive balls for $9.00. What is the price of the cheap golf balls?

25. Mr. Maclean and Mr. Hunter share a driveway. Mr. Maclean can clear it of snow with his snowblower in 30 min. Mr. Hunter, with a more powerful blower, can clear the driveway in 20 min. How long would it take them to clear the driveway working together?

26. It is estimated that to complete an excavation would take 2 days using 4 bulldozers and 3 steam shovels and 3 days using 5 bulldozers and 1 steam shovel. How many days would 1 bulldozer and 1 steam shovel take?

27. Muskoka Tire and Rubber Company produces two grades of radial tires. The cheaper tire requires 3 kg of rubber and takes 3 h of labor for a total production cost of $54.00. The higher grade tire requires 2 kg of rubber and 5 h of labor for a production cost of $63.00.
    a) What is the cost of 1 kg of rubber?
    b) What is the hourly labor cost?

C 28. A farmer sets out to travel on a 176 km stretch of lonely prairie highway at an average speed of 105 km/h. He is more than half way when his car breaks down and he has to complete the journey on foot. If he walks at 6 km/h and the whole trip takes 6 h 50 min,
    a) how far did he have to walk?
    b) how long was he driving?

29. A two-digit number is equal to four times the sum of its digits. Show that it is also equal to twelve times the difference of its digits.

30. Elizabeth travelled 618 km by airplane and bus to Calgary. The average speed of the airplane was 500 km/h, and the average speed of the bus was 60 km/h. If her total travelling time was 1.5 h, how far did she travel by bus?

31. Brian's average mark on three mathematics tests was 78. The mark on the first test was 86. His average for the first two tests was 3 more than the mark on the third test. What marks did Brian get on the second and third tests?

3 - 10 Solving Quadratic Equations

An equation of the form $ax^2 + bx + c = 0$, $a \neq 0$, is called a **quadratic equation**. Examples of quadratic equations are:

$$x^2 + 3x - 10 = 0 \qquad 3x^2 + 10x - 8 = 0$$

Many quadratic equations can be solved by factoring. Only this kind is considered in this section. The solution of a quadratic equation by factoring depends on an important property:

> If $A \cdot B = 0$, then $A = 0$, or $B = 0$, or both.

Example 1. Solve: a) $x^2 - x - 6 = 0$;
  b) $x^2 + 10x + 25 = 0$

Solution. a) $x^2 - x - 6 = 0$
Factor: $(x - 3)(x + 2) = 0$
If $x - 3 = 0$ $\qquad$ If $x + 2 = 0$
$\qquad x = 3$ $\qquad\qquad\qquad x = -2$

Check. If $x = 3$, $\qquad\qquad$ If $x = -2$,
$\qquad x^2 - x - 6$ $\qquad\qquad x^2 - x - 6$
$\qquad = (3)^2 - (3) - 6$ $\qquad = (-2)^2 - (-2) - 6$
$\qquad = 9 - 3 - 6$ $\qquad\qquad = 4 + 2 - 6$
$\qquad = 0$ $\qquad\qquad\qquad\quad = 0$

Both solutions are correct.

b) $x^2 + 10x + 25 = 0$
Factor: $(x + 5)(x + 5) = 0$
If $x + 5 = 0$
$\qquad x = -5$

Check. If $x = -5$,
$x^2 + 10x + 25 = (-5)^2 + 10(-5) + 25$
$\qquad\qquad\qquad\quad = 25 - 50 + 25$, or 0

The solution is correct.

If the solutions for a quadratic equation are known, the equation can be formed. This is a result of the following property:

> If $A = 0$ or $B = 0$, then $A \cdot B = 0$

Example 2. Write a quadratic equation with these solutions:

  a) -5 and 3 $\qquad$ b) $\frac{1}{2}$ and $-\frac{2}{3}$

Solution. a) If $x = -5$, then $x + 5 = 0$.
If $x = 3$, then $x - 3 = 0$.
And, $(x + 5)(x - 3) = 0$
$\qquad x^2 + 2x - 15 = 0$

b) If $x = \frac{1}{2}$, then $x - \frac{1}{2} = 0$.

If $x = -\frac{2}{3}$, then $x + \frac{2}{3} = 0$.

And, $(x - \frac{1}{2})(x + \frac{2}{3}) = 0$

$$x^2 + \frac{1}{6}x - \frac{1}{3} = 0$$

or, $6x^2 + x - 2 = 0$

Example 3. The height, h, in metres, of a football t seconds after a punt is given by the formula: $h = 20t - t^2$. How long after the kick is the football at a height of 15 m?

Solution. Substitute 15 for h in the formula:
$$15 = 20t - 5t^2$$
$$5t^2 - 20t + 15 = 0$$
$$t^2 - 4t + 3 = 0$$
$$(t - 1)(t - 3) = 0$$
$$t = 1, \text{ or } t = 3$$

The football is at a height of 15 m on the way up 1 s after the kick and on the way down 3 s after the kick.

Check. When $t = 1$, When $t = 3$,
$20t - 5t^2$ $\quad$ $20t - 5t^2$
$= 20(1) - 5(1)^2$ $\quad$ $= 20(3) - 5(3)^2$
$= 20 - 5, \text{ or } 15$ $\quad$ $= 60 - 45, \text{ or } 15$
The solution is correct.

Exercises 3-10

A 1. Solve:
   a) $x^2 + 8x + 15 = 0$ b) $x^2 - 7x + 12 = 0$
   c) $x^2 - x - 20 = 0$ d) $x^2 + 5x - 24 = 0$
   e) $x^2 + 8x + 12 = 0$ f) $x^2 - 5x - 36 = 0$
   g) $x^2 - 10x + 24 = 0$ h) $x^2 + 15x + 56 = 0$

2. Write a quadratic equation with these solutions:
   a) $9, -2$ b) $4, 10$ c) $-5, -\frac{1}{3}$ d) $\frac{7}{8}, -\frac{1}{4}$

B 3. Solve:
   a) $x^2 - 9x + 25 = 5$ b) $x^2 - 16x + 50 = -13$
   c) $x^2 - 6x - 20 = -4$ d) $x^2 + 10x + 25 = 4$
   e) $x^2 - 5x - 20 = -6$ f) $x^2 + 6x - 15 = 4x$
   g) $x^2 - 5x + 16 = 3x$ h) $x^2 - 10x + 16 = 4 - 2x$

3 - 10 Solving Quadratic Equations   **145**

4. Solve:
   a) $3x^2 + 15x + 18 = 0$ b) $2x^2 - 24x + 54 = 0$
   c) $4x^2 - 12x - 40 = 0$ d) $2x^2 - 26x + 60 = 0$
   e) $3x^2 + 6x - 72 = 0$ f) $5x^2 - 20x + 20 = 0$

5. The height, h, in metres, of an infield fly ball t seconds after being hit is given by the formula: $h = 30t - 5t^2$. How long after being hit is the ball at a height of 25 m?

6. The area, A, of a picture is given by the formula: $A = 28x - x^2$. Calculate the dimensions of a picture that has an area of
   a) 192 cm²;   b) 196 cm²;   c) 160 cm².

7. The sum, S, of the first n terms of the series:
   $$36 + 32 + 28 + 24 + \ldots.$$
   is given by the formula: $S = 2n(19 - n)$.
   a) If $S = 168$, find n.
   b) Why are two values of n possible?
   c) Find n for $S = 0$.

8. If the sides of a right triangle are x cm, $(x + 7)$ cm, and $(x + 9)$ cm, what is the actual length of the hypotenuse?

9. Two numbers differ by 6. The sum of their squares is 90. Find the numbers.

10. Two numbers have a sum of 12. The square of one number is double the other number. Find the numbers.

11. The ones' digit of a two-digit number is 1 less than the tens' digit. The sum of the squares of the digits is 85. Find the number.

12. A rectangular lawn measures 40 m by 30 m. If it is being cut from the outside in, how wide a strip has been cut when the job is half finished?

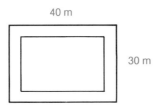

13. Solve:
    a) $5x^2 + 19x - 4 = 0$ b) $2x^2 + 15 = 11x$
    c) $6x^2 - 11x - 17 = 18$ d) $3x^2 + 5x = 2$

Consider Different Cases

Some problems, which may at first appear to be difficult, are easily solved by considering different cases in the solution. This method is particularly effective for problems which ask how many ways figures may be drawn or combined.

Example. How many different (non-congruent) triangles can be drawn on the 3 by 3 array of dots?

Solution. **Case 1.** Triangles with horizontal and vertical sides
There are three possibilities:

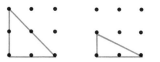

Case 2. Triangles with one side horizontal
There are four possibilities:

Since each of these triangles can be turned through 90° to make the horizontal side vertical, there are no additional triangles with one side vertical.

Case 3. Triangles with no sides horizontal or vertical
There is only one possibility:

Therefore, eight different triangles can be drawn on a 3 by 3 array of dots.

Exercises

1. On a 3 by 3 array of dots,
   a) how many line segments of different lengths can be drawn?
   b) how many squares of different areas can be drawn?

2. How many line segments with different non-negative slopes can be drawn on a 5 by 5 array of dots?

3. How many fractions of different value, in the form $\frac{m}{n}$, can be written if $0 < m \leq 4$, $0 < n \leq 4$, and m and n are integers?

4. A square is divided into a 4 by 4 array of small squares.

   a) How many different squares are in the figure?
   b) How many different rectangles are in the figure?

5. An equilateral triangle is divided into 16 congruent equilateral triangles. In the figure,
   a) how many triangles are there?
   b) how many rhombuses are there?
   c) how many parallelograms are there?
   d) how many trapezoids are there?

   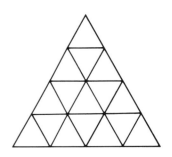

6. a) How many line segments of different lengths can be drawn joining two vertices of a cube?
   b) If the edge length of a cube is 1, find the length of each segment in (a).

7. How many different (non-congruent) triangles can be drawn joining three vertices of a cube?

8. The diagram shows eight unit cubes arranged to form a large cube. The vertices of the eight unit cubes form a lattice of 27 points.
   a) How many line segments of different lengths can be drawn to join points of the lattice?
   b) Find the lengths of the line segments identified in (a).

Review Exercises

1. Solve:
   a) $3(m - 5) = 2(5m - 11)$ b) $\frac{4}{3}m + \frac{5}{6} = \frac{1}{2}m - \frac{5}{3}$
   c) $3(x - 2) - 8 = 68 - 2(2x - 1)$
   d) $4(3y + 1) - 6(y - 3) = 4(2y - 7) + 34$

2. Find the solution set:
   a) $y - 2x = -8$ b) $3x + 5y = 20$ c) $4x - 24 = 3y$

3. If $4x + 3y = n$, find the value of n for these values of x and y:
   a) $(3, -2)$ b) $(-2, 3)$ c) $(-1, -5)$ d) $(4, 6)$

4. Solve by comparison:
   a) $x + y = -8$ b) $2x + y = 8$ c) $x + 2y = -2$
      $x - 2y = 7$ $4x - 9y = 5$ $-2x + y = 6$

5. Solve by substitution:
   a) $x + 2y = 4$ b) $2x + y = 9$ c) $2x + 3y = 9$
      $3x + 2y = 0$ $x - y = 3$ $x - y = 3$
   d) $4x + y = -6$ e) $2y + x + 10 = 0$ f) $y + 4x + 1 = 0$
      $-2x + 3y = 24$ $y - 4x = 13$ $y + 7 = 4x$

6. Solve by addition or subtraction:
   a) $3x - 4y = 1$ b) $3a + 2b = 5$ c) $3x - 4y = -2$
      $3x - 2y = -1$ $9a - 2b = 15$ $4x - 3y = -5$
   d) $3a + 2b = 5$ e) $x + 2y = 4$ f) $3a + 10b = -4$
      $2a + 3b = 0$ $\frac{5}{2}x - 13y = 10$ $4a - 5b = 13$

7. Represent each of the following algebraically:
   a) Two numbers have a sum of 80.
   b) One number is 7 less than another number.
   c) The sum of three times one number and four times another number is 30.
   d) Three years ago Ian was twice as old as Barbara.
   e) A number of nickels and dimes has a total value of $3.
   f) The perimeter of a rectangle with a length 5 cm greater than the width.

8. The sum of two integers is 36. Their difference is 4. Find the integers.

9. The sum of the digits of a two-digit number is 12. The ones' digit is 2 more than the tens' digit. Find the number.

10. The sum of two integers is 63. The smaller is 11 more than one-third the larger. Find the integers.

11. Lynn has $3.55 in dimes and quarters. If there are 25 coins altogether, how many dimes are there?

12. From his paper route, Andy collected $6.55 in nickels and dimes. The number of nickels was 6 more than one-half the number of dimes. How many nickels were there?

13. The sum of the digits of a two-digit number is 11. the number formed by reversing the digits is 45 more than the original number. What is the original number?

14. A 100 kg mixture of peanuts contains peanuts of two different kinds, one priced at $2/kg and the other at $2.40/kg. If the mixture is priced at $2.08/kg, how many kilograms of each kind of peanut does it contain?

15. One train leaves a station heading west. A second train, heading east, leaves the the same station 2 h later and travels 15 km/h faster than the first. They are 580 km apart 6 h after the second train departed. How fast is each train travelling?

16. A salmon-fishing boat put out to sea early in the morning travelling with the tide. It took 20 min to cover the 6 km to the captain's favorite fishing grounds. The return trip was against the tide and took 36 min. What was the still-water speed of the boat and what was the speed of the tide?

17. Sterling silver is 92.5% pure silver. How many grams of pure silver and sterling silver must be mixed to obtain 100 g of a 94% silver alloy?

18. A company hired four skilled and eight unskilled workers to install a solar-heated swimming pool, paying them a total of $300. To install a similar pool, they hired seven skilled and six unskilled workers paying them $375. What was each type of worker paid for the job?

19. Solve:
   a) $x^2 - 4x - 21 = 0$
   b) $x^2 + x - 56 = 0$
   c) $x^2 - 2x = 24$
   d) $x^2 - 24 = 5x$

20. Write a quadratic equation for each of these solutions:
   a) $(7, -5)$
   b) $(-3, 8)$
   c) $(4, -\frac{2}{3})$
   d) $(-\frac{1}{8}, \frac{3}{4})$

COMPUTER POWER

A Formula for Linear Systems

If there were many pairs of linear equations to be solved, it would be useful to have a formula to solve them. For a program could be written for this formula, and the equations solved by computer. Solving by computer is especially helpful when the coefficients in the equations are large or are decimal fractions.

To find such a formula, we solve the pair of equations:

$$Ax + By = C \quad \ldots \text{①}$$
$$Dx + Ey = F \quad \ldots \text{②}$$

① × E: $AEx + BEy = CE$
② × B: $BDx + BEy = BF$

Subtract: $AEx - BDx = CE - BF$
$x(AE - BD) = CE - BF$
$$x = \frac{CE - BF}{AE - BD}$$

① × D: $ADx + BDy = CD$
② × A: $ADx + AEy = AF$

Subtract: $BDy - AEy = CD - AF$
$y(BD - AE) = CD - AF$
$$y = \frac{CD - AF}{BD - AE}$$

$$\boxed{\frac{CD - AF}{BD - AE} \times \frac{-1}{-1}}$$

The solution of the above pair of equations is the ordered pair $\left(\dfrac{CE - BF}{AE - BD}, \dfrac{AF - CD}{AE - BD}\right)$, provided that $AE - BD \neq 0$. If $AE - BD = 0$, there may be more than one solution, or no solution.

A simple BASIC program for the above formula is as follows:

```
10  INPUT "WHAT ARE THE COEFFICIENTS FOR THE FIRST EQUATION"; A, B, C
20  INPUT "WHAT ARE THE COEFFICIENTS FOR THE SECOND EQUATION"; D, E, F
30  Z = A * E - B * D
40  IF Z = 0 THEN PRINT "NO UNIQUE SOLUTION": END
50  PRINT "THE SOLUTION OF THE SYSTEM IS"
60  PRINT (C * E - B * F) / Z, (A * F - C * D) / Z
```

Exercises

1. Use the program to solve:
   a) $3.3x - 4.2y = 12$
      $1.7x + 2.6y = 30$
   b) $86x + 49y = 97$
      $15x - 24y = -276$
   c) $9.3x + 1.6y = -8.2$
      $4.7x - 7.3y = 6.1$
   d) $26x - 34y = 105$
      $41x + 9y = 83$
   e) $243x + 155y = 528$
      $-62x + 417y = 166$
   f) $3.15x - 5.81y = 12.66$
      $8.69x + 4.07y = -19.22$

2. Acme Car Rentals calculates the amount it charges for rental cars on the basis of the length of time the car is rented and the distance travelled. When Lynne rented a car for 17 days and drove 850 km, she was charged $816.00. When Monique rented the same model car for 23 days and drove 1120 km, she was charged $1098.45. How much does Acme Car Rentals charge per day? per kilometre?

4 Graphing Equations and Inequalities

An 80 ha farm is to be planted with two crops, corn and wheat. Planting and harvesting costs, for which no more than a total of $12 000 is available, are $300/ha for corn and $100/ha for wheat. Draw a graph showing the number of hectares of each crop that can be planted. (See *Example 3* in Section 4-3.)

4-1 Graphing a Linear Equation

A linear equation such as $2x + y = 8$ has infinitely many solutions. Some of these are shown in the table.

x	0	1	2	3	4	5	6
y	8	6	4	2	0	-2	-4

Each solution is an ordered pair, (x, y), which may be represented by a point in the plane. Those listed in the table of values are shown plotted on a grid. The points appear to lie on a straight line. This line is the graph of the equation $2x + y = 8$.

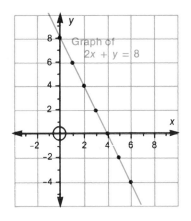

- The graph of a linear equation is a straight line.
- Every straight line is the graph of a linear equation.

The coordinates of only two points are needed to draw the graph of a linear equation. The two points easiest to find are usually those which have 0 as one coordinate.

Example 1. Draw the graph of $3x - 2y = 6$.

Solution. When $x = 0$, $3x - 2y = 6$ becomes
$$3(0) - 2y = 6$$
$$y = -3$$
The point $(0, -3)$ lies on the graph.
When $y = 0$, $3x - 2y = 6$ becomes
$$3x - 2(0) = 6$$
$$x = 2$$
The point $(2, 0)$ lies on the graph.
The line through the points $(2, 0)$ and $(0, -3)$ is the required graph.

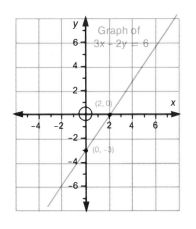

In the previous example, the solutions of $3x - 2y = 6$ can be expressed as a set of ordered pairs in the following manner.

$$\{(x, y) \mid 3x - 2y = 6, \ x, y \in R\}$$

The set / of all ordered pairs (x, y) / such that / $3x - 2y = 6$, / where x and y are real numbers.

The above is called "set-builder" notation.

Example 2 Draw the graph of $\{(x, y) \mid 4x + 3y = 21, \ x, y \in R\}$.

4-1 Graphing a Linear Equation 153

Solution. When $x = 0$, $4x + 3y = 21$ becomes
$$4(0) + 3y = 21$$
$$y = 7$$
The point $(0, 7)$ lies on the graph.
When $y = 0$, $4x + 3y = 21$ becomes
$$4x + 3(0) = 21$$
$$x = \frac{21}{4}, \text{ or } 5\frac{1}{4}$$

The point $(5\frac{1}{4}, 0)$ lies on the graph.
The line through the points $(0, 7)$ and $(5\frac{1}{4}, 0)$ is the required graph.

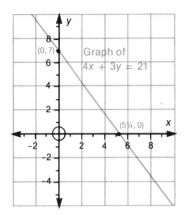

In the last example, a more accurate graph could have been obtained if, for the second point, a value of y had been chosen that yields an integral value for x. For example, when $y = -1$, $x = 6$; the point $(6, -1)$ is easier to plot than the point $(5\frac{1}{4}, 0)$.

Occasionally, the equation defining a set of ordered pairs contains only one variable. In such cases, the missing variable may represent any real number.

Example 3. Graph the sets $S = \{(x, y) \mid x = 5, y \in R\}$ and $T = \{(x, y) \mid y + 3 = 0, x \in R\}$.

Solution. S is the set of all points with first coordinate 5. The graph of S is a line parallel to the y-axis 5 units to the right of it.
T is the set of all points with second coordinate -3. The graph of T is a line parallel to the x-axis 3 units below it.

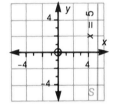

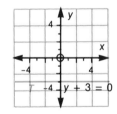

Example 4. A book publisher publishes a novel in both hardcover and paperback form. Printing and binding costs for each book are $8 for hardcover and $4 for paperback. Draw a graph showing the number of hardcover and paperback copies that the publisher can produce for $20 000.

Solution. Let x and y represent, respectively, the number of hardback and paperback copies that can be produced.
Since the total expenditure is $20 000,
$$8x + 4y = 20\,000.$$
Use the points corresponding to $x = 0$ and $y = 0$ to draw the graph.
When $x = 0$, $y = 5000$. When $y = 0$, $x = 2500$.
Since x and y represent numbers of books, the graph consists only of the points that have integral coordinates on the line segment shown.

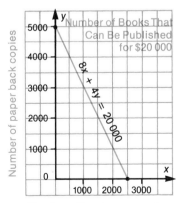

Exercises 4-1

A 1. Graph:
   a) $2x - y = 8$
   b) $2x + y = -8$
   c) $2x - y = -8$
   d) $3x + 2y = 6$
   e) $3x - 2y = -6$
   f) $3x + 2y = -6$
   g) $5x - 2y = 10$
   h) $4x + 3y = 12$
   i) $3x - 5y = 15$

2. Graph:
   a) $\{(x, y) \mid 3x + 2y = -12\}$
   b) $\{(x, y) \mid x - 2y = 10\}$
   c) $\{(x, y) \mid -5x - 3y = -30\}$
   d) $\{(x, y) \mid 4x + y = 7\}$
   e) $\{(x, y) \mid 3x - 2y = -8\}$
   f) $\{(x, y) \mid 3x + 2y = -9\}$
   g) $\{(x, y) \mid 5x - 10y = 40\}$
   h) $\{(x, y) \mid 4x - 8y = -36\}$

B 3. Graph:
   a) $\{(x, y) \mid 3x - 7y = 42\}$
   b) $\{(x, y) \mid 5x + 3y = 80\}$
   c) $\{(x, y) \mid x = -15\}$
   d) $\{(x, y) \mid 9x - 5y = -72\}$
   e) $\{(x, y) \mid 13y = 156\}$
   f) $\{(x, y) \mid 12x + 15y = -180\}$
   g) $\{(x, y) \mid 7x + 18y = 189\}$
   h) $\{(x, y) \mid 19x - 171 = 0\}$

4. Graph:
   a) $3x + 6y = 24$
   b) $12x - 15y = 60$
   c) $8x - 12y = -72$
   d) $9y + 117 = 0$
   e) $4x - 5y = 80$
   f) $9x + 15y = 90$
   g) $-10x + 15y = -120$
   h) $5x - 45 = 0$

5. A company makes radios and tape decks. The production time for each radio is 3 min and for each tape deck is 4 min. Draw a graph showing the number of radios and tape decks that can be produced in 1 h.

6. A farmer has $1680 for the rental of machinery to prepare the land for planting. A spring-tooth cultivator rents for $24/h while a disc-and-rake costs $30/h. Draw a graph showing the number of hours that the farmer can afford to rent each machine.

7. It will cost Naomi $600 to attend the Canadian Jamboree in Alberta. Her parents agreed to pay half the amount. Naomi earned the rest by cutting lawns at $5/h and weeding flower beds at $4 for each flower bed. Draw a graph showing how Naomi could have earned the amount she required.

8. Emmanuel intends to invest money in two different bonds to earn a total of $750 per year in interest. The first bond pays interest at the rate of 12% per year, the second at 15% per year. Draw a graph showing how he could invest his money.

C 9. Draw the graph of each equation:
   a) $y = |x|$
   b) $y = |x| + 3$
   c) $x^2 - y^2 = 0$
   d) $(x + 3)(y - 2) = 0$

10. Write the equation of the line that passes through:
    a) (5, 0) and (0, 5);
    b) (3, 0) and (0, −3);
    c) (2, 0) and (0, 1);
    d) (2, 0) and (0, 4);
    e) (2, 0) and (0, 3);
    f) (−1.5, 0) and (0, 3.5).

4-2 Graphing a Pair of Linear Equations

To solve a pair of linear equations such as $\begin{matrix} x - 2y = 6 \\ 3x + y = 11 \end{matrix}$ means to find all the ordered pairs (x, y), if any, that satisfy both equations. These may be found by graphing both equations on the same axes. If the lines are parallel, no ordered pair will satisfy both equations. If they intersect, one ordered pair will satisfy both equations. If they coincide, an infinite number of ordered pairs will satisfy them.

Example 1. Solve by graphing: $\begin{matrix} x - 2y = 6 \\ 3x + y = 11 \end{matrix}$

Solution. For the equation $x - 2y = 6$:
When $x = 0$, $y = -3$. When $y = 0$, $x = 6$.
For the equation $3x + y = 11$:
When $x = 0$, $y = 11$. When $y = 2$, $x = 3$.
The graph of each equation is a line. The point of intersection of the two lines, $(4, -1)$, is the only point that is common to both. Therefore, the graph of the solution set of the pair of equations is $(4, -1)$.

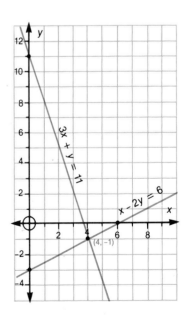

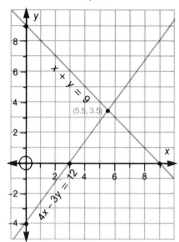

It is not always possible to obtain the exact coordinates of the point of intersection of two lines by graphing.

Example 2. Solve by graphing: $\begin{array}{l} x + y = 9 \\ 4x - 3y = 12 \end{array}$

Solution. For the equation $x + y = 9$:
  When $x = 0$, $y = 9$. When $y = 0$, $x = 9$.
For the equation $4x - 3y = 12$:
  When $x = 0$, $y = -4$. When $y = 0$, $x = 3$.
From the graph, the approximate solution is (5.5, 3.5).
The exact solution, $(5\frac{4}{7}, 3\frac{3}{7})$, can only be found by solving the equations algebraically.

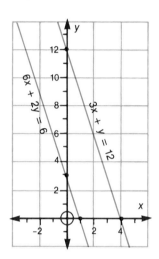

Example 3. Solve by graphing: $\begin{array}{l} 3x + y = 12 \\ 6x + 2y = 6 \end{array}$

Solution. For the equation $3x + y = 12$:
  When $x = 0$, $y = 12$. When $y = 0$, $x = 4$.
For the equation $6x + 2y = 6$:
  When $x = 0$, $y = 3$. When $y = 0$, $x = 1$.
Since the lines appear to be parallel, it would seem that the equations have no solution. If the second equation is divided by 2, the following pair of equations is obtained:
$$3x + y = 12$$
$$3x + y = 3$$
These equations have identical left sides but different numbers on the right side. Clearly, no ordered pair, (x, y), can satisfy both equations.

Example 4. Solve by graphing: $\begin{array}{l} 3x + y = 12 \\ 6x + 2y = 24 \end{array}$

Solution. If the second equation is divided by 2, it reduces to the first equation. Therefore, the graphs of the two equations are the same straight line. That is, there are infinitely many solutions, each solution being a point (x, y) on the line.

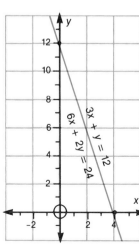

Exercises 4 - 2

A 1. Solve by graphing:

a) $x + y = 5$
   $3x - y = 3$

b) $x - y = -2$
   $4x + 2y = 16$

c) $x + y = 70$
   $3x + 4y = 240$

d) $x - y = 2$
   $3x + y = -14$

e) $x - y = 4$
   $2x + y = -4$

f) $5x + 4y = 40$
   $5x + 6y = 50$

2. Solve by graphing:
   a) $2x - y = 80$
      $x + 3y = -30$
   b) $x - 2y = 10$
      $3x - y = 0$
   c) $3x + 2y = 60$
      $3x - 5y = -150$
   d) $4x - 6y = 4$
      $3x + 2y = -12$
   e) $2x - y = 80$
      $3x + 6y = -120$
   f) $10x + 2y = -5$
      $4x + 6y = 2$

B 3. Solve by graphing:
   a) $x + y = 4$
      $x - y = 1$
   b) $x + 2y = 5$
      $x - 2y = 2$
   c) $2x - y = 8$
      $3x + 6y = 2$
   d) $2x - 3y = 0$
      $4x + 6y = 6$
   e) $3x - 6y = 180$
      $2x + 3y = 30$
   f) $2x + 3y = 2$
      $4x - 3y = 1$
   g) $5x + 4y = 2$
      $2x - 3y = -15$

4. Solve by graphing:
   a) $2x + 3y = 12$
      $3x - 2y = 5$
   b) $x + 2y = 6$
      $x + 2y = 2$
   c) $3x + 2y = 12$
      $6x + 4y = 24$
   d) $2x - 5y = 100$
      $-4x + 10y = -200$
   e) $6x + 10y = 18$
      $-3x - 5y = -9$
   f) $6x + 10y = 18$
      $3x + 5y = 9$
   g) $6x + 10y = 180$
      $3x + 5y = 300$

C 5. For which of the following pairs of equations is (2, −3) a solution?
   a) $x - y = 5$
      $3x + 4y = -6$
   b) $2x + y = 7$
      $x - 3y = 10$
   c) $4x - y = 11$
      $-12x + 3y = -33$
   d) $5x - 3y = 19$
      $-2x + 4y = -16$

6. For which of the equations in Exercise 5 is (2, −3) the only solution?

7. Under what conditions does a pair of equations have
   a) infinitely many solutions?
   b) no solution?
   c) one solution?

8. a) Let L represent the equation $x + 2y = 8$, and M represent the equation $3x - y = 3$. Form the equations represented by:
   i) $L + M$
   ii) $2L + M$
   iii) $L - M$
   iv) $3L - 2M$
   b) Draw the graphs corresponding to the six equations in (a).
   c) State a conclusion based on the results of parts (a) and (b).

THE MATHEMATICAL MIND

Diophantine Equations

People have always been interested in riddles and puzzles. The following example is typical of the arithmetic puzzles that are part of the folklore of times gone by.

Twenty measures of grain are distributed among some people so that each adult receives 3 measures and each child receives 2 measures. How many adults and children are there?

If there are x adults and y children, then this puzzle leads to the equation:

$$3x + 2y = 20$$
$$2y = 20 - 3x$$
$$y = \frac{20 - 3x}{2}$$

Since y must be a whole number, $20 - 3x$ must be divisible by 2. The only *non-negative* values of x for which this is true are: 0, 2, 4, 6. The table and graph show the solutions of the puzzle.

Number of adults x	Number of children y
0	10
2	7
4	4
6	1

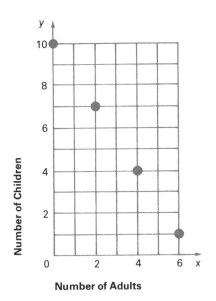

When the only valid solutions of an equation are integers, the equation is called a *diophantine equation*, after the Greek mathematician, Diophantus of Alexandria (c. 250 A.D.). His book, *Arithmetica*, is a collection of numerical problems, many of which have more than one answer. However, Diophantus did not study the problems systematically, and he usually gave only one of the answers.

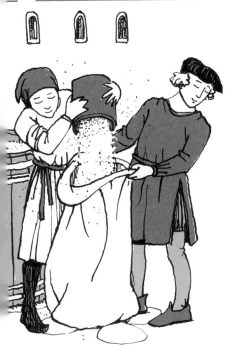

1. Solve and graph, if possible, the diophantine equations:
   a) $3x + 2y = 12$
   b) $x - 3y = 2$
   c) $4x + y = 17$
   d) $4x - 7y = 21$
   e) $2x + 5y = 9$
   f) $2x + 4y = 9$

2. Fay has $1.70 in quarters and dimes. How many of each does she have?

3. Ivor has $1.70 in quarters and nickels. How many of each does he have?

4. Spiders have 8 legs and flies have 6. In a collection of spiders and flies there are 138 legs. How many of each insect are there?

5. A store sells T-shirts for $2.50 and gym shorts for $1.50. If a day's receipts were $40, how many of each were sold?

6. Certain facts about Diophantus' life are found in a collection of problems dated about 500 A.D.

   "His boyhood lasted $\frac{1}{6}$ of his life; his beard grew after $\frac{1}{12}$ more; he married after $\frac{1}{7}$ more; his son was born 5 years later. The son lived to half his father's (final) age, and the father died 4 years after the son."

   How old was Diophantus when he died?

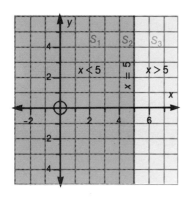

4-3 Graphing Linear Inequalities

The line defined by the equation $x = 5$ divides the plane into three sets of points:

- the set of points, S_1, to the left of the line,
- the set of points, S_2, which is the line,
- the set of points, S_3, to the right of the line.

In set-builder notation, these sets are:

$S_1 = \{(x, y) \mid x < 5\}$,
$S_2 = \{(x, y) \mid x = 5\}$,
$S_3 = \{(x, y) \mid x > 5\}$.

S_1 is the graph of the inequality $x < 5$ and S_3 is the graph of the inequality $x > 5$.

In general, the graph of any linear equation is a straight line which divides the plane into two **half-planes**. The half-planes are the graphs of the corresponding inequalities.

Example 1. Graph the inequality: $4x + 5y \leq 20$

Solution. The graph of the equation $4x + 5y = 20$ is drawn. The diagram shows the three sets of points into which the line divides the plane. The graph of the inequality is the line S_2 and one of the half-planes S_1 or S_3. Since $(0, 0)$ is a solution of $4x + 5y < 20$ and is in half-plane S_1, the graph of the inequality $4x + 5y \leq 20$ is the half-plane S_1 and the line S_2.

In *Example 1*, if the inequality had been simply $4x + 5y < 20$, the graph of $4x + 5y = 20$ would have been shown as a broken line to indicate that it was not part of the solution.

Example 1 shows that the set of points, or region, defined by a linear inequality is a half-plane. The next example shows the nature of the region defined by a *pair* of linear inequalities.

Example 2. Graph the solution set: $\begin{array}{l} 4x + 5y \leq 20 \\ 3x - 5y \geq 15 \end{array}$

Solution. Both equations are graphed on the same grid. The point $(0, 0)$ is a solution of $4x + 5y \leq 20$ but is not a solution of $3x - 5y \geq 15$. Therefore, the region to the left of the line representing $4x + 5y = 20$ defines the inequality $4x + 5y < 20$, and the region to the right of the line representing $3x - 5y = 15$ defines the inequality $3x - 5y > 15$. The graph of the solution set is the region where the graphs of the two inequalities overlap, and it includes the boundaries.

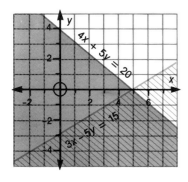

4 - 3   Graphing Linear Inequalities   **161**

Inequalities are used in many applications which have restrictions or limitations placed on the variables.

Example 3. An 80 ha farm is to be planted with two crops, corn and wheat. Planting and harvesting costs, for which no more than a total of $12 000 is available, are $300/ha for corn and $100/ha for wheat. Draw a graph showing the number of hectares of each crop that can be planted.

Solution. Let x and y represent respectively the number of hectares of corn and wheat that can be planted. x and y are subject to the following restrictions:
- The number of hectares cannot be negative.
$$x \geq 0 \ldots \text{①} \qquad y \geq 0 \ldots \text{②}$$
- No more than 80 ha can be planted.
$$x + y \leq 80 \ldots \text{③}$$
- Planting and harvesting costs cannot exceed $12 000.
$$300x + 100y \leq 12\,000$$
or $\qquad 3x + y \leq 120 \ldots \text{④}$

The shaded region shows the ordered pairs (x, y) which satisfy all four inequalities above. Any one of these ordered pairs gives a combination of amounts of corn and wheat that can be planted and harvested on the 80 ha farm for $12 000 or less.

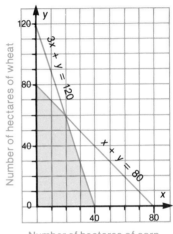

Exercises 4 - 3

A 1. Write the inequality that represents
   i) the shaded region;   ii) the unshaded region.

a) 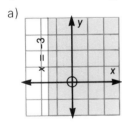   b) 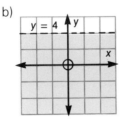   c)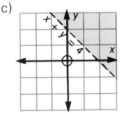

d) 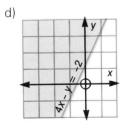   e) 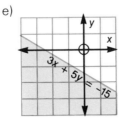   f)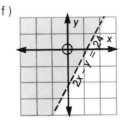

2. Graph the inequality:
   a) $x \leq -1$
   b) $y > 2$
   c) $x - y \leq -3$
   d) $x + 2y < 4$
   e) $3x - 2y \geq -6$
   f) $2x + y \geq 8$
   g) $2x + y \leq 5$
   h) $5x + 2y > -10$
   i) $2x - 7y \geq 14$

3. Write the pair of inequalities that represents the shaded region:

   a)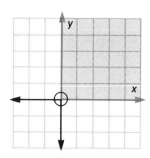
   b)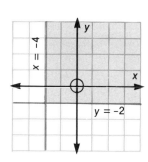
   c)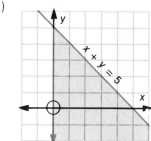
   d)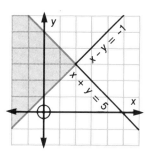
   e)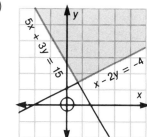
   f)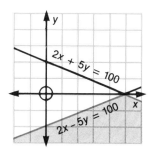

4. Graph the solution set:
   a) $x + y \leq 6$
      $x - 2y \geq 4$
   b) $2x + y \leq 6$
      $3x - 2y \leq 6$
   c) $x + 3y \leq -3$
      $4x - y \leq -4$
   d) $x - 4y \leq 8$
      $2x + y \geq -4$
   e) $3x - y \leq 6$
      $2x + y \geq -2$
   f) $6x + 3y \leq 12$
      $3x - 4y \geq 12$
   g) $5x + 2y \geq 10$
      $3x + 9y \geq 18$
   h) $4x - 5y \leq 20$
      $6x + 4y \geq -12$
   i) $8x - 3y \leq 24$
      $4x - 6y \geq 12$

4-3 Graphing Linear Inequalities   **163**

B 5. Graph the solution set:
   a) $2x + 3y \leq 8$
      $5x - 3y \leq 12$
   b) $3x - 4y \geq -9$
      $2x + 4y \geq 10$
   c) $6x + 5y \leq 120$
      $10x + 3y \leq 100$
   d) $5x - 3y \geq -10$
      $2x + 3y \leq -9$
   e) $4x - 7y \leq 140$
      $2x + 5y \leq -60$
   f) $4x + 3y \leq -6$
      $3x - 4y \geq -8$
   g) $12x - 5y \leq 48$
      $9x + 8y \leq 36$
   h) $7x + 4y \leq 280$
      $6x + 15y \geq 180$

6. Write the set of inequalities that represents the shaded region:

a)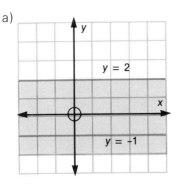

b)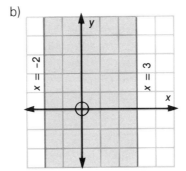

c)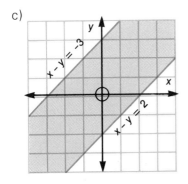

d)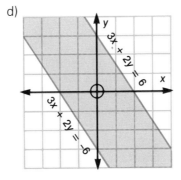

e)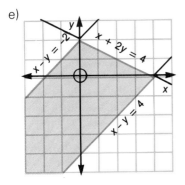

f)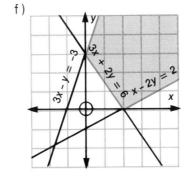

7. Graph the solution set:
   a) $x \leq 6$
      $x \geq 2$
   b) $y \leq -2$
      $y \geq -5$
   c) $x + y \leq 4$
      $x + y \geq -1$
   d) $2x - y \geq 2$
      $2x - y \leq 8$
   e) $x + 3y \leq 6$
      $x + 3y \geq -3$
   f) $3x - 4y \geq -12$
      $3x - 4y \leq 24$
   g) $x - y \geq -2$
      $y \leq 2$
      $x + y \leq 6$
   h) $5x + 2y \geq 10$
      $x + 3y \geq 6$
      $x - 2y \leq 4$
   i) $x + y \leq 7$
      $x + 3y \leq 12$
      $x \geq 0$
      $y \geq 0$

8. A company makes motorcycles and bicycles. The physical dimensions of the work area limit the number of both kinds that can be made in one day as follows:
   No more than 20 motorcycles can be made.
   No more than 30 bicycles can be made.
   No more than 40 vehicles in all can be made.
   Draw a graph of the ordered pairs showing the numbers of motorcycles and bicycles that can be made in one day.

9. A unit of film-processing equipment can develop a maximum of 80 rolls of film per hour provided there are not more than 50 rolls of color or of black-and-white. If color film is developed, there must be at least 15 rolls run at a time. Draw a graph showing the number of rolls of film of each type that can be processed in 1 h.

10. Anthony and Angela work in a plastics shop. Anthony's moulding machine can turn out 50 cups and 20 lids per hour. On her machine, Angela can produce 20 cups and 30 lids per hour. An order requiring more than 600 cups with lids is received. If no more than 30 man·hours can be spent on this order, draw a graph showing how many hours each person could work on this order.

11.
> **In China, men must be 22 to marry.**
>
> Peking. A Chinese law sets minimum legal ages for marrying. The minimum legal age is 22 for men and 20 for women.
>   In addition, couples are urged not to marry until the ages of the bride and groom total more than 52.

Use the information in the news item to draw a graph showing the ages at which Chinese men and women may marry.

4-4 Maximum and Minimum Values of Linear Expressions

A linear expression such as 7x − 3y + 2 may be evaluated for any ordered pair (x, y). If there is some restriction on the ordered pairs, then there may be both a maximum and minimum value of the expression.

Example 1. Find the maximum and minimum values of the expression 7x − 3y + 2 for the set of points marked on the graph.

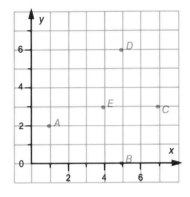

Solution. Evaluate the expression for each point.

Point	Value of 7x − 3y + 2
A(1, 2)	7(1) − 3(2) + 2 = 3
B(5, 0)	7(5) − 3(0) + 2 = 37
C(7, 3)	7(7) − 3(3) + 2 = 42
D(5, 6)	7(5) − 3(6) + 2 = 19
E(4, 3)	7(4) − 3(3) + 2 = 21

For the given points, the maximum value of 7x − 3y + 2 is 42 and the minimum value is 3.

If the ordered pairs are restricted to the points that lie on a given line segment, it is not possible to substitute the coordinates of every point to find the maximum and minimum values of an expression.

Example 2. Find the maximum and minimum values of the expression C = 2x + 5y for the points on the line segment AB shown on the graph.

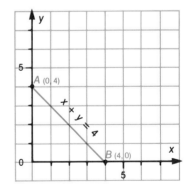

Solution. Since x + y = 4, then y = 4 − x, where 0 ≤ x ≤ 4 as indicated by the graph.
Substitute 4 − x for y in the expression C = 2x + 5y:
$$C = 2x + 5(4 − x)$$
$$= 20 − 3x$$
The maximum value of C occurs when x is least, at x = 0.
The maximum value is: C = 20 − 3(0), or 20.
The minimum value of C occurs when x is greatest, at x = 4. The minimum value is: C = 20 − 3(4), or 8.

Example 2 suggests that when a linear expression is evaluated for points on a line segment, the maximum and minimum values occur at the *endpoints* of the segment.

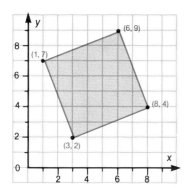

Often, the points for which an expression is evaluated are restricted to lie on the boundary of, or in, a given region.

Example 3. Find the maximum and minimum values of the expression $3x - 2y + 6$ for the points on or inside the square shown on the graph.

Solution. It can be shown that the maximum and minimum values of a linear expresson occur at the vertices of the region. Evaluate the expression at each vertex.

Vertex	Value of $3x - 2y + 6$
(3, 2)	$3(3) - 2(2) + 6 = 11$
(8, 4)	$3(8) - 2(4) + 6 = 22$
(6, 9)	$3(6) - 2(9) + 6 = 6$
(1, 7)	$3(1) - 2(7) + 6 = -5$

The maximum value of the expression is 22, and the minimum value is -5.

Exercises 4 - 4

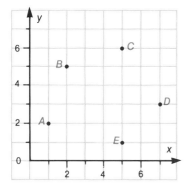

A 1. Find the maximum and minimum values of $5x - 3y + 4$ for the ordered pairs:
   a) (0, 6), (1, 3), (2, 4), (3, 5), (6, 0)
   b) (2, 7), (5, 5), (8, 3), (11, 1)

2. Find the maximum and minimum values of each expression for the points shown:
   a) $6x - y + 1$
   b) $8x - 3y + 4$
   c) $11x + 2y - 6$
   d) $3x - \frac{1}{2}y + 5$

3. Find the maximum and minimum values of each expression for the points on the line segment AB:
   a) $C = 3x - 2y$
   b) $C = x + 7y$
   c) $C = x + 4y - 2$
   d) $C = 2x + 5y - 7$

4. Find the maximum and minimum values of $C = 3x + 2y - 4$ for the points on line segments joining the following points:
   a) (0, 8) and (8, 0)
   b) (0, 3) and (3, 0)
   c) (3, 9) and (7, 1)
   d) (2, 11) and (11, 2)

4-4 Maximum and Minimum Values of Linear Expressions   167

B 5. Find the maximum and minimum values of these expressions
for the points shown on the graph:
   a) $C = x + 5y$
   b) $C = 2x - y + 3$
   c) $C = 3x + y - 2$
   d) $C = 5x - 3y + 2$

6. Find the maximum and minimum values of $C = 3x + y + 2$ for the points on the boundary of, or in, the regions shown:

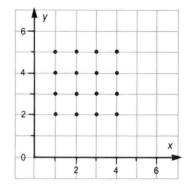

a)
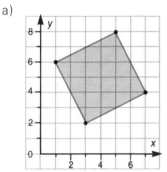

b)
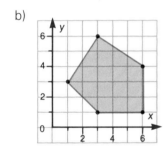

c)
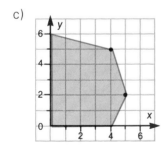

7. Find the maximum and minimum values of these expressions for the region in Exercise 6b:
   a) $C = 7x - 3y + 2$ b) $C = x + 5y - 1$
   c) $C = 8x - 3y - 2$ d) $C = 5x - y + 6$

Work Backwards to the Solution

In solving a word problem by an equation, we are working "backwards" from a known condition—the equation—to discover the unknown value of a variable. The process of working "backwards" from known conditions to unknown values is a useful strategy in the solution of a variety of mathematical problems.

Example. Louise rides her motorcycle home from school by a different route each day. Her home is 5 streets north and 5 streets east of school. How many different routes can Louise travel home if she must always travel north or east? (That is, she may not double back.)

Solution. Draw a 5 by 5 grid and label the intersections as shown. Work backwards from point A. How many different routes home are there from:
A? (Write your answer at point A.)
B? C? and E? (Write your answers at points B, C, and E.)
D? F? H? I? and M?
Continue the process until you have found the number of routes from school to home. If Louise lived 7 streets north and 7 streets east of the school, how many different routes could she travel?

Exercises

1. Sara keyed a number into a calculator. She then pressed the following keys:

and the final display showed 18. By working backwards, find the number that Sara keyed into the calculator.

2. A fungus growing on the surface of a pond doubles in area every day. In 30 days, the fungus grows from one spore to cover the pond completely. What fraction of the pond is covered after 26 days of growth?

3. When asked his height, Tony replied: "If I double my height in centimetres and add 50, divide the result by 2 and take the square root, then subtract 10 and take the positive square root, the result is a number that is 169 less than my height in centimetres." What is Tony's height?

Mathematics Around Us

The World's Tallest Building

The world's tallest building is the Sears Tower in Chicago. It is 440 m high and 69 m square at the base.

Questions

Use the information in the diagram to calculate the following:

1. the total floor area, in square metres
2. the total volume of the building in cubic metres

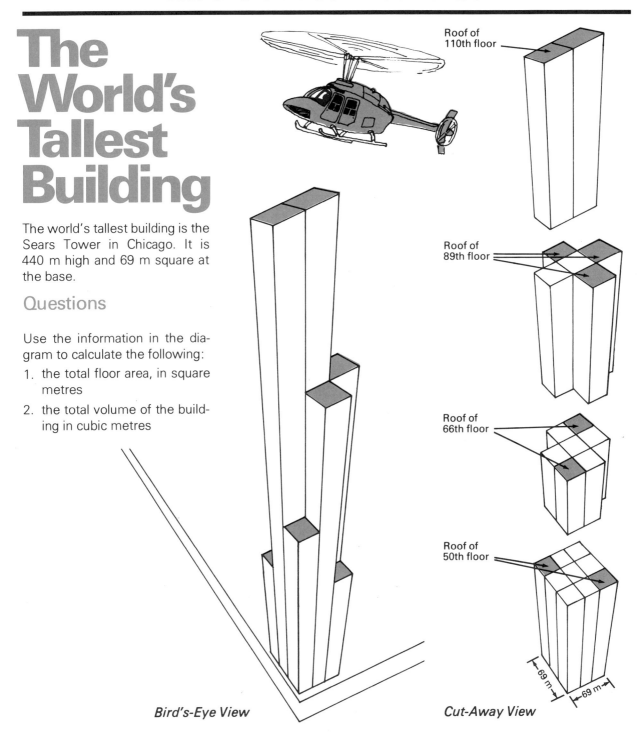

Bird's-Eye View

Roof of 110th floor
Roof of 89th floor
Roof of 66th floor
Roof of 50th floor

Cut-Away View

4-5 Linear Programming

In business and industry, considerable effort is spent in attempting to maximize profits and minimize costs. Many of the decisions made in such cases result from a mathematical analysis. Frequently, a branch of mathematics called **linear programming** is used.

The following examples show how to solve some problems using linear programming. Note the steps involved in each solution.

Example 1. A firm makes footballs for a profit of $10 each and soccer balls for a profit of $15 each. The length of time each ball requires on the cutting and stitching machines is shown in the table.

Type of ball	Time on cutting machine	Time on stitching machine
Football	2 min	1 min
Soccer ball	2 min	4 min

How many of each ball should be made per hour for the profit to be a maximum?

Solution. Step 1. **Identify two variables.**

Let x represent the number of footballs made per hour.
Let y represent the number of soccer balls made per hour.

Step 2. **Write an expression for the quantity to be maximized.**

The quantity to be maximized is the hourly profit:
$P = 10x + 15y$.

Step 3. **Write inequalities restricting the variables.**

x and y are subject to the following restrictions:
- The number of balls cannot be negative.
  $x \geqslant 0 \ldots$ ① $y \geqslant 0 \ldots$ ②
- The time on the cutting machine is not more than 60 min.
  $2x + 2y \leqslant 60$, or $x + y \leqslant 30 \ldots$ ③
- The time on the stitching machine is not more than 60 min.
  $x + 4y \leqslant 60 \ldots$ ④

Step 4. Graph the inequalities.
The shaded region shows the ordered pairs (x, y) which satisfy the four inequalities above. Any one of these ordered pairs gives a number of footballs and a number of soccer balls that can be made on the machines in 1 h.

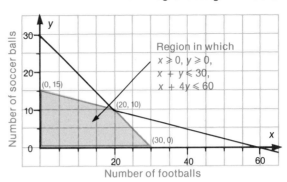

Step 5. Evaluate the expression to be maximized.
Since the maximum profit occurs at one of the vertices of the region defined by the inequalities, evaluate $P = 10x + 15y$ only for these points.

Vertex	Profit: $P = 10x + 15y$
(0, 0)	$10(0) + 15(0) = 0$
(30, 0)	$10(30) + 15(0) = 300$
(20, 10)	$10(20) + 15(10) = 350$
(0, 15)	$10(0) + 15(15) = 225$

Step 6. Answer the question.
Since (20, 10) yields the maximum profit, the firm should make 20 footballs and 10 soccer balls per hour. The maximum profit is $350/h.

Example 2. A farmer plans to allocate up to 45 ha for growing a mixed crop of corn and barley to pay off his fixed costs of $9000 for the year. He will get $250/ha for the corn and $150/ha for the barley. To get the required yield, he will need to prepare the land with fertilizer in the quantities 450 kg/ha for the corn and 150 kg/ha for the barley. How many hectares of each crop should he sow to keep the amount of fertilizer to a minimum?

Solution. **Step 1. Identify two variables.**
Let x represent the number of hectares to grow corn and y the number of hectares to grow barley.

Step 2. Write an expression for the quantity to be minimized.
The quantity to be minimized is the amount of fertilizer needed.
$$A = 450x + 150y$$

Step 3. Write inequalities restricting the variables.

x and y are subject to the following restrictions:

- The number of hectares cannot be negative.
  $x \geq 0 \ldots$ ① $y \geq 0 \ldots$ ②
- The number of hectares is not to exceed 45.
  $x + y \leq 45 \ldots$ ③
- The profit must equal or exceed $9000.
  $250x + 150y \geq 9000$, or
  $5x + 3y \geq 180 \ldots$ ④

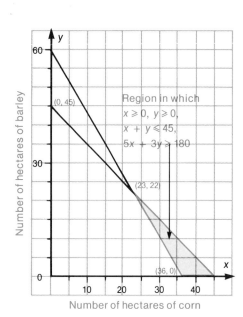

Step 4. Graph the inequalities.
The shaded region shows the ordered pairs (x, y) which satisfy the four inequalities above. Any one of these ordered pairs gives the number of hectares of corn and barley which will yield the required profit.

Step 5. Evaluate the expression to be minimized.
Since the minimum amount of fertilizer occurs at one of the vertices of the shaded region, evaluate $A = 450x + 150y$ only for these points.

Vertex	Fertilizer: $A = 450x + 150y$
(36, 0)	$450(36) + 150(0) = 16\,200$
(45, 0)	$450(45) + 150(0) = 20\,250$
(23, 22)	$450(23) + 150(22) = 13\,650$

Step 6. Answer the question.
Since (23, 22) requires the minimum amount of fertilizer, 13 650 kg, the farmer should plant approximately 23 ha with corn and 22 ha with barley.

Exercises 4 - 5

B 1. A firm makes footballs and soccer balls. Footballs take 2 min to cut and 1 min to stitch; soccer balls take 2 min to cut and 4 min to stitch. How many of each ball should be made per hour for the profit to be a maximum if
   a) the profit on a football is $20 and on a soccer ball is $4?
   b) the profit on a football is $4 and on a soccer ball is $20?

2. A farmer expects a profit of at least $8000 by planting 40 ha with wheat and barley. The ground requires fertilizer in the amounts 400 kg/ha for the wheat and 240 kg/ha for the barley. If the fertilizer used is to be kept to a minimum, how much land should he sow with each grain if he gets
   a) $240/ha for the wheat and $160/ha for the barley?
   b) $180/ha for the wheat and $240/ha for the barley?

3. An 80 ha farm is to be planted with corn and wheat. $12 000 is available for planting and harvesting costs which are $300/ha for corn and $100/ha for wheat. How many hectares of corn and wheat should be sown to maximize the profits if the profit is
   a) corn: $200/ha, wheat: $300/ha?
   b) corn: $300/ha, wheat: $200/ha?
   c) corn: $400/ha, wheat: $100/ha?

4. A company makes motorcycles and bicycles. A restricted work area limits the number of vehicles that can be made in one day.
   No more than 20 motorcycles can be made.
   No more than 30 bicycles can be made.
   No more than 40 vehicles of both kinds can be made.
   What should be the daily rate of production of both vehicles to maximize the profits if the profit per vehicle is
   a) motorcycle: $50, bicycle: $25?
   b) motorcycle: $25, bicycle: $50?

5. A unit of film-processing equipment can develop a maximum of 80 rolls of film per hour provided there are not more than 50 rolls of color or of black-and-white. If color film is developed, there must be at least 15 rolls run at a time. How many rolls of each kind of film should be processed per hour to maximize the profits if the profit per roll is
   a) black-and-white: 50¢, color: 90¢?
   b) black-and-white: 60¢, color: 40¢?

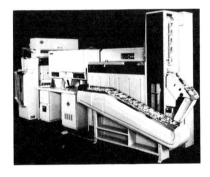

6. Ionic Electronics uses printed circuits and transistor blocks in its production of TV sets and stereo receivers. The numbers used in each piece of equipment and the profits are as follows.

	Circuits	Transistor blocks	Profit
TV set	2	1	$60
Stereo receiver	1	3	$75

If the supply of printed circuits and transistor blocks is limited to 100 and 150 per day respectively, how many TV sets and stereo units should be assembled per day to maximize profits?

7. A farmer's barn can accommodate not more than 75 head of cattle consisting of steers and heifers. Estimates of the feed and veterinary costs, and the profit, per head, are as shown.

	Feed costs	Veterinary costs	Profit
Steer	$90	$3	$50
Heifer	$60	$4	$40

If the farmer can spend no more than $5400 on feed and $240 on veterinary services, how many steers and how many heifers should she raise to maximize her profit?

C 8. The *World of Nuts* buys two different brands of nuts in bulk which it mixes, packages, and resells. The brands contain the same kinds of nuts but in different percentages, as follows:

	Brand A	Brand B	World of Nuts package
Almonds	20%	10%	at least $13\frac{1}{3}$%
Hazelnuts	10%	60%	at least 30%
Cashews	10%	20%	at least $16\frac{2}{3}$%
Peanuts	60%	10%	at most 40%
Cost	$4/kg	$6/kg	

The brands are mixed to give the percentage of each kind of nut in the last column of the table. In what ratio should the brands be mixed to minimize the cost?

9. The man·hours required to manufacture two models of ski—the *Cheetah* and the *Gazelle*—and the material cost are given in the table.

	Cheetah	Gazelle
Man·hours	6	10
Material cost	$15	$15

At least 30 pairs of *Cheetahs* and 20 pairs of *Gazelles* must be made each week. No more than 600 man·hours and $1200 of materials are available each week. How many pairs of each model must be made in order to maximize the profit if the profit on each model is

a) *Cheetah*: $20, *Gazelle*: $40?
b) *Cheetah*: $30, *Gazelle*: $40?

Review Exercises

1. Graph:
   a) $3x - y = 9$ b) $x + 3y = 6$ c) $3x - y = -15$

2. Graph:
   a) $\{(x, y) \mid 2x - 3y = 12\}$ b) $\{(x, y) \mid 2x + 5y = -10\}$

3. A company makes pots and pans. The production time for a pot is 3 min and for a pan is 2 min. Draw a graph showing the number of pots and pans that can be made in 1 h.

4. Josephine plans to invest a sum of money in two different stocks from which she will receive a total of $1000 a year in dividends. The first stock pays dividends at the rate of 12.5% per year, the second of 10% per year. Draw a graph to show how she could invest her money.

5. Solve by graphing:
   a) $x - 2y = -8$
      $3x + 2y = 16$
   b) $2x + y = 2$
      $-3x + y = 7$
   c) $4x - 3y = 18$
      $5x + 3y = 9$
   d) $2x - y = 5$
      $x + 2y = 10$
   e) $2x - 3y = -8$
      $6y - 4x = 16$
   f) $x - 5y = 3$
      $15y - 3x = 15$
   g) $3x + 7y = 11$
      $-2x + y = 4$
   h) $x + 5y = 9$
      $3x - 2y = -7$
   i) $-11x - 3y = 5$
      $-2x + 4y = 10$

6. Write the inequality that represents
   i) the shaded region;   ii) the unshaded region.

   a)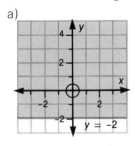
   b)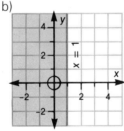
   c)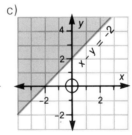

7. Graph the inequality:
   a) $x + 2y \geq 4$ b) $3x - 2y \leq -6$ c) $5x - 2y \leq -10$

8. Graph the solution set:
   a) $x + y \leq 3$
      $x - 2y \geq 6$
   b) $3x - y \geq 6$
      $2x + y \leq -4$
   c) $2x + y \leq 12$
      $3x - 4y \geq -12$
   d) $x + 3y \leq 5$
      $3x - y \geq -5$
   e) $2x - y \leq 6$
      $3x + 2y \geq 12$
   f) $x + 4y \leq 8$
      $2x - 3y \geq 6$

9. Write the inequalities that represent the shaded region:

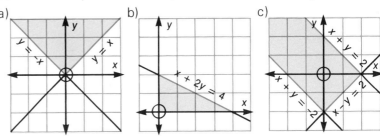

a) $y = -x$, $y = x$
b) $x + 2y = 4$
c) $x + y = 2$, $x - y = 2$

10. A company makes snowmobiles and powermowers. The physical dimensions of the work area limit the number of each kind of machine that can be made each day to the following:

   No more than 30 snowmobiles can be made.
   No more than 40 powermowers can be made.
   No more than a total of 50 machines can be made.

   Draw a graph which shows the number of snowmobiles and the number of powermowers that can be made in one day.

11. Find the maximum and minimum values of the expression $2x + 3y - 6$ for the points on line segments joining
   a) (0, 3) and (3, 0); b) (3, 8) and (8, 3);
   c) (2, 9) and (9, 2); d) (1, 7) and (6, 4).

12. Allsports Ltd. makes basketballs for a profit of $12 each and soccer balls for a profit of $15 each. The length of time each ball requires on the cutting and stitching machines is as follows:

Type	Cutting machine	Stitching machine
Basketball	3 min	2 min
Soccer ball	2 min	3 min

How many of each type of ball should be made per hour for the profit to be a maximum? What is the maximum profit per hour?

13. Sheila and Ken are employed making footballs and soccer balls and take times given in the table to make both kinds.
   An order is received for 8 dozen footballs and 5 dozen soccer balls. How many hours should each person work to minimize the payroll expense, and what will this expense be, if the rates of pay are
   a) Sheila: $4/h,   Ken: $4/h?
   b) Sheila: $6/h,   Ken: $3/h?
   c) Sheila: $3/h,   Ken: $6/h?

	Sheila	Ken
Football	2.5 min	5 min
Soccer ball	6 min	4 min

5 Coordinate Geometry and Relations

The sound of a thunderclap reaches you 15 s after you see the lightning which is 5 km away. If the time interval between the next thunderclap and the lightning is 2 s, how far away is the lightning? (See *Example 2* in Section 5-7.)

5 - 1 Distance in the Plane

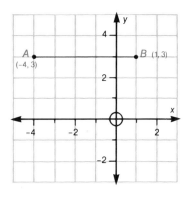

Coordinate systems are used to identify the locations of stars in the sky, the locations of cities on Earth's surface, and the locations of streets and buildings within cities. The Cartesian coordinate system is generally the most useful for describing the positions of points in the plane. One advantage of this system is that the distance between any two points is easily found.

In the diagram, line segment AB is parallel to the x-axis. Its length is the difference between the x-coordinates of A and B. That is, either

$$1 - (-4) = 5 \quad \text{or} \quad (-4) - 1 = -5$$

depending on the order of subtraction. If the distance is defined to be the absolute value of the difference, then the order of subtraction does not matter.

$$AB = |1 - (-4)| \qquad \text{or} \qquad AB = |(-4) - 1|$$
$$= |5| \qquad\qquad\qquad\qquad\quad = |-5|$$
$$= 5 \qquad\qquad\qquad\qquad\qquad = 5$$

This method is used for segments parallel to either axis.

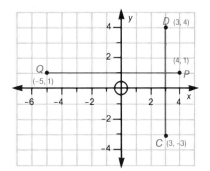

Example 1. Find the lengths of the line segments CD and PQ.

Solution. $CD = |4 - (-3)|$ $\qquad PQ = |-5 - 4|$
$\qquad\qquad\quad = |7| \qquad\qquad\qquad\quad = |-9|$
$\qquad\qquad\quad = 7 \qquad\qquad\qquad\quad\; = 9$

CD is 7 units long and PQ is 9 units long.

The Pythagorean theorem is used to determine the lengths of line segments that are not parallel to an axis.

Example 2. Find the length of line segment PQ.

Solution. Draw the right $\triangle PQR$ for which PQ is the hypotenuse. By the Pythagorean theorem:

$$PQ^2 = PR^2 + RQ^2$$
$$= |(-2) - 3|^2 + |4 - (-3)|^2$$
$$= 25 + 49$$
$$PQ = \sqrt{74}$$
$$\doteq 8.6$$

PQ is approximately 8.6 units long.

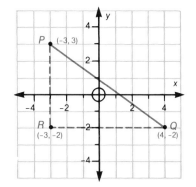

In general, the distance between two points, $P_1(x_1, y_1)$ and $P_2(x_2, y_2)$, is given by the formula:

$$P_1P_2 = \sqrt{(x_2 - x_1)^2 + (y_2 - y_1)^2}.$$

5 - 1 Distance in the Plane   179

Example 3. The points $L(1, 5)$, $M(-3, 1)$, and $N(6, -4)$ are the vertices of $\triangle LMN$. Is $\triangle LMN$ scalene, isosceles, or equilateral?

Solution. From the graph of $\triangle LMN$, LM is clearly seen to be shorter than either of the other two sides. Calculate the lengths of MN and LN.
$$MN = \sqrt{(6 + 3)^2 + (-4 - 1)^2}$$
$$= \sqrt{106}, \text{ or approximately } 10.30$$
$$LN = \sqrt{(6 - 1)^2 + (-4 - 5)^2}$$
$$= \sqrt{106}, \text{ or approximately } 10.30$$
Since $MN = LN$, $\triangle LMN$ is isosceles.

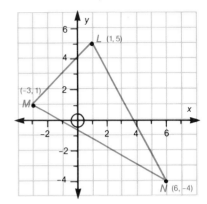

Exercises 5 - 1

A 1. Plot each pair of points and find the distance between them:
   a) (7, 3), (7, −2)
   b) (−3, −5), (−3, 2)
   c) (−4, 3), (7, 3)
   d) (6, −4), (−6, −4)
   e) (0, 0), (0, −8)
   f) (12.5, 5), (−1.5, 5)

2. Find the length of each line segment:

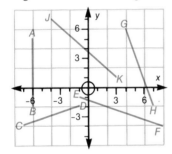

3. Plot each pair of points and find the distance between them:
   a) (5, 4), (1, 2)
   b) (−2, 3), (4, −1)
   c) (−8, 9), (−3, 4)
   d) (−3, 0), (8, −4)
   e) (0, 0), (−5, 2)
   f) (3.5, 5), (7.5, 8.5)

4. For each triangle shown on the grid, find
   a) the lengths of the sides;
   b) the area.

5. Find the length of the line segments with these endpoints:
   a) (5, 9), (−2, 2)
   b) (1, −6), (5, 2)
   c) (−3, 7), (5, 2)
   d) (−9, −3), (−3, 4)

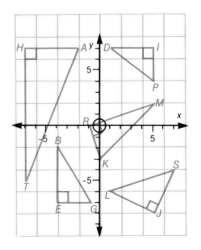

6. On a grid, draw the triangles with these vertices and classify them as scalene, isosceles, or equilateral:
   a) $A(-4, 3)$, $B(-2, -4)$, $C(3, 5)$
   b) $P(-1, 2)$, $Q(1, -5)$, $R(5, -2)$
   c) $J(-2, 5)$, $K(4, -1)$, $L(6, 7)$

B 7. On a grid, draw the rectangles with these vertices, and find:
   i) the lengths of the sides;
   ii) the perimeters;
   iii) the lengths of the diagonals;
   iv) the areas.
   a) $A(-6, -3)$, $B(3, -3)$, $C(3, 5)$, $D(-6, 5)$
   b) $J(-3, 3)$, $K(0, -6)$, $L(3, -5)$, $M(0, 4)$
   c) $P(-8, -1)$, $Q(-4, -7)$, $R(8, 1)$, $S(4, 7)$

8. a) Find the lengths of the line segments AB, BC, and AC:
   i) $A(9, 7)$, $B(4, 4)$, $C(-1, 1)$
   ii) $A(-5, -2)$, $B(3, 2)$, $C(7, 4)$
   iii) $A(8, -1)$, $B(-2, -6)$, $C(-4, -7)$

   b) Compare the lengths: $AB + BC$ and AC. What do you conclude about the points A, B, and C?

9. An ocean freighter sends a distress signal from a location given by (240, 160). A second freighter at (50, 420) and a coastguard cutter at (520, 100) hear the distress call. Which ship is closer to the ship in distress?

10. A surveyor establishes the corner points of a field to be $(-75, 375)$, $(-75, -100)$, $(150, -100)$, and $(250, 150)$, where the numbers represent the distances, in metres, from a pair of reference axes. Find the length of each diagonal, the perimeter, and the area of the field.

C 11. Which points are equidistant from (4, 0) and (0, 2)?
   $P(0, -3)$ $Q(3, 7)$ $R(5, 7)$ $S(2, 1)$

12. Find the coordinates of the points on the x-axis which are 5 units from (5, 4).

13. Find the coordinates of the point on the y-axis which is equidistant from the points
   a) (3,0) and (3,6);
   b) (4,0) and (2,6);
   c) (5,0) and (1,6).

14. Objects in space can be located using a three-coordinate system. A tracking station at (20, 50, 0) picks up a satellite at (70, 10, 40). How far away is the satellite?

5 - 2 Midpoint of a Line Segment

The coordinates of the midpoint of a line segment are related to the coordinates of its endpoints.

Example 1. Find the coordinates of the midpoint, M, of the line segment joining $A(2, 1)$ and $B(8, 5)$.

Solution. Complete the right triangle ABC for which AB is the hypotenuse. MD and ME are the perpendiculars onto BC and AC respectively. It is clear from the diagram that the coordinates of M are the means of the coordinates of A and B.

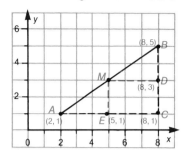

x-coordinate of M: $\dfrac{2 + 8}{2} = 5$

y-coordinate of M: $\dfrac{5 + 1}{2} = 3$

The coordinates of M are $(5, 3)$.

If M is the midpoint of a line segment having endpoints $A(x_1, y_1)$ and $B(x_2, y_2)$, then the coordinates of M are:
$$\left(\dfrac{x_1 + x_2}{2}, \dfrac{y_1 + y_2}{2}\right).$$

Example 2. Find the coordinates of the midpoint of the line segment joining $P(-5, 7)$ and $Q(2, -1)$.

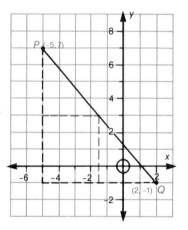

Solution. The x-coordinate is: $\dfrac{-5 + 2}{2} = -\dfrac{3}{2}$

The y-coordinate is: $\dfrac{7 + (-1)}{2} = 3$

The midpoint of PQ is $\left(-\dfrac{3}{2}, 3\right)$.

Exercises 5 - 2

A 1. Find the coordinates of the midpoint of the line segment with endpoints:
   a) $(0, 4)$, $(6, 4)$;
   b) $(-2, 5)$, $(-2, -1)$;
   c) $(1, 1)$, $(7, 9)$;
   d) $(5, 3)$, $(-3, 0)$;
   e) $(-6, 5)$, $(3, -2)$;
   f) $(0, 7)$, $(-4, -2)$;
   g) $(-3, -6)$, $(7, 5)$;
   h) $(-9, 0)$, $(5, -4)$.

2. Find the coordinates of the three points that divide line segment AB into four equal parts when the coordinates of A and B are:
   a) $(-4, 4)$, $(8, -12)$;
   b) $(5, -7)$, $(-9, 5)$;
   c) $(-2, -6)$, $(8, 4)$.
   d) $(-5, -3)$, $(5, 3)$.

3. A right triangle has vertices (−2, 8), (6, 4), and (4, 0).
   a) Draw the triangle on a grid and find the coordinates of the midpoint, M, of the hypotenuse.
   b) Show that M is equidistant from all three vertices.

4. A parallelogram has vertices (0, −3), (7, −1), (10, 4), and (3, 2). Find the coordinates of the midpoints of the diagonals. What conclusion can you draw?

B 5. M is the midpoint of the line segment AB. Find the coordinates of B if those of A and M are as follows:
   a) $A(-2, 4)$, $M(2, -1)$
   b) $A(8, -1)$, $M(5, -5)$
   c) $A(3, 5)$, $M(-1.5, 1.5)$
   d) $A(-6, -4)$, $M(-3, -2)$

6. Points P, Q, and R divide a line segment into four equal parts. Find the coordinates of the other endpoint if those of P and the adjacent endpoint are:
   a) $P(5, 2)$, $(2, 4)$;
   b) $P(-2, 1)$, $(-6, -2)$;
   c) $P(4, -1.5)$, $(8, -5)$;
   d) $P(-3, -6)$, $(-4, -8)$.

7. A triangle has vertices $A(2, 6)$, $B(-4, -2)$, and $C(8, -4)$. The midpoints of AB and AC are M and N respectively.
   a) Find the coordinates of M and N.
   b) Compare the lengths of MN and BC.

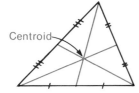

A **median** is the line segment from any vertex to the midpoint of the opposite side.

8. A triangle has vertices $P(-2, 4)$, $Q(-4, -4)$, and $R(6, 0)$. Find the lengths of its three medians.

C 9. A rectangle has vertices $A(0, -5)$, $B(12, 7)$ $C(6, 13)$ and $D(-6, 1)$.
   a) Plot the rectangle on a grid and find the coordinates of E, the midpoint of BC, and F, the midpoint of AD.
   b) Show that segments DE and BF trisect the diagonal AC.
   c) If G and H are the trisection points of AC, with G adjacent to C, calculate the ratio of the lengths:
      i) $EG:GD$;
      ii) $FH:FB$.

10. A square has vertices (4, 3), (11, 6), (8, 13), and (1, 10).
    a) Find the arithmetic mean of
       i) the x-coordinates of its vertices;
       ii) the y-coordinates of its vertices.
    b) Use the results of (a) as the coordinates of a point C and, by graphing, determine how C is related to the square.

11. Triangle ABC has vertices $A(-6, 2)$, $B(8, -2)$, $C(4, 6)$. Find the coordinates of its centroid.

5-3 Slope of a Line Segment

The pitch of a roof, the steepness of a ski run, or the gradient of a mountain road are all instances of a mathematical idea called **slope**.

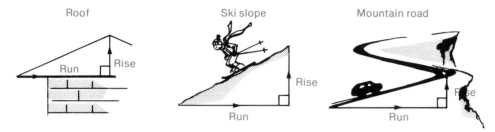

In each case, the slope is the ratio of the difference in the vertical position of the two points (rise) to the difference in the horizontal position (run).

$$\text{slope} = \frac{\text{rise}}{\text{run}}$$

In a coordinate system, the slope, m, of a line or line segment can be found if the coordinates of any two points, (x_1, y_1) and (x_2, y_2), on the line or segment are known.

The slope is the ratio: $\dfrac{\text{difference in } y\text{-coordinates}}{\text{difference in } x\text{-coordinates}}$

or, $m = \dfrac{y_2 - y_1}{x_2 - x_1}$

Example 1. Graph the line segments and find their slopes:
  a) $A(2, 1)\ B(5, 3)$ b) $M(-3, 4)\ N(-1, -2)$
  c) $R(-2, 4)\ S(5, 4)$ d) $J(1, 3)\ K(1, -2)$

Solution a) Slope of AB: $\dfrac{3 - 1}{5 - 2} = \dfrac{2}{3}$

The slope of AB is $\dfrac{2}{3}$.

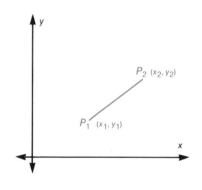

b) Slope of MN: $\dfrac{(-2) - (4)}{(-1) - (-3)} = -\dfrac{6}{2}$, or -3

The slope of MN is -3.

c) Slope of RS: $\dfrac{4 - 4}{5 - (-2)} = \dfrac{0}{7}$, or 0

The slope of RS is 0.

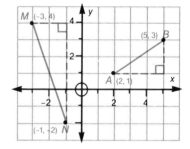

d) Slope of JK: $\dfrac{-2 - 3}{1 - 1}$

Since the denominator equals 0 and division by 0 is not defined, the slope of JK is not defined.

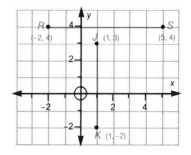

The conclusions from *Example 1* are:

- Lines *rising* to the right have *positive* slope.
- Lines *falling* to the right have *negative* slope.
- The slope of any horizontal line is zero.
- The slope of any vertical line is undefined.

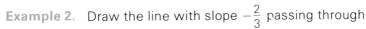

Example 2. Draw the line with slope $-\frac{2}{3}$ passing through

a) the point $(-2, 4)$; b) the point $(-1, -2)$.

Solution. a) Since $-\frac{2}{3} = \frac{-2}{3}$, for every difference of $+3$ in x there is a corresponding difference of -2 in y. Begin at $(-2, 4)$. Move 3 to the *right* and 2 *down*. This is a point on the line. Do the same again from this point to get a third point.

Also, since $-\frac{2}{3} = \frac{2}{-3}$, another point on the line is obtained by moving 3 to the *left* and 2 *up*.

b) Proceed as for (a) but begin at $(-1, -2)$.

Example 2 illustrates a fundamental property of slope:

> Two lines are parallel if, and only if, their slopes are equal.

Example 3. Determine whether or not the quadrilateral $A(0, -6)$ $B(2, -1)$ $C(-1, 5)$ $D(-3, 0)$ is a parallelogram.

Solution. Draw the quadrilateral on a grid.

Slope of AB: $\frac{-1 - (-6)}{2 - 0} = \frac{5}{2}$

Slope of DC: $\frac{5 - 0}{-1 - (-3)} = \frac{5}{2}$

Since the slopes are equal, $AB \parallel DC$.

Slope of AD: $\frac{0 - (-6)}{-3 - 0} = -2$

Slope of BC: $\frac{5 - (-1)}{-1 - 2} = -2$

Since the slopes are equal, $AD \parallel BC$.

Since both pairs of opposite sides are parallel, quadrilateral $ABCD$ is a parallelogram.

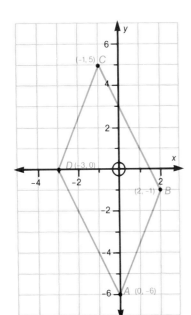

Exercises 5 - 3

A 1. Graph these line segments and find their slopes:
   a) $A(-2, 7)\ B(6, -4)$
   b) $C(3, -5)\ D(8, 10)$
   c) $E(1, 6)\ F(5, -4)$
   d) $G(-3, 7)\ H(-3, -7)$
   e) $J(-4, -3)\ K(8, 5)$
   f) $L(2, -7)\ M(7, -7)$

2. Find the pitch of the roof shown in the diagram.

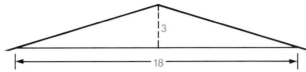

3. A section of roller-coaster track falls 25 m in a horizontal distance of 15 m. What is the slope of this section of track?

4. A new overpass is to be connected to an access road 250 m away by a road of uniform gradient. If the overpass is 8 m above the access road, what will be the slope of the connecting road?

5. On a grid, draw a line through
   a) $(-3, 5)$ with slope $\frac{-4}{7}$;
   b) $(-2, -6)$ with slope $\frac{3}{4}$;
   c) $(5, -4)$ with slope $-\frac{8}{3}$;
   d) $(6, 4)$ with slope undefined;
   e) $(-3, 1)$ with slope $\frac{7}{3}$;
   f) $(6, 4)$ with slope 0.

6. Find the slope of each line segment:
   a)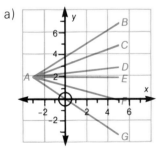
   b)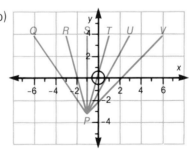

7. State the slope of a line that is
   a) parallel to: i) the x-axis,   ii) the y-axis;
   b) perpendicular to: i) the x-axis,   ii) the y-axis.

8. The coordinates of the vertices of a triangle are given. Graph the triangle and find the slope of each side.
   a) $(5, -1), (0, 4), (-2, -5)$
   b) $(-3, 4), (6, 7), (2, -3)$
   c) $(4, -2), (-4, 8), (4, 8)$
   d) $(-2, -1), (-1, -6), (5, 6)$

9. Graph the quadrilateral having vertices with the coordinates given, and determine whether or not it is a parallelogram.
   a) (5, 3), (−3, −3), (−2, −8), (6, −2)
   b) (−6, 1), (−2, −6), (10, 2), (7, 9)
   c) (−4, 5), (−2, −1), (6, −4), (4, 2)

B 10. A ladder, 2.6 m long, just reaches a window 2.4 m above the ground. What is its slope?

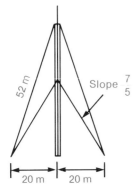

11. Guy wires supporting a broadcasting tower are fastened to the ground 20 m from its base. The first set of wires is attached to the tower part way up and have a slope of $\frac{7}{5}$. The second set is fastened to the top of the tower and each wire is 52 m in length.
    a) Find the length of the first set of guy wires.
    b) Find the slope of the second set of guy wires.

12. The following are slopes of parallel lines. Find the value of k.
    a) $\frac{2}{3}, \frac{4}{k}$ b) $-\frac{k}{5}, \frac{3}{2}$ c) $\frac{k}{-4}, \frac{3}{2}$ d) $\frac{1}{k}, -\frac{1}{2}$

13. The coordinates of the vertices of a triangle are given. Graph the triangle and compare the slopes of the line segments joining the midpoints of the sides with the slopes of the sides.
    a) (2, 6), (−2, 2), (6, −2)   b) (−2, 4), (4, −6), (6, 2)
    c) (−6, 3), (−2, −5), (4, −1)   d) (−4, 1), (3, −6), (2, 4)

14. A quadrilateral has vertices as given. Show that the midpoints of the sides are the vertices of a parallelogram.
    a) (−4, 3), (−1, 2), (5, 1), (6, 5)
    b) (−1, 4), (−3, −3), (1, −6), (5, −1)
    c) (−5, 0), (9, 0), (1, 4), (1, 12)

15. One endpoint of a line segment is (4, 6), the other is on the x-axis. Find the coordinates of the endpoint on the x-axis for the slope of the segment to be
    a) 1;   b) 2;   c) 3;   d) $\frac{1}{2}$;   e) −2.

16. Two line segments, with a common endpoint (0, 4), have slopes 2 and $-\frac{1}{2}$. If the other endpoints are on the x-axis, find their coordinates.

17. The coordinates of three vertices of a parallelogram are given. Find the coordinates of the fourth vertex.
    a) (−4, 1), (−3, −4), (5, 0)   b) (−4, 4), (1, −2), (8, −5)

18. a) Find the slopes of line segments AB, BC, and AC:
    i) $A(-3, -3)$, $B(2, 1)$, $C(11, 8)$,
    ii) $A(-6, 1)$, $B(-2, -1)$, $C(4, -4)$,
    iii) $A(8, -4)$, $B(-4, -1)$, $C(3, -3)$,
    iv) $A(-5, 4)$, $B(0, 2)$, $C(9, -2)$,
    b) Compare the slopes of AB, BC, and AC. What do you conclude about the points A, B, and C?

C 19. On April 14, 1981, the first American space shuttle, *Columbia*, returned to Earth. At one point in its reentry, it was travelling at approximately 1080 km/h and sinking at 4200 m/min. What was the slope of the reentry path?

20. The coordinates of the midpoints of the sides of a triangle are given. Find the coordinates of the vertices of the triangle.
    a) (5, 0), (2, 3), (7, 3)   b) (−3, 3), (−3, 8), (5, 5)

21. a) Can a triangle be drawn such that all three sides have positive slope?
    b) Can a quadrilateral be drawn such that
       i) three sides have positive slope?
       ii) all four sides have positive slope?

5-4 Slopes of Perpendicular Lines

Lines are parallel if they have equal slopes. There is also a relationship between the slopes of perpendicular lines.

In the diagram, right triangle ONA is rotated 90° counterclockwise about (0, 0) to $\triangle ON'A'$. The coordinates of A are (5, 3), the coordinates of A' are (−3, 5), and $OA \perp OA'$

Slope of OA: $\dfrac{3-0}{5-0} = \dfrac{3}{5}$

Slope of OA': $\dfrac{5-0}{-3-0} = -\dfrac{5}{3}$

The numbers $\dfrac{3}{5}$ and $-\dfrac{5}{3}$ are called **negative reciprocals**. The product of negative reciprocals is −1

$$\left(\dfrac{3}{5}\right)\left(-\dfrac{5}{3}\right) = -1$$

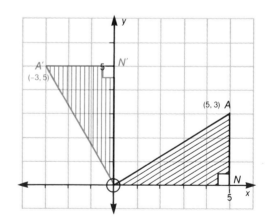

> Two lines are perpendicular if, and only if, their slopes are negative reciprocals.

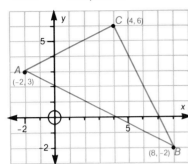

Example 1. A triangle has vertices $A(-2, 3)$, $B(8, -2)$, $C(4, 6)$. Determine whether or not it is a right triangle.

Solution. From the graph, $\angle C$ appears to be a right angle. Calculate the slopes of AC and BC.

Slope of AC: $\dfrac{6-3}{4-(-2)} = \dfrac{3}{6}$, or $\dfrac{1}{2}$

Slope of BC: $\dfrac{6-(-2)}{4-8} = \dfrac{8}{-4}$, or -2

Since $\left(\dfrac{1}{2}\right)(-2) = -1$, the slopes of AC and BC are negative reciprocals. Therefore, $AC \perp BC$, and $\triangle ABC$ is a right triangle.

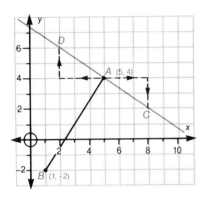

Example 2. A line segment has endpoints $A(5, 4)$, $B(1, -2)$. Draw a line through A which is perpendicular to segment AB.

Solution. Draw segment AB on a grid.

Slope of AB: $\dfrac{-2-4}{1-5} = \dfrac{-6}{-4}$, or $\dfrac{3}{2}$

The slope of a line perpendicular to AB is $-\dfrac{2}{3}$.

Begin at A. Points on the required line may be found by moving 3 to the right and 2 down, or 3 to the left and 2 up. Line CD is the required line.

Exercises 5 - 4

A 1. What is the slope of a line perpendicular to a line with slope
   a) $\dfrac{2}{3}$?   b) $\dfrac{5}{8}$?   c) $-\dfrac{3}{4}$?   d) $-\dfrac{1}{2}$?
   e) 3?   f) -2?   g) 0.2?   h) 1?

2. Which pairs of numbers are slopes of perpendicular lines?
   a) $\dfrac{3}{4}, -\dfrac{4}{3}$ b) $\dfrac{2}{3}, \dfrac{3}{2}$ c) $\dfrac{4}{5}, -\dfrac{4}{5}$
   d) $-4, \dfrac{1}{4}$ e) $3, \dfrac{1}{3}$ f) $2, -\dfrac{1}{2}$

3. Which triangles are right triangles?
   a) $A(3, 0)$ $B(-4, 4)$ $C(-1, -2)$ b) $P(-3, 1)$ $Q(3, -3)$ $R(7, 3)$
   c) $K(3, 2)$ $L(-5, -1)$ $M(-2, -8)$

4. Which quadrilaterals are rectangles?
   a) $A(5, 4)$ $B(-4, -2)$ $C(-2, -5)$ $D(7, 1)$
   b) $J(-3, 2)$ $K(-2, -3)$ $L(6, -2)$ $M(5, 3)$
   c) $P(5, 1)$ $Q(-4, 4)$ $R(-6, -2)$ $S(3, -5)$

5-4 *Slopes of Perpendicular Lines* **189**

5. Given the line through $P(-4, -2)$ and $Q(6, 4)$,
   a) draw the line through P perpendicular to PQ;
   b) draw the line through Q perpendicular to PQ.
   c) How are the lines constructed in (a) and (b) related?

B 6. If the following are slopes of perpendicular lines, find k.
   a) $\dfrac{k}{2}, \dfrac{1}{4}$
   b) $\dfrac{6}{k}, -\dfrac{2}{3}$
   c) $\dfrac{4}{3}, \dfrac{k}{2}$
   d) $\dfrac{k}{3}, -1$
   e) $\dfrac{3}{5}, \dfrac{k}{6}$
   f) $\dfrac{7}{12}, \dfrac{4}{k}$

7. A, B, and C are vertices of a rectangle. Find the coordinates of the fourth vertex.
   a) $A(1, 8), B(3, -2), C(6, 6)$ b) $A(2, -1), B(5, -3), C(7, 0)$
   c) $A(-4, 7), B(-6, 4), C(3, -2)$

8. In the diagram, $ABCD$ is a square. Find the coordinates of C and D if the coordinates of A and B are:
   a) $A(1, 0), B(0, 4)$;
   b) $A(3, 0), B(0, 7)$;
   c) $A(9, 0), B(0, 1)$;
   d) $A(a, 0), B(0, b)$.

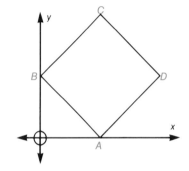

9. Two vertices of an isosceles triangle are $(-5, 4)$ and $(7, 8)$. The third vertex is on the x-axis. Find the possible coordinates of the third vertex.

10. a) What kind of quadrilateral has vertices $(-2, 7)$, $(-3, -1)$, $(4, -5)$, and $(5, 3)$?
    b) Find the midpoint and slope of each diagonal.
    c) What can you conclude from (a) and (b)?

C 11. The coordinates of two vertices of a square are given. Find the possible coordinates of the other two vertices.
   a) $(0, 0), (2, 5)$
   b) $(3, 1), (0, 5)$

12. The diagram shows right triangle OAB with squares drawn on the three sides. Points P, Q, and R are the centres of the three squares.
    a) Draw the diagram on a grid.
    b) Show that the segments AP and QR are perpendicular and equal in length.
    c) Show that the following line segments are also perpendicular and equal in length: i) BQ, PR ii) OR, PQ
    d) Investigate whether these relations are true for other right triangles.

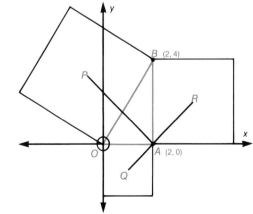

5-5 Using Slope to Graph a Linear Equation

One method of graphing a linear equation is to construct a table of values. Consider the equation $y = 2x + 3$.

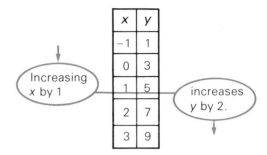

The table and graph suggest another method of graphing a linear equation. This method is based on two numbers:
- **the y-intercept.** This is the value of y when $x = 0$.
- **the slope.**

$$y = \underset{\text{slope}}{2}x + \underset{y\text{-intercept}}{3}$$

> In general, the graph of a linear equation in the form
> $$y = mx + b$$
> has a slope m and a y-intercept b.

The form, $y = mx + b$, is called the **slope y-intercept** form of a linear equation.

Example 1. Graph the equations:

a) $y = \frac{2}{3}x - 5$ b) $y = -2x + 4$

Solution. a) The slope is $\frac{2}{3}$ and the y-intercept is -5.

Begin at $(0, -5)$. Move 3 to the right and 2 up. This is a point on the line. Other points on the line can be obtained by continuing in this way, or by moving 3 to the left and 2 down.

b) The slope is -2 and the y-intercept is 4.
Begin at $(0, 4)$. Move 1 to the right and 2 down. This is a point on the line. Other points on the line can be obtained by continuing in this way, or by moving 1 to the left and 2 up.

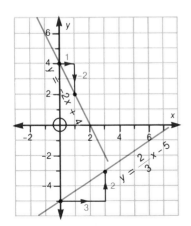

5-5 Using Slope to Graph a Linear Equation

Example 2. Find the equations of the lines shown on the grid.

Solution. The slope and the y-intercept of each line can be read from their graphs.
L_1 has a slope of 1 and a y-intercept of 2.
Its equation is $y = 1x + 2$
or, $y = x + 2$
L_2 has a slope of $-\frac{3}{2}$ and a y-intercept of -1.

Its equation is $y = -\frac{3}{2}x - 1$
or, $2y = -3x - 2$
$3x + 2y = -2$

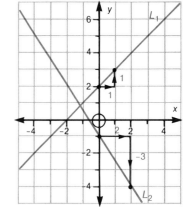

An equation that is not in the slope y-intercept form can be put into that form by solving for y.

Example 3. Graph the equation: $4x + 3y - 15 = 0$.

Solution. Solve for y: $4x + 3y - 15 = 0$
$3y = -4x + 15$
$y = -\frac{4}{3}x + 5$

The graph is drawn using the facts that the slope of the line is $-\frac{4}{3}$ and the y-intercept is 5.

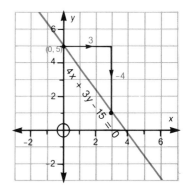

Exercises 5 - 5

A 1. Graph the equations:

a) $y = \frac{2}{5}x + 3$ b) $y = \frac{3}{4}x - 2$

c) $y = -\frac{1}{2}x + 1$ d) $y = -\frac{3}{2}x - 1$

e) $y = 2x - 3$ f) $y = -3x + 2$

2. Find the equations represented by the graphs:

a)

b)

c)

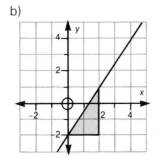
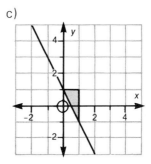

3. Find the equations represented by the graphs:

a) 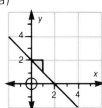 b) 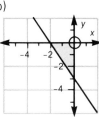 c)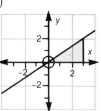

4. State the slope and the *y*-intercept:

   a) $y = 3x + 5$ b) $y = -2x + 3$ c) $y = \frac{2}{5}x - 4$

   d) $y = -\frac{1}{2}x + 6$ e) $y = -4x - 7$ f) $y = \frac{3}{8}x - \frac{5}{2}$

   g) $y = \frac{4}{3}x - 2$ h) $y = \frac{9}{5}x + 1$ i) $y = -\frac{2}{3}x$

B 5. a) Graph the following on one pair of axes:

   $y = 3x$ $\quad\quad$ $y = 2x$ $\quad\quad$ $y = 1x$

   $y = \frac{1}{2}x$ $\quad\quad$ $y = 0x$ $\quad\quad$ $y = -3x$

   $y = -2x$ $\quad\quad$ $y = -1x$ $\quad\quad$ $y = -\frac{1}{2}x$

   b) Describe how the graph of $y = mx$ changes as m changes.

6. a) Graph the following on one pair of axes:

   $y = x + 4$ $\quad\quad$ $y = x + 2$ $\quad\quad$ $y = x$
   $y = x - 4$ $\quad\quad$ $y = x - 2$

   b) Describe how the graph of $y = mx + b$ changes as b changes.

7. a) Graph the following on one pair of axes:

   $y = 3x + 4$ $\quad\quad$ $y = 2x + 4$ $\quad\quad$ $y = x + 4$

   $y = \frac{1}{2}x + 4$ $\quad\quad$ $y = 0x + 4$ $\quad\quad$ $y = -3x + 4$

   $y = -2x + 4$ $\quad\quad$ $y = -x + 4$ $\quad\quad$ $y = -\frac{1}{2}x + 4$

   b) Graph the following on one pair of axes:

   $y = 3x + 4$ $\quad\quad$ $y = 3x + 2$ $\quad\quad$ $y = 3x$
   $y = 3x - 4$ $\quad\quad$ $y = 3x - 2$

   c) Describe how the graph of $y = mx + b$ changes as:
   i) m changes, while b remains constant;
   ii) b changes, while m remains constant.

8. Write the equation of the line that has
   a) slope 2, y-intercept 3;
   b) slope −1, y-intercept 4;
   c) slope $\frac{2}{3}$, y-intercept −1;
   d) slope $-\frac{4}{5}$, y-intercept 8.

9. Write the equation of the line that passes through
   a) (0, 5) and (1, 7);
   b) (0, −2) and (3, 0).

10. State the slope and the y-intercept:
    a) $3x - 4y - 12 = 0$
    b) $5x - 2y + 10 = 0$
    c) $2x + y - 3 = 0$
    d) $3x + 5y + 20 = 0$
    e) $x + 2y - 5 = 0$
    f) $4x - 7y + 15 = 0$

11. Graph the equations:
    a) $3x - 2y = 6$
    b) $2x + y = 4$
    c) $5x - 2y = 10$
    d) $4x - 3y + 12 = 0$
    e) $x - 2y - 8 = 0$
    f) $5x + 3y - 6 = 0$
    g) $2x - y + 3 = 0$

12. Find pairs of: i) parallel lines; ii) perpendicular lines.
    a) $2x + 3y - 6 = 0$
    b) $5x - 2y + 12 = 0$
    c) $8x - 2y - 17 = 0$
    d) $2x + 5y - 8 = 0$
    e) $2x - 3y - 18 = 0$
    f) $x + 4y + 11 = 0$
    g) $5x - 2y = 20$
    h) $2x + 3y = 19$
    i) $4x - y = 12$

13. The three equations represent lines containing the sides of a triangle. Determine if the triangle is a right triangle.
    a) $y = 2x + 3$, $x + 2y - 2 = 0$, $3x + 2y + 7 = 0$
    b) $3x - 5y = 10$, $4x + 2y - 9 = 0$, $5x - 3y + 8 = 0$
    c) $4x + 7y - 12 = 0$, $3x - 6y - 11 = 0$, $2x + y - 17 = 0$

C 14. Find expressions for the slope and y-intercept of the line with equation $Ax + By + C = 0$.

15. Draw a line through the two points given and find its equation:
    a) (2, 2), (4, 3)
    b) (3, 4), (−3, 0)
    c) (−1, 5), (2, −1)
    d) (3, −3), (−6, 0)

16. The equations of the lines containing three sides of a square are as follows:
    $$y = -\frac{1}{2}x + 8, \quad y = -\frac{1}{2}x - 2, \quad y = 2x - 12.$$
    Find the equation of the line containing the fourth side.

Mathematics Around Us

Postal Rates Take a Licking

The graph shows Canadian first-class postal rates since 1900. For each rate, a photograph of a stamp which was in use at the time is shown.

Questions

1. a) In what years did the postal rates change?
   b) What was the increase in rates for each of those years?
   c) What was the percent increase in rate for each of those years?

2. a) In what year did the postal rate decrease?
   b) What was the percent decrease in rate that year?

3. During a recent five-year period, the philatelic value of the first ten stamps increased as shown in the table. What was the percent increase in value of each stamp?

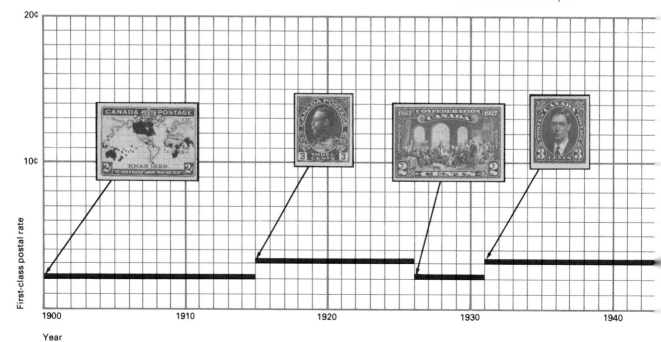

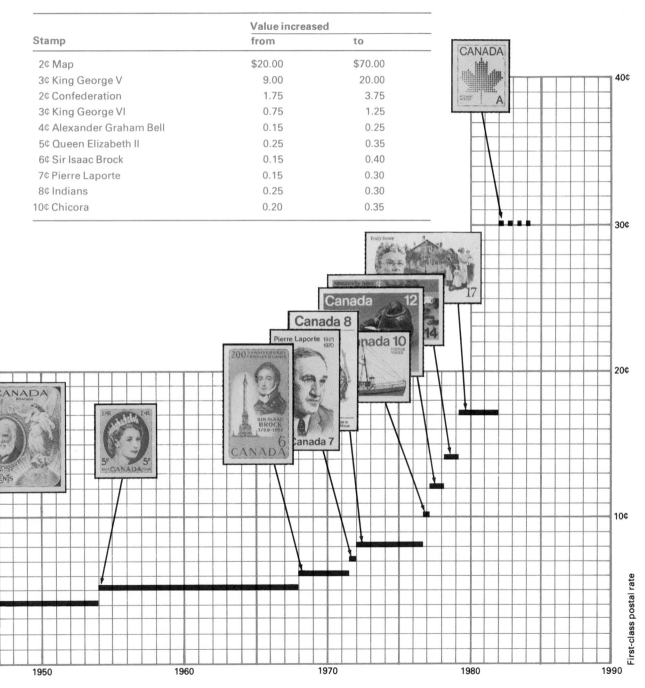

Stamp	Value increased	
	from	to
2¢ Map	$20.00	$70.00
3¢ King George V	9.00	20.00
2¢ Confederation	1.75	3.75
3¢ King George VI	0.75	1.25
4¢ Alexander Graham Bell	0.15	0.25
5¢ Queen Elizabeth II	0.25	0.35
6¢ Sir Isaac Brock	0.15	0.40
7¢ Pierre Laporte	0.15	0.30
8¢ Indians	0.25	0.30
10¢ Chicora	0.20	0.35

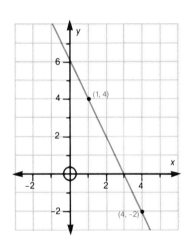

5-6 Finding the Equation of a Line

If both the slope, m, and y-intercept, b, of a line are known, its equation can be written using the form $y = mx + b$. If other information about the line is known instead, the equation of the line can still be found using the slope y-intercept form as a model.

Example 1. Find the equation of the line with slope 2 that passes through $(-1, 3)$.

Solution. Let $y = mx + b$ represent the equation of the line. Since the slope is 2, the equation becomes:
$$y = 2x + b.$$
Since $(-1, 3)$ is a point on the line:
$$3 = 2(-1) + b$$
$$b = 5$$
The y-intercept is 5. Therefore the equation of the line is $y = 2x + 5$.

Example 2. Find the equation of the line that passes through the points $(1, 4)$ and $(4, -2)$.

Solution. Slope of the line: $m = \dfrac{-2 - 4}{4 - 1}$
$$= \dfrac{-6}{3}, \text{ or } -2$$
Let $y = mx + b$ represent the equation of the line. Since $m = -2$, the equation becomes:
$$y = -2x + b.$$
Since $(1, 4)$ is a point on the line:
$$4 = -2(1) + b$$
$$b = 6$$
The y-intercept is 6. Therefore the equation of the line is $y = -2x + 6$.

Example 3. Find the equation of the line which passes through $(-1, 5)$ and is perpendicular to the line $3x - 2y = 12$.

Solution. Find the slope of the line $3x - 2y = 12$.
Solve for y: $-2y = -3x + 12$
$$y = \tfrac{3}{2}x - 6$$
The given line has a slope of $\tfrac{3}{2}$.

The slope of a line perpendicular to it must be $-\tfrac{2}{3}$.

Let $y = mx + b$ represent the equation of the required line.

Since $m = -\frac{2}{3}$: $y = -\frac{2}{3}x + b$

Since $(-1, 5)$ is a point on the line:
$$5 = -\frac{2}{3}(-1) + b$$
$$5 = \frac{2}{3} + b$$
$$b = \frac{13}{3}$$

The equation of the line is $y = -\frac{2}{3}x + \frac{13}{3}$

or, $3y = -2x + 13$
$2x + 3y = 13$

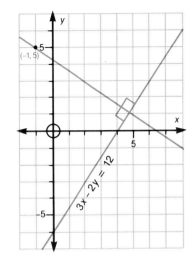

Exercises 5 - 6

A 1. The equation of a line is $y = 3x + b$. Find the value of b if the line passes through the point:
   a) (2, 1);
   b) (−1, 4);
   c) (3, −2);
   d) (0, 5);
   e) (0, −3);
   f) (2, 0).

2. The equation of a line is $y = -\frac{2}{3}x + b$. Find the value of b if the line passes through the point:
   a) (3, 2);
   b) (−6, −1);
   c) (5, 3);
   d) (0, 4);
   e) (2, 0);
   f) (0, 0).

3. Find the equation of each line, given the slope and a point on the line.
   a) 3, (2, 5);
   b) 7, (−4, 2);
   c) −4, (6, −8);
   d) 2, (−4, −5);
   e) $-\frac{3}{5}$, (−10, 3)
   f) $\frac{2}{3}$, (5, −2)
   g) $-\frac{7}{2}$, (−3, 7)
   h) $\frac{8}{3}$, (−5, −2)
   i) 0, $(\frac{1}{2}, \frac{3}{4})$

4. Find the equation of the line that passes through (−3, 2) and has the slope given. Graph all the lines on the same axes.
   a) 3
   b) 2
   c) 1
   d) $\frac{1}{2}$
   e) −3
   f) −2
   g) −1
   h) $-\frac{1}{2}$

5. Find the equation of the line with slope 2 that passes through the point given. Graph all the lines on the same axes.
   a) (1, 7)
   b) (1, 5)
   c) (1, 3)
   d) (1, 1)
   e) (1, −1)
   f) (1, −3)
   g) (1, −5)
   h) (1, −7)

6. Find the equation of the line that passes through:
   a) (2, 1) and (5, 7);
   b) (−3, 2) and (1, −10);
   c) (5, −2) and (7, 5);
   d) (−1, −3) and (4, 7);
   e) (4, −1) and (−2, −5);
   f) (−7, −12) and (−4, −4);
   g) (3, 7) and (7, −5);
   h) (−6, −2) and (−8, 2).

B 7. Find the equation of the line
   a) with slope 3 that passes through (2, 1);
   b) with slope $-\frac{2}{5}$ that passes through (−1, 4);
   c) that passes through (−1, 3) and (3, −1);
   d) that passes through (2, 5) and (−1, 3);
   e) with slope $\frac{3}{4}$ that passes through (4, 0);
   f) with slope −2 that passes through (0, 5);
   g) that passes through (0, 4) and (−2, 0);
   h) that has a slope of $-\frac{1}{3}$ and a y-intercept of 2.

8. Find the equation of the line that passes through (−2, 5) if
   a) the slope is 1;
   b) the y-intercept is 3;
   c) the line also passes through (3, −1);
   d) it is parallel to the line $y = -3x + 4$;
   e) it is perpendicular to the line $y = -3x + 4$.

9. A triangle has vertices (0, 5), (3, 0), and (5, 6). Draw the triangle, and find the equations of the lines containing the sides.

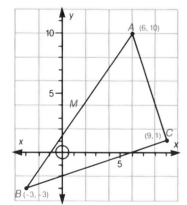

10. In △ABC, M is the midpoint of side AB. Find the equation of:
    a) the line BC;
    b) the line through A parallel to BC;
    c) the line through M parallel to BC;
    d) the line through M perpendicular to AB.

11. Find the equation of the line through (4, 1) that is
    a) parallel to the line $y = 3x + 2$;
    b) parallel to the line $y = -\frac{3}{4}x - 1$;
    c) perpendicular to the line $y = -\frac{3}{4}x - 1$;
    d) parallel to the line $2x + y - 4 = 0$;
    e) perpendicular to the line $5x - 3y + 12 = 0$;
    f) perpendicular to the line $3x + 8y = 7$.

12. A square has vertices (0, 4), (−6, 0), (−2, −6), and (4, −2). Draw the square, and find the equations of the lines containing
    a) the sides;
    b) the diagonals.

13. Find the equation of the perpendicular bisector of the segment joining (−3, 3) and (7, 7).

C 14. In the design on the cover of this book, the side lengths of the black and green squares are 1 and 2 respectively.
    a) Draw the design on a grid, as shown, starting with the black square with vertices (1, 1), (1, 2), (2, 2), and (2, 1).
    b) Find the equations of the following lines: AB, CD, EF, GH.
       What do they have in common?
    c) Find the equations of the following lines: AI, CJ, KL, MN. Show that they all pass through the same point. What are the coordinates of that point?

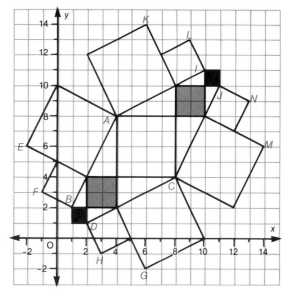

15. The equations of the lines containing the sides of a parallelogram are:
    $x + 2y = 8$ $x − 2y = 8$
    $x + 2y = 0$ $x − 2y = −8$.

    Find the equations of the lines containing the diagonals.

16. $A(−3, 1)$, $B(5, −1)$, and $C(9, 9)$ are the vertices of a triangle. Find the equation of the line containing
    a) the median from A;
    b) the altitude from C;
    c) the perpendicular from B onto side AC;
    d) the perpendicular bisector of AC.

17. $P(−6, −4)$, $Q(6, 2)$, $R(−2, 8)$ are the vertices of a triangle.
    a) Find the equations of the lines containing the sides.
    b) Find the equation of the perpendicular bisector of each side.
    c) Find the point of intersection, A, of the perpendicular bisectors, and the lengths of the segments AP, AQ, and AR.

18. $P(−1, 3)$, $Q(0, 6)$, $R(3, 5)$, $S(3, 0)$ are the vertices of a quadrilateral.
    a) Find the equations of the diagonals.
    b) Is PR the perpendicular bisector of QS?
    c) Is QS the perpendicular bisector of PR?
    d) Is $PQRS$: i) a rhombus?   ii) a parallelogram?

5-7 Direct Variation

If a ship travels at a steady speed of 50 km/h, then the distance, d, that is travelled is related to the elapsed time, t. The table and the graph show this relationship.

Distance travelled, d km	Elapsed time, t h
50	1
100	2
150	3
200	4

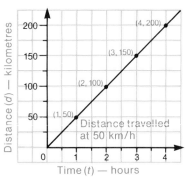

The table shows that when t is doubled, d is doubled. If t is tripled, d is tripled. We say that d **varies directly as** t. This is written:

$$d = kt, \text{ or } \frac{d}{t} = k.$$

where k is a number called the **constant of proportionality**.

From the table or graph above, it can be verified that, for all values of d and t,

$$d = 50t, \text{ or } \frac{d}{t} = 50.$$

The constant of proportionality is 50, and it represents the speed. It is also the slope of the line on the graph. The table, or the graph, represents ordered pairs which relate distance to time. The set of ordered pairs is called a **relation**.

> A relation is a set of ordered pairs.

Example 1. y varies directly as x. When $x = 10$, $y = 15$.
  a) Find the equation relating x and y.
  b) Find y when $x = 18$.
  c) Find x when $y = 60$.
  d) Draw a graph of the relation, and find its slope.

Solution. a) Since y varies directly as x, $y = kx$, where k is the constant of proportionality.
  Substitute 10 for x and 15 for y to find k:
  $$k = \frac{15}{10}, \text{ or } 1.5$$
  The equation relating x and y is $y = 1.5x$.

b) When $x = 18$: $y = 1.5(18)$, or 27

c) When $y = 60$: $60 = 1.5x$

$$x = \frac{60}{1.5}$$
$$= 40$$

d) The graph is drawn through the points representing the ordered pairs found in (b) and (c). It is a line through the origin with slope:

$$\frac{60 - 27}{40 - 18} = \frac{33}{22}, \text{ or } 1.5$$

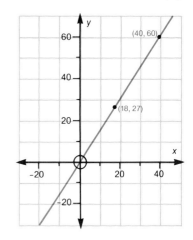

Example 2. The time interval between seeing a flash of lightning and hearing the thunderclap varies directly as your distance from the lightning. If the time interval is 15 s, you are 5 km from the lightning.

a) If the time interval is 2 s, how far away is the lightning?

b) If the lightning is 7 km away, how long will the sound of thunder take to reach you?

c) Graph the relation between the distance and the time interval.

Solution. a) Let d represent the distance to the lightning, in kilometres, and t represent the time interval between the flash and the sound, in seconds. Since d varies directly as t:

$$\frac{d}{t} = k, \text{ where } k \text{ is a constant.}$$

Substitute 5 for d and 15 for t, to find k:

$$k = \frac{5}{15}, \text{ or } \frac{1}{3}$$

The equation relating d and t is $d = \frac{1}{3}t$.

Substitute 2 for t: $d = \frac{1}{3}(2)$, or $\frac{2}{3}$

The lightning is about 0.67 km away.

b) Substitute 7 for d in the equation $d = \frac{1}{3}t$:

$$7 = \frac{1}{3}t$$
$$t = 21$$

It will be about 21 s before you hear the thunder.

c) The graph of the relation is a line through the origin with slope $\frac{1}{3}$.

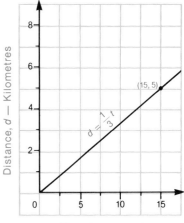

Exercises 5 - 7

A 1. In each of the following, y varies directly as x. Find the equation relating x and y, and the missing numbers.

a)
x	1	2	3	4	5
y			6		

b)
x	1	2	3	4	5
y		10			

c)
x	1	2	3	4	5
y				−12	

d)
x	1	2	3	4	5
y					1

2. y varies directly as x. When $x = 5$, $y = 20$.
   a) Find the equation relating x and y.
   b) Find y when $x = 12$.
   c) Find x when $y = 30$.
   d) Graph the relation and find its slope.

3. y varies directly as x. When $x = 12$, $y = 8$.
   a) Find the equation relating x and y.
   b) Find y when $x = 30$.
   c) Find x when $y = 15$.
   d) Graph the relation and find its slope.

4. At any given moment during daylight, the lengths of the shadows of trees vary directly as their heights. A tree 12 m tall casts a shadow 15 m long.
   a) Find the equation relating tree height to shadow length.
   b) How tall is a tree that casts a shadow 25 m long?
   c) Find the length of the shadow of a tree which is 9 m tall.
   d) Graph the relation between shadow length and height.

5. The depth of water in a bathtub varies directly as the length of time the taps are left on. If the taps are on for 2.5 min, the depth of water is 20 cm.
   a) Find the equation relating water depth to time.
   b) Find the depth of water if the taps are on for 4 min; for 1 min.
   c) Find the length of time the taps were left on if the depth of water is 15 cm.
   d) Graph the relation between water depth and time.

6. If y varies directly as x, how is y affected if
   a) x is doubled?
   b) x is halved?

5-7 Direct Variation

B 7. In each of the following, y varies directly as x. Find the equation relating x and y, and the missing numbers.

a)
x	3	6	9	12	15
y	4				

b)
x		10	18	8	
y			27		33	9

c)
x			28	42	70
y	5	10	20		

d)
x				85	95	105
y	2	10	34			

8. For the graph shown, find
   a) the equation relating x and y;
   b) the values of y when x = 12, 20, and 36;
   c) the values of x when y = 5, 12.5, 25;
   d) the slope of the line segment.

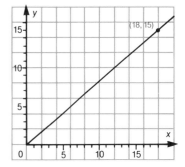

9. The amount of money an entrant in a "bike-a-thon" can collect varies directly as the distance covered. For the full distance of 32 km, the amount is $44.80.
   a) Find the equation relating the distance covered to the amount collected.
   b) How much would an entrant collect for covering 20 km?
   c) How far would an entrant have to ride to collect $40?
   d) Graph the relation between the amount collected and the distance covered.

10. The height to which a ball bounces varies directly as the distance through which it falls before hitting the ground. If it falls from a height of 4.0 m, it bounces to a height of 2.4 m.
    a) Find the height of bounce for a fall of 1.5 m.
    b) Find the distance of fall for a height of bounce of 3.0 m?

11. The number of tonnes of garbage collected in a city varies directly as the population. A city of 200 000 people generates 125 t of garbage per day. How many tonnes of garbage would you expect to be generated in a city with a population of
    a) 2 400 000?   b) 320 000?   c) 50 000?

C 12. For the diagram shown, find the equation relating x and y if the area of the inner circle is equal to the area of the shaded region.

13. A square with a side length of x is inscribed in a semicircle of radius r.
    a) Find the relation relating x and r.
    b) Find the ratio of the area of the square to the area of the semicircle.

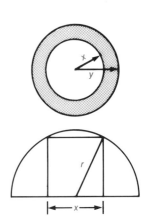

5-8 Partial Variation

The cost of publishing a school yearbook consists of two parts:

- **a fixed part**, which represents the cost of setting the type and making the printing plates. This cost is the same no matter how many copies of the yearbook are printed.
- **a variable part**, which represents the cost of the paper, ink, and the operation of the presses. This cost varies directly as the number of copies of the yearbook to be printed.

If the fixed cost is $1000 and the variable cost is $200 for every hundred copies, the table and graph below show how the total cost is related to the number of copies that are printed.

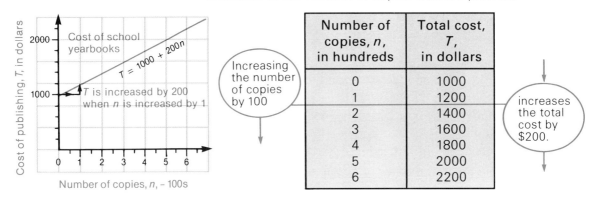

Number of copies, n, in hundreds	Total cost, T, in dollars
0	1000
1	1200
2	1400
3	1600
4	1800
5	2000
6	2200

Although the graph is a straight line, it does not represent direct variation. If the number of copies is doubled, the total cost is *not* doubled. However, from the given information we can write the equation which relates T and n.

$$T = 1000 + 200n$$

- 1000: Fixed cost
- 200n: Variable cost, $200 for every 100 yearbooks

T and n are in **partial variation**. In general, any two variables related by a linear equation with a non-zero constant term are in partial variation. Further, the graph showing the relation between the variables is a straight line which does not pass through (0, 0).

Example 1. A car's rate of fuel consumption averages 8.0 L/100 km.

a) If the fuel tank contains 60 L of gasoline to begin with, make a table of values relating n, the number of litres of gasoline left in the tank, to d, the distance travelled in hundreds of kilometres.

b) Draw a graph of the relation in (a).

c) Write the equation relating n and d.
d) When the car has travelled 280 km, about how much fuel is left in the tank?

Solution. a)

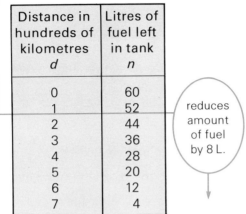

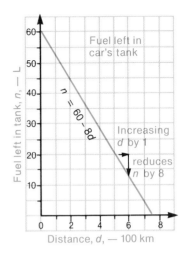

c) The equation relating n and d is:
$$n = 60 - 8d$$

d) Substitute 2.8 for d: $n = 60 - 8(2.8)$
$= 60 - 22.4$, or 37.6
There is about 37.6 L of fuel left in the tank after the car has travelled 280 km.

Example 2. The temperature of Earth's crust increases as the depth below the surface increases, as shown on the graph.
a) Find the equation relating the depth, d, in kilometres, to the temperature, T, in degrees Celsius.
b) What is the temperature at a depth of 3.2 km?
c) At what depth is the temperature 60°C?

Solution. a) The graph shows that the temperature at the surface ($d = 0$) is 20°C, and that the temperature increases 10°C for an increase in depth of 1 km. Therefore, the equation relating d and T is:
$$T = 10d + 20$$

b) Substitute 3.2 for d: $T = 10(3.2) + 20$
$= 52$
At a depth of 3.2 km, the temperature is about 52°C.

c) Substitute 60 for T: $60 = 10d + 20$
$40 = 10d$
$d = 4$
The temperature is 60°C at a depth of about 4 km.

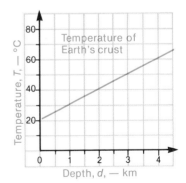

Exercises 5 - 8

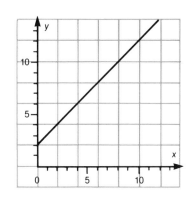

A 1. The relation between x and y is given by $y = 4x - 3$.
   a) Draw the graph of this relation.
   b) Find y when $x = 7$.
   c) Find x when $y = 45$.

2. The graph represents a partial variation between x and y.
   a) Construct a table of values.
   b) Find y when $x = 5$.
   c) Find x when $y = 12$.
   d) Find the equation relating x and y.

3. The graph represents the cost of printing election pamphlets.
   a) Construct a table of values for n and T, where n is the number of pamphlets, in hundreds, and T is the cost, in dollars.
   b) In this printing job, what is the fixed cost and what is the variable cost?
   c) What is the cost of printing 2500 pamphlets?
   d) How many pamphlets can be printed for $500?

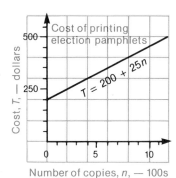

4. The fuel left (n litres) in the tank of a car after the car has been driven d hundreds of kilometres is given by the equation $n = 72 - 9d$.
   a) Construct a table of values for this relation.
   b) Graph the relation.
   c) How much fuel is left after 560 km?
   d) How far has the car been driven if 30 L of fuel remain?

B 5. Water rates in the Township of Woolwich consist of a service charge of $2.00 plus $0.56 for each 100 units used.
   a) Construct a table of values relating the water charge, C, to n, the number of units of water used.
   b) Draw the graph of this relation.
   c) Write the equation relating n and C.
   d) If the water bill is $6.70, how many units were used?

6. The cost, C, of photofinishing varies partially as the number of exposures, n, on a roll.
   a) Graph this relation.
   b) Find the equation relating n and C.

Photofinishing
12 exposures—$4.00
20 exposures—$6.00
24 exposures—$7.00
36 exposures—$10.00

5-8 Partial Variation **207**

7. In each of the following, x and y are in partial variation. Find the missing numbers.

   a)
x	1	2	3	4	5
y	5	8			

   b)
x	1	2	3	4	5
y			18	23	

   c)
x	2	4	6	8	10
y		2	6		

   d)
x	1	2	3	4	5
y			0	-2	

8. On a July day in Halifax, the sea level temperature is 20°C. The air temperature decreases 13°C for every 2000 m increase in altitude.
   a) Graph the relation between altitude and temperature.
   b) Find the equation that relates altitude and temperature.
   c) What is the temperature at 1500 m?
   d) At what altitude is the temperature 0°C?

9. The annual cost of owning a car depends partially on the distance driven. The cost is $2000 for 10 000 km, and $3800 for 25 000 km.
   a) Graph this relation.
   b) What is the fixed cost of owning a car?
   c) Determine the equation relating the total cost to the distance driven.
   d) Find the annual cost if 15 000 km are driven.

10. A set of rectangles are all 5 cm wide. Graph the relation between the perimeter and the length of these rectangles. Express the relation as an equation.

C 11. Weekly paid sales personnel at a furniture store have a choice of wage plans, as follows:

   Salary: $300 per week

   Salary + Commission: $100 + $3\frac{1}{3}$% commission on sales

   Commission: 5% commission on weekly sales

   a) How much is earned under each plan if sales are:
      i) $5000?   ii) $8000?
   b) Draw a graph showing the earnings under each plan for sales up to $10 000.
   c) For what sales are all three plans equivalent?

12. Christa wishes to rent a car for a weekend. Company A charges $60 plus 5¢/km, company B charges $30 plus 10¢/km, and company C charges 20¢/km.
    a) Graph each company's fee structure on the same grid.
    b) Under what circumstances should Christa rent the car from each company?
    c) Write equations for the relation between cost and distance.

13. For each of the diagrams below,
    a) find the equation relating x and y;
    b) draw a graph of the relation;
    c) describe the effect on the value of y as x varies.

    i) ii) iii)

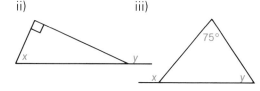

5-9 Inverse Variation

Police often catch speeders on a highway by measuring, from the air, the time it takes a car to travel over a marked portion of the road. The table and graph show how the speed v, in kilometres per hour, is related to the time t, in seconds, that a car takes to travel 0.5 km.

Time, t s	Speed, v km/h
20	90
40	45
60	30
80	22.5
100	18

The table shows that when t is multiplied by 2, v is divided by 2; when t is multiplied by 3, v is divided by 3; and so on. We say that v **varies inversely as** t, and write:

$$vt = k \quad \text{or} \quad v = \frac{k}{t}$$

where k is a constant.

From the table, it can be verified that, for all values of v and t,

$$vt = 1800 \quad \text{or} \quad v = \frac{1800}{t}.$$

5-9 Inverse Variation

Example 1. y varies inversely as x. When $x = 3$, $y = 6$.
  a) Find the equation relating x and y.
  b) Find y when $x = 9$.
  c) Find x when $y = 1$.
  d) Draw a graph of the relation between x and y.

Solution. a) Since y varies inversely as x, $xy = k$.
Substitute 3 for x and 6 for y to find k:
$$k = (6)(3), \text{ or } 18$$
The equation relating x and y is $xy = 18$.

b) When $x = 9$, $9y = 18$
$$y = 2$$

c) When $y = 1$, $x(1) = 18$
$$x = 18$$

d) Since the graph is not a straight line, several ordered pairs satisfying the equation $xy = 18$ must be plotted.

x	1	2	3	6	9	18
y	18	9	6	3	2	1

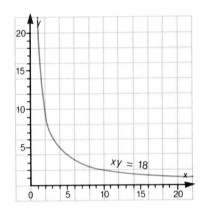

Example 2. After the first year, the value of a car varies inversely as its age.
  a) What is the value of a car at the end of 5 years if its value at the end of 2 years is $4000?
  b) Draw a graph of the relation between the value of the car and its age.

Solution. a) Let v represent the value of the car, in dollars.
Let a represent the age of the car, in years.
Since v varies inversely as a, $va = k$.
Substitute 4000 for v and 2 for a to find k:
$$k = (4000)(2), \text{ or } 8000$$
The equation relating v and a is $va = 8000$.
When $a = 5$, $v = \frac{8000}{5}$, or 1600.
At the end of 5 years, the car is worth $1600.

b) The graph is drawn from the following table of values.

a	1	2	3	4	5	6
v	8000	4000	2667	2000	1600	1333

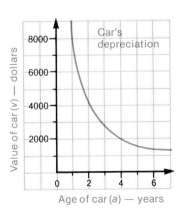

Car's depreciation

Exercises 5 - 9

A 1. If y varies inversely as x, find the equation relating x and y, and the missing numbers:

a)
x	2	3	4	6	8
y	12				

b)
x	10	12	15	20	30
y			4		

c)
x	4	8	12	16	24
y		6			

d)
x	44	33	22	12	11
y			6		

2. y varies inversely as x. When $x = 9$, $y = 4$.
   a) Find the equation relating x and y.
   b) Find: i) y when $x = 12$; ii) x when $y = 2$.
   c) Draw the graph of the relation.

3. y varies inversely as x. When $x = 7$, $y = 12$.
   a) Find the equation relating x and y.
   b) Find: i) y when $x = 6$; ii) x when $y = 4$.
   c) Draw the graph of the relation.

4. The cost per person of renting a van varies inversely as the number of people who rent it. When 4 people rent the van, the cost per person is $48.
   a) What is the cost per person if the van is rented by
      i) 2 people? ii) 6 people?
   b) How many people rent the van if the cost per person is
      i) $64.00? ii) $38.40?

5. The time required to empty a tank varies inversely as the rate of pumping. It takes 10 min to empty the tank at a rate of 4 L/min.
   a) How long does it take to empty the tank at a rate of 5 L/min?
   b) What is the rate of pumping if the tank is emptied in 2 min?

B 6. If y varies inversely as x, find the equation relating x and y and the missing numbers:

a)
x	6		9	36	
y		17	16		3

b)
x	6		9		15
y		24		18	12

c)
x			24	30	32
y	64	48			5

d)
x	15		9	7.5	2.5
y		22.5		30	

7. The time it takes to complete a bike-a-thon course varies inversely as the cyclist's average speed. At an average speed of 15 km/h, it takes 3.2 h to complete the course.
   a) How long does it take to complete the course at an average speed of: i) 12 km/h?   ii) 20 km/h?
   b) If the course is completed in 2 h, what is the average speed?
   c) Draw a graph of the relation between the time to complete the bike-a-thon course and the average speed.

8. The rotational speed of a gear varies inversely as the number of teeth on the gear. A gear with 24 teeth rotates at 40 r/min (revolutions per minute).
   a) What is the rotational speed of a gear with
      i) 32 teeth?   ii) 40 teeth?
   b) If a gear rotates at 60 r/min, how many teeth does it have?
   c) Draw a graph of the relation between rotational speed and number of teeth.

9. The number of hours required to construct a motion-picture set varies inversely with the number of workers if they work at the same rate. If the set can be constructed in 3 days by 20 workers, how many days would 12 workers require?

10. In a system of two pulleys turned by a single belt, the number of revolutions per minute of each pulley is inversely proportional to its radius. If one pulley has a radius of 50 cm and rotates at 120 r/min, what is the rotational speed of the other pulley if its radius is 80 cm?

11. If y varies inversely as x, how is y affected if
    a) x is increased by 25%?
    b) x is decreased by 20%

C 12. Each rectangle in a set of rectangles has an area of 360 cm².
    a) Graph the relation between the length and the width of the rectangles.
    b) Graph the relation between the length and the perimeter of the rectangles.
    c) Does the length vary inversely as:
       i) the width?   ii) the perimeter?   Why?

13. Twelve workers can do a job in 18 days.
    a) How many days will it take 18 workers to do the job?
    b) How many days will it take 9 workers to do the job?
    c) How many workers are needed to do the job in 6 days?

PROBLEM SOLVING STRATEGY

Reduce to a Simpler Problem

Sometimes a problem can be reduced to a simpler problem that can be more readily solved. Solving the simpler problem often suggests a solution to the original problem.

Example 1. At 08:00, when he was on his way to school, John noticed a man carrying a placard on which was written: "Beware—the world will end 23 997 hours from now." What time of day was the world supposed to end?

Solution. The problem would be much simpler if the placard had read: "Beware—the world will end 24 000 hours from now." As 24 000 is divisible by 24, the time of the predicted end of the world is 08:00. But 23 997 is 3 less than 24 000, so that the time of predicted end of the world is 05:00.

Many problems appear difficult until they are divided into two or more simpler problems.

Example 2. What fraction of the triangular region is shaded?

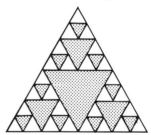

Solution. **Problem 1.** If the area of the triangular region is 1 unit^2, find the area of each of the three different sizes of shaded triangle.

Problem 2. Find the total number of each size of shaded triangle.

Problem 3. Find the total area of all the shaded triangles.

	Area	Number	Total area
Small △			
Medium △			
Large △			
Total area of all shaded triangles:			

Exercises

Solve each problem by reducing it to a simpler problem or dividing it into several simpler problems.

1. What number less than 2000 leaves a remainder of 1 when divided by 7, 11, and 13?

2. A hostess considered seating her guests in pairs but there was one guest left over. She considered groups of 3, 4, 5, and 6 but each time there was one guest left over. Finally, she seated them round small tables with 7 at each table and there were no guests left over nor seats empty. How many guests were at her party?

3. Complete the answer to *Example 2*.

4. A typist numbered the pages of the typescript of a book from 1 to 1250. How many digits did she type?

5. In a race, car *A* travels 3 laps of the track in the time it takes car *B* to travel 4 laps. Car *C* travels 6 laps in the time it takes car *D* to travel 9 laps. If car *C* travels four times as fast as car *A*, how many laps will car *B* have completed between consecutive simultaneous crossings of the timing line by all four cars?

5-10 Quadratic Relations

A **quadratic relation** is one that has at least one second degree term but no terms of a higher degree. The graph of a quadratic relation is usually a curve.

These are quadratic relations.
$$\begin{cases} \{(x, y) \mid x^2 + y^2 = 100\} \\ \{(x, y) \mid xy = 24\} \\ \{(t, d) \mid d = 4.9t^2\} \end{cases}$$

These are not quadratic relations.
$$\begin{cases} \{(x, y) \mid y = x^3 - 4x\} \\ \{(x, y) \mid y = 2^x\} \\ \{(x, y) \mid x^2y + xy^2 = 8\} \end{cases}$$

Example 1. The maximum distance a lighthouse beam can be seen depends on the height of the light above sea or lake level, and is given by the formula
$$d^2 = 13h$$
where d is the maximum distance of visibility in kilometres, and h is the height of the light, in metres, above sea or lake level.

a) Draw a graph to show the relationship between d and h for all possible heights up to 600 m.
b) Use the graph to find the maximum distance a light 250 m above sea level is visible at sea.

Solution. a) By giving values to h the corresponding values of d can be calculated and a table of values constructed. These points are plotted on a grid and joined by a smooth curve.

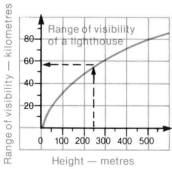

h m	0	100	200	300	400	500	600
d km	0	36	51	62	72	81	88

b) A light at a height of 250 m above sea level can be seen for about 57 km at sea.

The graph of *Example* 1 is part of a curve called a **parabola**.

Example 2. Draw the graph of H, where
$H = \{(x, y) \mid x^2 - y^2 = 9\}$.

Solution. To prepare a table of values, we first solve for y:
$$x^2 - y^2 = 9$$
$$x^2 - 9 = y^2$$
$$y = \pm\sqrt{x^2 - 9}$$

x	y
3	0
4	±2.65
5	±4.00
6	±5.20
7	±6.32
8	±7.42

x	y
−3	0
−4	±2.65
−5	±4.00
−6	±5.20
−7	±6.32
−8	±7.42

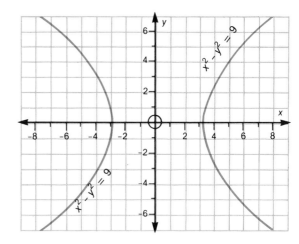

The graph of H is a **hyperbola**. This curve consists of two separate parts, or branches, as shown in the diagram. The inverse variation graphs which were drawn in the previous section showed one branch of a hyperbola.

Example 3. Draw the graph of E, where $E = \{(x, y) \mid 4x^2 + 9y^2 = 36\}$.

Solution. To prepare a table of values, we first solve for y:
$$4x^2 + 9y^2 = 36$$
$$9y^2 = 36 - 4x^2$$
$$y = \frac{\pm\sqrt{36 - 4x^2}}{3}$$

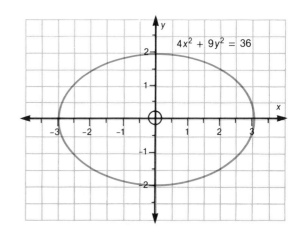

x	y
0	±2.00
0.5	±1.97
1.0	±1.89
1.5	±1.73
2.0	±1.49
2.5	±1.11
3.0	0

x	y
0	±2.00
−0.5	±1.97
−1.0	±1.89
−1.5	±1.73
−2.0	±1.49
−2.5	±1.11
−3.0	0

The graph of E is an **ellipse**.

Exercises 5 - 10

A 1. Graph the relation, and identify the curve:
a) $\{(x, y) \mid y = 2x^2\}$
b) $\{(x, y) \mid xy = 36\}$
c) $\{(x, y) \mid x^2 + y^2 = 25\}$
d) $\{(x, y) \mid 4x^2 + 6y^2 = 24\}$
e) $\{(x, y) \mid 4x^2 - 6y^2 = 24\}$

2. Graph the relation and identify the curve:
a) $\{(x, y) \mid x = 4y - y^2\}$
b) $\{(x, y) \mid x^2 - 4y^2 = 16\}$
c) $\{(x, y) \mid 7x^2 + 7y^2 = 49\}$
d) $\{(x, y) \mid xy = -60\}$
e) $\{(x, y) \mid 5x^2 + 2y^2 = 40\}$

3. In t seconds, and object falls a distance of h metres. The relation between t and h is: $t = 0.45\sqrt{h}$.
   a) Graph the relation. Is this a quadratic relation?
   b) If a pebble is dropped from a height of 196 m, how long does it take to reach the ground?
   c) If the pebble hits the ground after 2.7 s, from what height was it dropped?

4. The formula, $d = 3.6\sqrt{h}$, can be used to calculate the distance d, in kilometres, to the horizon from an altitude h, in metres.
   a) Draw a graph to show the relation between d and h for altitudes up to 10 000 m.
   b) How far away is the horizon seen from an aircraft at 7500 m?
   c) At what altitude would an an aircraft have to be for the horizon to be 300 km away?

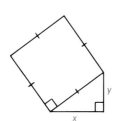

B 5. In the diagram, the area of the square on the hypotenuse of the right triangle is 65 units².
   a) Find the relation between x and y.
   b) Draw a graph of the relation.

6. A rectangle has a perimeter of 50 cm. If its length is x cm,
   a) find an expression in terms of x for:
      i) its width, w;   ii) its area, A;
   b) draw the graph of the relation between A and x.

7. A rectangle has an area of 36 cm². If its length is x cm,
   a) find an expression in terms of x for:
      i) its width, w;   ii) its perimeter, P;
   b) draw the graph of the relation between P and x.

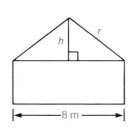

8. The diagram shows the cross section of a cottage.
   a) Find the relation between the height of the roof, h, and the length of a rafter, r.
   b) Draw the graph of the relation.

9. Graph the relation that gives the dimensions of all possible rectangles with a diagonal 4 units long.

C 10. The volume, V, in litres, of a sample of nitrous oxide gas is given by the formula $VP = 240$, where P is the pressure of the gas, in kilopascals.
   a) Graph the relation between P and V.
   b) Find the volume when the pressure is 7.5 kPa.
   c) Find the pressure when the volume is 9 L.

5-11 Other Non-Linear Relations

Many useful relations are neither linear nor quadratic.

Example 1. As a general rule, light does not penetrate much below 200 m into the ocean. The table shows the percent of surface light present at various depths.

Depth m	0	20	40	60	80	100
Percent of surface light present	100	63	40	25	16	10

a) Graph the relation between depth and percent of surface light present.

b) Use the graph to estimate:
  i) the percent of surface light present at a depth of 10 m;
  ii) the depth at which 50% of the surface light is present.

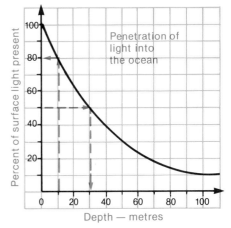

Solution. b) i) At a depth of 10 m, approximately 80% of the surface light is present.
  ii) 50% of the surface light is present at a depth of approximately 30 m.

If P is the percent of the surface light present, and d the depth, in metres, the equation for this relation is $P = 100e^{-0.023d}$, where e, like π, is an irrational number and is approximately equal to 2.718.

Example 2. Draw the graph of the relation:
$$\{(x, y) \mid y = \frac{5}{x^2 + 1}, -3 \leqslant x \leqslant 3\}$$

Solution. Use several values of x between -3 and $+3$ to construct a table.

x	y
0	5
±0.5	4
±1.0	2.5
±1.5	1.54
±2.0	1
±2.5	0.69
±3.0	0.5

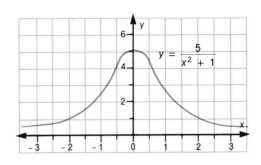

Example 3. Lynn and Lisa take a ride on a Ferris wheel. The graph shows how their height above the ground varies during the first 40 s of the ride.

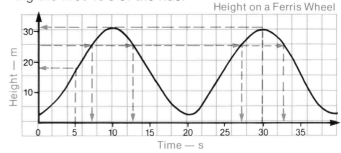

a) What is their height after 5 s? after 30 s?
b) When is their height 25 m?
c) How long does it take the Ferris wheel to make one rotation?

Solution. a) After 5 s, their height is about 17 m.
After 30 s, their height is about 32 m.

b) Their height is 25 m at about 7 s, 13 s, 27 s, and 33 s after the start.

c) Their height is 2 m at 0 s and 20 s. Therefore, it takes the Ferris wheel 20 s to make one rotation.

The up-and-down motion of the two girls is called **simple harmonic motion**. If h is their height in metres after t seconds, the equation of the relation is:

$$h = 15 \sin 2\pi \left(\frac{t-5}{20} \right) + 2$$

Exercises 5-11

A 1. Classify each of these graphs as linear, quadratic, or others:

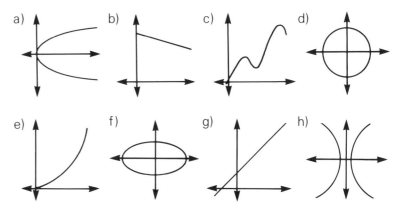

2. Graph the relation for $-3 < x < 3$:
   a) $\{(x, y) \mid y = x^3\}$
   b) $\{(x, y) \mid y = x^3 + 9x\}$
   c) $\left\{(x, y) \mid y = \dfrac{10}{x^2 + 5}\right\}$

B 3. If $1250 is invested at 12% per annum compound interest, the amount, A, to which it grows after n years, is given by: $A = 1250(1.12)^n$.
   a) Graph the relation.
   b) What is the amount, A, after 4 years?
   c) How long does it take for the investment to double in value?

4. When calculating drug dosages, doctors sometimes need to estimate the total area of a patient's skin. The graph shows how this area varies with a person's mass.
   a) What is the skin area of a person whose mass is
      i) 50 kg?   ii) 90 kg?
   b) About how heavy is a person who has a skin area of
      i) 1 m²?   ii) 2 m²?

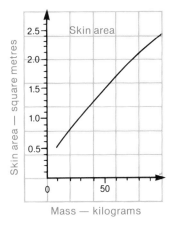

5. The graph shows how the height of a Skywheel passenger above the ground varies with the time elapsed since starting.

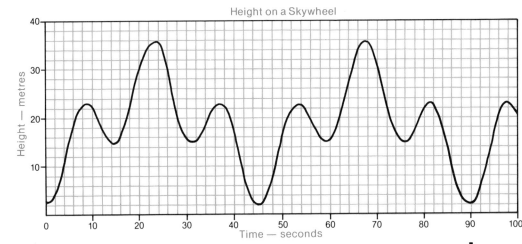

   a) Estimate a passenger's height after
      i) 10 s;   ii) 20 s;   iii) 30 s.
   b) When is the passenger's height
      i) 10 m?   ii) 20 m?   iii) 36 m?
   c) How long does it take for the ride to make one complete cycle?

6. If a ball is dropped from a height of 3 m and allowed to bounce freely, the height, h, in metres, to which it bounces is given by the formula $h = 3(0.9)^n$, where n is the number of bounces.
   a) Graph the relation for the first 8 bounces.
   b) What height does the ball reach after the third bounce? the tenth bounce?
   c) After what bounce does the ball reach a height of approximately 1.6 m?

7. Data relating to people's forgetfulness are given in the table.

Number of days	Percent forgotten
1	14
5	29
15	39
30	44
60	46

   a) Show this information on a graph.
   b) After how many days will a person have forgotten 25% of information previously known?
   c) How much are you likely to remember after the summer holidays?

C 8. a) State the relation, in terms of x and y, between all numbers that have a sum equal to their product.
   b) Draw a graph of the relation.
   c) Name five pairs of numbers that exhibit this property.

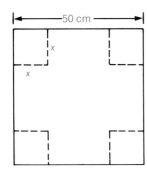

9. A square box, x cm deep, is to be made from a piece of cardboard, 50 cm square, by cutting equal squares from the corners and turning up the sides.
   a) Find an expression for the volume, V, of the box.
   b) Draw a graph of this relation for reasonable values of x.
   c) For what value of x will the volume be a maximum?

10. The cost per hour, C, in dollars, of operating a river steamer is given by the formula $C = \dfrac{100v^2}{v - 4}$, where v is the steamer's speed, in knots, relative to the water.
    a) Graph this relation for values of v from 4.5 to 10 knots.
    b) Find the speed of the steamer which results in the lowest value of C.

Review Exercises

1. Find the length of the line segment with endpoints:
   a) $(-6, 3), (9, 3)$;   b) $(4, 11), (4, -3)$;   c) $(3, 7), (-7, -3)$;
   d) $(-8, 9), (-3, 4)$;   e) $(5, 2), (1, -6)$;   f) $(-5, 3), (3, -5)$.

2. A rectangle has vertices $(-2, -4), (6, -4), (6, 2), (-2, 2)$. Find
   a) the lengths of the sides;
   b) the perimeter;
   c) the lengths of the diagonals;
   d) the area.

3. Find the coordinates of the midpoint of the line segment with endpoints:
   a) $(3, -8), (3, 12)$;
   b) $(5, -1), (-11, -1)$;
   c) $(2, 6), (-10, -12)$;
   d) $(5, -3), (13, -7)$;
   e) $(-2, 8), (5, 4)$;
   f) $(3, 7), (4, -2)$.

4. A triangle has vertices $K(3, 7), L(-5, 1), M(7, -5)$.
   a) Find the coordinates of the midpoint of each side.
   b) Find the length of the median from L to KM.

5. Find the slope of the line segment with endpoints:
   a) $(3, 8), (-1, -2)$;
   b) $(8, -6), (-3, -6)$;
   c) $(-3, 8), (4, -5)$;
   d) $(-7, 2), (5, 10)$;
   e) $(6, -3), (-8, 7)$;
   f) $(2, -6), (-3, 9)$.

6. A triangle has vertices $A(5, 7), B(-3, 4), C(2, -5)$.
   a) Find the slopes of AB and AC.
   b) Find the slope of a line through A parallel to BC.
   c) Find the slope of the median from B to AC.

7. On a grid, draw a line through
   a) $(4, 7)$ with slope $\frac{3}{5}$;
   b) $(3, -5)$ with slope $-\frac{4}{3}$.

8. What is the slope of a line perpendicular to a line with slope
   a) $-\frac{3}{5}$?
   b) 4?
   c) $-\frac{7}{4}$?
   d) 0.3?

9. A triangle has vertices $P(-4, -2), Q(6, 4), R(-7, 3)$. Show that $\angle PQR$ is a right angle.

10. State the slope and the y-intercept:
    a) $y = 4x - 3$
    b) $y = -\frac{5}{3}x + 7$
    c) $y = -\frac{9}{4}x - 3$
    d) $2x + 5y = 15$
    e) $3x - 4y = 16$
    f) $5x - 3y - 18 = 0$

11. Write the equation of the line that has
    a) slope $-\frac{1}{2}$, y-intercept -4;
    b) slope $\frac{4}{3}$, y-intercept -6.

12. Find the equation of the line that passes through
    a) (3, 5) and (−5, −3);   b) (−4, 7) and (5, −4);
    c) (2, 10) and (−2, −6);   d) (−8, 3) and (6, −5).

13. Find the equation of the line that passes through (5, 5) and
    a) is parallel to the line $3x + 4y = -16$;
    b) is perpendicular to the line $5x + 2y = 10$.

14. If y varies directly as x, find the equation relating x and y and the missing numbers:

    a)
x	3	5	9	12	4	20
y			36			

    b)
x	−3	3	6	12	15
y			−2		

15. y varies directly as x. When $x = 20$, $y = 15$.
    a) Find the equation relating x and y.
    b) Find y when $x = 30$.   c) Find x when $y = 60$.

16. If x and y are in partial variation, find the missing numbers:

    a)
x	4	6	8	10	14
y		11			27

    b)
x	−3	0	3	6	9
y	−4			14	

17. The relation between x and y is given by $y = 2x - 3$.
    a) Draw the graph of this relation.
    b) Find y when $x = 40$.   c) Find x when $y = 37$.

18. If y varies inversely as x, find the equation relating x and y, and the missing numbers:

    a)
x	4	6	12	18	24
y			6		

    b)
x	72		36	24	
y		4		9	12

19. The time required to empty a tank varies inversely as the rate of pumping. It takes 12 min to empty the tank at a rate of 5 L/min.
    a) How long does it take to empty the tank at a rate of 6 L/min?
    b) What is the pumping rate if the tank is emptied in 5 min?

Cumulative Review (Chapters 3 - 5)

1. Solve:
   a) $13x - 12 = 49x + 24$
   b) $15 - 2(3 + 2a) = 4 + 3(2a - 5)$

2. Solve:
   a) $3x - y = 5$
   b) $4x + 3y = 24$
   c) $5x + 2y = 20$

3. Solve by comparison:
   a) $2x - y = 1$
      $x + 2y = 8$
   b) $3x - y = -7$
      $2x + 3y = 10$
   c) $x - 4y = 11$
      $3x + y = -6$

4. Solve by substitution:
   a) $2x - y = 2$
      $2x + 3y = -2$
   b) $3x + y = 7$
      $x - 4y = 24$
   c) $x + 3y = 18$
      $4x - y = 7$

5. Solve:
   a) $3x - 5y = -3$
      $4x + 3y = 25$
   b) $4x - 7y = 15$
      $5x + 3y = 7$
   c) $6x - 3y = -5$
      $12x + 15y = 4$

6. The sum of two numbers is 39 and their difference is 3. Find the numbers.

7. The sum of three consecutive odd numbers is 273. Find the numbers.

8. The total receipts for 120 tickets were $280. If student tickets cost $2 each and adult tickets $3 each, how many of each type were sold?

9. The units' digit of a two-digit number is 1 more than twice the tens' digit. If the sum of the digits is 10, find the number.

10. Solve
    a) $x^2 - x - 30 = 0$
    b) $x^2 + 3x - 28 = 0$

11. Graph the inequality:
    a) $3x + 2y \leq 12$
    b) $2x + 7y \geq -14$

12. Solve by graphing:
    a) $2x - y = -4$
       $3x + 5y = 7$
    b) $x + 4y = 7$
       $2x - 5y = -12$
    c) $8x + 3y = -2$
       $-3x + y = 5$

13. Graph the solution set:
    a) $2x - y \geq -4$
       $x + 4y \leq 7$
    b) $-3x + y \leq 5$
       $2x - 5y \geq -12$
    c) $3x + 5y \leq 7$
       $8x + 3y \geq -2$

14. Find the length of the line segment with endpoints:
    a) (5, 8), (−3, 2);
    b) (−5, 7), (−5, −11);
    c) (4, 8), (−8, −4);
    d) (2, −3), (9, 4).

15. Find the coordinates of the midpoint of the line segment with endpoints:
    a) (5, −3), (9, −5);
    b) (−6, 3), (10, −7);
    c) (−1, 7), (4, −2);
    d) (2, −3), (9, 4).

16. Find the slope of the line segment with endpoints:
    a) (−3, 7), (7, −3);
    b) (2, 8), (−6, 4);
    c) (8, −2), (−1, 6);
    d) (−5, −5), (10, 4).

17. A triangle has vertices $A(3, 5)$, $B(1, −6)$, $C(−5, 4)$.
    a) Find the slope of AB.
    b) Find the coordinates of the midpoint of BC.
    c) Find the length of the median from A to BC.

18. Find the equation relating x and y, and the missing numbers:
    a)

y varies directly as x.				
x	5	10	40	80
y	−2		−8	

    b)

y varies inversely as x.				
x	2	3	4	6	12
y			27		

19. State the slope and y-intercept:
    a) $y = -2x + 7$
    b) $3x - 4y = 12$
    c) $5x + 2y = -6$
    d) $5 - 4x - 2y = 0$

20. Write the equation of the line that has
    a) slope $-\frac{2}{3}$, y-intercept 4;
    b) slope $\frac{6}{5}$, y-intercept -3.

21. Find the equation of the line that passes through
    a) (4, 8) and (−3, 6);
    b) (−5, 10) and (4, −6).

22. Find the equation of the line that passes through (−1, 2) and
    a) is parallel to the line $2x + 5y = 10$;
    b) is perpendicular to the line $3x - 5y = 15$.

23. The time required to empty a tank varies inversely as the rate of pumping. It takes 24 min to empty a tank at a rate of 10 L/min.
    a) How long does it take to empty the tank at a rate of 15 L/min?
    b) What is the pumping rate if the tank is emptied in 1 h?

6 Transformations

Ore from Rock Island, R, is to be shipped to the countries of Martinesia and New Hampton. Where should the unloading ports, M in Martinesia and N in New Hampton, be built so that the round trip, $R \to M \to N \to R$, is a minimum? (See *Example 2* in Section 6-8.)

6-1 Transformations as Mappings

Whenever the shape, size, or position of a figure is changed, it has undergone a **transformation**. Under a transformation, some of the characteristics of a figure may be changed while others remain the same. Those characteristics that are unchanged are said to be **invariant**. Some common transformations are illustrated below.

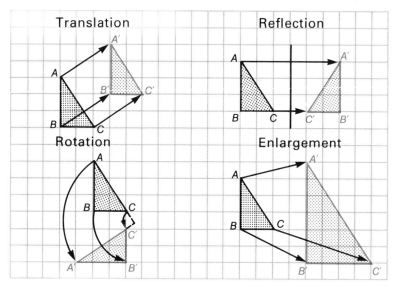

In each of the four diagrams there is a relation between corresponding points of the two triangles.

A maps onto A': $A \to A'$
B maps onto B': $B \to B'$
C maps onto C': $C \to C'$

A', B', and C' are the **images** of the points A, B, and C.

> A transformation is a relation that maps every point P onto an image point P'.

Transformations can be represented as mappings of ordered pairs. A **mapping rule** such as

$$(x, y) \to (x + 5, y - 2)$$

can be used to find the image of any point (x, y). This mapping rule is: first coordinates are increased by 5, second coordinates are decreased by 2.

Example 1. Draw the image of the trapezoid with vertices $A(-2, 2)$, $B(0, 2)$, $C(0, 6)$, $D(-2, 4)$, under the transformation: $(x, y) \to (x + 5, y - 2)$. Describe the transformation.

Solution. The table shows the coordinates of the images of the vertices of the trapezoid.

Point (x, y)	Image → (x + 5, y − 2)
A(−2, 2) →	A′(3, 0)
B(0, 2) →	B′(5, 0)
C(0, 6) →	C′(5, 4)
D(−2, 4) →	D′(3, 2)

From the graph, the transformation appears to be a translation 5 units to the right and 2 units down.

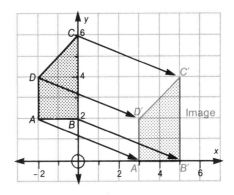

The next example shows how the coordinates of a point can be found given the coordinates of its image.

Example 2. Under the transformation $(x, y) \to (-x, -y)$, the image of $\triangle ABC$ has vertices $A'(-3, 2)$, $B'(3, 6)$, $C'(-5, 6)$. Draw $\triangle ABC$ and $\triangle A'B'C'$, and describe the transformation.

Solution. Let (x, y) be the coordinates of each of the vertices, in turn, of $\triangle ABC$. Then $(-x, -y)$ represent the coordinates of the corresponding vertices of $\triangle A'B'C'$.

For A′: $-x = -3$ and $-y = 2$
          $x = 3$ $y = -2$
For B′: $-x = 3$ and $-y = 6$
          $x = -3$ $y = -6$
For C′: $-x = -5$ and $-y = 6$
          $x = 5$ $y = -6$

$\triangle ABC$ has vertices $A(3, -2)$, $B(-3, -6)$, and $C(5, -6)$. From the graph, the transformation appears to be a $\frac{1}{2}$-turn about $(0, 0)$.

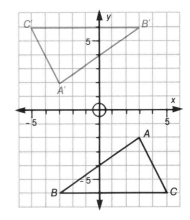

Exercises 6-1

A 1. Find the image of the point (5, 2) under the transformation:
     a) $(x, y) \to (x + 3, y - 2)$;   b) $(x, y) \to (-x, y)$;
     c) $(x, y) \to (2x, y)$;   d) $(x, y) \to (x - 2, 3y)$;
     e) $(x, y) \to (-x + 10, y)$;   f) $(x, y) \to (-3x, 1 - y)$;
     g) $(x, y) \to (2x + 1, 2y - 1)$;   h) $(x, y) \to (3x - 4, 5y + 1)$.

2. a) Draw the square that has vertices $A(-1, 1)$, $B(3, 1)$, $C(3, 5)$, $D(-1, 5)$.
   b) Draw its image under the transformation: $(x, y) \rightarrow (x + 5, y + 2)$.
   c) Describe the transformation.

3. a) Draw the parallelogram that has vertices $A(-2, 1)$, $B(4, 3)$, $C(5, 6)$, $D(-1, 4)$.
   b) Draw its image under the transformation: $(x, y) \rightarrow (x, -y)$.
   c) Describe the transformation.

4. A triangle has vertices $P(1, -1)$, $Q(7, 2)$, $R(4, 6)$.
   a) Draw its image under the transformation: $(x, y) \rightarrow (-x, -y)$.
   b) Describe the transformation.

5. Find the coordinates of the point that has the image point $(0, 6)$ under the transformation:
   a) $(x, y) \rightarrow (x, -y)$;
   b) $(x, y) \rightarrow (x - 2, y - 2)$;
   c) $(x, y) \rightarrow (x, -y + 12)$;
   d) $(x, y) \rightarrow (x, 2y)$;
   e) $(x, y) \rightarrow (2x, y - 1)$;
   f) $(x, y) \rightarrow (x + 4, 3y)$.

B 6. Find the image of the point $(2, 3)$ under the transformation:
   a) $(x, y) \rightarrow (x + 1, y - 4)$;
   b) $(x, y) \rightarrow (y, x)$;
   c) $(x, y) \rightarrow (2x, -y)$;
   d) $(x, y) \rightarrow (3 - x, 2y + 1)$;
   e) $(x, y) \rightarrow (y + 3, x - 3)$;
   f) $(x, y) \rightarrow (y - 1, -x + 5)$.

7. A quadrilateral has vertices $J(1, -2)$, $K(5, -2)$, $L(5, 3)$, $M(2, 1)$. Draw its image under the transformation: $(x, y) \rightarrow (-y + 2, x + 4)$.

8. A triangle has vertices $A(-2, -1)$, $B(4, -1)$, $C(5, 2)$. Draw its image under the transformation: $(x, y) \rightarrow (x + y, 2x - y)$.

9. Find the coordinates of the point that has the image point $(-4, 8)$ under the transformation:
   a) $(x, y) \rightarrow (x + 2, y - 3)$;
   b) $(x, y) \rightarrow (2x, 2y)$;
   c) $(x, y) \rightarrow (x, 4y)$;
   d) $(x, y) \rightarrow (y, x)$;
   e) $(x, y) \rightarrow (y, -x)$;
   f) $(x, y) \rightarrow (5 - x, y)$.

C 10. Draw the image figure $ABCDEO$ under each transformation and describe the transformation:
   a) $(x, y) \rightarrow (x, y + 5)$;
   b) $(x, y) \rightarrow (x - 5, y)$;
   c) $(x, y) \rightarrow (x, -y)$;
   d) $(x, y) \rightarrow (-x, -y)$;
   e) $(x, y) \rightarrow (-y, x)$;
   f) $(x, y) \rightarrow (3x, y)$.

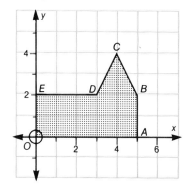

6-2 Translations

When points and their corresponding images under a given transformation are plotted on a grid, certain properties of the transformation often emerge. *Example 1* illustrates two properties of line segments joining points and their images under a **translation**.

Example 1. a) Plot the points $(-6, 3), (-3, 5), (-1, 6), (1, 0), (4, 4)$, and $(5, 1)$ and their images under the transformation $(x, y) \to (x + 3, y - 2)$ on a grid.

b) Join each point to its image with a line segment. What do you notice?

Solution. a)

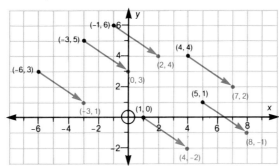

b) The transformation is a translation 3 units to the right and 2 units down. The line segments appear to be equal in length and parallel.

Example 2. A square has vertices $A(6, 1), B(8, 3), C(6, 5), D(4, 3)$.

a) On a grid, draw the image of the square under the transformation: $(x, y) \to (x - 5, y - 2)$.

b) Compare the square and its image with respect to
 i) side length;
 ii) slopes of sides;
 iii) measures of angles.

Solution. a) The image of square $ABCD$ is figure $A'B'C'D'$

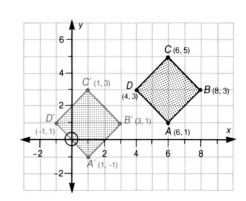

b) i) Length of BC:
$$\sqrt{(6 - 8)^2 + (5 - 3)^2}$$
$$= \sqrt{8}, \text{ or } 2\sqrt{2}$$
Length of $B'C'$:
$$\sqrt{(1 - 3)^2 + (3 - 1)^2}$$
$$= \sqrt{8}, \text{ or } 2\sqrt{2}$$
Therefore, line segment BC has the same length as its

image, $B'C'$. Similarly, the other sides of the square can be shown to have the same lengths as their images. Under this translation, length is invariant.

ii) Slope of BC: $\frac{5 - 3}{6 - 8} = -1$

Slope of $B'C'$: $\frac{3 - 1}{1 - 3} = -1$

Therefore, line segment BC is parallel to its image, $B'C'$. Similarly, the other sides of the square can be shown to be parallel to their images. Under this translation, slope is invariant.

iii) Slope of AB: $\frac{3 - 1}{8 - 6} = 1$

Slope of $A'B'$: $\frac{1 - (-1)}{3 - 1} = 1$

Slope of BC: -1 Slope of $B'C'$: -1

Therefore, both $\angle B$ and $\angle B'$ are right angles. Similarly, the other angles and their images can be shown to be right angles. Under this translation, angle measure is invariant.

Example 2 illustrates these properties of translations:

> A translation preserves:
> - length
> - direction, or slope
> - angle measure

Example 3. Graph the line $2y = 3x + 4$ and its image under the translation: $(x, y) \rightarrow (x, y - 3)$.

Solution. Two points on the line are $A(0, 2)$ and $B(-2, -1)$.
Under the translation: $A(0, 2) \rightarrow A'(0, -1)$
$B(-2, -1) \rightarrow B'(-2, -4)$
Using these coordinates, the line and its image can now be drawn.

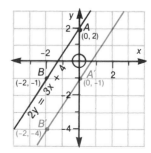

Exercises 6 - 2

A 1. Copy the diagram and draw the translation image:

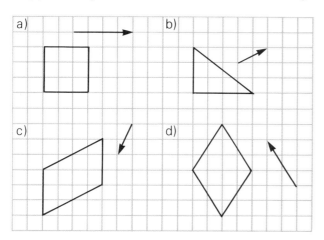

2. Find the missing points or mapping rule:

Mapping rule	Point	Image
$(x, y) \to (x - 2, y)$	$(5, -1)$	
$(x, y) \to (x, y + 3)$	$(-2, 1)$	
$(x, y) \to (x + 4, y)$		$(7, 2)$
$(x, y) \to (x - 3, y - 6)$		$(-2, -3)$
	$(0, 0)$	$(4, -1)$
$(x, y) \to (x + 5, y + 1)$	$(4, -3)$	
	$(-2, 1)$	$(3, -1)$
$(x, y) \to (x, y - 2)$		$(-3, 0)$
	$(4, -2)$	$(1, 1)$

3. Consider the translation: $(x, y) \to (x - 2, y + 5)$.
   a) Find the images of the points $(0, 0)$, $(3, 1)$, and $(2, -6)$.
   b) Find the points that have $(0, 3)$, $(1, 7)$, and $(-3, 1)$ as their images.
   c) Graph all points and their images from (a) and (b), and join matching points with line segments.
   d) Determine the length and slope of each segment.

4. In the diagram, which of the images indicated are the result of a translation of the shaded figure?

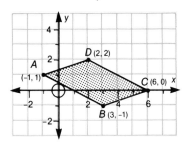

5. Copy the graph, and draw the image of parallelogram ABCD under each of the following translations:
   a) $(x, y) \rightarrow (x + 5, y + 2)$ b) $(x, y) \rightarrow (x - 8, y - 1)$
   c) $(x, y) \rightarrow (x + 2, y - 4)$ d) $(x, y) \rightarrow (x - 4, y + 2)$

6. A triangle has vertices $A(-4, 2)$, $B(2, 2)$, $C(-4, 5)$. Draw the graph of the triangle and its image under the transformation:
   a) $(x, y) \rightarrow (x + 5, y + 2)$ b) $(x, y) \rightarrow (x - 3, y - 4)$
   c) $(x, y) \rightarrow (x + 6, y - 3)$ d) $(x, y) \rightarrow (x - 3, y + 1)$

B 7. A triangle has vertices $P(-3, 0)$, $Q(5, 4)$, $R(2, -2)$.
   a) Graph the triangle and its image under the translation: $(x, y) \rightarrow (x + 8, y + 1)$.
   b) Compare the triangle and its image with respect to
      i) side length;   ii) slopes of sides;   iii) area.

8. A quadrilateral has vertices $J(1, 3)$, $K(3, -1)$, $L(6, -2)$, $M(6, 4)$.
   a) Graph the quadrilateral and its image under the translation: $(x, y) \rightarrow (x - 3, y - 2)$.
   b) Compare the quadrilateral and its image with respect to
      i) side length;   ii) slopes of sides;   iii) area.

9. A parallelogram has vertices $A(-4, 0)$, $B(1, -1)$, $C(7, 2)$, $D(2, 3)$.
   a) Graph the parallelogram and its image under the translation: $(x, y) \rightarrow (x + 2, y + 1)$.
   b) Compare the parallelogram and its image with respect to
      i) slopes of sides;   ii) side length.

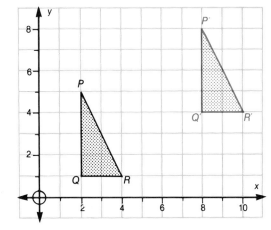

10. a) Copy the diagram and draw the line segments joining matching points PP', QQ', RR'.
    b) Compare these segments with respect to length and slope.
    c) Describe a property of translations that (b) illustrates.

11. Graph the line $y = 2x + 3$ and its image under the translation: $(x, y) \rightarrow (x, y + 2)$.

12. Graph the line $x + y = 4$ and its image under the translation: $(x, y) \rightarrow (x + 1, y + 2)$.

C 13. Find the equation of the image of the line $3x - 2y = 12$ under the translation: $(x, y) \rightarrow (x + 5, y + 2)$.

6-3 Rotations

In this section, the properties of a second kind of transformation, **rotations**, are investigated.

Example 1. A triangle has vertices $O(0, 0)$, $P(7, 1)$, $Q(6, -2)$.
  a) Graph the triangle and its image under the transformation: $(x, y) \rightarrow (-y, x)$.
  b) Compare the triangle and its image with respect to
      i) side length;
      ii) slopes of sides;
      iii) measures of angles.

Solution. a) Under this transformation, the coordinates of points are interchanged, and then the sign of the new first coordinate is reversed.

Point (x, y)	Image $(-y, x)$
$P(7, 1)$	$P'(-1, 7)$
$Q(6, -2)$	$Q'(2, 6)$
$O(0, 0)$	$O'(0, 0)$

The translation is a rotation with turn centre O, which is an invariant point.

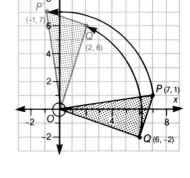

b) i) Length of PQ: $\sqrt{(7 - 6)^2 + (1 + 2)^2}$
$= \sqrt{10}$
Length of $P'Q'$: $\sqrt{(-1 - 2)^2 + (7 - 6)^2}$
$= \sqrt{10}$
Line segment PQ has the same length as its image, $P'Q'$. Similarly, the other sides of the triangle can be shown to have the same lengths as their images. Under this rotation, length is invariant.

ii) Slope of PQ: $\dfrac{1 - (-2)}{7 - 6} = 3$
Slope of $P'Q'$: $\dfrac{7 - 6}{-1 - 2}, = -\dfrac{1}{3}$
Therefore, the side PQ is perpendicular to its image, $P'Q'$. Similarly, the other sides can be shown to be perpendicular to their images. Direction, or slope, is not necessarily preserved under a rotation.

Line segments are said to be perpendicular to each other if the lines containing them are perpendicular.

iii) Since the sides of △POQ and the corresponding sides of △P'OQ' are equal in length, the triangles are congruent. Therefore, their respective angles are equal. Under this transformation, angle measure is invariant.

Example 1 illustrates these properties of rotations:

> A rotation preserves:
> - length
> - location of the turn centre
> - angle measure
>
> A rotation does not necessarily preserve direction, or slope.

Example 2. A triangle has vertices $A(1, -1)$, $B(8, 3)$, $C(5, 5)$. On the same grid, graph △ABC and its image under the transformations: $(x, y) \to (-x, -y)$ and $(x, y) \to (y, -x)$. Describe each transformation.

Solution.

Point (x, y)	Image (−x, −y)	Image (y, −x)
(1, −1)	(−1, 1)	(−1, −1)
(8, 3)	(−8, −3)	(3, −8)
(5, 5)	(−5, −5)	(5, −5)

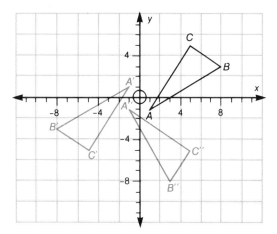

The first transformation represents a $\frac{1}{2}$-turn about $(0, 0)$, or a 180° rotation. The second transformation represents a $\frac{3}{4}$-turn about $(0, 0)$, or a 270° rotation.

All rotations are considered to be counterclockwise.

Exercises 6 - 3

A 1. Copy and complete this table.

Object	2						
After a $\frac{1}{4}$-turn	ᴎ		4		6		
After a $\frac{1}{2}$-turn	ᴢ	ε				8	
After a $\frac{3}{4}$-turn	ᴎ			5		7	
After a full turn	2						

2. In the diagram, which of the images indicated are the result of a rotation of the shaded figure?

3. Consider the rotation $(x, y) \to (-y, x)$.
   a) Find the images of the points $(4, 1)$, $(0, 5)$, and $(-3, 2)$.
   b) Find the points that have $(2, 6)$, $(3, 0)$, and $(-1, -4)$ as their images.
   c) Graph all points and their images from (a) and (b).

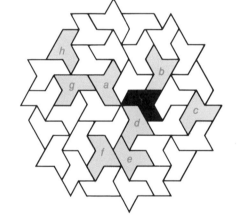

4. A triangle has vertices $A(4, 1)$, $B(3, 4)$, $O(0, 0)$. On the same grid, graph the triangle and its images under these rotations:
   a) $(x, y) \to (-y, x)$;
   b) $(x, y) \to (-x, -y)$;
   c) $(x, y) \to (y, -x)$.

5. A quadrilateral has vertices $P(-6, 5)$, $Q(-2, 1)$, $R(4, 1)$, $S(3, 5)$. On the same grid, graph the quadrilateral and its images under these rotations:
   a) $(x, y) \to (-y, x)$;
   b) $(x, y) \to (-x, -y)$;
   c) $(x, y) \to (y, -x)$.

B 6. Copy these diagrams and, with O as the turn centre, draw
   i) the $\frac{1}{4}$-turn image;   ii) the $\frac{1}{2}$-turn image.
   a)   b)   c)

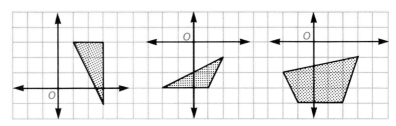

7. A triangle has vertices P(2, 5), Q(4, 5), R(4, 1).
   a) Graph the triangle and its image under the rotation:
      $(x, y) \rightarrow (-y, x)$.
   b) Compare the triangle and its image with respect to
      i) side length;   ii) slopes of sides;   iii) area.

8. A rectangle has vertices $A(-1, -2)$, $B(7, 2)$, $C(5, 6)$, $D(-3, 2)$.
   a) Graph the rectangle and its image under the rotation:
      $(x, y) \rightarrow (y, -x)$.
   b) Compare the rectangle and its image with respect to
      i) side length;   ii) slopes of sides;   iii) area.

9. A triangle has vertices J(5, 0), K(7, 0), and L(5, 3). Graph the triangle and its image under the transformation:
   $(x, y) \rightarrow (-y + 3, x - 3)$. Describe the transformation.

10. A trapezoid has vertices $A(-3, 1)$, $B(-2, -2)$, $C(1, -1)$, $D(3, 3)$.
    a) Graph the trapezoid and its image under the transformation:
       i) $(x, y) \rightarrow (-y + 6, x)$;
       ii) $(x, y) \rightarrow (-x + 6, -y + 6)$;
       iii) $(x, y) \rightarrow (y, -x + 6)$.
    b) Describe each transformation.

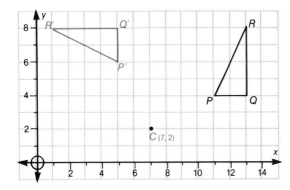

11. In the diagram, $\triangle P'Q'R'$ is the rotation image of $\triangle PQR$ under a $\frac{1}{4}$-turn about C(7, 2).
    a) Copy the diagram, and draw the segments PP', QQ', RR' and their perpendicular bisectors.
    b) Describe the property of rotations that is illustrated by (a).

12. Graph the line $y = x$ and its image under the rotation: $(x, y) \rightarrow (-y, x)$.

13. Graph the line $2x + y = 6$ and its image under the rotation: $(x, y) \to (-x, -y)$.

C 14. Find the equation of the image of the line $2x + 3y = 6$ under the rotation:
$(x, y) \to (-y, x)$.

15. A triangle has vertices $A(3, 6)$, $B(2, 1)$, $C(5, 4)$.
    a) Draw the triangle and its image under rotations about $(0, 0)$ of $90°$, $180°$, and $270°$.
    b) How are the coordinates of the vertices affected by these rotations?

16. Because Polaris, the north star, is virtually in the line of Earth's rotational axis, the other stars seem to revolve around it as Earth rotates. If the Plough (Big Dipper) was in position A at 10 p.m. relative to Polaris, and later the same night was in position B,
    a) through what angle had Earth rotated?
    b) what was the time when it was in position B?

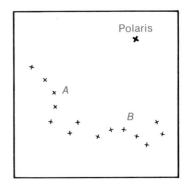

17. a) Draw the triangle with vertices $J(-1, 4)$, $K(5, -4)$, $L(7, -1)$ and its $\frac{1}{4}$-turn image about $(0, 0)$, $\triangle J'K'L'$.
    b) Draw the perpendicular bisectors of JJ', KK', and LL'. What point do these lines have in common?
    c) By extending the sides if necessary, measure the angles between JK and $J'K'$, JL and $J'L'$, KL and $K'L'$. What property do these angles have in common?

18. A line segment has endpoints $P(-2, 3)$, $Q(3, 1)$.
    a) Graph the segment and its image under the transformation:
       $(x, y) \to (-y - 2, x + 4)$.
    b) Find the coordinates of the turn centre.

19. A triangle has vertices $A(-4, 4)$, $B(-3, 1)$, $C(2, 4)$.
    a) Graph the triangle and its image under the transformation:
       $(x, y) \to (y + 1, -x + 3)$.
    b) Find the coordinates of the turn centre.

20. A square has vertices $A(2, 3)$, $B(6, 3)$, $C(6, 7)$, $D(2, 7)$.
    a) Graph the square and its image under the transformation:
       $(x, y) \to (-y + 1, x + 3)$.
    b) Find the coordinates of the turn centre.

6-4 Reflections

The properties of a third kind of transformation, **reflections**, are illustrated by the examples that follow.

Example 1. A triangle has vertices $A(3, 7)$, $B(-4, 3)$, $C(3, 3)$.
    a) Graph the triangle and its image under the transformation: $(x, y) \rightarrow (y, x)$.
    b) Compare the triangle and its image with respect to
        i) side length;
        ii) slopes of sides;
        iii) measures of angles.

Solution. a) Under this transformation, the coordinates of points are interchanged.

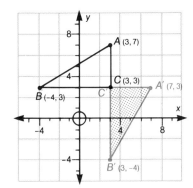

Point (x, y)	Image (y, x)
$A(3, 7)$	$A'(7, 3)$
$B(-4, 3)$	$B'(3, -4)$
$C(3, 3)$	$C'(3, 3)$

b)   i) Length of AB:
$$\sqrt{(7-3)^2 + (3+4)^2}$$
$$= \sqrt{65}$$
Length of $A'B'$:
$$\sqrt{(3+4)^2 + (7-3)^2}$$
$$= \sqrt{65}$$
Line segment AB has the same length as its image, $A'B'$. Similarly, the other sides of the triangle can be shown to have the same lengths as their images. Under this rotation, length is invariant.

  ii) Slope of AB: $\dfrac{7-3}{3-(-4)} = \dfrac{4}{7}$
Slope of $A'B'$: $\dfrac{3-(-4)}{7-3} = \dfrac{7}{4}$
The side AB is not parallel to its image. Direction, or slope, is not preserved under this reflection.

  iii) Since the sides of $\triangle ABC$ and the corresponding sides of $\triangle A'B'C'$ are equal in length, the triangles are congruent. Therefore, their respective angles are equal. Under this reflection angle measure is invariant.

6-4 Reflections

In the diagram of *Example 1*, to move from A to B to C is to move in a counterclockwise direction; to move from A' to B' to C' is to move in a clockwise direction. That is, a reflection reverses **orientation**. This reversal does not occur with translations or rotations.

The transformation $(x, y) \rightarrow (y, x)$. is a reflection in the line $y = x$. This line is called the **reflection line**.

Example 1 illustrates these properties of reflections:

> A reflection preserves:
> - length
> - angle measure
> - location of all points on the reflection line
>
> A reflection does not preserve:
> - direction, or slope
> - orientation

In the next example, the reflection line is other than the line $y = x$.

Example 2. A rectangle has vertices $A(4, 1)$, $B(6, 5)$, $C(4, 6)$, $D(2, 2)$.

    a) Graph the rectangle and its image under the transformation: $(x, y) \rightarrow (-x, y)$.

    b) Describe the transformation.

Solution. a)

Point (x, y)	Image (−x, y)
A(4, 1)	A'(−4, 1)
B(6, 5)	B'(−6, 5)
C(4, 6)	C'(−4, 6)
D(2, 2)	D'(−2, 2)

    b) The transformation is a reflection in the y-axis.

In *Example 2*, what would be the mapping rule for a reflection in the x-axis?

Exercises 6 - 4

A 1. Copy the diagram and draw the reflection image:

    a)   b)   c)

2. Copy the diagram and draw the reflection image:

a) b) c)

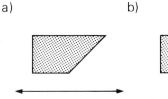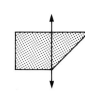

d) e) f)

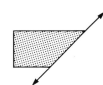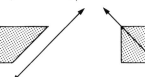

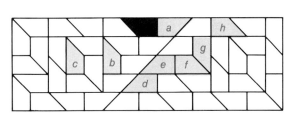

3. In the diagram, which of the images indicated are a result of a reflection of the shaded figure?

4. Consider the reflection: $(x, y) \rightarrow (-x, y)$.
   a) Find the images of the points $(1, 3)$, $(4, -2)$, and $(-1, 5)$.
   b) Find the points that have $(2, -3)$, $(-1, 2)$, and $(3, 0)$ as their images.
   c) Graph all points and their images from (a) and (b).

5. A triangle has vertices $A(3, 1)$, $B(7, 1)$, $C(7, 3)$. Graph the triangle and its image under the reflection:
   a) $(x, y) \rightarrow (-x, y)$;
   b) $(x, y) \rightarrow (x, -y)$;
   c) $(x, y) \rightarrow (y, x)$;
   d) $(x, y) \rightarrow (-y, -x)$.

6. Draw the image of $\triangle PQR$, shown on the graph, under the reflection:
   a) $(x, y) \rightarrow (y, x)$;
   b) $(x, y) \rightarrow (-y, -x)$;
   c) $(x, y) \rightarrow (-x, y)$;
   d) $(x, y) \rightarrow (x, -y)$.

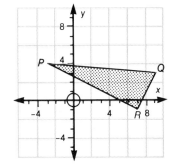

7. A triangle has vertices $J(2, 2)$, $K(10, 6)$, $L(0, 6)$.
   a) Graph the triangle and its image under the reflection: $(x, y) \rightarrow (x, -y)$.
   b) Compare the triangle and its image with respect to
      i) side length; ii) slopes of sides; iii) area.

B 8. A square has vertices $A(0, 2)$, $B(-5, 0)$, $C(-3, -5)$, $D(2, -3)$.
   a) Graph the square and its image under the reflection: $(x, y) \rightarrow (-y, -x)$.

b) Compare the square and its image with respect to
   i) side length;   ii) slopes of sides;   iii) area.

9. A triangle has vertices $P(2, 4)$, $Q(-1, 3)$, and $R(1, -1)$. Graph the triangle and its images under the transformation:
$(x, y) \rightarrow (-x + 6, y)$.

10. A parallelogram has vertices $E(-2, 1)$, $F(-3, -2)$, $G(3, 0)$, $H(4, 3)$. Graph the parallelogram and its image under the transformation: $(x, y) \rightarrow (y + 3, x - 3)$.

11. A quadrilateral has vertices $P(2, 3)$, $Q(4, -1)$, $R(8, 1)$, $S(7, 4)$. Graph the quadrilateral and its image under the transformation: $(x, y) \rightarrow (-y - 1, -x - 1)$.

12. The vertices of a triangle are $A(1, 4)$, $B(3, -2)$, $C(5, 1)$.
    a) Draw the image of $\triangle ABC$ under a reflection in
       i) the line $x + 2 = 0$. Call it $\triangle A'B'C'$.
       ii) the line $y + 3 = 0$. Call it $\triangle A''B''C''$.
    b) Draw the image of $\triangle A'B'C'$ in the line $y + 3 = 0$.
    c) Draw the image of $\triangle A''B''C''$ in the line $x + 2 = 0$.

13. In the diagram, $\triangle P'Q'R'$ is the reflection image of $\triangle PQR$ in the line $y = -2x + 17$.
    a) Copy the diagram and draw the segments joining matching points PP' and QQ'.
    b) Show that the reflection line
       i) is perpendicular to both PP' and QQ';
       ii) bisects both PP' and QQ'.
    c) Describe the property of reflections that is illustrated in (b).

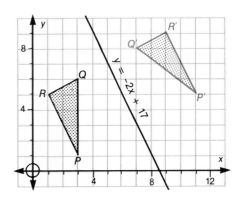

14. In the diagram, $\triangle A'B'C'$ is the reflection image of $\triangle ABC$.

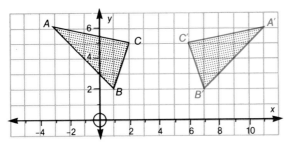

a) Copy the diagram and locate the reflection line.
b) Extend AB and $A'B'$ to meet at P, AC and $A'C'$ to meet at Q, BC and $B'C'$ to meet at R,
c) What property of reflections is illustrated in (b)?

15. Copy the diagrams. Use the property of reflections illustrated in Exercise 14 to locate the reflection line for each object and its image.

a) b)

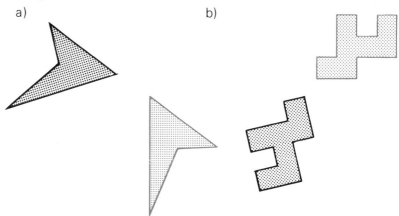

16. Graph the line $y = 2x + 1$ and its image under the reflection: $(x, y) \rightarrow (-x, y)$.

17. Graph the line $x + y = 3$ and its image under the reflection: $(x, y) \rightarrow (x, -y)$.

C 18. Find the equation of the image of the line $2x + 3y = 6$ under the reflection: $(x, y) \rightarrow (y, x)$.

19. A quadrilateral has vertices $A(1, -1)$, $B(2, -5)$, $C(5, -5)$, $D(6, -3)$.
    a) Graph the quadrilateral and its image under the tranformation: $(x, y) \rightarrow (x + 9, -y)$.
    b) Show that the transformation is neither a translation, nor a reflection, nor a rotation.
    c) What properties does this new transformation possess?

20. In the diagram, $\triangle A'B'C'$ is a reflection image of $\triangle ABC$.
    a) Copy the diagram and draw the reflection line.
    b) Find the equation of the reflection line.
    c) Find the image of $P(-3, -2)$ under this reflection.

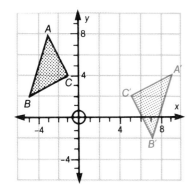

21. A reflection maps the point $(-3, -1)$, onto $(3, 5)$.
    a) Draw the reflection line and find the images of the points $(1, 5)$, $(2, -1)$, and $(4, -2)$ under the same reflection.
    b) Find the points that have $(3, 2)$, $(-4, 6)$, and $(-1, -3)$ as images under the same reflection.
    c) Write the equation of the reflection line.

6-5 Successive Transformations

Egg

Larva

Pupa

Adult Monarch Butterfly

Each of the successive transformations in the life cycle of a butterfly is necessary. Successive transformations can also occur in mathematics. However, unlike transformations in nature, when the final image is compared with the original object, an equivalent single transformation can be found.

Combining Two Reflections in Parallel Lines

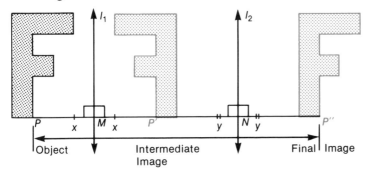

The diagram shows the results of two successive reflections in the parallel lines l_1 and l_2. By comparing the final image with the object, we see that the successive reflections are equivalent to a single translation.

By the properties of a reflection: $PM = P'M$, and $P'N = P''N$
Let $PM = x$ and $P'N = y$.
The length of the translation arrow is:
$$PP'' = PM + MP' + P'N + NP''$$
$$= x + x + y + y$$
$$= 2(x + y)$$
$$= 2MN$$
But MN is the distance between the reflection lines.

> The result of two successive reflections in parallel lines is a translation in a direction perpendicular to both lines. The length of the translation arrow is twice the distance between the reflection lines.

Combining Two Reflections in Non-Parallel Lines

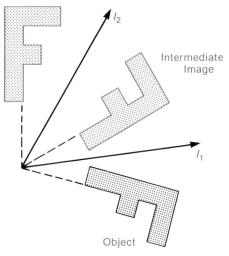

The diagram shows the result of two successive reflections in the non-parallel lines l_1 and l_2. By comparing the final image with the object, we see that the successive reflections are equivalent to a rotation. Furthermore, the angle of rotation is twice the angle between the reflection lines.

> The result of two successive reflections in non-parallel lines is a rotation. The turn centre is the point of intersection of the reflection lines, and the angle of rotation is twice the angle between the reflection lines.

Combining a Translation and a Reflection

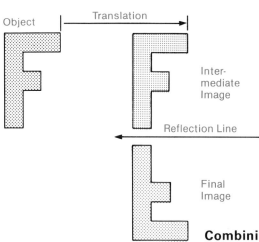

The diagram shows the result of a translation and a reflection in a line parallel to the direction of the translation. This kind of transformation is called a **glide reflection**.

> The result of a translation and a reflection in a line parallel to the direction of the translation is a glide reflection.

Since lengths, angles, and areas are invariant under both a translation and a reflection, they must also be invariant under a glide reflection.

Combining Transformations to Form Patterns

Successive transformations of a basic figure can result in interesting patterns. When all the images cover the entire plane without gaps or overlapping, the pattern is called a *tessellation*.

The examples of successive transformations in this section suggest that every translation and every rotation can be represented by two combined reflections. Consequently, any of the basic figures in a tessellation can be mapped onto any other congruent figure by two or more combined reflections in suitably chosen reflection lines.

Exercises 6 - 5

B 1. A triangle has vertices $A(-3, 7)$, $B(2, 4)$, $C(4, 7)$.
   a) Graph $\triangle ABC$ and its image, $\triangle A'B'C'$, under a reflection in the line $y = 2$.
   b) Draw the image of $\triangle A'B'C'$ under a reflection in the line $y = -3$.
   c) Describe the result of these successive reflections.

2. A triangle has vertices $P(3, 1)$, $Q(7, 1)$, $R(7, 3)$.
   a) Graph $\triangle PQR$ and its image, $\triangle P'Q'R'$, under the reflection: $(x, y) \to (y, x)$.
   b) On the same grid, graph the image of $\triangle P'Q'R'$ under the reflection: $(x, y) \to (-x, y)$.
   c) Describe the result of these successive reflections.

3. A triangle has vertices $A(-4, 1)$, $B(-2, 2)$, $C(-4, 6)$.
   a) Graph $\triangle ABC$ and its image, $\triangle A'B'C'$, under a reflection in the line: $y = x + 2$.
   b) Graph the image of $\triangle A'B'C'$ under a reflection in the line: $y = x - 8$.
   c) Describe the result of these successive reflections.

4. A triangle has vertices $K(1, 4)$, $L(1, 6)$, $M(-3, 6)$.
   a) Graph $\triangle KLM$ and its image, $\triangle K'L'M'$, under a reflection in the line: $y = x + 2$.
   b) Graph the image of $\triangle K'L'M'$ under a reflection in the line: $y = -x$.
   c) Describe the result of these successive reflections.

5. A quadrilateral has vertices $P(2, 3)$, $Q(4, 1)$, $R(6, 1)$, $S(6, 4)$.
   a) Graph quadrilateral $PQRS$ and its image $P'Q'R'S'$ under the translation: $(x, y) \to (x - 8, y)$.
   b) Graph the image of quadrilateral $P'Q'R'S'$ under a reflection in the x-axis.
   c) Describe the result of these successive transformations.

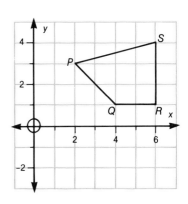

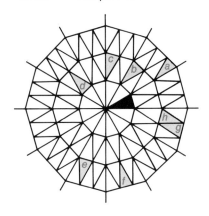

6. In the diagram, the images indicated are a result of successive transformations of the shaded figure. Name the transformations and say which are glide reflections.

7. Complete the following:

   a) If 2 becomes ᑌ , then P becomes ? .

   b) If 🌳 becomes 🌳(flipped) , then ⊟ becomes ? .

   c) If 6 becomes e , then 3 becomes ? .

   d) If 8 becomes ထ , then N becomes ? .

8. Identify the transformation, or series of transformations, that produce the following:

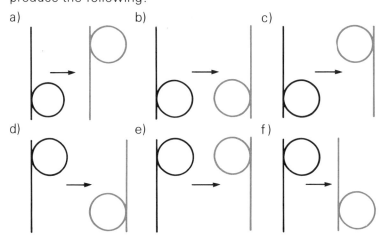

9. State how transformations are used to make each pattern from the colored basic figure:

   a)   b)   c)   d)

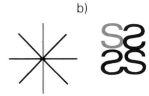

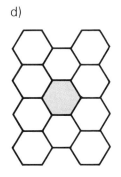

6-5 *Successive Transformations* **247**

e)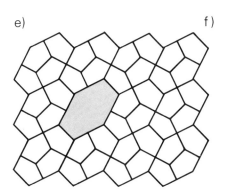

f)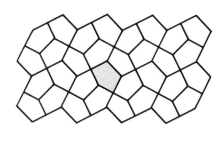

10. i) Copy each diagram and draw the images of the figure under successive reflections first in l_1 and then in l_2.
    ii) Describe a single equivalent transformation.
    iii) Repeat parts (i) and (ii) but reflect first in l_2 and then in l_1.
    iv) Does it matter which reflection line is used first?

    a)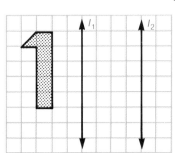
    b)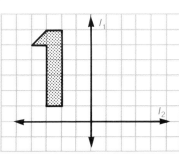
    c)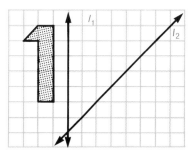

11. i) Copy each diagram and draw the images of the figure under successive reflections first in l_1 and then in l_2.
    ii) Describe a single equivalent transformation.
    iii) Compare the results with those of Exercise 10a. Does it matter if the figure being transformed lies on one of, or between, the reflection lines?

    a)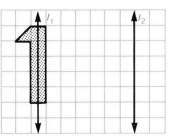
    b)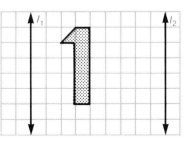
    c)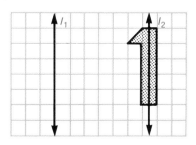

12. i) Copy each diagram and draw the images of the figure under successive rotations through the angles shown, first about P_1 and then about P_2.
    ii) In each case, describe a single equivalent transformation.

    a)   b)

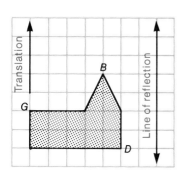

13. a) Copy the diagram and draw the image, $G'B'D'$, of the figure under a glide reflection consisting of a translation followed by a reflection.
    b) Find the midpoint of each line segment joining the matching points BB', GG', DD'.
    c) What property do the line segments joining the matching points have?
    d) Would it make any difference if the reflection were done first?

C 14. A triangle has vertices $P(2, -5)$, $Q(10, -9)$, $R(11, -2)$.
    a) Graph $\triangle PQR$ and its image, $\triangle P'Q'R'$, under a reflection in the line $x + 2y = 2$.
    b) The image of $\triangle P'Q'R'$ in a second line of reflection is a triangle with vertices $P''(2, 5)$, $Q''(-6, 9)$, $R''(-7, 2)$. Find the equation of this line of reflection.
    c) Find the point of intersection of the two lines of reflection.
    d) Describe the single transformation that maps $\triangle PQR$ onto $\triangle P''Q''R''$.

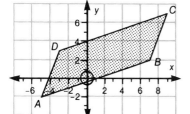

15. a) Copy the graph and draw the images of the quadrilateral under rotations about $(0, 0)$ of $90°$, $180°$, and $270°$.
    b) On another copy of the graph, draw the images of the quadrilateral under successive reflections first in the y-axis, then in the x-axis, and then again in the y-axis.
    c) List as many properties of the resulting figures in (a) and (b) as you can.

6-6 Isometries and Congruence

The four transformations discussed in this chapter—translations, rotations, reflections, and glide reflections—and combinations of them, all preserve length. That is, matching segments of the object and the image have the same length. For this reason, each is called an isometry, a word derived from two Greek words: *isos* (same) and *metria* (measure). Any transformation that preserves length is called an **isometry**.

Translations, rotations, reflections, and glide reflections are the only isometries in the plane. Their properties, which have been verified in the examples and exercises of previous sections, are summarized in the following table.

	Properties of the Four Isometries			
	Translation	Rotation	Reflection	Glide Reflection
Length	invariant	invariant	invariant	invariant
Angle measure	invariant	invariant	invariant	invariant
Area measure	invariant	invariant	invariant	invariant
Orientation	invariant	invariant	reversed	reversed
Invariant points	none	centre of rotation	all points on the reflection line	none
Segments joining matching points	equal and parallel	turn centre lies on perpendicular bisector of segments	reflection line is perpendicular bisector of all segments	midpoints lie on reflection line
Perpendicular bisectors of line segments joining matching points	perpendicular to translation arrow	pass through the centre of rotation	coincide with the reflection line	have no particular pattern

Figures that have the same shape and size are said to be **congruent**. True congruence rarely occurs in the physical world, but many things are so nearly identical that they may be considered to be so.

> Two figures are said to be congruent if there is some isometry that maps one onto the other.

If two figures are congruent, one is the image of the other under one of the four isometries. Translations and reflections are usually easy to see, but rotations and glide reflections are not. The following tests are based on the perpendicular-bisector properties presented in the preceding table. They can be used to determine which isometry maps a given shape onto a congruent image.

Tests for Isometries	
Test for Translations Perpendicular bisectors are all parallel.	**Test for Reflections** Perpendicular bisectors are all the same line.
Test for Rotations Perpendicular bisectors pass through the turn centre.	**Test for Glide Reflections** Perpendicular bisectors have no particular pattern.
Figures related by a translation or a rotation are called *directly congruent* because their orientation is the same.	Figures related by a reflection or a glide reflection are called *oppositely congruent* because their orientation is reversed.

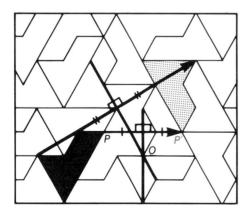

Example. In the tessellation shown, describe the isometry that maps the shaded figure onto its colored image.

Solution. Since line segments joining matching points are not parallel, the isometry cannot be a translation. Since the orientation is not reversed, it cannot be a reflection or a glide reflection. Therefore, the isometry must be a rotation. To find the turn centre, O, construct the perpendicular bisectors of two line segments joining matching points. To find the angle of rotation,

6 - 6 *Isometries and Congruence* **251**

join any point, *P*, and its image, *P'*, to the turn centre. Measured with a protractor, the angle is found to be 120°.

Exercises 6 - 6

A 1. State the isometry that maps each polygon onto its image:

a)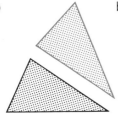
b)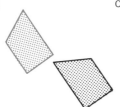
c)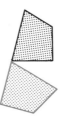

d)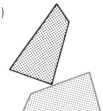
e)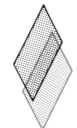
f)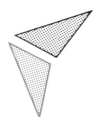

2. Identify the isometry that maps:
   a) ① → ②   b) ① → ⑤
   c) ⑥ → ③   d) ⑧ → ③
   e) ① → ③   f) ⑦ → ⑥
   g) ⑧ → ⑥   h) ⑦ → ⑧
   i) ⑤ → ③   j) ① → ⑥

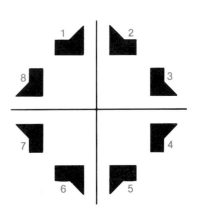

3. Identify the isometry that maps:
   i)   Figure I → Figure II
   ii) Figure II → Figure III
   iii) Figure I → Figure III

   a)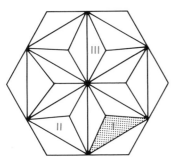
   b)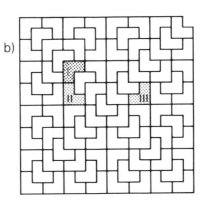

4. i) Identify the pairs of figures that are directly congruent.
   ii) Name the isometry that maps the object onto the image for each pair of congruent figures.

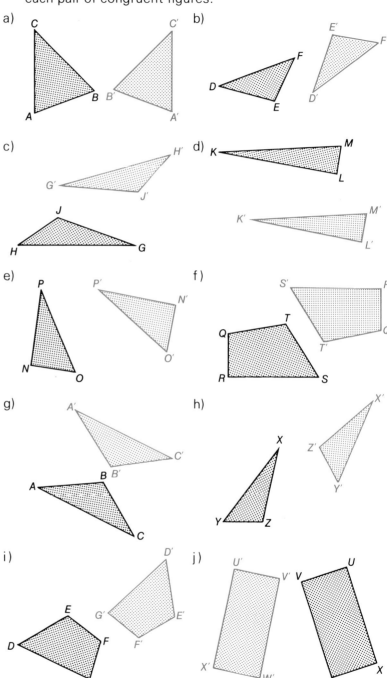

6-6 *Isometries and Congruence*   **253**

5. Copy each figure and divide it into two congruent parts. State the isometry that maps one part onto the other.

a)   b)   c)   d)   e)

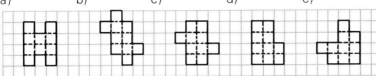

6. By drawing two straight lines through a square, divide it into four congruent:
   a) squares;   b) triangles;   c) quadrilaterals.

B 7. A puzzle consists of five pairs of congruent shapes which, when assembled, fit into a rectangular tray. Two solutions are shown. Copy each solution on graph paper and
   a) identify an isometry which maps each shape to its matching shape;
   b) explain why there are different answers in (a).
   Cut the pieces out of cardboard and try the puzzle yourself. There are many solutions. Try to find one having half-turn symmetry.

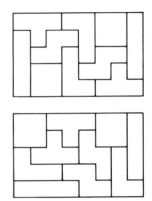

8. The diagram shows a parallelogram divided into 32 congruent pentagons.
   a) Identify the isometry that maps
      i) Figure I → Figure II;
      ii) Figure I → Figure III;
      iii) Figure II → Figure III.
   b) How many pentagons are
      i) directly congruent to Figure I?
      ii) oppositely congruent to Figure I?

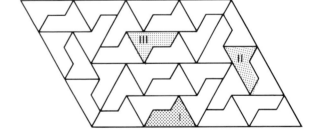

9. i) Divide a tracing of each figure into the number of congruent shapes indicated.
   ii) What isometries map the congruent parts onto each other?

a)   b)   c)

2 congruent parts
3 congruent parts
4 congruent parts

2 congruent parts

2 congruent parts
4 congruent parts

6-7 Dilatations

The transformations up to now have all been isometries. The simplest transformation that does not preserve length is a dilatation. A **dilatation** is a transformation that enlarges or reduces all dimensions of a figure by a factor k, called the *scale factor*.

$k < 1$

$k = 1$ $k > 1$

Example   A triangle has vertices $A(1, 3)$, $B(5, 5)$, $C(6, 3)$.

a) Graph $\triangle ABC$ and its image under the transformation: $(x, y) \rightarrow (2x, 2y)$.

b) Compare the triangle and its image with respect to
    i) side length;   ii) slopes of sides;
    iii) measures of angles;   iv) area.

c) Investigate the line segments joining matching points.

Solution.   b)   i) Length of AB:

$$\sqrt{(5 - 1)^2 + (5 - 3)^2}$$
$$= \sqrt{16 + 4}$$
$$= \sqrt{20}, \text{ or } 2\sqrt{5}$$

Length of $A'B'$:
$$\sqrt{(10 - 2)^2 + (10 - 6)^2}$$
$$= \sqrt{64 + 16}$$
$$= \sqrt{80}, \text{ or } 4\sqrt{5}$$

Clearly, $A'B' = 2AB$
Similarly, $B'C' = 2BC$
    and $A'C' = 2AC$

ii) Slope of AB: $\frac{5 - 3}{5 - 1} = \frac{1}{2}$

    Slope of $A'B'$: $\frac{10 - 6}{10 - 2} = \frac{1}{2}$

Therefore, $A'B' \parallel AB$

Similarly, $B'C' \parallel BC$
    and $A'C' \parallel AC$

a)

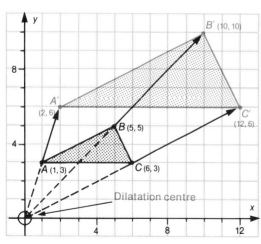

Lengths are doubled.

Direction is invariant.

iii) Slope of BC: $\frac{5-3}{5-6} = -2$

Slope of AB: $\frac{1}{2}$

Therefore, $\angle B$ is a right angle.

Slope of B'C': $\frac{10-6}{10-12} = -2$

Slope of A'B': $\frac{1}{2}$

Therefore, $\angle B'$ is a right angle and $\angle B = \angle B'$.   Angle measure is invariant.
Using a protractor, it can be shown that
$\angle A = \angle A'$ and $\angle C = \angle C'$.

iv) Area of $\triangle ABC$: $\frac{1}{2}bh = \frac{1}{2}(5)(2)$
$= 5$ units^2

Area of $\triangle A'B'C'$: $\frac{1}{2}bh = \frac{1}{2}(10)(4)$ Area is multiplied by 4.
$= 20$ units^2

c) Slope of OA: $\frac{3-0}{1-0} = 3$

Slope of OA': $\frac{6-0}{2-0} = 3$

Therefore O, A, A' are collinear. Similarly, O, B, B' are collinear, and O, C, C' are collinear. This shows that lines containing matching points intersect at the same point. This point is called the **dilatation centre**.

The transformation $(x, y) \rightarrow (kx, ky)$ is a dilatation with scale factor k and dilatation centre $(0, 0)$.

A dilatation preserves: • direction, or slope
• angle measure
• location of the dilatation centre

A dilatation does not preserve length except when $|k| = 1$.

Exercises 6 - 7

A 1. A triangle has vertices $A(-1, 2)$, $B(4, 2)$, $C(3, 5)$.
   a) Graph $\triangle ABC$ and its image under the transformation:
      $(x, y) \rightarrow (2x, 2y)$.
   b) Compare the triangle and its image with respect to
      i) slopes of sides   ii) lengths of sides;   iii) area.
   c) Determine the scale factor and the dilatation centre.

2. A triangle has vertices $A(2, 2)$, $B(4, 2)$, $C(2, -2)$.
   a) Graph $\triangle ABC$ and its image, $\triangle A'B'C'$, under the transformation: $(x, y) \to (3x, 3y)$.
   b) Join AA', BB', CC' and extend to meet at $O(0, 0)$.
   c) Calculate
      i) the lengths of the segments OA', OA, OB', OB, OC', OC;
      ii) the ratios: $\dfrac{OA'}{OA}$, $\dfrac{OB'}{OB}$, $\dfrac{OC'}{OC}$.
   d) Calculate
      i) the lengths of the segments $A'B'$, AB, $B'C'$, BC, $A'C'$, AC;
      ii) the ratios: $\dfrac{A'B'}{AB}$, $\dfrac{B'C'}{BC}$, $\dfrac{A'C'}{AC}$.
   e) What is the scale factor?

3. A triangle has vertices $P(3, 6)$, $Q(12, 6)$, $R(9, 12)$.
   a) Graph $\triangle PQR$ and its image, $\triangle P'Q'R'$, under the dilatation: $(x, y) \to (\tfrac{1}{3}x, \tfrac{1}{3}y)$.
   b) What is the geometric effect of this dilatation?
   c) Calculate
      i) the lengths of the segments OP', OP, OQ', OQ, OR', OR;
      ii) the ratios: $\dfrac{OP'}{OP}$, $\dfrac{OQ'}{OQ}$, $\dfrac{OR'}{OR}$.
   d) Compare $\triangle PQR$ and its image with respect to
      i) slopes of sides;
      ii) lengths of sides.
   e) What is the scale factor?

4. A rectangle has vertices $A(2, 0)$, $B(2, 3)$, $C(4, 3)$, $D(4, 0)$.
   a) Graph the rectangle and its dilatation image with $(0, 0)$ as the dilatation centre and 3 as the scale factor.
   b) What is the geometric effect of this dilatation?
   c) i) Measure the lengths of the line segments OB', OB, OC', OC.
      ii) Calculate the ratios: $\dfrac{OB'}{OB}$, $\dfrac{OC'}{OC}$.
   d) Compare rectangle $ABCD$ and its image with respect to:
      i) lengths of sides;
      ii) measures of angles;
      iii) area.

6-7 Dilatations   257

B 5. Copy the diagram. With (0, 0) as the dilatation centre, draw the dilatation image of △ABC with a scale factor of
   a) 1.5;   b) 0.5;   c) −1.5;   d) −0.5.

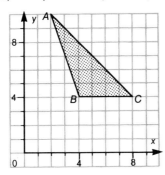

6. In the diagram below, determine the scale factor, k, for each of the dilatations:
   a) $A \to B$ b) $A \to C$ c) $B \to C$
   d) $C \to A$ e) $C \to B$ f) $B \to A$

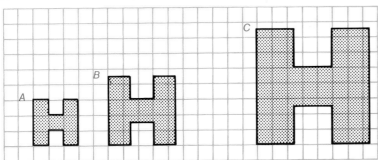

7. An object, a dilatation centre, and an image point are given. Copy each diagram and complete the image.
   a)   b)

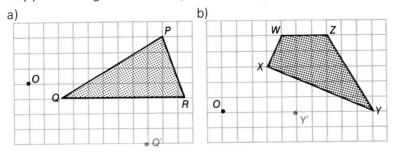

8. A trapezoid has vertices $A(2, 2)$, $B(8, 2)$, $C(8, 6)$, $D(4, 6)$.
   a) Graph the trapezoid and its image under the dilatation:
      $(x, y) \to (-x, -y)$.
   b) What is the geometric effect of this dilatation?
   c) What is the scale factor?

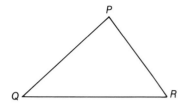

9. a) Trace △PQR and draw its dilatation image using a scale factor of 2 and a dilatation centre at: i) P; ii) Q; iii) R.
   b) For each part of (a), what is the ratio of the area of the image triangle to the area of △PQR?

10. The coordinates of the vertices of four polygons are given. Draw each polygon on a separate grid and show the positions of its dilatation images for the scale factors indicated. In each case, (0, 0) is the dilatation centre.
    a) (−1, 4), (2, 1), (5, 2)   i) $k = 2$ ii) $k = 3$
    b) (4, −2), (2, 4), (6, 3)   i) $k = \frac{3}{2}$ ii) $k = -\frac{1}{2}$
    c) (0, 0), (9, 0), (0, 6)   i) $k = \frac{2}{3}$ ii) $k = \frac{4}{3}$
    d) (0, 0), (4, 0), (6, 6), (0, 4)   i) $k = \frac{1}{2}$ ii) $k = -\frac{3}{2}$

11. The diagram shows a tessellation using a trapezoid. The tessellation is unusual because dilatations of the trapezoid can be seen in it.

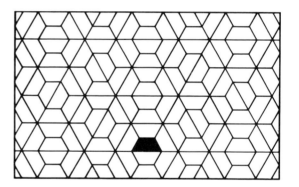

   a) Find three successive images of the shaded trapezoid and compare each image and the shaded trapezoid with respect to
      i) their perimeters;   ii) their areas.
   b) Explain how the tessellation illustrates the properties of a dilatation.

12. Find other tessellations containing dilatations of the basic figure in earlier sections of this chapter.

13. Graph the line $x + y = 3$ and its image under the dilatation: $(x, y) \rightarrow (2x, 2y)$.

C 14. Find the equation of the image of the line $y = 2x + 3$ under the dilatation: $(x, y) \rightarrow (2x, 2y)$.

Mathematics Around Us

The Pantograph

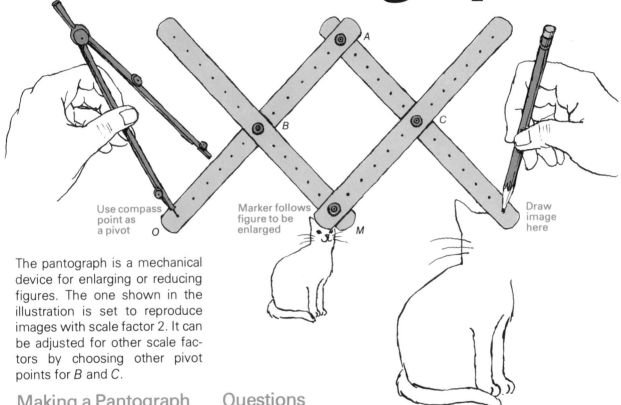

Use compass point as a pivot O

Marker follows figure to be enlarged M

Draw image here

The pantograph is a mechanical device for enlarging or reducing figures. The one shown in the illustration is set to reproduce images with scale factor 2. It can be adjusted for other scale factors by choosing other pivot points for B and C.

Making a Pantograph

A working model can be made from four strips of thin cardboard pivoted together with dressmakers' snap fasteners. Convenient dimensions for the cardboard strips are 13 cm by 1 cm. To allow the pantograph to move freely at the pivots, use only the part of the snap fastener shown. To make the marker, M, fix the head of a pin to a snap fastener with epoxy cement and cut off the excess when the cement is dry.

Questions

1. How would you adjust the pantograph to make enlargements with scale factors other than 2 ?
2. How would you make reductions with the pantograph ?
3. Investigate the effect of interchanging the positions of the compass point, O, and the marker, M.
4. Explain the pantograph's principle of operation.

Use for points A, B, and C.

Use for marker M.

6-8 Applications of Transformations

Transformations are a powerful tool for solving certain types of problems. Often, transformations can reduce a problem to a simpler form which is more readily solved. The following examples use transformations to solve problems which would be difficult to solve by other methods.

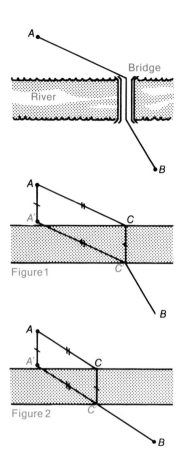

Figure 1

Figure 2

Example 1. A university students' residence is located at A. The athletic complex is located on the other side of the river at B. If a bridge is to be built perpendicular to both banks of the river, where should it be located so that the length of the path from A to B is a minimum?

Solution. Let C and C' denote the ends of the bridge. Locate A', the image of A under a translation that maps C onto C'. Since length is invariant under a translation,
$$AC = A'C'.$$
Also, $AA' = w$ and $CC' = w$, where w is the width of the river.
The length of the path from A to C to C' to B is
$$AC + w + C'B.$$
This length is also equal to $A'C' + w + C'B$. The path is shortest when $A'C' + C'B$ is a minimum. This occurs when C' is located such that $AC'B$ is a straight line, as shown in Figure 2. That is, the position of the bridge is found by translating A a distance w perpendicular to the river banks to obtain A'. Then C' is the point where $A'B$ intersects the opposite bank.

When solving problems by means of transformations, use is made of their invariance properties. *Example 1* involved the invariance of length and direction under a translation. *Example 2* involves the invariance of length under two reflections.

Example 2. Ore from Rock Island, R, is to be shipped to the countries of Martinesia and New Hampton. Where should unloading ports, M in Martinesia and N in New Hampton, be built so that the round trip, $R \rightarrow M \rightarrow N \rightarrow R$, is a minimum?

Solution. Let R' and R'' denote the images of R when reflected in the shorelines of Martinesia and New Hampton respectively. The round trip distance, $RM + MN + NR$, is equal to $R'M + MN + NR''$. Clearly, this is a minimum when R', M, N, and R'' lie on a straight line. The locations of the unloading ports are found by first finding R' and R'' and then joining them. M and N are the points of intersection of $R'R''$ with the shoreline.

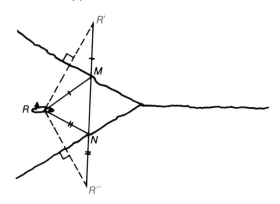

Exercises 6 - 8

A 1. A contractor must install a fountain at B. The nearest water supply is at A. Copy the diagram and locate the position of the perpendicular cut that must be made across the sidewalk for the length of piping required to connect A and B to be a minimum.

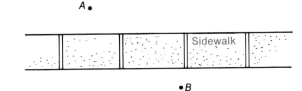

2. Consider the points $A(2, 9)$ and $B(12, 6)$. If N is a point on the x-axis, find
   a) the coordinates of N for the distance $AN + NB$ to be a minimum;
   b) the minimum distance from A to N to B.

3. a) Plot the points $A(2, 1)$ and $B(12, 8)$ on a grid and shade the strip: $4 \leq y \leq 6$. Points $C(x, 4)$ and $D(x, 6)$ are the endpoints of a segment parallel to the y-axis.
   b) Find x for the length of the path $\triangle ACDB$ to be a minimum.
   c) Find the minimum length of the path $ACDB$.

4. In a race, runners must start at post P, run to the fence and then finish at post Q. Find the point on the fence for the total distance run to be a minimum.

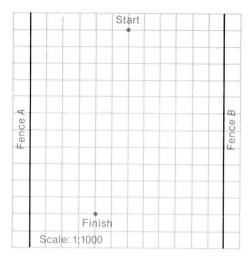

B 5. a) Plot the points A(3, 6) and B(6, 3) on a grid.
   b) Calculate the length of the shortest path from A
      i) to the x-axis to B;
      ii) to the y-axis to B;
      iii) to the y-axis to the x-axis to B.

6. The diagram shows the layout of a race at an athletic event. Runners must touch both fences before running toward the finish.
   a) Which fence should be touched first?
   b) Use the scale to calculate the shortest possible route from start to finish.

7. In a race, runners start from a post P, touch fence l_1 then fence l_2 and return to post P. Copy the diagram below (left) and locate the points on l_1 and l_2 for the total distance run to be a minimum.

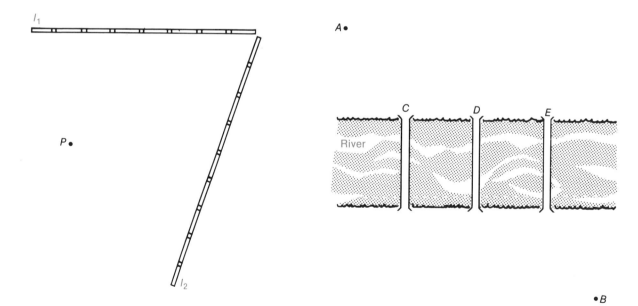

C 8. A race from A to B involves crossing a river by one of three bridges (See diagram above, right.).
   a) Which bridge should be used for the distance run to be a minimum?
   b) Using a scale of 1 : 10 000, how long is the race?

Finding Minimum Distances

Given two fixed points, $P_1(x_1, y_1)$ and $P_2(x_2, y_2)$, the problem of finding the coordinates of a point $P(x, 0)$ on the x-axis such that the total distance $P_1P + PP_2$ is a minimum can be solved using a computer. One method employs the computer's ability to try successive values for x and print a table of corresponding distances.

In the following program, the computer will print a table of values showing the total distance $P_1P + PP_2$ for several values of x. For most problems, the least value of the total distance in the table will be close to the actual minimum distance.

```
10  INPUT "WHAT ARE THE COORDINATES OF THE FIRST POINT"; A, B
20  INPUT "WHAT ARE THE COORDINATES OF THE SECOND POINT"; C, D
30  PRINT "VALUE OF X", "TOTAL DISTANCE"
40  F = (C − A) / 10
50  FOR I = 0 TO 10
60  X = A + I * F
70  PRINT X, SQR ((A − X) ↑ 2 + B ↑ 2) + SQR ((X − C) ↑ 2 + D ↑ 2)
80  NEXT I
```

Example. Use the preceding program to find the minimum total distance between P_1, the x-axis, and P_2 when the coordinates of the given points are $P_1(2, 5)$ and $P_2(8, 2)$.

Solution. WHAT ARE THE COORDINATES OF THE FIRST POINT?   2, 5
WHAT ARE THE COORDINATES OF THE SECOND POINT? 8, 2

VALUE OF X	TOTAL DISTANCE
2.0	11.3245553
2.6	10.7943434
3.2	10.3419841
3.8	9.96601345
4.4	9.66442192
5.0	9.43650318
5.6	9.28526859
6.2	9.2206559
6.8	9.26347014
7.4	9.44740911
8.0	9.81024968

The minimum distance is approximately 9.22.

Exercises

1. Draw a graph of the output of the example to show how the total distance $P_1P + PP_2$ varies as the value of x varies. How can you determine the minimum distance from the graph?

2. The coordinates of two points P_1 and P_2 are given. Find the minimum distance $P_1P + PP_2$ where P is a point on the x-axis.
   a) $P_1(1, 4)$, $P_2(6, 8)$
   b) $P_1(-2, 6)$, $P_2(5, 9)$
   c) $P_1(-3, 5)$, $P_2(7, 5)$
   d) $P_1(4, 7)$, $P_2(4, 2)$

Review Exercises

1. Find the image of the point (4, 6) under the transformation:
   a) $(x, y) \to (x + 4, y - 3)$;
   b) $(x, y) \to (-y, -x)$;
   c) $(x, y) \to (-x + 8, -y + 12)$;
   d) $(x, y) \to (y + 3, x - 3)$.

2. Find the coordinates of the point having the image (3, 8) under the transformation:
   a) $(x, y) \to (x - 3, y - 3)$;
   b) $(x, y) \to (2x, y - 5)$;
   c) $(x, y) \to (y - 4, x + 2)$;
   d) $(x, y) \to (y + 1, -x + 3)$.

3. A triangle has vertices $A(-3, 3)$, $B(4, 3)$, $C(-3, 6)$. Graph the triangle and its image under the transformation:
   a) $(x, y) \to (x + 3, y + 2)$;
   b) $(x, y) \to (x - 2, y - 3)$.

4. Graph the line $2y = 2x + 1$ and its image under the transformation: $(x, y) \to (x + 1, y - 1)$.

5. A triangle has vertices $P(5, 2)$, $Q(2, 5)$, $R(0, 0)$.
   a) On the same grid, graph the triangle and its images under the transformations:
      i) $(x, y) \to (-y, x)$;
      ii) $(x, y) \to (-x, -y)$.
   b) Describe each transformation.

6. Graph the line $x + 2y = 3$ and its image under the transformation: $(x, y) \to (-y, -x)$.

7. A triangle has vertices $A(2, 2)$, $B(6, 2)$, $C(6, 5)$.
   a) On the same grid, graph the triangle and its images under the transformations:
      i) $(x, y) \to (-x, y)$;
      ii) $(x, y) \to (-y, -x)$.
   b) Describe each transformation.

8. Graph the line $x - y = 4$ and its image under the transformation: $(x, y) \to (-x, y)$.

9. A triangle has vertices $K(3, 5)$, $L(-2, 4)$, $M(2, -4)$. Graph the triangle and its image under the transformation: $(x, y) \to (-x + 4, y)$.

10. A triangle has vertices $A(5, 1)$, $B(6, 5)$, $C(10, 3)$.
    a) Draw its image, $\triangle A'B'C'$, under a reflection in the line $x = 3$.
    b) Draw the image of $\triangle A'B'C'$, under a reflection in the line $y = -2$.
    c) Describe the result of these successive reflections.

11. A triangle has vertices $P(3, 5)$, $Q(6, 2)$, $R(7, 6)$. Graph the triangle and its image under the transformation: $(x, y) \to (x - 6, -y)$.

12. Identify the isometry that maps:
    a) Figure I onto Figure II;
    b) Figure I onto Figure III;
    c) Figure II onto Figure III.

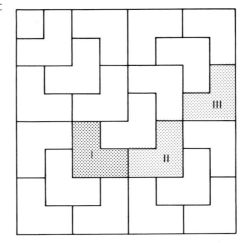

13. i) Identify the pairs of figures that are directly congruent.
    ii) State the isometry that maps each object onto its image.

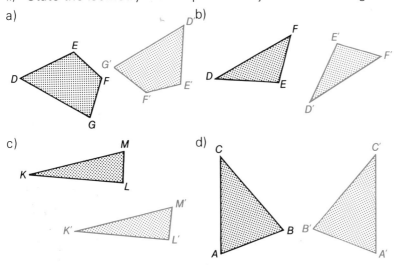

14. A triangle has vertices $A(-3, 2)$, $B(1, 4)$, $C(4, 1)$. Graph $\triangle ABC$ and its image under the dilatation: $(x, y) \to (2x - 3, 2y + 1)$.

15. Graph the line $2x + 3y = 4$ and its image under the transformation: $(x, y) \to \left(\frac{x}{2}, \frac{y}{2}\right)$.

16. A fly lands at the point $(3, 8)$ on a grid. It walks to the y-axis, then to the x-axis, and finally stops at the point $(6, 4)$. What is the shortest distance the fly could have walked?

THE MATHEMATICAL MIND

A Genius From India

Srinivasa Ramanujan 1887–1920

The brilliant Indian mathematician, Srinivasa Ramanujan, developed many new and complicated results about numbers. He had an amazing memory and an unusual ability to see patterns in and relations between numbers quickly.

Once, when a friend was visiting Ramanujan in a hospital where he was being treated for tuberculosis, the friend mentioned that he had come in a taxi with the number 1729. He remarked that the number seemed to be a dull one, and hoped that this was not an unfavorable omen.

"On the contrary," Ramanujan replied. "1729 is a very interesting number. It is the smallest number that can be expressed as the sum of two cubes in two different ways."

1. Find natural numbers a, b, c, d such that $1729 = a^3 + b^3$ and $1729 = c^3 + d^3$.
2. What is the smallest number that can be expressed as the sum of two perfect squares in two different ways?
3. Find numbers which can be expressed as the sum of three perfect squares in two different ways.

7 Geometry

A trawler and a cruise ship leave port together, the trawler to sail due east and the cruise ship to sail due south. After 3 h, the trawler captain, wishing to check his course, radios the cruise ship for information. He is told that the cruise ship has maintained a steady 20 km/h due south and that the ships are now 75 km apart. If the trawler has maintained a steady 15 km/h, is it on course? (See *Example 1* in Section 7-9.)

7-1 What Are Deductions?

A succession of statements in which the last statement follows from the others by logical reasoning is called a **deduction**.

Deduction 1.	Statements:	Any four-legged animal is a quadruped.
		Juno is a four-legged animal.
	Conclusion:	Juno is a quadruped.
Deduction 2.	Statements:	Maria is 3 years older than Gerry.
		Gerry is 13 years old.
	Conclusion:	Maria is 16 years old.
Deduction 3.	Statements:	Andy and Bill have the same weight.
		Cindy and Donna have the same weight.
	Conclusion:	Andy and Cindy together weigh the same as Bill and Donna together.
Deduction 4.	Statement:	$3x - 12 = 9$
	Conclusion:	$x = 7$

The statements of *Deductions 1* and *2* lead to obvious conclusions. The conclusion in *Deduction 3* is less apparent and results from applying the principle, or axiom:

> If equals are added to equals, the sums are equal.

An **axiom** is a statement that is assumed to be true without question.

This and another axiom are used in *Deduction 4* to obtain the conclusion $x = 7$. First, 12 is added to the equal quantities $3x - 12$ and 9 to obtain $3x$ and 21. Then, both quantities are divided by 3. The second axiom is:

> If equals are divided by equals, the quotients are equal.

Example 1. State a conclusion that follows logically from the given statements:
  a) All students are perfect. My sister is a student.
  b) A square is a rectangle. A rectangle has four right angles.

Solution. a) My sister is perfect.
  b) A square has four right angles.

Axioms are usually considered to be so obvious that they are often used without being mentioned. However, it is important to be aware of their use.

7-1 What Are Deductions?   269

Example 2. In the following deductions, what axioms are needed to establish the conclusions?

a) Statements: $x < 5$. $y > 5$.
Conclusion: $x < y$.

b) Statements: $AB = 2$ cm, $BC = 3$ cm, $CD = 2$ cm.
Conclusion: $AD = 7$ cm.

Solution. a) If $a < b$ and $b < c$, then $a < c$.

b) The whole is equal to the sum of its parts.

What axioms are used in this next example?

Example 3. Prove that when two lines intersect, the vertically opposite angles are equal.

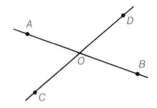

Proof. AB and CD are two lines that intersect at O.
Since $\angle AOD + \angle AOC = 180°$,
and $\angle AOD + \angle DOB = 180°$
then $\angle AOD + \angle AOC = \angle AOD + \angle DOB$
Therefore, $\angle AOC = \angle DOB$

In a similar manner, it can be proved that $\angle AOD = \angle COB$. That is, when two lines intersect, the vertically opposite angles are equal.

Deductions that are used in proving other deductions are called **theorems**. The deduction of *Example 3*—**When two lines intersect, the vertically opposite angles are equal**—is used to solve problems involving intersecting lines. The deduction is therefore called the **Opposite Angle theorem**.

Example 4. Find the value of x:

a)    b)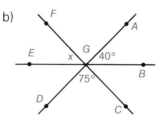

Solution. a) $x = 71°$ ... Opposite Angle theorem

b) $\angle EGD = 40°$ ... Opposite Angle theorem
Since $\angle FGC$ is a straight angle,
$x + 40° + 75° = 180°$
$x = 65°$

Exercises 7-1

A 1. State a conclusion that follows logically from the given statements:
   a) All students in this school are teenagers.
      Lynn is a student in this school.
   b) If you jog regularly you will stay fit.
      Manuel jogs regularly.
   c) All musicians have long hair.
      Sharon is a musician.
   d) When a certain number is multiplied by 5 and 4 is added, the result is 39.
   e) A square is a rhombus. A rhombus has four equal sides.
   f) A square is a parallelogram.
      A parallelogram has both pairs of opposite sides parallel.
   g) Some rectangles are squares. A square has four equal sides.
   h) In the figure, $\angle ABC = \angle ABD$
   i) $x^2 = 25$ and $x < 0$.
   j) There is only one even prime number.
      2 is a prime number.

2. Name two pairs of vertically opposite angles:
   a)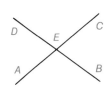
   b)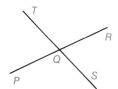

3. Find the values of x and y:
   a)
   b)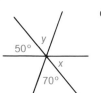
   c)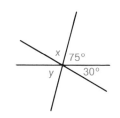

B 4. Decide if the conclusion follows logically from the given statements:
   a) All mathematics students can compute. Lisa is a mathematics student. Therefore, Lisa can compute.
   b) Some students like pizza. Sophie is a student. Therefore, Sophie likes pizza.

c) Some apples are green. All apples are fruit. Therefore, some fruit is green.
d) Some teachers teach Mathematics and Science. Some teachers teach Mathematics and English. Therefore, some teachers teach Science and English.
e) Some prime numbers are even. Some even numbers are multiples of 5. Therefore, some prime numbers are multiples of 5.
f) An isosceles triangle has at least two equal sides. An equilateral triangle has three equal sides. Therefore, an equilateral triangle is isosceles.
g) $x < y$. $y = z$. Therefore, $x < z$.
h) $a < 10$. $10 > b$. Therefore, $a < b$.
i) Three lines, $l_1, l_2,$ and l_3, are such that $l_1 \perp l_2$, and $l_2 \perp l_3$. Therefore, $l_1 \perp l_3$.
j) The coordinates of one point on the line $3x + y = k$ are $(2, 5)$. Therefore, $k = 11$.

5. In the diagram, $\angle ABC = 90°$. Prove that each of the other three angles equals 90°.

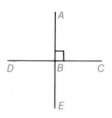

6. In the diagram below (left), $\angle ABC = \angle PQR$. Prove that $\angle ABD = \angle PQS$.

Two angles with a sum of 180° are **supplementary**. Two angles with a sum of 90° are **complementary**.

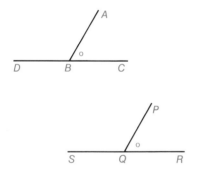
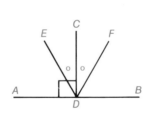

7. In the diagram above (right), $\angle CDA = 90°$ and $\angle EDC = \angle FDC$. Prove that $\angle ADE = \angle BDF$.

8. Find the value of x:

a)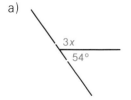
b)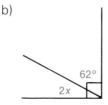
c)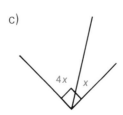

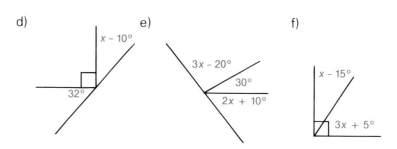

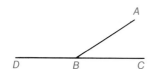

C 9. In the diagram, if $\angle ABD = n \times \angle ABC$, express $\angle ABC$ and $\angle ABD$ in terms of n.

10. Alec, Bob, Carl, Dan, and Eddie are on the school basketball team. List the boys in order of increasing height if
    i) there are at least two boys shorter than Alec,
    ii) Dan is shorter than Carl,
    iii) Bob is not the shortest boy, and
    iv) Dan is taller than Alec.

11. The capacities of two pails are 5 L and 3 L. How can the pails be used to obtain exactly 1 L of water?

12. State the axioms needed to reach the conclusions:

    a) $\frac{1}{3}x + 2 = 10$
       $x = 24$

    b) $3a \geq 18$
       $2b \leq 12$
       $b \leq a$

    c) Right $\triangle ONA$ is rotated 90° counterclockwise about (0, 0) to $ON'A'$. The coordinates of A' are $(-3, 5)$.

    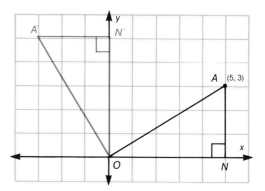

    d) The product of the binomials $(x + 3)$ and $(x + 4)$ is $x^2 + 7x + 12$.

    e) The equation of one line is $3x - y = 2$. The equation of another line is $x + y = 6$. The lines intersect at the point (2, 4).

7-2 Congruent Triangles: SAS

If two triangles are congruent,
  i) their corresponding angles are equal; and
  ii) their corresponding sides are equal.
However, it is not necessary to know this much information to show that two triangles are congruent. The minimum conditions under which triangles are congruent are called **congruence axioms**.

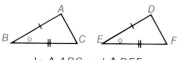

In △ABC and △DEF,
if AB = DE
  ∠B = ∠E
  BC = EF
then △ABC ≅ △DEF.

> **SAS Congruence Axiom:** If two sides and the contained angle of one triangle are equal to two sides and the contained angle of another triangle, the triangles are congruent.

Example 1. In △JKL and △PQR, JL = PR, KL = QR, and ∠L = ∠R.
  a) Explain why △JKL ≅ △PQR.
  b) List pairs of equal angles and equal sides.

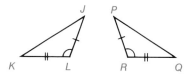

Solution. a) △JKL ≅ △PQR by the SAS congruence axiom.
  b) Equal angles: Equal sides:
    ∠J = ∠P   JK = PQ
    ∠K = ∠Q   KL = QR
    ∠L = ∠R   JL = PR

The SAS congruence axiom is used to prove the next theorem.

Isosceles Triangle Theorem: In an isosceles triangle, the angles opposite the equal sides are equal.

Given: △ABC in which AB = AC.
To Prove: ∠B = ∠C.
Proof: Let the bisector of ∠A meet BC in D.
  In triangles BAD and CAD: AB = AC
                ∠BAD = ∠CAD
                AD = AD
        Therefore, △BAD ≅ △CAD ... SAS
        and   ∠B = ∠C.

This theorem can now be used in the proofs of other theorems and deductions.

Example 2. In the figure, AB = AC. Prove that ∠ABE = ∠ACD.

Proof. ∠ABC = ∠ACB   ... Isosceles Triangle theorem
  Therefore, 180° − ∠ABC = 180° − ∠ACB,
    and   ∠ABE = ∠ACD.

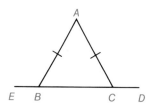

Exercises 7 - 2

A 1. Explain why the triangles are congruent, and list pairs of equal sides and equal angles:

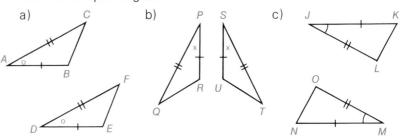

2. State all the facts you can about each figure:

   a)
   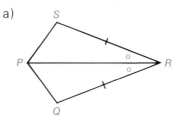

   b)
   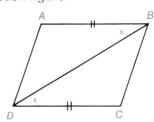

   c)
   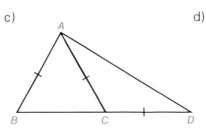

   d)
   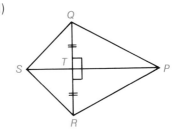

3. In the diagram below (left), $AE = BE$ and $CE = DE$. Prove that $AD = BC$ and $\angle D = \angle C$.

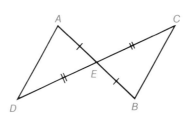
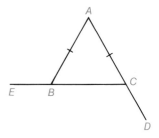

4. In the diagram above (right), sides AC and CB of isosceles $\triangle ABC$ are extended to D and E respectively. Prove that $\angle ABE = \angle BCD$.

7-2 Congruent Triangles: SAS **275**

B 5. A and B are two points on a circle with centre O.
   a) In △OAB, prove that ∠A = ∠B.
   b) Is the result of (a) true for any two points, A and B, on the circle?

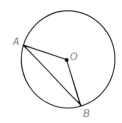

6. Prove that an equilateral triangle is equiangular.

7. Prove that any point on the perpendicular bisector of a line segment is equidistant from the endpoints of the segment. That is, in the diagram below (left), prove that PA = PB.

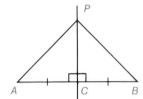 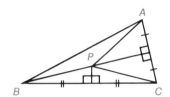

8. In △ABC (above, right), P is the point of intersection of the perpendicular bisectors of BC and AC. Prove that P is equidistant from the three vertices.

9. In △PQR, S is a point on QR such that SP = SQ = SR. Prove that ∠QPR = ∠Q + ∠R.

10. Isosceles triangles ABC and DBC (below, left) share a common base, BC.
    a) Prove that ∠ABD = ∠ACD.
    b) Would the result in (a) be true if both triangles were on the same side of the common base?

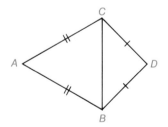 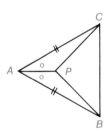

11. In △ABC (above, right), AB = AC. P is any point on the bisector of ∠BAC.
    a) Prove that △PBC is isosceles.
    b) Would the result of (a) still be true if P is outside △ABC?

C 12. Side BC of isosceles △ABC is extended to D and E such that BE = CD. Prove that AE = AD.

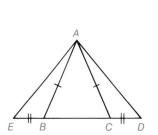

13. In the diagram below (left), $AC = AE$ and $DC = BE$. Prove that $BC = DE$.

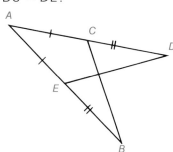
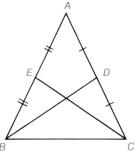

14. In $\triangle ABC$ (above, right), $AB = AC$. Prove that the medians BD and CE are equal in length.

15. Draw two triangles that illustrate why SSA is not a congruence axiom. That is, if two sides and a non-contained angle of one triangle are equal to two sides and a non-contained angle of a second triangle, the triangles are not necessarily congruent.

7-3 Congruent Triangles: SSS

Another important congruence axiom is presented in this section.

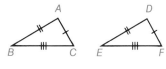

In $\triangle ABC$ and $\triangle DEF$,
if $\quad AB = DE$
$\quad\quad BC = EF$
$\quad\quad AC = DF$
then $\triangle ABC \cong \triangle DEF$.

> **SSS Congruence Axiom:** If three sides of one triangle are equal to three sides of another triangle, the triangles are congruent.

The SSS congruence axiom is used to prove this next theorem.

Perpendicular Bisector Theorem: Any point that is equidistant from the endpoints of a line segment lies on the perpendicular bisector of the segment.

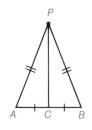

Given: Line segment AB and point P such that $PA = PB$.
To Prove: P is on the perpendicular bisector of AB.
Proof: Join P to the midpoint, C, of AB.
In $\triangle PCA$ and $\triangle PCB$: $\quad PC = PC$
$\quad\quad\quad\quad\quad\quad\quad\quad\quad\quad\quad CA = CB$
$\quad\quad\quad\quad\quad\quad\quad\quad\quad\quad\quad PA = PB \quad$ ...given
Therefore, $\quad\quad\quad\quad\quad\triangle PCA \cong \triangle PCB \quad$ ... SSS
and $\quad\quad\quad\quad\quad\quad\angle ACP = \angle BCP$.
Since $\angle ACP + \angle BCP = 180°$
$\quad\quad\quad\quad\angle ACP = 90°$
$PC \perp AB$, and P is on the perpendicular bisector of AB.

7-3 *Congruent Triangles:* SSS   **277**

The Perpendicular Bisector theorem is used in the next deduction.

Example. Prove that the diagonals of a rhombus bisect each other at right angles.

Proof. ABCD is any rhombus.
Since $AB = BC$,
    B is on the perpendicular bisector of AC.
        ... Perpendicular Bisector theorem
Since $AD = DC$,
    D is on the perpendicular bisector of AC.
        ... Perpendicular Bisector theorem

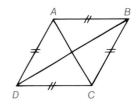

That is, BD is the perpendicular bisector of AC.
Similarly, AC is the perpendicular bisector of BD.
Therefore, each diagonal bisects the other at right angles.

Exercises 7 - 3

A 1. Explain why the triangles are congruent, and list pairs of equal sides and equal angles:

a) 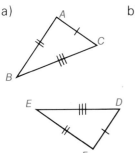   b) 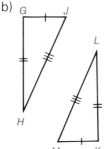   c)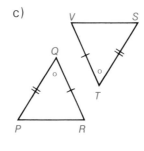

2. State all the facts you can about each figure:

a)    b) 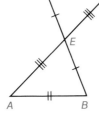   c)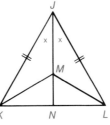

3. In quadrilateral ABCD, $AB = DC$ and $BC = AD$. Prove that $\angle B = \angle D$.

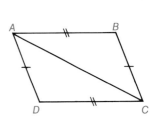

4. In △PQR (below, left), PQ = PR and S is the midpoint of QR. Prove that
   a) PS ⊥ QR;
   b) PS bisects ∠QPR.

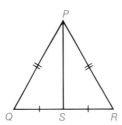
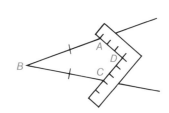

B 5. The diagram above (right) shows how to use a carpenter's square to bisect an angle. Prove that BD bisects ∠B.

6. AB and CD are chords of equal length in a circle with centre O. Prove that ∠AOB = ∠COD.

7. If XY and WZ are diameters of a circle, prove that XW = YZ.

8. Isosceles triangles ABC and DBC share a common base, BC.
   a) Prove that
      i) AD is the perpendicular bisector of BC;
      ii) AD bisects ∠A and ∠D.
   b) Would the results in (a) be true if both triangles were on the same side of the common base?

9. Prove:
   a) the opposite angles of a rhombus are equal;
   b) the diagonals of a rhombus bisect the angles at the vertices.

10. Two circles with centres A and B intersect at C and D. Prove that
    a) AB is the perpendicular bisector of CD.
    b) AB bisects ∠CAD and ∠CBD.

11. The diagonals of quadrilateral ABCD bisect each other at right angles. Prove that ABCD is a rhombus.

12. Draw two triangles to illustrate why AAA is not a congruence axiom. That is, if three angles of one triangle are equal to three angles of another triangle, the triangles are not necessarily congruent.

13. In quadrilateral PQRS, PQ = QR and PS = SR. T is any point on the diagonal QS. Prove that PT = TR.

7-4 Congruent Triangles: ASA

In this section, a third axiom establishing congruence is introduced.

> **ASA Congruence Axiom:** If two angles and the contained side of one triangle are equal to two angles and the contained side of another triangle, the triangles are congruent.

In △ABC and △DEF,
if ∠B = ∠E
BC = EF
∠C = ∠F
then △ABC ≅ △DEF.

In the next example, the ASA congruence axiom is used in a deduction.

Example. In △ABC, the bisector of ∠A is perpendicular to BC. Prove that △ABC is isosceles.

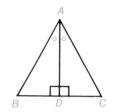

Proof. In the diagram, the bisector of ∠A meets BC in D, and AD ⊥ BC.
In △ABD and △ACD: ∠BAD = ∠CAD
AD = AD
∠BDA = ∠CDA
Therefore, △ABD ≅ △ACD   ... ASA
and   AB = AC.
Therefore, △ABC is isosceles.

Exercises 7-4

A 1. State why the triangles are congruent, and list pairs of equal angles and equal sides:

a) 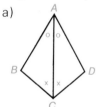   b) 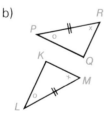   c)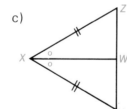

2. State all the facts you can about each figure:

a) 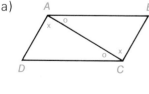   b)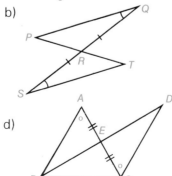

c) 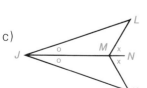   d)

3. In the diagram below (left), $BC = EC$ and $\angle B = \angle E$. Prove that $\angle A = \angle D$.

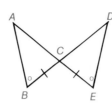
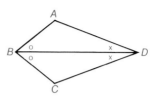

4. In quadrilateral $ABCD$ (above, right), diagonal BD bisects $\angle ABC$ and $\angle ADC$. Prove that $AB = CB$ and $AD = CD$.

5. Find pairs of congruent triangles and state the required congruence axiom:

a) i) ii) iii) b) i) ii) iii)

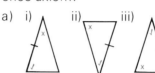
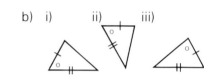

c) i) ii) iii) d) i) ii) iii)

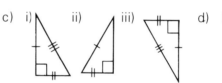
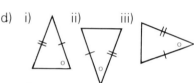

e) i) ii) iii) iv)

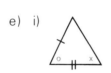

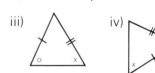

f) i) ii) iii) iv) v)

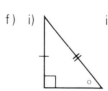
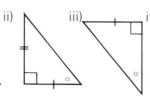
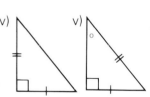

B 6. Find pairs of congruent triangles and state the required congruence axiom:

a) b) c)

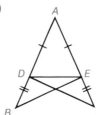
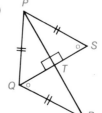
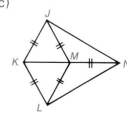

7-4 Congruent Triangles: ASA

7. In the figure given, $\angle KLN = \angle NMK$ and $\angle KML = \angle NLM$. Prove that $KL = NM$ and $KM = NL$.

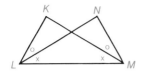

8. In $\triangle ABC$ (below, left), $AB = AC$. D and E are points on AC and AB respectively such that $\angle ABD = \angle ACE$. Prove that $BD = CE$.

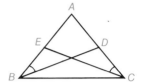
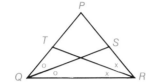

9. In $\triangle PQR$ (above, right), $PQ = PR$. The bisectors of $\angle R$ and $\angle Q$ meet PQ and PR in T and S respectively. Prove that $QS = RT$.

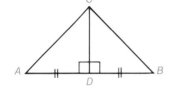

10. In $\triangle ABC$, D is the midpoint of AB and $CD \perp AB$. Prove that $\triangle ABC$ is isosceles.

11. XY and WZ are diameters of a circle. Prove that $\angle WXY = \angle WZY$.

12. In $\triangle PQR$, $PQ = PR$. $PS \perp QR$. Prove that $\angle QPS = \angle RPS$.

13. P, R are points on the inner of two concentric circles and Q, S are points on the outer circle such that PS and QR intersect at the centre, O (below, left). Prove that $PQ = RS$.

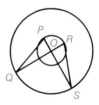
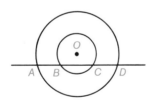

C 14. A line intersects two concentric circles, centre O, in A, B, C, D (above, right). Prove that $AB = CD$.

15. In quadrilateral $PQRS$, $PQ = RS$ and $\angle Q = \angle R$. Prove that $\angle P = \angle S$.

16. GH, JK, LM are three diameters of a circle. Prove that $\triangle GJL \cong \triangle HKM$.

17. XY and WZ are two line segments of equal length. The perpendicular bisectors of XW and YZ meet at O. Prove that $\triangle OXY \cong \triangle OWZ$.

18. In $\triangle ABC$, $AB = AC$. Sides BC and AB are extended to E and D respectively, such that $CE = BC$ and $BD = AB$. Prove that $AE = DC$.

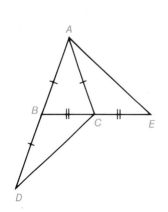

THE MATHEMATICAL MIND

A Best Seller From Way Back

In 300 B.C., the city of Alexandria was the intellectual and cultural centre of ancient Greek civilization. A university was established there, and Euclid, who may have come from Athens, was chosen to head the mathematics department.

Euclid

Euclid's *Elements* is one of the world's most famous books. It has been used and studied more than any other book except the Bible. More than 1000 editions have appeared in the last 500 years. In the *Elements*, Euclid organized the results in geometry, number theory, and algebra of mathematicians from 600 to 300 B.C. He showed that the theorems of what is now called Euclidean geometry can be deduced from the following ten axioms.

1. Things which are equal to the same thing are equal to each other.
2. If equals be added to equals, the sums are equal.
3. If equals be subtracted from equals, the remainders are equal.
4. Things which coincide with one another are equal to one another.
5. The whole is greater than the part.
6. It is possible to draw a straight line from one point to any other point.
7. It is possible to extend a straight line indefinitely.
8. It is possible to describe a circle with a given centre and radius.
9. All right angles are equal to one another.
10. If the interior angles between a transversal and two straight lines total less than 180°, the two straight lines are not parallel.

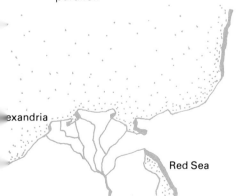

Two pages of an edition of Elements *published in 1254.*

Because it was the first substantial source of mathematical knowledge, Euclid's *Elements* has probably had more influence on mathematical thinking than any other work. However, it does contain certain logical errors and unwarranted assumptions. That is, assumptions that are true but which do not follow from the axioms. Some examples are:
i) He assumed that a straight line which intersects one side of a triangle necessarily intersects another side.
ii) He proved two triangles congruent by moving one to coincide with the other. That is, he assumed that in moving a figure its properties do not change.

Despite these obvious defects, the *Elements* survived and dominated the study of geometry for more than two thousand years.

1. What is the difference between an axiom and a theorem?
2. Euclid assumed that there is only one straight line joining two points. Is this assumption justified by the axioms?
3. Euclid used the word "elements" to refer to the important, or most useful, theorems in the subject. From the examples and exercises of this chapter, make a list of the theorems which might be considered to be the "elements" of geometry.

7-5 Parallel Lines

Two lines in the same plane must either intersect or never meet. If they never meet, they are called **parallel lines**. A **transversal** is a line that intersects two or more lines. When a transversal intersects two parallel lines, certain pairs of angles are important.

There are two pairs of **alternate angles**.

There are four pairs of **corresponding angles**.

i)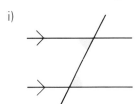

i) ii)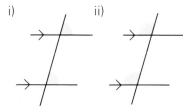

ii)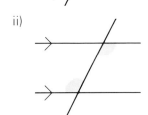

iii) iv)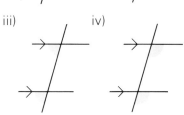

To prove theorems and deductions involving parallel lines, the following axiom is used.

> **Parallel Lines Axiom**:
> If l_1 and l_2 are parallel lines, then
> $x + y = 180°$
>
>

The above axiom can be used to prove important theorems about alternate and corresponding angles.

Parallel Lines Theorem: When a transversal intersects two parallel lines, the alternate angles are equal and the corresponding angles are equal.

Given: Parallel lines l_1 and l_2 intersected by a transversal. w, x, y, z, represent the measures of the angles indicated.

To Prove: $y = z$ and $y = w$

Proof: a) $\quad\quad\quad\quad\quad x + y = 180°\quad$ ... Parallel Lines axiom
But, $\quad\quad x + z = 180°\quad$ ... straight angle
Therefore, $\quad y = z$.
That is, the alternate angles are equal.

b) $z = w$ ... vertically opposite angles
But, $y = z$ ... proved in part (a)
Therefore, $y = w$.
That is, corresponding angles are equal.

Example. Find the values of x, y, and z:

a)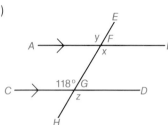

b)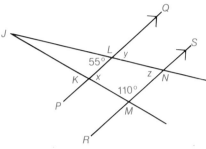

Solution. a) $\angle BFG = \angle FGC$ ... alternate angles.
That is, $x = 118°$
$\angle EFA = \angle FGC$ ... corresponding angles.
That is, $y = 118°$
$\angle HGD = \angle FGC$ ... vertically opposite angles.
That is, $z = 118°$

b) $\angle LKM + \angle KMN = 180°$ ... Parallel Lines axiom.
$$x + 110° = 180°$$
$$x = 70°$$
$\angle JLK = \angle QLN$ ... vertically opposite angles.
$$55° = y$$
$\angle QLN = \angle LNM$ ... alternate angles.
$$55° = z$$

Parallelogram's Sides Theorem: The opposite sides of a parallelogram are equal.

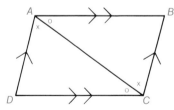

Given: Parallelogram $ABCD$
To Prove: $AB = CD$ and $BC = DA$
Proof: Draw AC.
In $\triangle ABC$ and $\triangle CDA$:
             $\angle BAC = \angle DCA$ ... alternate angles
             $AC = CA$
             $\angle ACB = \angle CAD$ ... alternate angles
Therefore, $\triangle ABC \cong \triangle CDA$ ... ASA
and $AB = CD$, and $BC = DA$.

Exercises 7-5

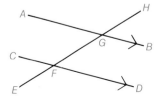

A 1. In the figure given, name two pairs of alternate angles and four pairs of corresponding angles.

2. Find the values of x and y:

a)    b)    c)

3. Find the values of w, x, y, and z:

a) 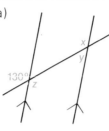   b) 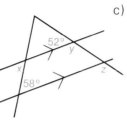   c)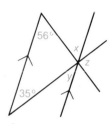

d) 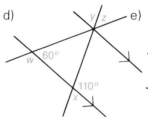   e)   f)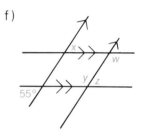

4. State all the facts you can about each figure:

a) 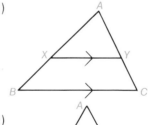   b)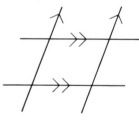

c) 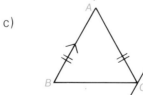   d)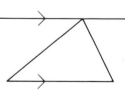

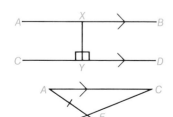

5. In the diagram given, $AB \parallel CD$ and $XY \perp CD$. Prove that $XY \perp AB$.

6. Prove that the opposite angles of a parallelogram are equal.

B 7. In the diagram given, E is the midpoint of AB and $AC \parallel DB$. Prove that E is also the midpoint of CD.

8. In △ABC, AB = AC. Side BA is extended to D, and AE is drawn parallel to BC. Prove that AE bisects ∠DAC.

9. The measure of one angle of a parallelogram is given. Find the measures of the other three angles:
   a) 50°   b) 70°   c) 90°

10. In quadrilateral PQRS, PS ∥ QR and PS = QR. Prove that PQ = SR.

11. Prove that the diagonals of a parallelogram bisect each other.

12. In parallelogram ABCD (below, left), P and Q are the midpoints of sides AB and CD respectively. Prove that PQ and AC bisect each other.

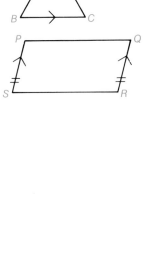

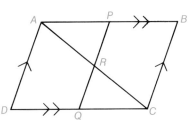 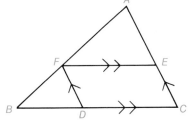

C 13. In △ABC (above, right), FE ∥ BC and FD ∥ AC. Prove that the three angles of △AFE are respectively equal to the three angles of △FBD.

7-6 Angles in a Triangle

A triangle has three sides and three angles. Although triangles can have many different sizes and shapes, the sum of the angles in every triangle is 180°.

Angle Sum Theorem: The sum of the angles in a triangle is 180°.

Given: △ABC
To Prove: ∠A + ∠B + ∠C = 180°
Proof: Through A, draw DE parallel to BC.
  ∠DAB = ∠B and ∠EAC = ∠C   ... alternate angles
  But, ∠BAC + ∠DAB + ∠EAC = 180°
  Therefore, ∠A + ∠B + ∠C = 180°

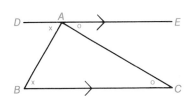

288 Chapter 7

Example 1. In the figure, ∠AED = ∠ABC. Prove that ∠ADE = ∠ACB.

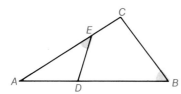

Solution. In △ADE: ∠ADE = 180° − ∠A − ∠AED
...Angle Sum theorem
In △ABC: ∠C = 180° − ∠A − ∠B
...Angle Sum theorem
         = 180° − ∠A − ∠AED
Therefore, ∠ADE = ∠ACB.

Example 2. Find the value of x:

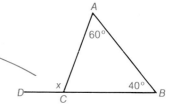

This is called an exterior angle.

Solution. ∠ACB + 40° + 60° = 180°   ...Angle Sum theorem
                    ∠ACB = 80°
Since ∠BCD is a straight angle,
                80° + x = 180°
                      x = 100°

The result of the last example may be generalized as follows:

Exterior Angle Theorem: The exterior angle of a triangle is equal to the sum of the two opposite interior angles.

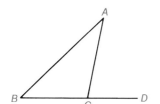

Given: △ABC with BC extended to D
To Prove: ∠ACD = ∠A + ∠B
Proof: ∠A + ∠B + ∠ACB = 180°   ... Angle Sum theorem
        ∠ACD + ∠ACB = 180°   ... straight angle
Therefore, ∠ACD + ∠ACB = ∠A + ∠B + ∠ACB
That is,   ∠ACD = ∠A + ∠B.

Exercises 7 - 6

A 1. Find the value of x:

a)

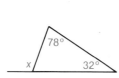

b)

c)

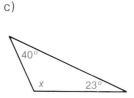

7-6 Angles in a Triangle 289

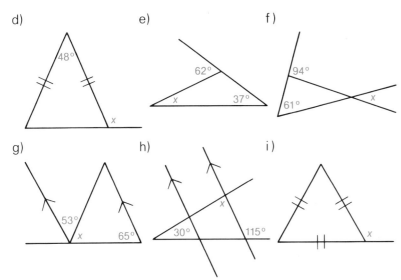

2. State all the facts you can about each figure:

a)  b)  c)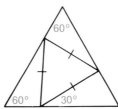

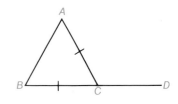

3. Side BC of $\triangle ABC$ is extended to D. If $BC = AC$, prove that $\angle ACD = 2 \angle A$.

B 4. Find the values of x and y:

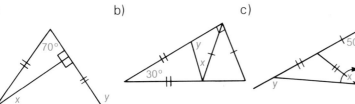

d) 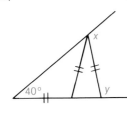 e)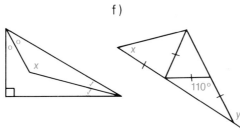

5. Find the value of *x*:

a)   b)  c)

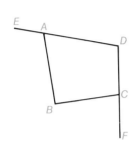

6. Sides *DA* and *DC* of quadrilateral *ABCD* are extended to *E* and *F* respectively. Prove that ∠ *EAB* + ∠ *BCF* = ∠ *B* + ∠ *D*.

7. Any polygon can be divided into triangles by joining vertices. The diagram shows a pentagon divided into three triangles.

a) Copy and complete this table:

Polygon	Number of Sides	Number of Triangles	Sum of Angles
Triangle	3	1	180°
Quadrilateral	4		
Pentagon	5	3	
Hexagon			
Octagon			
Decagon			

b) Write a simple formula for the sum of the angles in a polygon with *n* sides.

8. A **regular polygon** is one that has all sides the same length and all angles equal.

a) Make an additional column to your table of Exercise 7 giving the measure of each angle of a regular polygon.

b) Write a simple formula for the measure of each angle in a regular polygon with *n* sides.

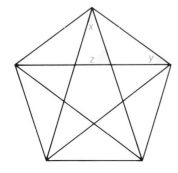

9. The diagram shows a regular pentagon and its five diagonals. Find the values of *x*, *y*, and *z*.

10. Prove that the sum of the exterior angles of a triangle, one at each vertex, equals 360°.

11. For each of the following figures,
    i) find the equation relating x and y;
    ii) draw a graph of the relation between x and y for reasonable values of the variables;
    iii) describe the effect of varying x on the value of y.

    a)   b)   c)

    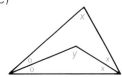

12. a) Find the sum of the shaded angles:
    i)   ii)   iii)

    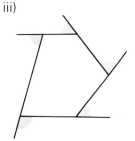

    b) State a probable conclusion based on the results of (a).

13. Find the sum of the shaded angles:
    a)   b)

    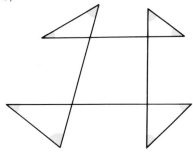

14. From the data given in the diagram, prove that $\angle BEC = 90°$.

15. Prove that $\triangle ABC \cong \triangle DEF$.

    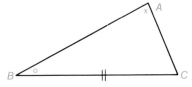
    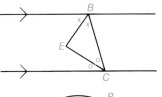

C 16. AB is a diameter of circle with centre O. If P is any point on the circle, prove that $\angle APB = 90°$.

    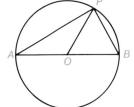

7-7 Congruent Triangles: AAS

The ASA congruence axiom states that if two angles and the contained side of one triangle are equal to two angles and the contained side of another triangle, the triangles are congruent. However, the triangles are congruent whether or not the equal sides are contained by the given angles, because the remaining angles of the triangles must also be equal.

AAS Congruence Theorem: If two angles and any side of one triangle are equal to two angles and the corresponding side of another triangle, the triangles are congruent.

Given: Triangles *ABC* and *DEF* with two angles and a non-contained side equal as indicated.

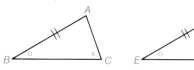

To Prove: △*ABC* ≅ △*DEF*

Proof: $AB = DE$
        ∠*A* = 180° − (∠*B* + ∠*C*)   ... Angle Sum theorem
        ∠*D* = 180° − (∠*E* + ∠*F*)   ... Angle Sum theorem
        But, ∠*B* = ∠*E* and ∠*C* = ∠*F*   ... given
        Therefore, ∠*A* = ∠*D*
        and △*ABC* ≅ △*DEF*   ... ASA

Angle Bisector Theorem: Any point on the bisector of an angle is equidistant from the sides of the angle.

Given: ∠*ABC* and its bisector, *BD*

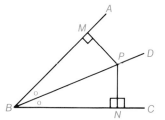

To Prove: Any point, *P*, on *BD* is equidistant from the sides of ∠*ABC*.

Proof: From *P*, draw perpendiculars *PM* and *PN* onto *BA* and *BC* respectively.
      In △*PMB* and △*PNB*:   ∠*PMB* = ∠*PNB*
                             ∠*PBM* = ∠*PBN*
                             PB = *PB*
Therefore,   △*PMB* ≅ △*PNB*   ... AAS
    and   *PM* = *PN*.
That is, *P* is equidistant from the sides of ∠*ABC*.

7-7   Congruent Triangles: AAS   **293**

Exercises 7-7

A 1. State why the triangles are congruent, and list pairs of equal sides and equal angles:

a)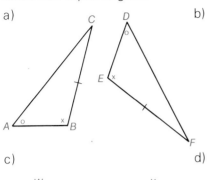
b)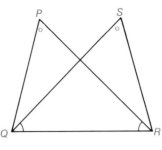

c)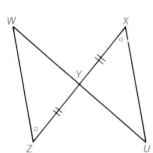
d)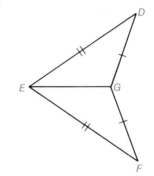

2. State a third condition necessary for congruence:

a)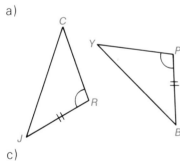
b)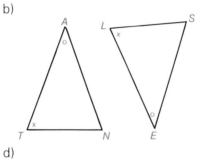

c)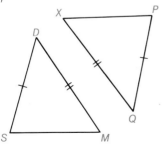
d)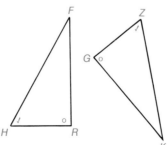

3. Explain why △ABC ≇ △DEF.

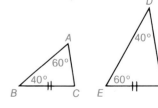

4. State all the facts you can about each figure:

a)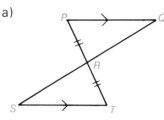

b)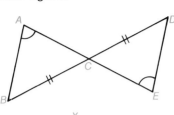

c)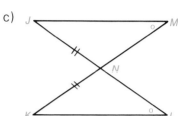

d)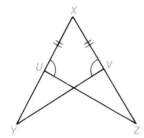

B 5. From the data given in the diagram below (left), prove that $BC = EC$.

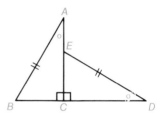
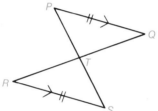

6. Given that $PQ = RS$ and $PQ \parallel RS$ in the diagram above (right), prove that line segments PS and QR bisect each other.

7. In $\triangle ABC$ (below, left), $AB = AC$. Prove that altitudes BD and CE are equal.

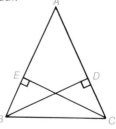
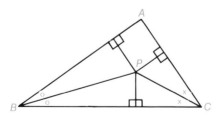

8. In $\triangle ABC$ (above, right), P is the point of intersection of the bisectors of $\angle B$ and $\angle C$. Prove that P is equidistant from the three sides.

9. Prove that the altitudes drawn from corresponding vertices of congruent triangles are equal.

10. In $\triangle ABC$, $AE = BD$, $BD \perp AC$, and $AE \perp BC$. Prove that $\triangle ABC$ is isosceles.

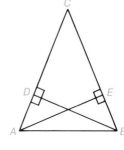

Use Indirect Proof

In deductions, when a desired conclusion cannot be obtained directly from given statements and theorems, the conclusion may often be reached by a deductive process called **indirect proof**.

In an indirect proof, the alternatives to the desired conclusion are assumed, in turn, to be true. If, as expected, they lead to the contradiction of a known fact, then the desired conclusion is the only remaining possibility.

Example 1. In $\triangle ABC$, M is a point on BC such that $BM \neq CM$ and AM bisects $\angle A$. Prove: $AB \neq AC$.

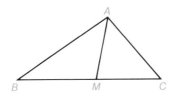

Proof. **Step 1.** State the possible alternatives.
Either $AB \neq AC$, or $AB = AC$.

Step 2. Assume that the opposite of the desired result is true.
Assume that $AB = AC$.

Step 3. Show that a conclusion can be reached which contradicts the known facts.
In $\triangle ABM$ and $\triangle ACM$,
$$AB = AC \quad \ldots \text{assumption}$$
$$\angle BAM = \angle CAM$$
$$AM = AM$$
Therefore, $\triangle ABM \cong ACM \ldots$ SAS
and $BM = CM$
This contradicts the known fact that $BM \neq CM$.

Step 4. State the conclusion.
The assumption — $AB = AC$ — was false. Therefore, $AB \neq AC$ is the only remaining possibility.

The next example is an illustration of indirect proof in algebra.

Example 2. If n is an integer such that n^2 is odd, prove that n is odd.

Proof. **Step 1.** Either n is odd or n is even.

Step 2. Assume that n is even.

Step 3. Then $n = 2k$, where k is an integer;
and $n^2 = (2k)^2$
$= 4k^2$.
Since n^2 is a multiple of 4, n^2 is even. This contradicts the known fact that n^2 is odd.

Step 4. Therefore, the assumption that n is even is false; and n must be odd.

Exercises

Prove the following using indirect proof.

1. In $\triangle PQR$, $\angle Q = 50°$ and $\angle R = 60°$. Prove: $PQ \neq PR$.

2. Prove that a triangle cannot have
   a) two right angles;
   b) two obtuse angles.

3. Prove that only one perpendicular can be drawn from the point, P, to the line, l.

4. If n is an integer such that n^2 is even, prove that n is even.

5. If m and n are positive integers such that their product, mn, is odd, prove that both m and n are odd.

6. In $\triangle ABC$, AM is the median from A to BC, and $\angle AMC = 60°$. Prove: $AB \neq AC$.

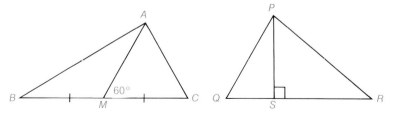

7. In $\triangle PQR$, PS is the altitude from P to QR, and $QS \neq RS$. Prove: $PQ \neq PR$.

8. In $\triangle DEF$, G and H are the midpoints of DE and DF respectively, and $DF \neq DE$. Prove that EH and FG cannot bisect each other.

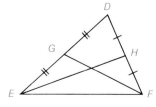

7-8 Some Theorems and Their Converses

The statements:
> If the car has no gasoline, then it will not start.
> If the car will not start, then it has no gasoline.

are called *converse* statements. To form the converse of a statement, interchange the clauses following the words "if" and "then". Other examples of statements and their converses are as follows.

Statement	Converse
1. If it is raining, then the ground is wet.	1. If the ground is wet, then it is raining.
2. If $2x - 3 = 5$, then $x = 4$.	2. If $x = 4$, then $2x - 3 = 5$.
3. If two triangles are congruent, then their areas are equal.	3. If two triangles have equal areas, then they are congruent.
4. All whales are mammals.	4. All mammals are whales.
5. Parallel lines are lines in the same plane that do not intersect.	5. Lines in the same plane that do not intersect are parallel.
6. Every square is a rectangle.	6. Every rectangle is a square.

Because statements 4, 5, and 6 are not written in the "if... then" form, their converses are formed simply by interchanging subject and complement.

It is clear, from the above examples, that the converse of a true statement is not necessarily true. The converses of statements 2 and 5 are true, but the converses of statements 1 and 3 are not necessarily true, while those of 4 and 6 are false.

The table that follows shows three theorems (proved in earlier sections of this chapter) and their converses. Although the converses of these theorems are true, not all theorems have true converses.

Theorem	Converse
1. If two sides of a triangle are equal, then the angles opposite those sides are equal.	1. If two angles of a triangle are equal, then the sides opposite those angles are equal.

Theorem	Converse
2. Any point on the perpendicular bisector of a line segment is equidistant from the endpoints of the segment.	2. Any point which is equidistant from the endpoints of a line segment is on the perpendicular bisector of the segment.
3. If a transversal intersects two parallel lines, the alternate angles are equal.	3. If a transversal intersects two lines and the alternate angles are equal, the lines are parallel.

The converse of the second of the three theorems was proved in *Example 1* of Section 7-3. The converses of the other two theorems are proved in the next two examples.

Example 1. Prove that if two angles of a triangle are equal, the sides opposite those angles are equal.

Proof. In $\triangle ABC$, $\angle B = \angle C$.
Let the bisector of $\angle A$ meet BC in D.
In $\triangle BAD$ and $\triangle CAD$: $\angle B = \angle C$
$\angle BAD = \angle CAD$
$AD = AD$
Therefore, $\triangle BAD \cong \triangle CAD$ ... AAS
and $AB = AC$.

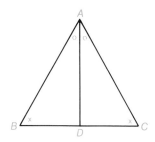

Figure 1

Example 2. Prove that if a transversal intersects two lines and the alternate angles are equal, the lines are parallel.

Proof. l_1 and l_2 are two lines cut by a transversal t. Equal alternate angles are indicated by x (Figure 1). Using indirect proof, either $l_1 \parallel l_2$, or l_1 intersects l_2. If l_1 intersects l_2 at a point P, $\triangle PQR$ is formed (Figure 2).
$x + y + z = 180°$ ... Angle Sum theorem
also, $x + y = 180°$ ... straight angle
Therefore, $z = 0°$
This means that one of the angles of $\triangle PQR$ is $0°$, which is impossible. Therefore, the assumption that l_1 intersects l_2 is false, and $l_1 \parallel l_2$ is the only other possibility.

It is also the case that if a transversal makes equal corresponding angles with two lines, the lines are parallel.

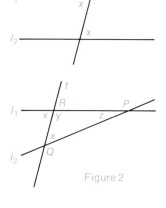

Figure 2

If a theorem and its converse are both true, they can be combined

7-8 Some Theorems and Their Converses

into a single theorem using the phrase "if, and only if". The following combined theorems have been proved in this chapter.

Isosceles Triangle Theorem: A triangle has two equal angles if, and only if, it has two equal sides.

Perpendicular Bisector Theorem: A point is on the perpendicular bisector of a line segment if, and only if, it is equidistant from the endpoints of the segment.

Parallel Lines Theorem: When a transversal intersects two lines, the lines are parallel if, and only if,
- the alternate angles are equal; or
- the corresponding angles are equal.

Exercises 7-8

A 1. Write the converse of each statement:
   a) If you live in Vancouver, then you live in Canada.
   b) Ice is frozen water.
   c) If $x = 2$, then $x^3 = 8$.
   d) If $x^2 > 25$, then $|x| > 5$.
   e) If two triangles are congruent, then their angles are equal.
   f) If a right triangle has a 30° angle, then it also has a 60° angle.
   g) Every student in this room takes mathematics.
   h) All fish can swim.
   i) Perpendicular lines intersect at right angles.
   j) Every rhombus is a parallelogram.

2. In Exercise 1, each statement is true.
   a) Which converse statements are true?
   b) Rewrite the statements which have a true converse in the "if, and only if" form.

3. State all the facts you can about each figure:

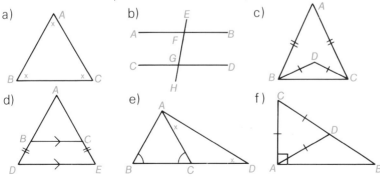

B 4. Write in the "if... then" form:
   a) A dog is a four-legged animal.
   b) All students like hamburgers.
   c) All prime numbers greater than 2 are odd.
   d) A perfect square has an odd number of factors.
   e) An equilateral triangle is isosceles.
   f) Every right triangle has two acute angles.

5. a) Write the converse of each statement in Exercise 4, and decide whether it is true.
   b) Rewrite the statements which have a true converse in the "if, and only if" form.

6. Write a sentence and its converse that can be obtained from each of the following, and decide whether the given statement is true.
   a) You can drive a car if, and only if, you have insurance.
   b) Shadows are seen if, and only if, the sun is shining.
   c) $x + 5$ is an even number if, and only if, x is an odd number.
   d) A line is perpendicular to the line $y = 2x + 3$ if, and only if, its slope is $-\frac{1}{2}$.
   e) A quadrilateral is a rectangle if, and only if, it has four right angles.
   f) Two rectangles have equal areas if, and only if, they have equal lengths.

7. In the figure, $\angle ABE = \angle ACD$. Prove that $AB = AC$.

8. In the figure below (left), $\angle ACD = \angle CBE$. Prove that $\triangle ABC$ is isosceles.

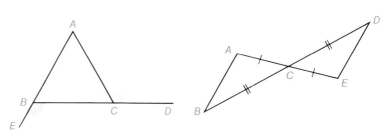

9. In the figure above (right), $AC = EC$ and $BC = DC$. Prove that $AB \parallel DE$.

10. In quadrilateral $ABCD$, $AB = AD$ and $\angle B = \angle D$. Prove that $BC = DC$.

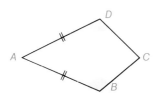

11. In quadrilateral ABCD (below, left), the diagonal BD bisects ∠B and BC = DC. Prove that AB ∥ DC.

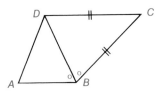
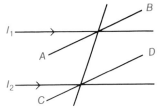

12. In the figure above (right), l_1 is parallel to l_2 and AB and CD bisect the angles through which they pass. Prove that AB ∥ CD.

13. In the figure, $l_1 \parallel l_2$ and AB and AC bisect the angles through which they pass. Prove that ∠A = 90°.

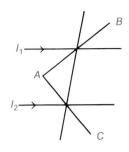

14. a) Prove the theorem: The opposite sides of a parallelogram are equal.
    b) Write the converse of the theorem in (a). If the converse appears to be true, write a proof; if it does not appear to be true, give an example to show why.

15. a) Prove the theorem: The diagonals of a rectangle bisect each other.
    b) Write the converse of the theorem in (a). If the converse appears to be true, write a proof; if it does not appear to be true, give an example to show why.

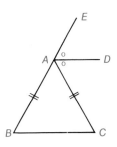

C 16. In △ABC, AB = AC and AD bisects the exterior angle ∠CAE. Prove that AD ∥ BC.

17. In each figure, △ABC, and △ADE are isosceles. Prove that BC ∥ DE.
    a)
    b)

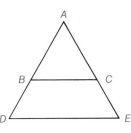
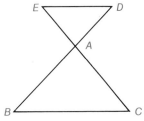

18. Give an example of each of the following:
    a) A statement and its converse which are both true
    b) A statement and its converse which are both false
    c) A true statement which has a false converse
    d) A false statement which has a true converse

Mathematics Around Us

How Fast Do Glaciers Move?

The Columbia Icefield in Jasper National Park is the largest accumulation of ice in the Rocky Mountains. It is estimated that about 6–10 m of snow falls in this region each year. Not all of this snow melts in the summer and, over the centuries, the accumulated snow has been pressed into ice.

The Athabasca Glacier "flows" from the rim of the Columbia Icefield, and can be seen by travellers from the Banff-Jasper highway. This glacier is about 5.3 km long, 1.2 km wide, and has a maximum thickness of approximately 300 m. However, the glacier is presently receding, or becoming shorter, because the ice is melting at the end at a faster rate than it can be resupplied from above.

There are two rates associated with the Athabasca Glacier which have been measured:

- The rate of flow of the ice down the mountain—about 10 cm per day;
- The rate of recession of the end of the glacier—about 12 m per year.

These rates depend on the season or on the location on the glacier where the measurements are made.

Questions

1. a) About how many years might it take ice at the top of the glacier to "flow" to the end?
   b) About how long would it take to "flow" the length of your classroom?
   c) About how far has it "flowed" since you were born?

2. a) About how far has the end of the glacier receded since you were born?
   b) About how long would it take to recede the length of your classroom?
   c) About how long might it take for the glacier to disappear?

3. If the ice stopped "flowing" down the mountain, what would be the rate of recession of the end of the glacier?

4. What important assumptions did you make in answering the above questions? Discuss the validity of these assumptions.

7-9 The Pythagorean Theorem and Its Converse

The Pythagorean theorem is one of the most important of all mathematical theorems because of its many applications in real-world problems as well as in mathematics. As a result, over the last 2500 years, many mathematicians have sought new and original proofs of this relationship. One of the simplest is shown below.

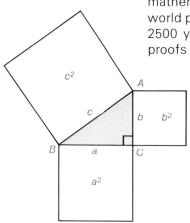

The Pythagorean Theorem: The area of the square on the hypotenuse of a right triangle is equal to the sum of the areas of the squares on the other two sides.

Given: Right $\triangle ABC$ with side lengths a, b, c.
To Prove: $a^2 + b^2 = c^2$

Proof: Draw square $CDEF$ with sides of length $a + b$, and vertices A and B on CF and CD respectively. Locate points G, H, as shown, and draw quadrilateral $ABGH$. By the SAS congruence axiom, the four corner triangles are all congruent. Therefore, the sides of the quadrilateral $ABGH$ are all c units long. If x and y represent the measures of the angles indicated,

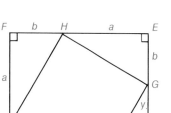

$x + y = 90°$ ... Angle Sum theorem
Therefore, $\angle ABG = 90°$
... $\angle CBD = 180°$,
and quadrilateral $ABGH$ is a square.

Area of square $CDEF$ = Area of square $ABGH$ + 4(Area of $\triangle ABC$)

$$(a + b)^2 = c^2 + 4(\tfrac{1}{2}ab)$$
$$a^2 + 2ab + b^2 = c^2 + 2ab$$
$$a^2 + b^2 = c^2$$

The converse of the Pythagorean theorem is also true, and can be proved by using congruent triangles.

$c^2 = a^2 + b^2$

Converse of the Pythagorean Theorem: If the area of the square on the longest side of a triangle is equal to the sum of the areas of the squares on the other two sides, the triangle is a right triangle.

Given: $\triangle ABC$ with side lengths a, b, c such that $c^2 = a^2 + b^2$.
To Prove: $\angle C = 90°$

7-9 The Pythagorean Theorem and Its Converse

Proof: Draw △DEF with ∠F = 90°, FE = a, FD = b.
In right △DEF: $DE^2 = a^2 + b^2$ ... Pythagorean theorem
$ = c^2$
$DE = c$
Therefore,   △ABC ≅ △DEF ... SSS
and   ∠C = ∠F, or 90°.

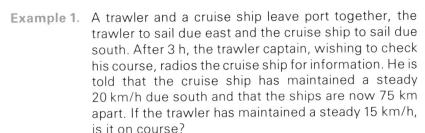

The ancient Egyptians used the converse of the Pythagorean theorem to construct right angles, and this technique is still in use.

Example 1. A trawler and a cruise ship leave port together, the trawler to sail due east and the cruise ship to sail due south. After 3 h, the trawler captain, wishing to check his course, radios the cruise ship for information. He is told that the cruise ship has maintained a steady 20 km/h due south and that the ships are now 75 km apart. If the trawler has maintained a steady 15 km/h, is it on course?

Solution. Distance travelled by the cruise ship in 3 h:
3 × 20 km, or 60 km
Distance travelled by the trawler in 3 h:
3 × 15 km, or 45 km
Since the intended courses of the two ships are at right angles, their distance apart after 3 h should be:

$$\sqrt{60^2 + 45^2} = \sqrt{3600 + 2025}$$
$$= \sqrt{5625}$$
$$= 75 \text{ km}$$

This is the distance given by the cruise ship. Therefore, the trawler is on course.

The Pythagorean theorem can be used to prove another congruence theorem.

HS Congruence Theorem: If the hypotenuse and one side of a right triangle are equal to the hypotenuse and one side of another right triangle, the triangles are congruent.

Given: △ABC and △DEF with ∠C = ∠F = 90°, AB = DE, AC = DF

To Prove: △ABC ≅ △DEF

Proof: $BC^2 = AB^2 - AC^2$ ... Pythagorean
and, $EF^2 = DE^2 - DF^2$ theorem
$ = AB^2 - AC^2$
That is, $BC^2 = EF^2$, and BC = EF.
Therefore, △ABC ≅ △DEF   ... SSS

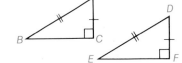

Example 2. Prove that any point which is equidistant from the sides of an angle is on the bisector of the angle.

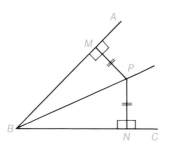

Proof. ABC is any angle and P any point such that the perpendiculars PM and PN onto BA and BC respectively are equal.

In △PMB and △PNB: PB = PB
PM = PN
∠PMB = ∠PNB = 90°

Therefore, △PMB ≅ △PNB   ... HS congruence
and ∠PBM = ∠PBN.   theorem

That is, BP bisects ∠ABC.

Therefore, any point which is equidistant from the sides of an angle is on the bisector of the angle.

Exercises 7-9

A 1. Use the Pythagorean theorem to find the value of x rounded to two decimal places:

a) 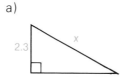   b) 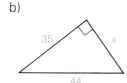   c)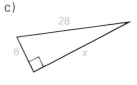

2. Which sets of three numbers represent the lengths of the sides of right triangles?

   a) 2, 3, 4   b) 4, 5, 6   c) 6, 8, 10
   d) 3, 4, 5   e) 4, 6, 8   f) 5, 12, 13
   g) 10, 24, 36   h) 8, 15, 17   i) 24, 45, 51

3. A window frame is to be 135 cm long and 94 cm wide. What distance between opposite corners will assure that the frame is rectangular?

4. A demolition ball, swinging on a chain 10 m long, must hit a wall 3 m above the ground. To get the necessary momentum, the pivot of the chain is 6 m from the wall. How high is the pivot above the ground?

B 5. Determine if ∠BAC is a right angle:

a) 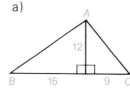   b) 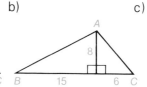   c)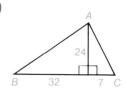

6. In the figure below (left), each colored square has sides of 2 cm.
   a) Find the lengths of the sides of the outer square.
   b) What percent of the outer square is covered by the colored squares?

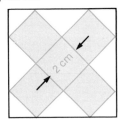
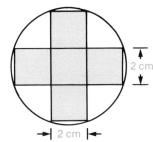

7. In the figure above (right), each colored square has sides of 2 cm.
   a) Find the radius of the circle.
   b) What percent of the circle is covered by the five squares?

8. a) If $s > t$, prove that a triangle with sides of length $s^2 + t^2$, $s^2 - t^2$, and $2st$ is a right triangle.
   b) If s and t represent positive integers, the expressions in (a) yield sets of three integers which represent the lengths of the sides of right triangles. Such sets of integers are called Pythagorean triples. Find five Pythagorean triples.

9. If AD is the perpendicular bisector of BC, prove that $\angle BAE = 90°$.
   a)
   b)

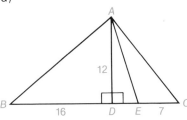
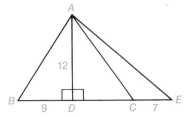

10. Prove that $\angle A = 90°$:
    a)
    b)
    c)

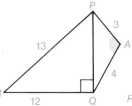
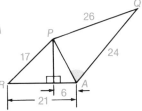
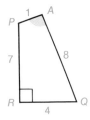

11. In quadrilateral PQRS, ∠Q and ∠S are right angles and PQ = PS. Prove that RQ = RS.

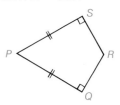

12. Prove that AB ∥ CD:
    a)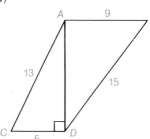
    b)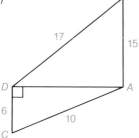

C 13. Use each figure given to prove that $c^2 = a^2 + b^2$.
    a)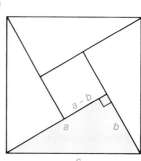
    b)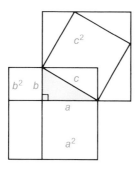

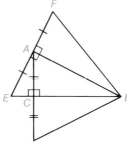

14. In the figure given, BC is the perpendicular bisector of AD, and BA is the perpendicular bisector of EF. Prove that the area of △BAF is equal to the sum of the areas of △BCD and △ACE.

15. a) If the lengths of the three sides of a triangle are known, find a method of deciding if the triangle is right, acute, or obtuse.
    b) Use your method to classify these triangles as right, acute, or obtuse:
    i)   ii)   iii)

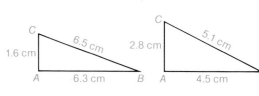 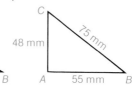

7-10 Proofs Using Transformations

The following properties of transformations have been demonstrated in earlier sections.

1. Length and angle measure are invariant under translations, rotations, and reflections.

2. If two lines are parallel, one may be mapped onto the other by a translation, a $\frac{1}{2}$-turn, or a reflection.

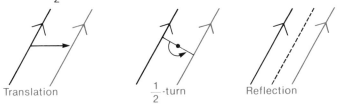

Translation $\frac{1}{2}$-turn   Reflection

3. A line may be mapped onto itself by a translation, a $\frac{1}{2}$-turn, or a reflection.

Translation $\frac{1}{2}$-turn   Reflection

4. The perpendicular bisector of any line segment AB is the reflection line of the reflection which maps $A \to B$ and $B \to A$.

5. If two lines intersect, one may be mapped onto the other by a reflection or a rotation.

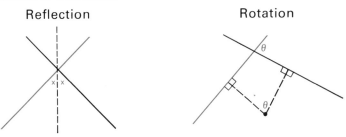

Reflection

The bisector of the angle formed by the line and its image is the reflection line.

Rotation

The angle formed by the line and its image is equal to the angle of rotation.

If properties 1-5 above are assumed as axioms, theorems and deductions similar to those seen earlier, can be proved.

A Proof Using Reflection

Example 1. Prove that the angles opposite the equal sides of an isosceles triangle are equal.

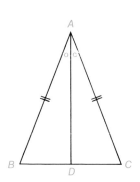

Proof. In $\triangle ABC$, $AB = AC$ and the bisector of $\angle A$ meets BC in D. Under a reflection in AD, the line containing segment AB maps onto the line containing segment AC.
Since $AB = AC$, $B \to C$.
Therefore, $\angle B \to \angle C$.
Since angle measure is invariant, $\angle B = \angle C$.
That is, the angles opposite the equal sides of an isosceles triangle are equal.

A Proof Using Translation

Example 2. Prove that if a transversal intersects two parallel lines, the corresponding angles are equal.

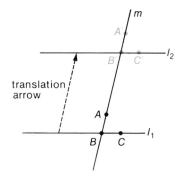

Proof. Label the figure as shown.
Under a translation parallel to the transversal, m, that maps line l_1 onto line l_2,
$$A \to A', \quad B \to B', \quad C \to C'.$$
Therefore, $\angle ABC \to \angle A'B'C'$.
That is, corresponding angles are equal.

The next example illustrates a problem that is easier to solve by transformations than by other methods.

A Proof Using Rotation

Example 3. $\triangle ABC$ and $\triangle ECD$ are equilateral. Prove that $AD = BE$ and that $\angle AFB = 60°$.

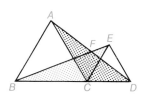

Proof. Under a 60° rotation about C,
$$\triangle ACD \to \triangle BCE$$
Therefore, $AD \to BE$ and AD intersects BE at 60°.
That is, $AD = BE$ ... length invariant under a rotation and $\angle AFB = 60°$.

Exercises 7-10

Proofs Using Translations

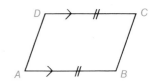

B 1. In quadrilateral *ABCD*, *AB* || *DC* and *AB* = *DC*. Prove that *AD* || *BC* and *AD* = *BC*.

2. Line segments *AD*, *BE*, and *CF* (below, left) are equal and parallel. Prove that △*ABC* ≅ △*DEF*.

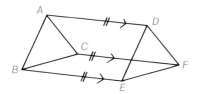
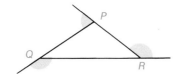

3. In △*PQR* (above, right), prove that the sum of the exterior angles is 360°.

Proofs Using Reflections

4. Prove that any point on the perpendicular bisector of a line segment is equidistant from the endpoints of the segment.

5. In the figure below (left), *PR* is the perpendicular bisector of *QS*. Prove that ∠*SPR* = ∠*QPR*.

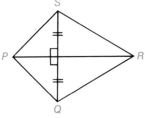
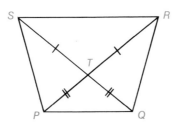

6. In the figure above (right), *PT* = *QT* and *RT* = *ST*. Prove that △*PQS* ≅ △*QPR*.

7. In square *ABCD*, prove that
   a) *CE* = *CF*;   b) *BE* = *DF*.

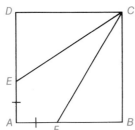
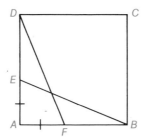

Proofs Using Rotations

8. Two lines intersect. Prove that the vertically opposite angles are equal.

9. Prove that when a transversal intersects two parallel lines, the alternate angles are equal.

10. The diagonals of quadrilateral ABCD bisect each other. Prove that ABCD is a parallelogram.

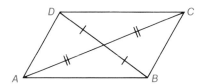

11. △ABC (below, left) is an equilateral triangle, and BD = CE. Prove that AD = BE and that ∠BFD = 60°.

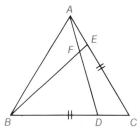
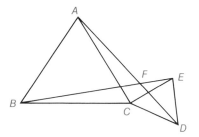

12. △ABC and △ECD (above, right) are equilateral triangles. Prove that BE = AD and that ∠AFB = 60°.

Proofs Using Dilatations

13. In △PQR (below, left), S and T are the midpoints of PQ and PR respectively. Prove that ST ∥ QR and $ST = \frac{1}{2}QR$.

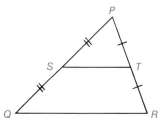
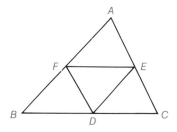

14. In △ABC (above, right), D, E, F, are the midpoints of the sides. Prove that △AFE, △FBD, △EDC, and △DEF are all congruent.

Review Exercises

1. State a conclusion that follows logically from the given statements:
   a) Every artist is poor. Jeffrey is an artist.
   b) The opposite angles of a parallelogram are equal. *PQRS* is a parallelogram.
   c) A triangle with two equal sides is isosceles. △*KLM* has two equal sides.

2. Decide if the conclusion follows logically from the given statements:
   a) All musicians have long hair. Roger is a musician. Therefore, Roger has long hair.
   b) All artists are poor. Michael is poor. Therefore, Michael is an artist.
   c) Everytime someone enters the backyard, the dog barks. The dog is barking. Therefore, someone is in the backyard.

3. a) Prove: $PS = PT$ b) Prove: $\angle KML = \angle K + \angle L$

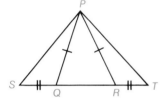 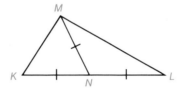

4. a) Prove: $\angle M = \angle K$ b) Prove: AD bisects $\angle A$

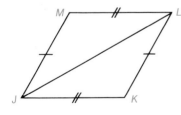 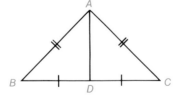

5. In △*ABC*, $AB = BC$ and $\angle BCE = \angle BAD$. Prove: $EB = DB$.

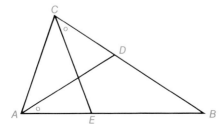

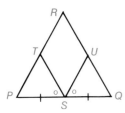

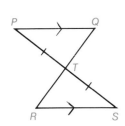

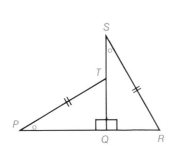

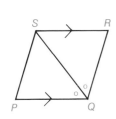

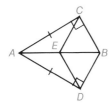

6. In $\triangle PQR$, $PR = QR$, S is the midpoint of PQ, and $\angle TSP = \angle USQ$. Prove: $ST = SU$.

7. Find the values of w, x, y, and z:
   a) 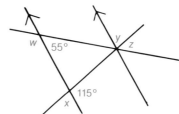   b)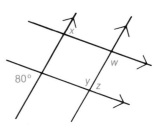

8. In the figure, $PQ \parallel RS$ and T is the midpoint of PS. Prove that T is the midpoint of QR.

9. Find the values of x and y:
   a) 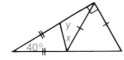   b) 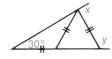   c)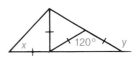

10. In the figure, $\angle S = \angle P$, $PT = SR$, and $SQ \perp PR$. Prove: $SQ = PQ$.

11. Write in the "if...then" form:
    a) All right angles are equal.
    b) Congruent triangles have equal corresponding angles.
    c) Perpendicular lines form right angles.
    d) Congruent triangles have equal corresponding sides.

12. a) Write the converse of each statement in Exercise 11, and decide whether it is true.
    b) Rewrite the statements that have a true converse in the "if, and only if" form.

13. In quadrilateral $PQRS$, diagonal SQ bisects $\angle Q$ and $SR \parallel PQ$. Prove: $SR = QR$.

14. Which sets of numbers represent the lengths of the sides of right triangles?
    a) 15, 20, 25   b) 7, 8, 11   c) 9, 40, 41

15. In the figure, $\angle ACB$ and $\angle ADB$ are right angles and $AC = AD$. Prove: $CE = DE$.

8 Statistics and Probability

Each box of Barley Puff cereal contains one International Action hockey card. Assuming the cards are randomly distributed in equal numbers among the boxes, estimate the number of Barley Puff cereal boxes it would normally take to obtain the entire set of six different cards. (See *Example 2* of Section 8-9.)

8-1 Interpreting Graphs

Statistics is the branch of mathematics which deals with the collection, interpretation, and analysis of data for the purpose of drawing inferences and making predictions. This section presents some common data displays and graphs and shows how to interpret them.

Stem-and-Leaf Diagrams

A **stem-and-leaf** diagram is a useful way of displaying such data as the marks obtained by a class on a test. It has the advantage that all the individual marks can be shown. The stem is formed by the tens' digits of the marks and the leaf by the units' digits. This is illustrated in the following example.

On a mathematics test a class obtained these marks:

43	57	82	68	70	93	63
85	79	77	62	73	75	98
81	63	56	59	69	41	77
93	70	80	84	78	89	82

The diagram below on the left shows the first column of the test marks—43, 85, 81, 93. On the right is the completed stem-and-leaf diagram for all the marks.

```
Tens   Units           Stem-and-Leaf Diagram for the
                       Mathematics Test Marks
  4  |  3               4 | 3 1
  5  |                  5 | 7 6 9
  6  |                  6 | 3 8 2 9 3
  7  |                  7 | 9 0 7 0 3 8 5 7
  8  |  5 1             8 | 5 1 2 0 4 9 2
  9  |  3               9 | 3 3 8
```

(Means 85 and 81.)

- Use the stem-and-leaf diagram to determine how many students obtained a mark less than 75.

The above diagram has intervals of 10 marks. A diagram with intervals of 5 marks can be drawn by splitting each interval of 10 into two intervals of 5. Thus,

```
  6 | 3 8 2 9 3      becomes      6 | 3 2 3
                                  6 | 8 9
```

Histograms

A graph that records the number of times a variable has a value in a particular interval is called a **histogram**. The one at the top of the next page shows the distribution of incomes among the employees of Ajax Canada Ltd.

8 - 1 Interpreting Graphs   **317**

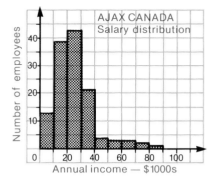

- What percent of the workers at Ajax earn less than $30 000 per year?

Broken-line Graphs

The only points on a **broken-line graph** that represent actual data are the endpoints of the segments. The broken-line graph opposite shows the sales in one shop of the hit album, "Denim Dolly Sings." The sales are recorded at the end of each month and the cumulative total represented by points on the graph. Since these are the only known data, the points are joined by line segments.

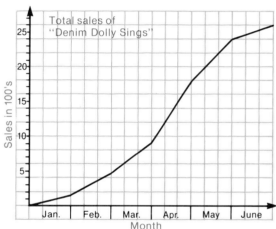

- About how many albums were sold in March?
- What was the peak month for sales?

Continuous-line Graphs

A graph which displays the value of one variable for all values of another variable over an interval is called a **continuous-line graph**. The graph opposite shows the temperature of a cup of coffee during the first 6 min of cooling. The graph was obtained by measuring the temperature every minute. Since the temperature can be expected to fall steadily, not erratically, between successive measurements, the points corresponding to actual data were joined by a smooth curve.

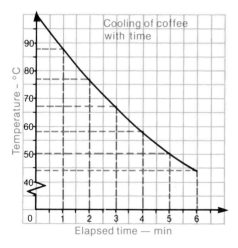

- How long did it take the temperature of the coffee to fall to 60°C?

Unlike a broken-line graph, *all* points on a continuous-line graph are considered to represent actual values of the variables, unless a statement to the contrary is made.

Exercises 8-1

A 1. This stem-and-leaf diagram shows the marks obtained by students on a mathematics test.

```
4 | 7 5
5 | 6 3 9
6 | 7 2 6 4 8 7 5 7
7 | 2 4 7 0 3 8 0
8 | 1 6 4 7
9 | 2
```

   a) How many students received a mark of:
      i) 59?   ii) 67?   iii) 70?
   b) How many students received a mark of 75 or higher?
   c) What was the most frequently-occurring mark?
   d) How many students took the test?

2. At a bicycle rodeo the participants received the scores shown.

```
2 | 4 7 2
3 | 2 0 6 1 7 6 9 6
4 | 8 3 5 9 5 1
5 | 0 0 0
```

   a) How many received a score of:
      i) 32?   ii) 35?   iii) 45?
   b) How many received a score below 33?
   c) How many received a perfect score?
   d) How many scored in the top 40%?

3. The masses, in kilograms, of the 30 students in a grade 10 class are shown in the accompanying diagram.

```
4 | 4 2 3
4 | 6 9
5 | 4 2 4
5 | 7 5 8 5 7
6 | 3 1 3 2 2 4 3
6 | 6 5 8 8 7
7 | 2 4 2
7 | 8 5
```

   a) How many students weigh:
      i) 54 kg?   ii) 56 kg?   iii) 63?   iv) 66 kg?
   b) How many students weigh:
      i) more than 70 kg?   ii) less than 55 kg?
   c) What is the total mass of the 30 students?

4. In a ski race, the contestants' times, in seconds, were as shown.

```
23 | 5
24 | 2 1 3
24 | 6 9 5 8 6
25 | 0 4 1 3 3 4 4
25 | 5 6 8 5
26 | 1 0 0
26 | 6 5
```

   a) What was the winning time?
   b) How many completed the course in:
      i) 249 s?   ii) 253 s?   iii) 264 s?
   c) How many skiers finished in less than 4 min 18 s?
   d) How many contestants were in the top 20%

5. The heights, in centimetres, of 27 students in Elmwood Collegiate are listed below.

   150 162 171 159 168 148 175 163 157
   150 168 170 172 155 164 153 160 149
   176 161 155 149 156 168 172 158 173

   Construct a stem-and-leaf diagram using:
   a) 10 cm intervals;   b) 5 cm intervals.

B 6. From the histogram shown below (left),
   a) determine how many students received marks in Science;
   b) determine how many students had a mark between 60 and 69;
   c) make a stem-and-left diagram of the marks.

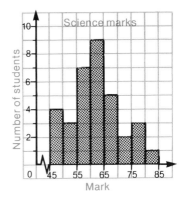
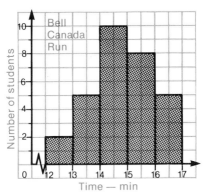

7. A group of 13-14 year old students completed the Bell Canada Run in the times indicated in the histogram above (right).
   a) How many students ran?
   b) What percent of the students completed the run in 13 min?

8. A typical Life Insurance Mortality graph is shown.
   a) What is the death rate per 1000 at age:
      i) 25?   ii) 50?   iii) 70?
   b) At what age is the death rate per 1000:
      i) 5?   ii) 40?
   c) Predict the death rate at age 80.

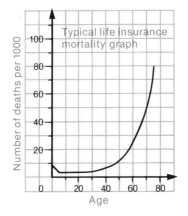

9. The graph below shows the wind-chill temperature on a day when the still-air temperature is −12°C.
   a) What is the wind-chill temperature when the wind speed is:
      i) 20 km/h?   ii) 35 km/h?
   b) What wind-speed gives a wind-chill temperature of −35°C?

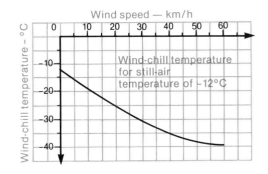

10. The graph below (left) shows how the distance to the horizon varies with the height from which it is seen.
    a) How far is the horizon when seen from a tower:
       i) 40 m high?   ii) 60 m high?
    b) How high is an observer if the horizon is:
       i) 15 km away?   ii) 30 km away?

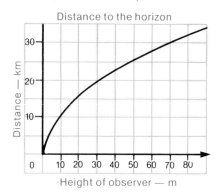
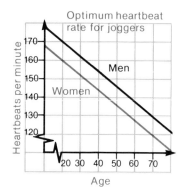

11. The graph above (right) shows the optimum rate of heartbeat for joggers according to age and sex.
    a) What is the optimum heartbeat rate for those whose age is: i) 30?   ii) 45?   iii) 57?
    b) How old is a person whose optimum heartbeat rate is:
       i) 145 beats/min?   ii) 135 beats/min?

8-2 Measures of Central Tendency: Simple Data

The table gives the lengths of the drives of two golfers in a driving contest. Who hits a longer drive, on the average, Lise or Monique?

	Length of Drive—metres						
Lise	260	302	268	260	250	288	291
Monique	285	257	282	252	290	280	250

The answer to the question depends on which of the following meanings of "average" is chosen.

8-2 Measures of Central Tendency: Simple Data

- The **mean** is the arithmetical average of the numbers in a set—the sum of all the numbers divided by the number of numbers.

- The **median** is the middle number when the numbers are arranged in order. If there is an even number of numbers, the median is the mean of the two middle numbers.

- The **mode** of a set of numbers is the most frequently occurring number. There may be more than one mode, or no mode.

Since the mean, median, and mode of a set of numbers are single values around which the numbers cluster, they are called **measures of central tendency**.

Arranged in order, the lengths of the drives, in metres, and their sums are:

Lise: 250, 260, 260, 268, 288, 291, 302.   Sum = 1919

Monique: 250, 252, 257, 280, 282, 285, 290.   Sum = 1896

The average distances for the drives of each golfer are:

	Mean	Median	Mode
Lise	$\frac{1919}{7} \doteq 274$ m	268 m	260 m
Monique	$\frac{1896}{7} \doteq 271$ m	280 m	None

If the average used is the mean, Lise's average drive is longer. If the median is used, then Monique's average drive is longer.

Averages are usually calculated using larger sets of data than those for the golfers. The following example shows how to use a stem-and-leaf diagram to find the median of a set of numbers.

Example. A school's grade 10 students obtained the following marks on a mathematics examination. Find the median.

```
63  57  82  80  71  93  44  53  58  61  55
80  66  69  63  72  77  81  50  58  70  62
85  83  74  42  57  62  68  81  58  74  71
79  80  60  67  95  41  65  73  82  77  60
67  63  72  78  50  57  60  65  55  82  86
79  69  60  63  71  78  68  60  58  70
```

Solution. Form the stem-and-leaf diagram.

```
4 | 4 2 1
5 | 5 7 7 8 5 8 3 7 8 0 8 0
6 | 3 5 0 6 2 5 8 9 0 3 0 7 0 3 7 9 2 0 1 8 3
7 | 0 1 0 3 8 4 1 7 1 9 2 4 7 9 2 8
8 | 1 0 2 2 0 5 3 2 0 6 1
9 | 3 5
```

Since there are 65 marks, the median is the 33rd mark when the marks are ordered—there are 32 lower marks and 32 higher marks. It is seen from the stem-and-leaf diagram that there are 36 marks in the 40's, 50's, and 60's. The median is therefore the fourth highest mark in the 60's. The highest four marks in this row, in order, are 68, 68, 69, 69. Therefore the median is 68.

Exercises 8-2

A 1. Find the mean, median, and mode:
   a) 5, 9, 4, 6, 2, 8, 7, 6, 3
   b) 24, 33, 25, 29, 32, 37, 25, 40, 38, 25, 33
   c) 3.7, 4.2, 7.1, 5.8, 6.3, 4.8, 5.2, 6.3, 5.4, 3.9
   d) 114, 92, 126, 85, 94, 109, 111, 88, 96, 107, 100, 105, 95, 90, 97, 99, 101
   e) 65, 52, 73, 86, 58, 47, 65, 78, 69, 58, 71, 67, 56, 57, 55, 58, 54, 59, 60, 70, 75, 55, 59, 61, 71, 85, 80, 81, 79, 66

2. The heights, in centimetres, of 13 players on a junior basketball team are as follows: 172, 182, 178, 187, 183, 176, 182, 182, 178, 173, 177, 181, 176. Calculate the mean, median, and mode.

3. Bobolink Sports Equipment employs 16 people at a weekly salary of $200, 3 people at a weekly salary of $150, and 2 people at a salary of $750 per week. Calculate the mean, median, and modal salary paid by the company.

4. Find the mean, median, and mode:

a)
```
3 | 7
4 | 6 3 9
5 | 5 2 8 5
6 | 3 2
```

b)
```
17 | 8 6
18 | 3 0 2 3
18 | 5 9 7 8 5
19 | 2 1 4 3 2 2
19 | 7 5 8
```

8-2 Measures of Central Tendency: Simple Data

5. The batting averages of the Tillson Giants were: 0.263, 0.309, 0.350, 0.207, 0.256, 0.278, 0.378, 0.283, 0.274, and 0.229.
   a) Find the median.
   b) Find the mean of these averages.
   c) Would the mean necessarily be the team's batting average?

B 6. From the clipping, determine:
   a) the total value of the prizes;
   b) the total number of prizes;
   c) the mean value of the prizes.

> **Hello, young lovers. Did you win $5000?**
>
> MONTREAL (CP)—A total of 770 winning numbers were drawn in a special Loto Canada "Sweetheart Draw" on Valentine's Day.
>
> In a departure from its usual million-dollar lottery, Loto Canada awarded 70 prizes of $5000 each and 700 prizes of $500 each. The winners were part of a Valentine's Day bonus program.

7. The prizes in a lottery were as follows:

       5 prizes of $100 000
       2 prizes of $75 000
     102 prizes of $7 500
     102 prizes of $750
   53 000 prizes of $25

   Find:
   a) the mean value of the prizes;
   b) the median value;
   c) the mode.

8. A bowler had these scores after eight games of 5-pin bowling: 299, 321, 317, 396, 245, 390, 340, 272.
   a) Find: i) the mean score; ii) the median score.
   b) In two more games, the bowler scored 173 and 216. Find, for the ten games: i) the mean score; ii) the median score.

9. In 1980, George Brett of the Kansas City Royals almost became the first major league baseball player since 1941 to have a 0.400 batting average. By September 22, he had 162 hits in 411 times at bat. During the remaining games of the season, he had 13 hits in 38 times at bat.
   a) What was his batting average on September 22?
   b) What was his batting average for the entire season?
   c) How many more hits did he need to reach a 0.400 batting average?
   d) In 1941, Ted Williams had 185 hits in 456 times at bat. What was his batting average?

10. In a set of data, each number is multiplied by 3. What changes occur in the measures of central tendency?

8-3 Measures of Central Tendency: Grouped Data

Sometimes, data has already been grouped and displayed, as in a histogram. In such cases, other methods are required to determine the measures of central tendency. The following example illustrates one such method.

Example. 1000 Tiger Track tires were road tested until treadwear indicators became visible. The distance travelled by each tire was recorded and the results summarized in a histogram. The same test was performed on 1000 Puma Paw tires. Use the histograms to determine which brand of tire has

a) the longer median lifetime;
b) the longer mean lifetime.

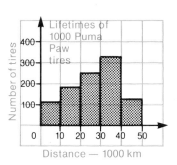
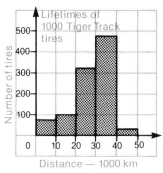

Solution. a) From the histograms, it is evident that the 500th (and 501st) longest wearing tire in the Tiger Track set lasted somewhere between 20 000 km and 30 000 km. Therefore, the median lifetime of Tiger Track tires is between 20 000 km and 30 000 km. Similarly, the median lifetime of Puma Paw tires is between 30 000 km and 40 000 km.

b) Since the exact lifetimes of individual tires cannot be determined from a histogram, the mean lifetime is calculated as if all the tires in each interval had the same lifetime, the middle value of the interval. Mean lifetime of Tiger Track tires:

$$\frac{110(5) + 192(15) + 248(25) + 325(35) + 125(45)}{1000}$$

= 26.63 thousand kilometres

Mean lifetime of Puma Paw tires:

$$\frac{75(5) + 102(15) + 320(25) + 476(35) + 27(45)}{1000}$$

= 27.78 thousand kilometres

The Puma Paw tires have longer median and mean lifetimes than Tiger Track tires.

In the last example, although the average lifetime of Puma Paw tires is longer than that of Tiger Track tires, it is evident from the histograms that a Tiger Track tire is more likely to run for more than 40 000 km. The manufacturers might use the data in their advertising as follows.

> **Puma Paw Tires go 4% farther than those of our leading competitor.**

> **Tiger Track Tires are over 4 times more likely to last 40 000 km than the tires of our leading competitor.**

Is each advertisement correct based on the test data?

Exercises 8 - 3

A 1. Find the mean, median, and mode:
a)

b)

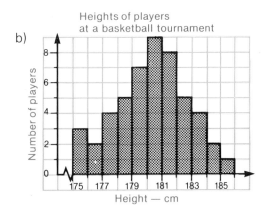

2. Find the mean, median, and mode:
a)

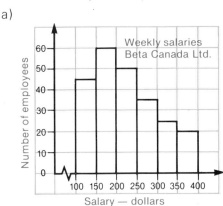

b)

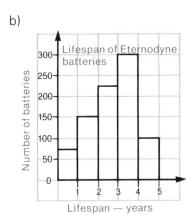

3. Find the mean, median, and mode:

a)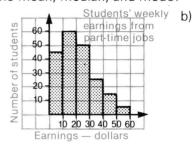
b)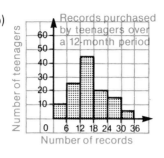

4. Find the mean, median, and mode:

a)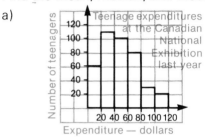
b)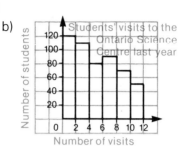

5. Find the mean, median, and mode:

a) Class marks on a quiz

Mark	Frequency
6	//
7	/
8	////
9	𝍷𝍷𝍷𝍷𝍷 ///
10	𝍷𝍷𝍷𝍷𝍷 𝍷𝍷𝍷𝍷𝍷
11	𝍷𝍷𝍷𝍷𝍷 /
12	///
13	/

b) Sample expenditures for lunch

Cost	Frequency
$1.00	𝍷𝍷𝍷𝍷𝍷 𝍷𝍷𝍷𝍷𝍷 ///
1.25	𝍷𝍷𝍷𝍷𝍷 𝍷𝍷𝍷𝍷𝍷 𝍷𝍷𝍷𝍷𝍷 ///
1.50	𝍷𝍷𝍷𝍷𝍷 𝍷𝍷𝍷𝍷𝍷 𝍷𝍷𝍷𝍷𝍷 𝍷𝍷𝍷𝍷𝍷
1.75	𝍷𝍷𝍷𝍷𝍷 𝍷𝍷𝍷𝍷𝍷 𝍷𝍷𝍷𝍷𝍷 𝍷𝍷𝍷𝍷𝍷 ///
2.00	𝍷𝍷𝍷𝍷𝍷 𝍷𝍷𝍷𝍷𝍷 //
2.25	𝍷𝍷𝍷𝍷𝍷 ////
2.50	𝍷𝍷𝍷𝍷𝍷

B 6. If your marks (out of 25) on six tests are 19, 18, 22, x, 17, and 23, find the value of x to give you a mean mark of 20.

7. In a set of data, the smallest number is decreased by 3 and the largest number is increased by 3. What changes occur in the measures of central tendency?

C 8. Find 7 numbers that have
a) a mean of 12 and a median of 13;
b) a mean of 62 and a median of 65;
c) a mean of 8, a median of 10, and a mode of 5;
d) a mean of 15, a median of 12, and a mode of 15;

8-4 Measures of Dispersion

In Section 8-2, measures of central tendency were used to determine which of two golfers was the longer driver based on a sample of 7 drives. The same data can be used to determine which golfer is more consistent. That is, which golfer has the least variation in distance when she drives the ball. Numbers for expressing this variation are called **measures of dispersion**. Two common measures of dispersion are range and mean deviation.

The **range** of a set of numbers is the difference between the greatest and the least number.

The **deviation** of a number, x, from the mean, m, of a set of numbers is $|x - m|$. The **mean deviation** of a set of numbers is the mean of all such deviations.

Example 1. Based on the data in the previous section, which golfer, Lise or Monique, is more consistent in the distance of her drives?

Solution.

Lise		Monique	
Distance m	Deviation from the mean, 274	Distance m	Deviation from the mean, 271
260	$\|260 - 274\| = 14$	285	$\|285 - 271\| = 14$
302	$\|302 - 274\| = 28$	257	$\|257 - 271\| = 14$
268	$\|268 - 274\| = 6$	282	$\|282 - 271\| = 11$
260	$\|260 - 274\| = 14$	252	$\|252 - 271\| = 19$
250	$\|250 - 274\| = 24$	290	$\|290 - 271\| = 19$
288	$\|288 - 274\| = 14$	280	$\|280 - 271\| = 9$
291	$\|291 - 274\| = 17$	250	$\|250 - 271\| = 21$
	Total 117		Total 107

Mean deviation: $\frac{117}{7} \doteq 16.7$ Mean deviation: $\frac{107}{7} \doteq 15.3$

Range: $302 - 250 = 52$ Range: $290 - 250 = 40$

Since the mean deviation for the lengths of Lise's drives is 16.7 compared with a mean deviation of 15.3 for Monique's drives, we conclude that Monique's drives tend to be more consistent. Furthermore, the range of Lise's drives is 54 compared with a range of 40 for Monique's drives. This also indicates a smaller variation in the lengths of Monique's drives.

To compute the mean deviation of a set of numbers, the frequency of occurrence of each number must be considered, just as in computing the mean.

Mark	Frequency
14	//
15	////
16	//// ///
17	//// /
18	/

Example 2. Find the range and mean deviation for the set of marks shown.

Solution. The range of the marks is $18 - 14$, or 4.
The mean is:

$$\frac{2(14) + 4(15) + 8(16) + 6(17) + 1(18)}{21}$$

$$= \frac{28 + 60 + 128 + 102 + 18}{21}$$

$$= \frac{336}{21}, \text{ or } 16$$

The mean deviation is:

$$\frac{2|14 - 16| + 4|15 - 16| + 8|16 - 16| + 6|17 - 16| + 1|18 - 16|}{21}$$

$$= \frac{4 + 4 + 0 + 6 + 2}{21}$$

$$= \frac{16}{21}, \text{ or approximately } 0.76$$

Exercises 8 - 4

A 1. Find the range and mean deviation for each set of data:
  a) 8, 4, 6, 12, 10
  b) 3, 4, 8, 6, 7, 4, 8, 10, 13
  c) 3, 5, 7, 9, 18, 12, 15, 11
  d) 2, 9, 5, 9, 2, 9, 7, 10, 10
  e) 9, 5, 6, 7, 8, 3, 18
  f) 3, 2, 1, 3, 6, 5, 5, 4, 7

2. Find the range and mean deviation for each set of data:
  a) 12, 19, 22, 24, 28   b) 4, 19, 22, 24, 36
  c) 4, 13, 22, 30, 36   d) 12, 16, 22, 27, 28
  e) 17, 19, 22, 23, 24   f) 10, 14, 22, 29, 30

3. In Madrid, the average temperatures for the days in June were:

30	24	22	25	19
28	24	27	28	25
28	26	27	28	25
25	20	27	26	26
27	23	20	19	22
29	21	24	26	29

Find the range and the mean deviation.

B 4. Find the range and mean deviation:

a)
Mark	Frequency
20	///
30	ЖЖ /
40	ЖЖ ЖЖ
50	ЖЖ
60	/

b)
```
3 | 7
4 | 5 2
5 | 8 6 4 7
6 | 2 6
7 | 3
```

c)
```
12 | 8
13 | 4 1
13 | 7 5 6
14 | 2 0 1 1 3
14 | 9 6 5
15 | 2
```

d)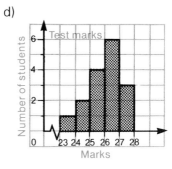

e)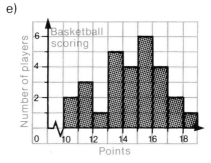

5. The marks of 23 driver-education students on a test out of 35 were:

9	29	18	30	20	35
13	21	18	20	19	19
26	28	28	17	34	27
24	32	21	22	19	

Find the measures of dispersion.

6. Find the measures of central tendency and the measures of dispersion:

a)
17	20	15	22	17
15	18	24	12	8
24	15	22	17	15
10	8	12	15	20

b)
Mark	0	1	2	3	4	5	6	7	8	9	10
Number of students	1	3	4	3	2	9	5	12	8	6	3

c)
```
8  10 11 13  8
4   6  8  5 11
5  13 14 13  6
```

d)
Salary	Frequency
$250	///
280	ЖЖ /
300	ЖЖ ЖЖ //
325	ЖЖ
350	///
895	/

e)
```
13 14 16 18 14
17 22 21 15 16
15 14 13
```

Mathematics Around Us

Estimating Wildlife Populations

One of the world's largest surviving buffalo herds lives in Wood Buffalo National Park which is on the boundary between Alberta and the Northwest Territories. It is estimated that there are about 12 000 buffaloes in the park. This number is determined by scientists who use a "capture-recapture" sampling technique.

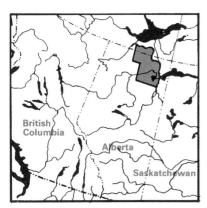

Wood Buffalo National Park

The "capture-recapture" sampling technique is based on the assumption that if enough members of a wildlife population are captured, tagged, and released, any sufficiently large sample of that population will contain the same proportion of tagged individuals as the entire population. One year, scientists captured, tagged, and released 300 buffaloes in Wood Buffalo National Park. The next year they captured 1000 buffaloes and found that 25 had tags. From this data, they estimated the number of buffaloes in the park as follows:

Let n represent the number of buffaloes in the park.
Proportion of tagged buffaloes in the entire population:
$$\frac{25}{1000}$$
Proportion of tagged buffaloes in the sample:
$$\frac{300}{n}$$
If the sample is representative of the population, then:
$$\frac{25}{1000} = \frac{300}{n}$$
$$25n = 300\ 000$$
$$n = 12\ 000$$

There are about 12 000 buffaloes in the park.

Questions

1. a) What assumptions are made about the buffalo in the park, and their movements, in the above calculation?

   b) What is the least number of buffaloes there could be?

2. A game warden nets and tags 250 lake trout in Fairy Lake. Two months later she nets 150 lake trout and finds that 30 of them are tagged. Estimate the number of lake trout in Fairy Lake.

3. Biologists, studying the migratory patterns of Canada geese, tag 1425 of the birds in the month of October. The next year, they capture 760 Canada geese and find that 285 are tagged. About how many Canada geese are in that region in October?

8-5 Sampling and Predicting

One of the principal uses of statistics is in making predictions. For example, suppose it is necessary to know the number of Canadians with type O blood. It is clearly not possible to check the blood types of all Canadians. Instead, a representative portion of the population, called a **sample**, is tested. From the results, the number of Canadians with blood type O can be predicted.

If, in a sample of 2000 Canadians, 900 have type O blood, the **relative frequency** of type O blood is $\frac{900}{2000}$, or 0.45.

In a population of 25 000 000, the number of Canadians with type O blood would be about:
$0.45 \times 25\,000\,000$, or $11\,250\,000$.

> If an outcome, A, occurs r times in n repetitions of an experiment, then the relative frequency of A is $\frac{r}{n}$.

In statistics, **population** is taken to mean the whole of anything of which a sample is being taken.

The accuracy of a prediction depends on the sample being sufficiently large and truly representative of the population as a whole. A representative sample is one chosen in such a way that each element of the population has an equal chance of being included. Such a sample is called a **random sample**.

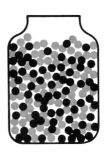

Example. A jar contains 3000 white, black, and red beads. The beads are thoroughly mixed and a sample of 60 taken. The sample is found to contain 17 white beads, 32 black beads, and 11 red beads. Estimate the number of beads of each color in the jar.

Solution. White: Relative frequency: $\frac{17}{60}$

Estimated number in jar: $\frac{17}{60} \times 3000$, or 850

Black: Relative frequency: $\frac{32}{60}$

Estimated number in jar: $\frac{32}{60} \times 3000$, or 1600

Red: Relative frequency: $\frac{11}{60}$

Estimated number in jar: $\frac{11}{60} \times 3000$, or 550

The answers given in the above example are only estimates of the number of beads of each color. If other samples are taken, the results may differ. For a more reliable estimate, it would be necessary to take several samples and find the mean of the results.

Exercises 8 - 5

A 1. A class of mathematics students tossed pennies a total of 36 000 times. Heads occurred 17 563 times. What was the relative frequency of heads?

2. A die was rolled 7200 times. The frequency of each outcome is shown in the table. What was the relative frequency of each outcome? Do you think it was a fair die? Explain.

Outcome	⚀	⚁	⚂	⚃	⚄	⚅
Frequency	1175	1225	1142	1168	1273	1217

3. A pair of dice was rolled 5350 times. A pair of 6's occurred 140 times. What was the relative frequency of a pair of 6's?

4. a) Toss a coin 50 times and record the frequency of heads.
   b) Calculate the relative frequency of heads.
   c) Combine your results with those of other students to obtain the relative frequency of heads for a greater number of tosses.
   d) How does the relative frequency of heads compare with 0.5 as the number of tosses increases?

5. When a thumbtack is tossed, there are two possible outcomes: If the relative frequency of "point up" is found to be 0.62, what should be the relative frequency of "point down"?

point up   point down

B 6. Choose 20 consecutive lines from a magazine story or newspaper article. Count the number of words and the number of syllables in each word. What is the relative frequency of words containing:
   a) one syllable?   b) more than two syllables?

7. a) Use a paper cup to shake and toss 5 coins. Record the number of heads.
   b) Repeat part (a) 19 times.
   c) Calculate the relative frequency of each outcome.
   d) Combine your results with those of other students to obtain more accurate values.
   e) Based on these results, if 5 coins were tossed 1280 times, about how many times would you expect to get:
      i) 1 head?   ii) 3 heads?   iii) 5 heads?

8. a) Toss two dice 25 times and record the results.
   b) Combine your results with those of at least three other students.
   c) Based on these results, if two dice were tossed 750 times, how many times would you expect them to show:
      i) a sum of 7?   ii) a sum of 11?   iii) a product of 12?

9. A dental survey of 360 students in Glentown High School revealed that 135 of them had two or more cavities. If the total school enrolment is 1656, about how many students would you expect to have two or more cavities?

10. The table shows the blood types of a random sample of Inuits.

Blood type	O	A	B	AB
Number of Inuits	75	59	14	8

    Estimate the number of Inuits with each blood type in a population of 1850.

C 11. In a random survey, 240 homeowners in each of towns A and B were asked if they used a water softener. 80 in town A and 66 in town B said they did. In which town should a manufacturer of water softeners concentrate his sales efforts if the populations of the towns are:
   a) A—4620, B—5265?   b) A—8430, B—7856?

12. A jar contains 1250 red, blue, and yellow marbles. Ten marbles are taken out at random and their colors recorded. This is done eight times, the marbles being returned to the jar after each draw. From the following table of results, estimate the number of marbles of each color in the jar.

Draw	1st	2nd	3rd	4th	5th	6th	7th	8th
Red	5	3	3	1	3	4	2	4
Blue	2	2	3	2	3	1	2	0
Yellow	3	5	4	7	4	5	6	6

13. a) Count the number of photographs on a random sample of 20 pages in this book.
    b) Use the result of (a) to predict the number of photographs in the whole book.
    c) Combine your results with those of others to obtain a more accurate prediction of the total number of photographs.

8-6 Probability

Every day we encounter statements such as:
- There is a 20% chance of rain.
- The probability of winning a prize in a draw is 0.02.
- Sheri's chances of living past age 60 are 70%.
- The probability of drawing a spade from a well-shuffled deck of cards is 0.25.

This last statement follows from the fact that 13 of the 52 cards in a deck of cards are spades. If a card is drawn, there are 52 equally likely outcomes of which 13 are favorable to spades. We say that the probability of drawing a spade is $\frac{13}{52}$, and we write:

$$P(\text{spade}) = \frac{13}{52}$$
$$= \frac{1}{4}, \text{ or } 0.25$$

Any outcome, or set of outcomes, of an experiment is called an **event**.

> If an experiment has n equally likely outcomes of which r are favorable to event A, then the **probability** of event A is: $P(A) = \frac{r}{n}$.

By permission of Johnny Hart and Field Enterprises, Inc.

Example 1. Give the meaning of each statement without using the words "chance" or "probability".
  a) There is a 20% chance of rain.
  b) The probability of winning a prize in a draw is 0.02.
  c) Sheri's chances of living past age 60 are 70%.

Solution.
  a) In the past, when weather conditions have resembled present conditions, there has been rain in 20% of the cases.
  b) The ratio of the number of winning tickets to the total number of tickets is $\frac{2}{100}$, or $\frac{1}{50}$.
  c) About 70% of the girls Sheri's age and in similar health are expected to live beyond age 60.

Example 2. A card is drawn from a well-shuffled deck. Find the probability that the card is:
  a) a club;
  b) an ace;
  c) a red card;
  d) the 7♣.

Solution. There are 52 cards and each has the same chance of being drawn.

a) There are 13 clubs. $P(\text{club}) = \frac{13}{52}$
$= \frac{1}{4}$, or 0.25

b) There are 4 aces. $P(\text{ace}) = \frac{4}{52}$
$= \frac{1}{13}$, or about 0.077

c) There are 26 red cards. $P(\text{red card}) = \frac{26}{52}$
$= \frac{1}{2}$, or 0.5

d) There is one 7♣. $P(7♣) = \frac{1}{52}$, or about 0.019

Example 3. If a die is thrown 300 times, about how many times should it show 5?

Solution. Since $P(5) = \frac{1}{6}$, the die should show 5 about $\frac{1}{6} \times 300$, or 50 times.

Exercises 8-6

A 1. A card is drawn from a well-shuffled deck of cards. What is the probability the card drawn will be:
   a) a heart?
   b) black?
   c) a 5?
   d) a red Jack?
   e) a black 3, 6, or 9?

2. If one letter is selected at random from the word "mathematics", what is the probability it will be:
   a) an "m"?
   b) an "e"?
   c) a vowel?
   d) a "t" or an "h"?
   e) one of the first ten letters of the alphabet?

3. Nashila buys 3 tickets for a Student-Council Christmas draw. What is the probability of her winning if the number of tickets sold is:
   a) 360?
   b) 600?
   c) 945?

4. A lottery issues 100 000 tickets. What is the probability of your winning if the number of tickets you hold is:
   a) 1?
   b) 10?
   c) 120?
   d) none?
   e) 100 000?

5. A die is tossed. What is the probability that the die shows:
   a) a 3?
   b) an even number?
   c) a perfect square?
   d) a prime number?

6. When the wheel is spun, what is the probability it will stop with the arrow pointing to:
   a) a 4?
   b) an odd number?
   c) a prime number?
   d) a number divisible by 3?
   e) a number greater than 5?
   f) a number less than 8?
   g) a two-digit number?
   h) a one-digit number?

7. The game of euchre is played with the 9, 10, Jack, Queen, King, and Ace of each suit. If one card is drawn from those cards, what is the probability that it will be:
   a) a 9 or 10?
   b) a face card?
   c) a red ace?
   d) the queen of hearts?

8. About how many times should a die show ⊡ if it is tossed:
   a) 50 times?
   b) 500 times?
   c) 5000 times?

B 9. If 26 cards are dealt from a well-shuffled deck of 52 cards, about how many should be:
   a) hearts?
   b) face cards?
   c) aces?

10. Ibrahim says that the probability of passing a test is 0.5 because there are two possible outcomes, pass or fail. Do you agree? Why?

11. A bag contains 40 marbles—12 red, 10 yellow, and 18 blue. If one is taken out at random, what is the probability it will be:
    a) red?
    b) yellow?
    c) blue?
    d) not blue?
    e) red or blue?
    f) green?

12. Bags A, B, and C contain green and red counters in the numbers shown. From which bag would you stand the best chance of selecting a green counter in one draw?

  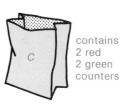

A contains 5 red 7 green counters

B contains 4 red 6 green counters

C contains 2 red 2 green counters

13. Slips of paper, numbered from 1 to 30, are placed in a bowl. If one is selected at random, what is the probability that it bears a number with one or both digits a 2?

14. A card is drawn from a deck of cards, replaced, and the deck shuffled. If this is done 1000 times, about how many times should the card drawn be:
    a) black?
    b) a Queen?
    c) a diamond?
    d) the ace of spades?

15. In the game "In Between", the value of the third card dealt must be in between the first two in order to win. What is the probability of winning if the two cards already dealt are:
    a) a 2 and a 6?
    b) a 5 and a Queen?
    c) a 7 and an 8?
    d) the Jack and a King?

16. The table shows the distribution of blood types among Canadians.

Blood type	O	A	B	AB
Percent of Canadians	45%	40%	11%	4%

    a) Determine the probability that a person selected at random will have:
       i) type AB blood;
       ii) either type A or type B blood;
       iii) neither type A nor type O blood.
    b) About how many Canadians in a sample of 100 000 would you expect to have type B blood?
    c) A hospital tries to keep 35 bottles of type AB blood on hand. About how many bottles of type A should be kept on hand?

C 17. The table shows how the blood types of donor and patient must be matched. The check marks indicate compatible combinations of blood types.

	Patient			
Donor	O	A	B	AB
O	✓	✓	✓	✓
A		✓		✓
B			✓	✓
AB				✓

    Use this table and the one in the previous exercise to find the probability that a person selected at random may give blood to a person with:
    a) type A blood;   b) type B blood;   c) type AB blood.

8-7 The Probability of Two or More Events

Suppose that a coin and a die are tossed. What is the probability that the coin shows heads and the die shows an even number?

The **tree diagram** shows that there are 12 possible outcomes. The event that the coin shows heads and the die shows an even number has three favorable outcomes: $H2$, $H4$, $H6$.

P (heads and even number) $= \dfrac{3}{12}$

$= \dfrac{1}{4}$

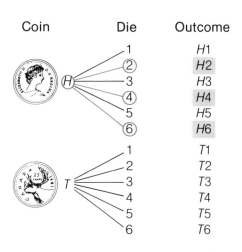

Another way of finding the above probability is by noticing that the number of outcomes involving heads is one-half the total number of outcomes, and the number of outcomes involving an even number is also one-half the total number of outcomes.

P (heads and even number) $= \dfrac{1}{2} \times \dfrac{1}{2}$, or $\dfrac{1}{4}$

The probability of two or more events is the product of the probability of each event.

Example 1. A die is tossed three times. What is the probability that it shows 5 or greater each time?

Solution. For each toss, $P(5 \text{ or greater}) = \dfrac{2}{6}$, or $\dfrac{1}{3}$

P (5 or greater on all 3 tosses) $= \dfrac{1}{3} \times \dfrac{1}{3} \times \dfrac{1}{3}$, or $\dfrac{1}{27}$

When a die is tossed several times, the outcomes are independent of each other. Care must be taken in calculating the probability of two or more events when the outcomes are related.

Example 2. A bag contains 5 black balls and 5 red balls. Find the probability of drawing 2 red balls in succession if

a) the first ball is replaced before drawing the second;

b) the first ball is not replaced.

Solution. a) First draw: There are 10 balls of which 5 are red.

$P(\text{red}) = \dfrac{5}{10}$, or $\dfrac{1}{2}$

Second draw: There are 10 balls of which 5 are red.

$P(\text{red}) = \dfrac{5}{10}$, or $\dfrac{1}{2}$

$$P(\text{2 reds in succession}) = \frac{1}{2} \times \frac{1}{2}$$
$$= \frac{1}{4}, \text{ or } 0.25$$

b) $\quad P(\text{red on first draw}) = \frac{1}{2}$

Second draw: There are 9 balls of which 4 are red.
$$P(\text{red on second draw}) = \frac{4}{9}$$
$$P(\text{2 reds in succession}) = \frac{1}{2} \times \frac{4}{9}$$
$$= \frac{2}{9},$$
$$\doteq 0.22$$

Exercises 8-7

A 1. A coin and a regular tetrahedron (with faces marked 1, 2, 3, 4) are tossed. Draw a tree diagram to find the probability of getting:
   a) a head and a 2;
   b) a tail and a perfect square.

2. What is the probability of two tossed dice showing a sum of 12?

3. What is the probability of tossing four coins and getting four heads?

4. If it is equally likely that a child be born a girl or a boy, what is the probability that:
   a) a family of three children will be all girls?
   b) a family of six children will be all boys?

B 5. A baseball player has a batting average of 0.300.
   a) What is the probability that the player has a hit each time he comes to bat?
   b) If he comes to bat four times in the next game, what is the probability that: i) he has 4 hits?   ii) he has no hits?

6. A weather report gives the chance of rain on both days of the weekend as 80%. If this is correct, what is the probability that:
   a) there is rain on both days?
   b) it does not rain on either day?

7. A multiple-choice test has 4 questions. Each has 5 responses, only one of which is correct. If all the questions are attempted by guessing, what is the probability of getting all 4 right?

8. A bag contains 3 red balls and 5 green balls. Find the probability of drawing 2 green balls in succession if:
   a) the first ball is replaced before drawing the second;
   b) the first ball is not replaced.

9. Repeat the previous exercise for a bag containing:
   a) 4 red balls and 6 green balls;
   b) 5 red balls and 4 green balls;
   c) 8 red balls and 12 green balls.

10. A meal at a fast-food outlet has three items:
    i) a hamburger, cheeseburger, or hot dog
    ii) a soft drink or shake
    iii) a sundae, a piece of pie, or cookies.
    If a choice is made at random, what is the probability that a meal will include:
    a) a hamburger?   b) a hot dog and a shake?
    c) a cheeseburger, a shake, and cookies?

11. Two cards are drawn from a well-shuffled deck. If
    i) the first card is replaced before the second is drawn,
    ii) the first card is not replaced,
    find the probability that they are both:
    a) spades;   b) aces;   c) face cards.

12. In the SCRABBLE® Brand Crossword Game, the letters of the alphabet are distributed over 100 tiles as shown in the table. Two tiles are selected simultaneously from a full bag of tiles. What is the probability that they are both:
    a) "B"?   b) "E"?   c) vowels?

13. The word "Mathematics" is spelled out with SCRABBLE® Brand Crossword tiles and the tiles are put in a bag. What is the probability that two tiles drawn simultaneously will be:
    a) both vowels?   b) both consonants?   c) both "M"?

14. In basketball, a player given a one-and-one foul shot is given a second shot only if the first is successful. Thus, the player can score 0, 1, or 2 points in this situation. If the player shoots with 75% accuracy, find the probability that she will score:
    a) 0 points;   b) 1 point;   c) 2 points.

C 15. What is the probability that two persons selected at random would both have birthdays in:
    a) February?   b) the same month?
    (What assumption are you making?)

Distribution of Tiles		
A—9	J—1	S—4
B—2	K—1	T—6
C—2	L—4	U—4
D—4	M—2	V—2
E—12	N—6	W—2
F—2	O—8	X—1
G—3	P—2	Y—2
H—2	Q—1	Z—1
I—9	R—6	Blank—2

16. Five dice are tossed simultaneously. Find the probability that:
    a) they all show 6;
    b) no die shows 6;
    c) no die shows 5 or 6;
    d) they all show the same number.

17. Cards are dealt until a Jack appears. What is the probability that the first Jack is:
    a) the second card?
    b) the third card?

18. Cards are dealt until a heart appears. What is the probability that the first heart is:
    a) the second card?
    b) the third card?

8 - 8 Monte Carlo Methods

For some problems, advanced mathematical techniques are needed to calculate probabilities precisely. However, good estimates of probabilities can still be obtained without these techniques by simulating problems. This can be done in two ways:
- by designing an experiment;
- by using random numbers.

The techniques used in the following examples are called **Monte Carlo Methods**. The study of such methods forms an important branch of statistics.

Example 1. Design a classroom experiment to estimate the probability that a student, guessing the answers on a 10-question, true-false test, will get 6 or more correct.

Solution. Since the probability of guessing the correct answer for each question is $\frac{1}{2}$, the probability can be simulated by tossing coins. Toss 10 coins many times, each time recording if 6 or more heads show. The fraction of times they do show gives an estimate of the required probability.

The results of one such experiment were 21 favorable outcomes in 50 trials. Therefore, an estimate of the probability of getting 6 or more correct answers in a 10-question, true-false test by guessing is $\frac{21}{50}$, or 0.42. For a closer estimate, many more trials are needed.

Example 2. Design a classroom experiment to estimate the probability that in a group of five people at least two were born in the same month.

Solution. Assuming that the probability of being born in a particular month is $\frac{1}{12}$, the situation can be simulated by using a coin and die. Assign the outcomes to months as follows:

H ⚀ January		H ⚀ February
H ⚁ March		H ⚁ April
H ⚂ May		H ⚂ June
T ⚃ July		T ⚃ August
T ⚄ September		T ⚄ October
T ⚅ November		T ⚅ December

Toss a coin and a die five times and record whether any month occurred more than once. Repeat the experiment many times.

In one such experiment, consisting of 25 trials, a month was duplicated 12 times. Therefore, an estimate of the required probability is $\frac{12}{25}$, or 0.48.

Exercises 8-8

For each exercise, design an experiment to arrive at an estimate of the probability.

A 1. Design a classroom experiment to estimate the probability that a student, guessing the answers on a 7-question, true-false test, will get 5 or more correct.

  2. Design a classroom experiment to estimate the probability that in a group of 7 people at least 3 were born under the same sign of the zodiac.

B 3. What is the probability that a family with 3 children will have 2 girls and a boy?

  4. According to a news report, 1 in every 6 railroad cars is defective. What is the probability that a 7-car train contains no defective cars?

  5. A survey indicates that 75% of all consumers prefer Brand *A* cola to Brand *B*. If a sample of 10 consumers is chosen, what is the probability that 8 or more of them prefer Brand *A*?

  6. Lynn and Lisa are playing tennis and have reached a score of 40-40, or deuce. Play continues until one player has a margin of two points. Estimate the probability that Lisa will win the game if the probability of winning each point played is 0.5.

2250 Random Numbers

1	67983	60852	09916	43596	20363	53315	37287	07662	26401	28650
2	19010	91956	31795	41845	25190	06991	66521	93755	02166	79003
3	41830	13963	52289	51633	77785	31712	93500	19449	77822	36645
4	50115	21246	09195	09502	53413	26357	63992	52872	42570	80586
5	22712	09067	51909	75809	16824	41933	97621	68761	85401	03782
6	82806	82277	88300	29832	22806	92486	36042	34590	55743	85297
7	68885	23670	25151	14619	33069	05296	14748	43282	62802	30626
8	41971	29316	23695	60065	62854	01237	72575	98475	61743	66763
9	86818	10485	28018	57382	70220	77420	94651	05024	24716	63746
10	61411	17729	56740	10634	56007	05873	36764	41765	97918	49916
11	20240	11618	52392	19715	20334	01124	39338	73458	63616	72057
12	47935	98490	99047	20071	81921	13627	99672	26523	53766	01219
13	28555	86201	62668	98919	54425	52470	21863	38900	96199	02418
14	59767	56647	35868	12109	29037	72768	45163	69121	72091	48070
15	99784	67224	77465	88593	61371	05036	41838	02224	34532	14840
16	69823	11868	50659	64782	20491	11303	41774	80579	09599	00703
17	89585	58666	00566	73433	67326	86922	42271	45800	59208	94299
18	93707	06735	84194	51810	19421	68021	05152	06217	57168	95760
19	04063	28256	83450	70758	00038	24278	55795	30155	78395	65622
20	89294	03751	09422	22965	09888	95835	80131	65972	16145	59876
21	60301	16519	51348	36322	70572	48637	05309	08369	79567	67699
22	74006	15355	95718	91467	30481	31576	84764	67417	19343	01920
23	86117	80403	42385	64085	70178	07265	87005	48570	25755	81223
24	87860	70624	75971	40430	43435	34945	70220	32445	18369	01990
25	10484	39599	04817	06980	22037	43080	52425	77667	67793	92230
26	45635	80376	17981	83957	91343	18249	85861	90149	59239	10040
27	89884	99155	65450	31432	60782	51442	31091	91187	81633	54164
28	08854	49077	20318	73772	85867	61524	78601	92812	34536	97897
29	43755	12282	84744	58693	25640	66247	58618	40854	85560	00699
30	36381	94203	18050	28540	97769	63915	65191	04638	76462	13106
31	05478	49611	27465	72222	56456	82646	09667	43683	33611	15020
32	30900	37036	68577	43276	57609	88486	16952	46799	49171	19846
33	87097	50134	42000	51378	60900	70086	51319	51408	85037	15608
34	84951	45154	20051	46979	79305	46375	16686	96475	54604	14795
35	52243	19460	67237	95379	78426	75457	05919	05828	13052	51831
36	48397	03688	27314	19086	58154	56293	78283	87702	17610	97741
37	36108	73699	01494	88477	18706	86938	40590	38087	22757	04249
38	73798	17752	23699	42632	77518	34777	66590	12061	35079	14551
39	26577	40103	74102	06328	43037	77254	78000	61577	41810	85898
40	81004	57367	28642	02357	23267	76973	29206	69086	42603	49297
41	27346	40553	51131	02911	76081	81510	32041	12592	53089	16876
42	75820	17172	54381	10928	63101	20817	08819	00264	38508	94741
43	06940	77259	57591	56640	35761	81718	40979	88063	50261	66925
44	17772	34298	05693	26833	91346	57941	76006	78275	85258	45102
45	01458	47852	91481	69165	93122	30194	92266	29754	80002	19151

8-9 Monte Carlo Methods With Random Numbers

Tossing a coin or die is awkward and time-consuming when it is necessary to make many trials. A more efficient method is to use random numbers. A very large set of single-digit numbers in which each number has an equal chance of occurring each time is called a set of **random numbers**. A set of random numbers is given on page 344, or you can generate your own by taking the last digits of the telephone numbers on any page of a telephone directory.

The next two examples show how random numbers can be used in Monte Carlo simulations.

Example 1. Use the table of random numbers on page 344 to estimate the probability that a student, guessing the answers on a 10-question, true-false test, will get 6 or more correct.

Solution. Start anywhere in the table and select a sequence of ten numbers. Record whether six or more of the digits are even. Repeat many times using other sequences of ten digits. Five typical results are as follows:

Starting at:	Random numbers	Number of even digits
6th number, 4th row	2 1 2 4 6 0 9 1 9 5	5
11th number, 15th row	7 7 4 6 5 8 8 5 9 3	4
26th number, 17th row	8 6 9 2 2 4 2 2 7 1	7
27th number, 21st row	8 6 3 7 0 5 3 0 9 0	5
34th number, 29th row	1 8 4 0 8 5 4 8 5 5	6

(0 is considered to be an even number.)

In 2 out of 5 sequences, 6 or more digits were even. Therefore, an estimate of the required probability is $\frac{2}{5}$, or 0.4.

In some problems, some of the numbers in the set of random numbers are disregarded.

Example 2. Each box of Barley Puff cereal contains one International Action hockey card. Assuming the cards are randomly distributed in equal numbers among the boxes, estimate the number of boxes of Barley Puff cereal it would normally take to obtain the entire set of six different cards.

Solution. Start anywhere in the table and record a sequence of numbers until each number from 1 to 6 occurs at least once. Note the number of numbers in the sequence excluding 7, 8, 9, 0. Repeat the procedure using other sequences. The following are five typical results:

Starting at:	Random numbers (disregarding 0, 7, 8, 9)	Number of numbers
11th number, 3rd row	5 2 2 5 1 6 3 3 5 3 1 1 2 3 5 1 4	17
1st number, 10th row	6 1 4 1 1 1 2 5 6 4 1 6 3	13
21st number, 17th row	6 3 2 6 6 2 2 4 2 2 1 4 5	13
36th number, 24th row	3 2 4 4 5 1 3 6	8
1st number, 36th row	4 3 3 6 2 3 1 4 1 6 5	11
	Total	62

An estimate of the mean number of boxes needed is $\frac{62}{5}$, or about 12. For a more reliable estimate, many more sequences of random numbers must be examined.

Exercises 8-9

For each exercise, use the table of random numbers to arrive at an estimate of the probability.

A 1. About how many times must a die be tossed until a 6 appears?

2. What is the probability that a family with 3 children will have 2 girls and a boy?

3. When a thumbtack is tossed, there are two possible outcomes: point up or point down. By experiment, it is found that the probability that the thumbtack lands with point up is 0.6. When two thumbtacks are tossed, what is the probability that one lands with the point up and the other lands with the point down?

B 4. If a red die and a white die are tossed, what is the probability that the red die shows a greater number than the white die?

5. According to a news report, 1 in every 5 railroad cars is defective. What is the probability that a 7-car train contains no defective cars?

6. Ten coins are tossed. What is the probability that there are 5 heads and 5 tails?

7. A survey indicates that 70% of all consumers prefer Brand A cola to Brand B. If a sample of 10 consumers is chosen, what is the probability that 8 or more of them prefer Brand A?

8. In a dice game, a player tosses a die three times. To score, the number appearing on the second toss must be in between the numbers appearing on the other tosses. What is the probability of scoring?

9. Lynn and Lisa are playing tennis and have reached a score of 40-40, or deuce. Play continues until one player has a margin of two points. Estimate the probability that Lisa will win the game, assuming that the probability of her winning each point played is: a) 0.6; b) 0.7.

10. A baseball player has a 0.300 batting average. Assuming that he will be at bat four times in the next game, estimate:
   a) the probability that he has 4 hits;
   b) the probability that he has no hits.

11. The championship series in the American and National Baseball Leagues are best-of-five series. In order to win the series, a team must win three games. A sports announcer remarked that in previous years the team which won the second game usually won the series. Estimate the probability that this is true, assuming that the teams are evenly matched.

12. The Stanley Cup hockey final is a best-of-seven series. To win the cup, a team must win four games. Assuming that the two teams are evenly matched, estimate the probability that:
   a) the series will last 7 games;
   b) the team which wins the first game wins the Cup.

COMPUTER POWER

Monte Carlo Methods With Computers

Computers are capable of generating random numbers which can be used to simulate experiments. The computer can scan thousands of numbers and record results in a matter of seconds. The computer programs that follow can be used to simulate the accompanying experiments. The programs will produce reasonably good estimates of the probabilities being investigated.

Example 1. What is the probability that a student guessing randomly on a 10-question true-false test will get at least 6 correct?

Solution. Estimate the probability, then run this program.

```
10   PRINT "THE EXPERIMENT IS BEING RUN."
20   PRINT "PLEASE WAIT."
30   FOR J = 1 TO 100:C = 0
40   FOR I = 1 TO 10
50   IF INT (2 * RND (1)) = 1 THEN 70
60   C = C + 1
70   NEXT I
80   IF C < 6 THEN 100
90   T = T + 1
100  NEXT J
110  PRINT "ON 100 TESTS, RANDOM GUESSING YIELDED AT LEAST 6 CORRECT ";
120  PRINT "ANSWERS ON "T" OCCASIONS."
130  PRINT "THE RELATIVE FREQUENCY WAS "T / 100
```

Run this program several times. Compare the relative frequencies with your estimate.

Example 2. What is the probability that at least two people in a group of five have birthdays in the same month? (Assume all months are equally likely.)

Solution. Estimate how many groups of five out of 10 such groups, and 100 such groups, contain at least two people with birthdays in the same month. Key in the following program and run it.

```
10   M = 10
20   PRINT "THE EXPERIMENT IS BEING RUN."
30   PRINT "PLEASE WAIT."
40   FOR I = 1 TO 5
50   X(I) = INT (12 * RND (1) + 1)
60   NEXT I
70   FOR J = 2 TO 5
80   FOR K = 1 TO J − 1
90   IF X (K) = X(J) THEN 130
100  NEXT K
110  NEXT J
```

```
120  N = N + 1: GOTO 140
130  S = S + 1:N = N + 1
140  IF N < M THEN 40
150  PRINT "IN "M" GROUPS OF 5 PEOPLE"
160  PRINT "THERE WERE "S" CASES OF COINCIDENCE."
170  PRINT "THE RELATIVE FREQUENCY OF COINCIDENCE IS "S / N
180  IF M = 10 THEN PRINT "PLEASE WAIT FOR NEXT RESULT."
190  IF M = 100 THEN END
200  M = 100: GOTO 40
```

Compare your estimates with the actual number of cases of coincidence. How close were your estimates? Run the program several times. Compare the results with the first run of the experiment.

Example 3. Each box of Barley Puff cereal contains one International Action hockey card. Assuming the cards are randomly distributed in equal numbers among the boxes, how many boxes of Barley Puff cereal would it normally take to get the complete set of six different cards?

Solution. Estimate the number of boxes of cereal it would take to get all six different cards. Then key in and run the following program.

```
10  N = 0
20  N = N + 1:S = 0
30  M = INT (6 * RND (1) + 1)
40  A(M) = 1
50  FOR I = 1 TO 6
60  S = S + A(I)
70  NEXT I
80  IF S < 6 THEN 20
90  PRINT "IT TOOK "N" BOXES TO GET ALL 6 CARDS."
```

Compare the number of boxes with your estimates. Run the program several times and compare the results. How would you get a good estimate of the average number of boxes that you would have to purchase to get all six cards?

Review Exercises

```
4 | 8
5 | 7 3 8 7
6 | 0 4 2 7 6 7 3 8 7
7 | 8 6 0 5 9 3 4 4 2
8 | 0 7 3 8 6
9 | 1 4
```

1. This stem-and-leaf diagram shows the marks obtained by students on a science test.
   a) How many students received a mark of:
      i) 57?   ii) 67?   iii) 77?
   b) How many students received a mark of 76 or higher?
   c) What was the most frequently-occurring mark?
   d) How many students took the test?

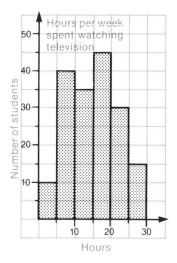

2. The histogram shows the results of a survey of student television-viewing habits. From the histogram, determine how many students:
   a) took part in the survey;
   b) watch television 15 h per week or less;
   c) watch more than 20 h.

3. Find the mean, median, and mode:
   a) 8, 3, 5, 6, 6, 4, 7, 5, 6, 8
   b) 36, 45, 42, 40, 37, 45, 48, 50, 45, 41, 39
   c) 5.4, 3.8, 5.2, 4.9, 3.9, 4.0, 4.1, 3.8, 4.2, 3.8

4. Find the mean, median, and mode:
   a)
   ```
   3 | 9
   4 | 5 2 2 6
   5 | 7 3 6 5 4
   6 | 3 2
   ```
   b)

Science Marks	
Mark	Frequency
8	/
9	𝍫 ///
10	𝍫 𝍫 //
11	𝍫 //
12	//

5. For each set of data in Exercise 3, find the range and mean deviation.

6. For each set of data in Exercise 4, find the range and mean deviation.

7. A pair of dice was rolled 5400 times. A pair of 6's occurred 144 times. What was the relative frequency of a pair of 6's?

8. A card is drawn from a well-shuffled deck of cards. What is the probability that the card drawn will be:
   a) a diamond?   b) red?   c) a 7?
   d) a black Queen?   e) a red 2, 5, or 8?

9. A bag contains 30 marbles—6 red, 9 yellow, and 15 blue. If one is taken out at random, what is the probability it will be:
   a) red?
   b) yellow?
   c) blue?
   d) not yellow?
   e) yellow or blue?
   f) green?

10. A die is tossed. What is the probability that the die shows:
    a) a 4?
    b) a number greater than 4?
    c) a number less than 4?
    d) a number divisible by 3?

11. A bag contains 5 red balls and 7 yellow balls. Find the probability of drawing 2 yellow balls in succession if:
    a) the first ball is replaced before drawing the second;
    b) the first ball is not replaced.

12. Two cards are drawn from a well-shuffled deck. If
    i) the first card is replaced before the second is drawn,
    ii) the first card is not replaced,
    find the probability that they are both:
    a) clubs;
    b) Kings;
    c) red face cards.

13. The letters in "Manitoba" are written on separate slips of paper and placed in a bag. What is the probability that two slips drawn simultaneously will both show:
    a) vowels?
    b) consonants?
    c) "a"?

14. A case of 150 flash cubes is known to contain 5 defective ones. If 2 cubes are taken from the box, what is the probability that both are defective?

15. The faces of two regular tetrahedrons are numbered 1 to 4. If they are tossed, what is the probability of the bottom faces:
    a) being a pair?
    b) having a total of 6?
    c) having a difference of 1?

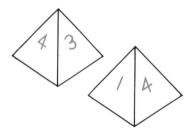

Cumulative Review (Chapters 6-8)

1. Find the image of the point (5, 7) under the transformation:
   a) $(x, y) \to (x + 2, y - 1)$;
   b) $(x, y) \to (-y, -x)$.

2. A triangle has vertices $A(-2, 2)$, $B(5, 2)$, $C(-2, 5)$. Graph the triangle and its image under the transformation:
   a) $(x, y) \to (x + 4, y + 3)$;
   b) $(x, y) \to (x - 1, y - 2)$.

3. Graph the line $x + 3y = 6$ and its image under the transformation: $(x, y) \to (-y, -x)$.

4. A triangle has vertices $K(3, 3)$, $L(7, 3)$, $M(7, 5)$.
   a) On the same grid, graph the triangle and its images under the transformations:
      i) $(x, y) \rightarrow (-x, y)$;   ii) $(x, y) \rightarrow (-y, -x)$.
   b) Describe each transformation.

5. A triangle has vertices $A(-3, 2)$, $B(1, 4)$, $C(4, 1)$. Graph $\triangle ABC$ and its image under the dilatation: $(x, y) \rightarrow (2x + 1, 2y - 3)$.

6. Decide if the conclusion follows logically from the statements:
   a) All musicians have long hair. Evlyn has long hair. Therefore, Evlyn is a musician.
   b) All artists are poor. Andy is an artist. Therefore, Andy is poor.

7. a) Prove: $\angle YXZ = \angle WXZ$ b) Prove: $MN \perp KL$

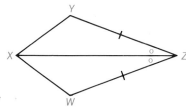 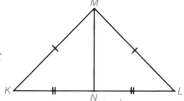

8. a) If $AC = BC$ and $\angle EAB = \angle DBA$, prove: $AE = BD$.

   b) If $\angle C = \angle D$, $AC = DE$, and $CB \perp AD$, prove: $CB = BD$.

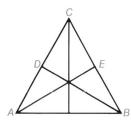 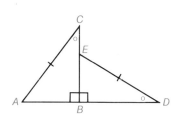

9. Find the values of w, x, y, and z:
   a)   b)

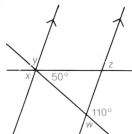 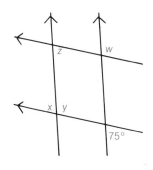

10. Find the values of x and y:

a)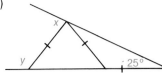
b)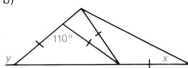

11. Write in the "if...then" form:
    a) A rectangle has four right angles.
    b) Congruent triangles are equal in area.
    c) The measure of an obtuse angle is between 90° and 180°.

12. a) Write the converse of each statement in Exercise 11, and decide whether it is true or false.
    b) Rewrite the statements that have a true converse in the "if, and only if" form.

13. Which sets of numbers represent the lengths of the sides of right triangles?
    a) 21, 28, 35   b) 8, 15, 17   c) 5, 12, 14

14. This stem-and-leaf diagram shows the marks obtained by students on a science test.
    a) How many students took the test?
    b) How many received a mark higher than 73?
    c) Find the mean, median, and mode for this set of data.
    d) Find the range and mean deviation for the marks.

5	3 0 7 8 8
6	9 7 6 5 0 3 6 6 8
7	5 3 6 7 8 6 3 4
8	3 8 5 7 7 6
9	3 1

15. A pair of dice was rolled 6300 times. A pair of 1's occurred 171 times. What was the relative frequency of a pair of 1's?

16. A card is drawn from a well-shuffled deck of cards. What is the probability that the card drawn will be:
    a) a heart?   b) black?   c) a 5?
    d) a red ace?   e) a black 4, 7, or 10?

17. A bag contains 50 marbles—28 red, 12 yellow, and 10 blue. If one is taken out at random, what is the probability it will be:
    a) red?   b) yellow?   c) blue?
    d) not red?   e) red or yellow?   f) black?

18. A bag contains 6 white balls and 9 black balls. Find the probability of drawing 2 white balls in succession if:
    a) the first ball is replaced before drawing the second;
    b) the first ball is not replaced.

Mathematics Around Us

Gwennap Pit is an historical amphitheatre in Cornwall, England. It was converted from a tin-mining pit in the 16th century. The famous clergyman, John Wesley, delivered sermons on 17 different occasions to crowds that overflowed the amphitheatre.

Gwennap Pit

The lowest ring, or stage, of the amphitheatre has a radius of about 3 m. Each successive ring of seats is 1 m wide.

Questions

1. What is the total seating area of Gwennap Pit, including the ring of seats at the top?

2. What is the seating area of
   a) the second ring?
   b) the fifth ring?
   c) the tenth ring?

3. If a person requires a seating area of 0.25 m², how many people will the amphitheatre seat?

4. How many people will be able to sit on
   a) the second ring?
   b) the fifth ring?
   c) the tenth ring?

5. If Gwennap Pit is completely full and a person is randomly chosen, what is the probability that that person will be sitting in
   a) the second ring?
   b) the fifth ring?
   c) the tenth ring?

9 Trigonometry

Two office towers are 25 m apart. From the 15th floor (40 m up) of the shorter tower, the angle of elevation of the top of the other tower is 72°. Find the angle of depression of the base of the taller tower from the 15th floor, and the height of the taller tower. (See *Example 3* in Section 9-5.)

9-1 Angle of Inclination

The steepness of a road, a roof, or a ski run can be determined by finding the slope. It can also be determined by measuring the angle of inclination.

The angle of inclination of the hill is 20°.

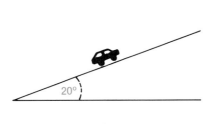

The angle of inclination of the roof is 45°.

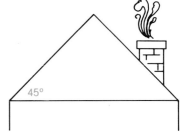

The **angle of inclination** of a line is the angle between the line and a horizontal line which meets it.

If either the slope or the angle of inclination of a line is known, the other can be found. One way to do this is to draw an accurate diagram.

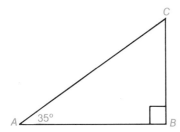

Example 1. Find the slope of a line with an angle of inclination of 35°.

Solution. With ruler and protractor construct any right $\triangle ABC$ such that $\angle A = 35°$. Measure AB and BC. In the diagram, $AB \doteq 4.0$ cm and $BC \doteq 2.8$ cm.

Slope of AC: $\frac{BC}{AB} \doteq \frac{2.8}{4.0}$, or 0.70

A line with an angle of inclination of 35° has a slope of approximately 0.70.

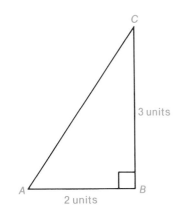

Example 2. Find the angle of inclination of a line with slope 1.5.

Solution. Since $1.5 = \frac{3}{2}$, construct any right $\triangle ABC$ such that $AB = 2$ units and $BC = 3$ units. Measure $\angle A$. In the diagram, $\angle A \doteq 56°$.

A line with slope 1.5 has an angle of inclination of approximately 56°.

Examples 1 and *2* suggest that there is a unique value of the slope for each angle of inclination less than 90°. The slope is called the **tangent** of the angle of inclination.

9-1 Angle of Inclination

The symbol, tan θ, represents the slope of a line with an angle of inclination of θ degrees, $0° \leq θ < 90°$.

$$\tan θ = \frac{\text{rise}}{\text{run}}$$

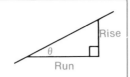

It is neither efficient nor accurate to determine the tangent of a given angle or the angle corresponding to a given tangent by drawing and measuring, as in the first two examples. A better method is to use a table, such as the one on page 403 or a scientific calculator.

Example 3. Find the slope of a line that has an angle of inclination of 24°.

Solution. The slope of the line is tan 24°. From the table of tangents, tan 24° ≐ 0.445. This value can also be obtained with a scientific calculator (in degree mode) by entering 24 and pressing the [tan] key.
A line with an angle of inclination of 24° has a slope of approximately 0.445.

Example 4. Find the angle of inclination of a line with slope 3.7.

Solution. If θ is the angle of inclination, then tan θ = 3.7.
From the table, θ is closest to 75°.
If 3.7 is entered on a scientific calculator and the keys [inv] [tan] pressed, the more accurate value of 74.88° is obtained.
A line with slope 3.7 has an angle of inclination of approximately 75°.

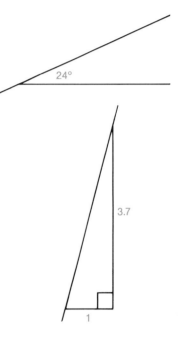

Exercises 9-1

A 1. Draw each right triangle on a grid and measure the angle of inclination:

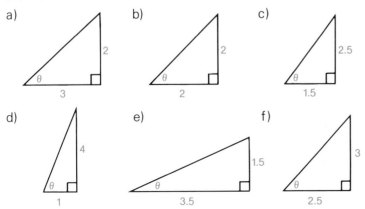

2. Use a grid and protractor to find the angle of inclination of a line with slope
   a) $\frac{3}{4}$; b) $\frac{2}{5}$; c) $\frac{3}{5}$; d) $\frac{4}{3}$; e) $\frac{5}{2}$; f) $\frac{5}{3}$.

3. Use ruler and protractor to find the slope of a line that has an angle of inclination of
   a) 40°; b) 65°; c) 75°;
   d) 20°; e) 30°; f) 55°.

4. Use the table on page 403 or a scientific calculator to find the angle of inclination:

a) 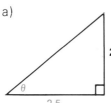   b) 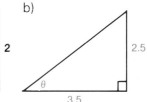   c)

d) 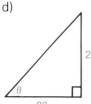   e) 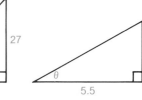   f)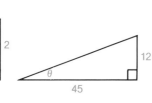

5. Find, as accurately as possible, the angle of inclination of a line with slope
   a) $\frac{2}{3}$; b) $\frac{5}{8}$; c) $\frac{4}{9}$; d) 2.4; e) $\frac{2}{7}$; f) $\frac{41}{12}$.

6. Find, as accurately as possible, the slope of a line that has an angle of inclination of
   a) 25°; b) 69°; c) 5°;
   d) 45°; e) 60°; f) 89°.

B 7. A rectangle measures 6 cm by 3 cm. Find the measures of the two acute angles formed at A by the diagonal AC.

8. An isosceles $\triangle PQR$ has a base, QR, of length 10 cm and a height of 4 cm. Find the measures of the three angles of the triangle.

9. In isosceles $\triangle ABC$, $AB = AC$, $\angle B = 70°$, and $BC = 4$ cm. Find the height of the triangle to the nearest millimetre.

10. Find the acute angle formed by the line and the x-axis:

a)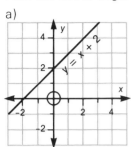
b)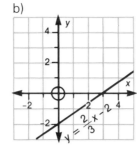
c)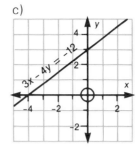

11. Find the obtuse angle formed by the line and the x-axis:

a)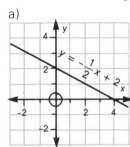
b)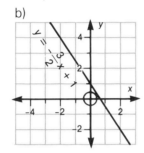
c)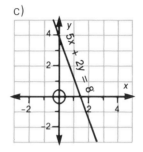

C 12. A rectangle measures 12 cm by 8 cm. Find the acute angle at the intersection of the two diagonals.

13. The diagram shows squares arranged side by side to form two rectangles at right angles. The rectangles can be lengthened by adding more squares.

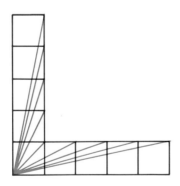

a) Find the angles of inclination of the diagonals shown.
b) How many squares would be needed on the vertical column to have an angle of inclination greater than
  i) 80°? ii) 85°? iii) 88°? iv) 89°?
c) How many squares would be needed on the horizontal rectangle to have an angle of inclination smaller than
  i) 10°? ii) 5°? iii) 2°? iv) 1°?

9-2 The Tangent Ratio in Right Triangles

The definition of the tangent of an angle of inclination, given in Section 9-1, can be extended to any acute angle.

For the acute angle shown:

$$\tan \theta = \frac{\text{length of side opposite } \theta}{\text{length of side adjacent to } \theta}$$

The side adjacent to θ is always taken to be the arm of θ which is not the hypotenuse.

The tangent ratio is often used to determine the unknown angles or sides of a right triangle.

Example 1. In right $\triangle ABC$, find

a) $\tan A$ and $\angle A$;   b) $\tan C$ and $\angle C$.

Solution. a) The side opposite $\angle A$ is BC. The adjacent side is AB.

Therefore, $\tan A = \frac{5}{12}$, or approximately 0.417.

From the table, or a scientific calculator, $\angle A \doteq 23°$.

b) The side opposite $\angle C$ is AB. The adjacent side is BC.

Therefore, $\tan C = \frac{12}{5}$, or 2.4

From the table, or a scientific calculator, $\angle C \doteq 67°$.

Check. The sum of the angles in $\triangle ABC$ is $23° + 67° + 90°$, or $180°$.

Example 2. In right $\triangle PQR$, $\angle R = 90°$ and $PR = 7$.

a) Find QR if $\angle Q = 25°$.   b) Find $\angle Q$ if $QR = 9$.

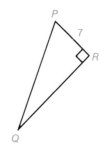

Solution. a) Let x represent the length of QR.

Then $\frac{7}{x} = \tan 25°$

$7 = x \tan 25°$

$7 \doteq x(0.466)$

$x \doteq \frac{7}{0.466}$, or 15.0

If $\angle Q = 25°$, then $QR \doteq 15.0$.

b) $\tan Q = \frac{PR}{QR}$

$= \frac{7}{9}$, or 0.778

$\angle Q \doteq 38°$

If $QR = 9$, then $\angle Q \doteq 38°$.

9 - 2 The Tangent Ratio in Right Triangles   361

Example 3. When the foot of a ladder is 2 m from a wall, the angle
formed by the ladder and the ground is 68°.
  a) How high up the wall does the ladder reach?
  b) How long is the ladder?

Solution. a) Let d represent the distance, in metres, the ladder
reaches up the wall.
Then, $\frac{d}{2} = \tan 68°$
$$d = 2 \tan 68°$$
$$\doteq 2(2.475), \text{ or } 4.95$$
The ladder reaches approximately 4.95 m up the wall.

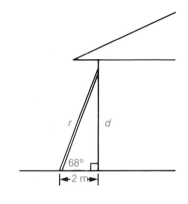

  b) Let r represent the length of the ladder, in metres.
Then, using the Pythagorean theorem,
$$r^2 = 2^2 + (4.95)^2$$
$$\doteq 28.5025$$
$$r \doteq \sqrt{28.5025}, \text{ or } 5.34$$
The length of the ladder is approximately 5.34 m.

Exercises 9 - 2

A 1. Calculate tan A and tan B:

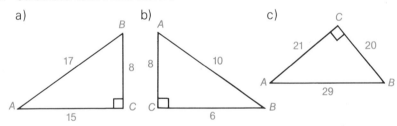

2. Find the measures of $\angle A$ and $\angle B$ in Exercise 1.

3. Find the value of x:

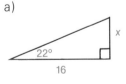
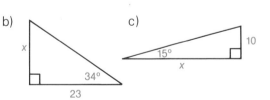
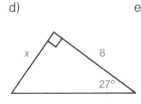
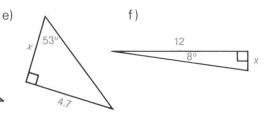

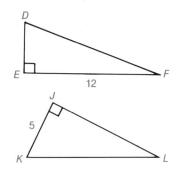

4. In △DEF, find
   a) DE if ∠F is: i) 18°, ii) 36°, iii) 54°;
   b) ∠F if DE is: i) 3, ii) 6, iii) 9.

5. In △JKL, find
   a) JL if ∠K is: i) 20°, ii) 30°, iii) 40°;
   b) ∠K if JL is: i) 8, ii) 12, iii) 16.

B 6. A guy wire fastened 40 m from the base of a television tower makes an angle of 60° with the ground.
   a) How high up the tower does the guy wire reach?
   b) How long is the guy wire?
   c) Repeat part (a) if the angle of the guy wire is 45°.

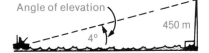

7. A communications tower is 450 m high. From a ship at sea, its angle of elevation is 4°.
   a) How far is the ship from the tower?
   b) What would be the angle of elevation if the ship were 25 km from the tower?

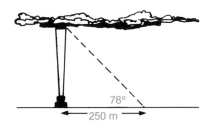

8. To measure the height of clouds, or "ceiling", at night, airport controllers can beam a light vertically and measure the angle of elevation of the spot of light made on the clouds.
   a) How high are the clouds in the diagram?
   b) What would be the angle of elevation of clouds 500 m high?

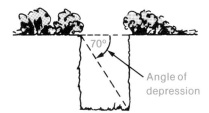

9. A gorge of rectangular cross section is 65 m wide. The angle of depression of a bottom corner when viewed from the opposite edge is 70°.
   a) How deep is the gorge?
   b) What would be the angle of depression if the gorge were 100 m deep?

10. A helicopter hovers directly over a landing pad on the top of a 125 m high building. The angle of elevation of the helicopter to an observer 145 m from the base of the building is 58°. How high is the helicopter above the landing pad?

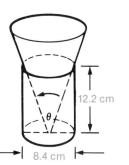

11. The diagram shows a cone inverted in a cylindrical can with the vertex of the cone just touching the bottom of the can. Find the vertex angle, θ, of the cone.

C 12. In △ABC, ∠B = 90°, AB = 1, and ∠A = θ. Find expressions, in terms of θ, for BC and AC.

9-3 The Sine and Cosine Ratios in Right Triangles

Although the tangent ratio is useful for solving problems involving right triangles, there are many problems in which it does not apply. For example, suppose a guy wire, 20 m long, is used to support a tower and forms an angle of 62° with the ground. The tangent ratio, $\tan 62° = \frac{BC}{AB}$, cannot be used to find either the distance AB or the length of BC because both are unknown. The ratio used must involve the one length given, that of the hypotenuse. Ratios involving the hypotenuse are the **sine** ratio and the **cosine** ratio.

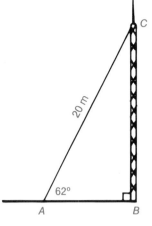

If θ is an acute angle in right triangle ABC,

$$\sin \theta = \frac{\text{length of side opposite } \theta}{\text{length of hypotenuse}}$$

and $$\cos \theta = \frac{\text{length of side adjacent to } \theta}{\text{length of hypotenuse}}.$$

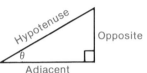

The sine, cosine, and tangent ratios are called trigonometric ratios because each is the ratio of the lengths of two sides of a right triangle.

Example 1. In right $\triangle ABC$, find

    a) $\sin A$, $\cos A$, $\tan A$;   b) $\sin C$, $\cos C$, $\tan C$.

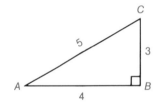

Solution. a) The side opposite $\angle A$ is BC. The adjacent side is AB, and the hypotenuse is AC. Therefore,

$$\sin A = \frac{3}{5} \qquad \cos A = \frac{4}{5} \qquad \tan A = \frac{3}{4}$$
$$= 0.600 \qquad\quad = 0.800 \qquad\quad = 0.750$$

b) The side opposite $\angle C$ is AB. The adjacent side is BC. Therefore,

$$\sin C = \frac{4}{5} \qquad \cos C = \frac{3}{5} \qquad \tan C = \frac{4}{3}$$
$$= 0.800 \qquad\quad = 0.600 \qquad\quad \doteq 1.333$$

Sines and cosines of angles may be found from tables, such as those on page 403, or with a scientific calculator.

Example 2. In right $\triangle PQR$, find the measures of $\angle P$ and $\angle R$ and the length of PQ.

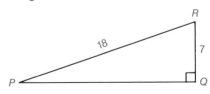

Solution. $\quad \sin P = \dfrac{7}{18} \qquad\qquad \cos R = \dfrac{7}{18}$

$\qquad\qquad\qquad \doteq 0.389 \qquad\qquad\qquad \doteq 0.389$

From the tables, or a scientific calculator,

$\qquad\qquad P \doteq 23° \qquad\qquad\qquad\qquad R \doteq 67°$

$\cos P = \dfrac{PQ}{PR}$

$PQ = PR \cos 23°$

$\qquad \doteq 18 \times 0.920,$ or approximately 16.6

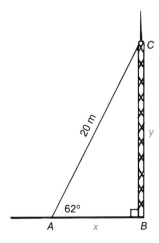

Example 3. A guy wire, 20 m long, supports a tower and forms an angle of 62° with the ground.

a) How high up the tower is the guy wire attached?

b) How far from the base of the tower is the guy wire attached to the ground?

Solution. a) Let y represent the length of BC.

Then, $\dfrac{y}{20} = \sin 62°$

$y = 20 \sin 62°$

$\quad \doteq 20 \times 0.883,$ or 17.66

The guy wire is attached about 17.7 m up the tower.

b) Let x represent the distance AB.

Then, $\dfrac{x}{20} = \cos 62°$

$x = 20 \cos 62°$

$\quad \doteq 20 \times 0.469$

$\quad \doteq 9.38$

The guy wire is attached to the ground about 9.4 m from the base of the tower.

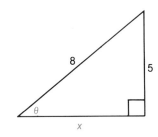

Example 4. If $\sin \theta = \dfrac{5}{8}$, find values for $\cos \theta$ and $\tan \theta$.

Solution. Let θ be an acute angle in a right triangle where the length of the hypotenuse is 8 units and the length of the side opposite θ is 5 units. Let x represent the length of the adjacent side.

By the Pythagorean theorem:

$\qquad\qquad x^2 = 8^2 - 5^2,$ or 39

$\qquad\qquad x = \sqrt{39}$

Then $\cos \theta = \dfrac{\sqrt{39}}{8} \quad$ and $\quad \tan \theta = \dfrac{5}{\sqrt{39}}$

$\qquad\qquad\qquad \doteq 0.781 \qquad\qquad\qquad \doteq 0.801$

9-3 The Sine and Cosine Ratios in Right Triangles

Exercises 9 - 3

A 1. In right △ABC, find the value of
   i) sin A, cos A, tan A;   ii) sin C, cos C, tan C.
   a) 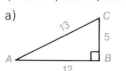 b) 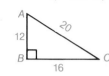 c)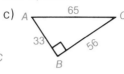

2. Find the measures of ∠A and ∠C in each triangle in Exercise 1.

3. Determine the measures of the acute angles:
   a) 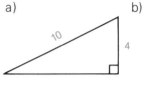 b) 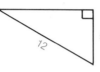 c)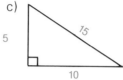
   d)  e)   f)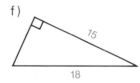

4. Determine the lengths of the other two sides:
   a)  b) 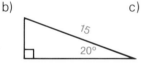 c)
   d)  e)   f)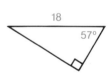

5. Determine the length of the hypotenuse:
   a)  b)   c)
   d) 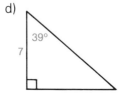 e) 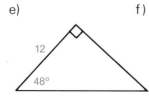 f)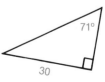

6. In △ABC, ∠B = 90° and AC = 25.
   a) Find AB and BC if: i) ∠A = 31°;   ii) ∠C = 78°.
   b) Find ∠A and ∠C if: i) BC = 10;   ii) AB = 17.

7. In △PQR, ∠Q = 90° and PQ = 15.
   a) Find PR and QR if: i) ∠P = 58°;   ii) ∠R = 42°.
   b) Find ∠P and ∠R if: i) PR = 28;   ii) QR = 25.

8. A 10 m ladder leans against a vertical wall at an angle of 73°. Find
   a) the height the ladder reaches up the wall;
   b) the distance from the foot of the ladder to the wall.

9. A storm causes some 15 m telephone poles to lean over.
   a) One pole leans at an angle of 72° to the ground. How high is the top of the pole from the ground?
   b) The top of another pole is 12 m above the ground. What is the pole's angle to the ground?

B 10. If $\sin \theta = \frac{7}{12}$, find values for $\cos \theta$ and $\tan \theta$.

11. If $\cos \theta = \frac{17}{25}$, find values for $\sin \theta$ and $\tan \theta$.

12. If $\tan \theta = 1.42$, find values for $\sin \theta$ and $\cos \theta$.

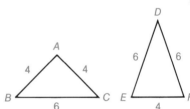

13. Find all the angles in isosceles triangles ABC and DEF.

C 14. A 20 m telephone pole was broken in the manner shown in the diagram. How high is the stump of the pole?

15. An equilateral triangle is inscribed in a circle. If the radius of the circle is 10 cm, calculate the length of a side of the triangle.

16. Consider the trigonometric tables on page 403.
   a) Explain why none of the values of $\sin \theta$ and $\cos \theta$ are greater than 1.
   b) Explain why the values of $\tan \theta$ are greater than 1 when θ is greater than 45°.
   c) Compare: i) sin 30° and cos 60°; ii) sin 80° and cos 10°. Explain the relationship between the sine and cosine of complementary angles.
   d) Verify that, for any value of θ, $\sin^2\theta + \cos^2\theta = 1$

17. In △ABC, ∠B = 90°, ∠A = θ, and AC = 1.
   a) Find expressions, in terms of θ, for: i) AB; ii) BC.
   b) Show that $\sin^2\theta + \cos^2\theta = 1$.

9-4 Solving Right Triangles

Finding the unknown sides and angles of a triangle is called *solving the triangle*. In a right triangle, this is possible if either of the following are known:

- the lengths of any two sides;
- the length of one side and an acute angle.

If the lengths of two sides of a right triangle are known, the Pythagorean theorem can be used to find the length of the third side, and any trigonometric ratio can be used to find one of the acute angles.

Example 1. Solve $\triangle ABC$, given that $AB = 5$, $BC = 2$, and $\angle B = 90°$.

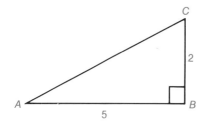

Solution. By the Pythagorean theorem:
$$AC^2 = AB^2 + BC^2$$
$$= 5^2 + 2^2$$
$$= 29$$
$$AC = \sqrt{29}, \text{ or approximately } 5.4$$

$\angle A$ can be found by the tangent ratio:
$$\tan A = \frac{2}{5}$$
$$= 0.400$$
$$\angle A \doteq 22°$$
And $\angle C \doteq 90° - 22°$, or $68°$

Example 2. Solve $\triangle PQR$, given that $PQ = 7$, $PR = 9$, and $\angle Q = 90°$.

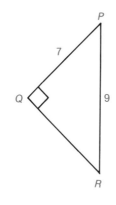

Solution. $QR^2 = PR^2 - PQ^2$
$$= 9^2 - 7^2$$
$$= 32$$
$$QR = \sqrt{32}, \text{ or approximately } 5.7$$

$\angle R$ can be found by the sine ratio:
$$\sin R = \frac{7}{9}$$
$$\doteq 0.778$$
$$\angle R \doteq 51°$$
And $\angle P \doteq 90° - 51°$, or $39°$

If the length of one side and one acute angle are known, a trigonometric ratio can be used to find the length of a second side. The length of the third side can then be found using either the Pythagorean theorem or another trigonometric ratio.

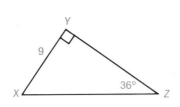

Example 3. Solve $\triangle XYZ$, given that $XY = 9$, $\angle Y = 90°$, and $\angle Z = 36°$.

Solution. $\angle X = 90° - 36°$, or $54°$

Trigonometric ratios of either $\angle X$ or $\angle Z$ can be used to find the lengths of YZ and XZ.

$\dfrac{9}{YZ} = \tan 36°$ $\qquad\qquad$ $\dfrac{9}{XZ} = \sin 36°$

$\qquad \doteq 0.727$ $\qquad\qquad\qquad \doteq 0.588$

$YZ \doteq \dfrac{9}{0.727}$, or 12.4 $\qquad$ $XZ \doteq \dfrac{9}{0.588}$, or 15.3

In applied problems, it is not usually necessary to find all the unknown lengths and angles.

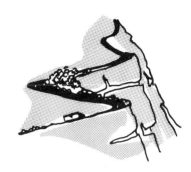

Example 4. A truck travels 6 km up a mountain road. If the change in height is 1250 m, what is the angle of inclination of the road?

Solution. Let θ represent the angle of inclination of the road.

Then, $\sin \theta = \dfrac{1250}{6000}$, or approximately 0.208

$\theta \doteq 12°$

The angle of inclination of the road is approximately $12°$.

Exercises 9 - 4

A 1. Find all unknown angles and sides:

a) b) c)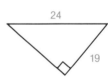

d) e) f)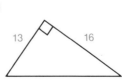

2. Solve each triangle:

a)   b)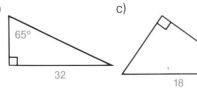

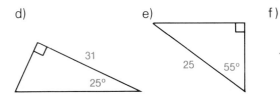
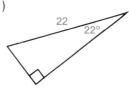

3. Solve △ABC, in which ∠B = 90°, with the data given:
   a) ∠A = 27°, AC = 10
   b) ∠A = 35°, AB = 15
   c) ∠C = 50°, AC = 20
   d) ∠C = 62°, BC = 100
   e) AB = 40, BC = 27
   f) AC = 48, AB = 35

4. From a distance of 80 m, the angle of elevation of the top of a flagpole is 18°. Determine the height of the flagpole, to the nearest tenth of a metre.

5. A radio tower is 350 m high. If the sun's rays make an angle of 39° with the ground, determine the length of the tower's shadow, to the nearest metre.

6. Sharon is flying a kite on a string 130 m long. Determine the height of the kite, to the nearest metre, if the string is at an angle of 34° to the ground. What assumptions are you making?

7. The diagonal of a rectangle is 12 cm long and makes an angle of 32° with the longer side. Find the length of the rectangle.

8. The diagonal of a rectangle is 15 cm long and makes a 20° angle with the longer side. Find the width of the rectangle.

9. An airplane is flying at an altitude of 6000 m over the ocean directly towards an island. When the angle of depression of the coastline from the airplane is 14°, how much farther does the airplane have to fly before it crosses the coast?

10. Find the measure of the acute angle formed by the intersection of the diagonals of a rectangle which measures
    a) 8 cm by 6 cm;
    b) 12 cm by 9 cm.

B 11. The inclination of a warehouse conveyor can be set between 14° and 40°. If the length of the conveyor is 12 m, what are the minimum and maximum heights the conveyor can reach?

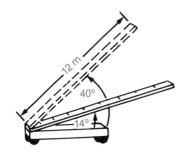

12. Determine the perimeter of a regular pentagon which is inscribed in a circle of 10 cm radius.

13. The length-to-width ratio of a child's kite in the shape of a kite is 5 : 3. Its length is 2 m and the angle at the tail is 40°. What is the angle at the nose?

Mathematics Around Us

Gondola Lifts in Banff and Jasper National Parks

Gondola lifts may generally be assumed to follow the hypotenuse of a right triangle, beginning at the lower terminal and ending at the upper terminal. If the elevations of these terminals and the length of the cable, or track, are known, the average angle of inclination of the lift can be found.

Data for the three gondola lifts in Banff and Jasper National Parks are given below.

Question

- Calculate the average angle of inclination for each gondola lift.

Banff Sulphur Mountain Gondola Lift

Lower Terminal: 1583 m
Upper Terminal: 2286 m
Length of Track: 1561 m

Lake Louise Gondola

Lower Terminal: 1532 m
Upper Terminal: 2036 m
Length of Track: 3353 m

Jasper Tramway

Lower Terminal: 1408 m
Upper Terminal: 2500 m
Length of Track: 2000 m

Investigating Repeating Decimals

When a calculator or a computer is used to divide two natural numbers and the quotient is a non-terminating decimal, only a small number of the decimal digits are shown. Since a computer is capable of repeating a sequence of steps very rapidly and accurately, it can be programmed to perform a division to any desired number of decimal places. Therefore, a computer is an ideal tool for investigating the patterns which arise when rational numbers are expressed as repeating decimals. The following program will cause the computer to print as many decimal digits as desired.

```
10   INPUT "WHAT IS THE NUMERATOR"; N
20   INPUT "WHAT IS THE DENOMINATOR"; D
30   F = N / D
40   I = INT (F)
50   R = N − I * D
60   IF N < D THEN 80
70   PRINT "THE INTEGRAL PART IS "I
80   A = INT ((F − I) * 1000000)
90   PRINT "THE FIRST 6 DECIMAL DIGITS ARE 0." A
100  INPUT "DO YOU WANT MORE DIGITS"; A$
110  IF A$ = "NO" THEN END
120  R = R * 1000000 − D * A
130  A = INT (R * 1000000 / D)
140  PRINT "THE NEXT 6 DIGITS ARE "A: GOTO 100
```

Example. Express $\frac{39}{17}$ as a repeating decimal using the above program.

Solution. WHAT IS THE NUMERATOR? 39
WHAT IS THE DENOMINATOR? 17
THE FIRST 6 DECIMAL DIGITS ARE 0.294117
DO YOU WANT MORE DIGITS? YES
THE NEXT 6 DIGITS ARE 647058
DO YOU WANT MORE DIGITS? YES
THE NEXT 6 DIGITS ARE 823529
DO YOU WANT MORE DIGITS? YES
THE NEXT 6 DIGITS ARE 411764
DO YOU WANT MORE DIGITS? NO

The result shows that $\frac{39}{17} = 2.\overline{294\ 117\ 647\ 058\ 823\ 5}$.

Exercises

1. Express as a repeating decimal:

   a) $\frac{23}{79}$ b) $\frac{187}{84}$ c) $\frac{38}{23}$ d) $\frac{424}{757}$

2. Investigate the patterns in the repeating decimals for:

   a) $\frac{1}{17}, \frac{2}{17}, \frac{3}{17}, \ldots$;   b) $\frac{1}{41}, \frac{2}{41}, \frac{3}{41}, \ldots$

9-5 Applications of the Trigonometric Ratios

Trigonometry was originally developed to solve problems in navigation. Since then, it has been applied to a wide range of problems in mathematics, science, and industry.

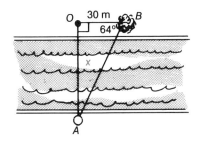

Example 1. In the diagram, an observer, O, is directly opposite a hydro pole, A, on the other side of a canal. A tree, B, is 30 m from O. If $\angle B = 64°$ and $\angle O = 90°$, what is the width of the canal?

Solution. Let the width of the canal, AO, be x metres.
Then, $\frac{x}{30} = \tan 64°$
$x \doteq 30 \times 2.050$,
$\doteq 61.5$
The width of the canal is approximately 61.5 m.

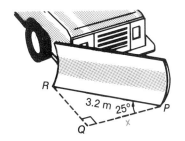

Example 2. A snow plow has a 3.2 m blade set at an angle of 25°. How wide a path will the snow plow clear?

Solution. Let the width of the path, PQ, be x metres.
Then, $\frac{x}{3.2} = \cos 25°$
$x \doteq 3.2 \times 0.906$,
$\doteq 2.90$
The snow plow will clear a path about 2.9 m wide.

Example 3. Two office towers are 25 m apart. From the 15th floor (40 m up) of the shorter tower, the angle of elevation of the top of the other tower is 72°. Find
a) the angle of depression of the base of the taller tower from the 15th floor;
b) the height of the taller tower.

Solution. a) Let the angle of depression be θ.
Then, $\tan \theta = \frac{40}{25}$, or 1.6
$\theta \doteq 58°$
The angle of depression is approximately 58°.
b) Let the height of the taller tower above the 15th floor be x metres.
Then, $\frac{x}{25} = \tan 72°$
$x \doteq 25 \times 3.078$, or 76.95
The height of the taller tower is approximately 77 + 40, or 117 m.

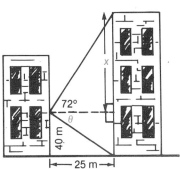

Exercises 9 - 5

A 1. A guy wire, supporting a TV tower 50 m tall, joins the top of the tower to an anchor point 25 m from the base. Find the length of the wire and the angle it makes with the ground.

2. A tree casts a shadow 42 m long when the sun's rays are at an angle of 38° to the ground. How tall is the tree?

3. A kite string, 140 m long, makes an angle of 40° with the ground. Determine the height of the kite.

4. The diagram shows a rod, 10 m long, in a well with 1.4 m of the rod protruding. How deep is the well?

5. The diagram below (left) shows a house designed for solar heating. Determine the length, l, of the heating panels.

Well

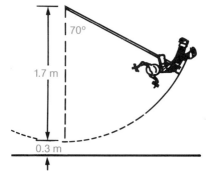

B 6. In the diagram above (right), how high above the ground is the child on the swing?

7. When a Ferris wheel, 20 m in diameter, is in the position shown, how high are its seats above the ground?

8. A 30 cm ruler rests at an angle of 20° to the horizontal in a 25 cm diameter hemispherical bowl. How far along the ruler is its point of contact with the rim of the bowl?

9. Water in a hemispherical bowl begins to pour out when the bowl is tilted through an angle of 35°. How deep is the water in the bowl?

10. When the foot of a ladder is 2.0 m from a wall, the angle formed by the ladder and the ground is 60°.
    a) How long is the ladder?
    b) How high up the wall does the ladder reach?

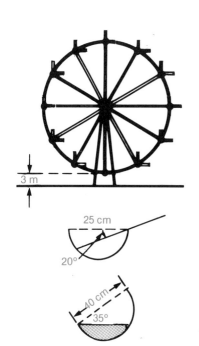

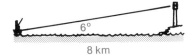

11. The top of a communications tower has an angle of elevation of 6° when observed by a ship 8 km from the base of the cliff below the tower. How high is the top of the tower above sea level?

12. An airplane flying at a speed of 52 m/s increases its altitude at the rate of 400 m/min. At what angle is it climbing?

C 13. From an apartment window 24 m above the ground, the angle of depression of the base of a nearby building is 38° and the angle of elevation of the top is 63°. Find the height of the nearby building.

14. To an observer at A, the angle of elevation of a church spire is 31°. To an observer at B, the angle of elevation is 35°. If the observers are 65 m apart with the spire directly between them, what is the height of the spire?

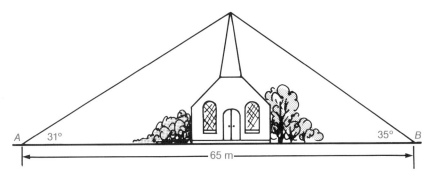

15. A man whose eyes are 1.8 m above the ground notes that the angle of elevation of the top of a building is 65°. He walks 30 m farther away and finds the angle of elevation to be 55°. How tall is the building?

16. Two trees are 100 m apart. From a point midway between them, the angles of elevation of their tops are 12° and 16°. How much taller is one tree than the other?

Review Exercises

1. Find the angle of inclination of a line with slope
   a) $\frac{2}{5}$; b) $\frac{3}{8}$; c) $\frac{3}{2}$; d) $\frac{5}{4}$; e) 2; f) 2.7.

2. Find the slope of a line that has an angle of inclination of
   a) 12°; b) 27°; c) 30°; d) 55°; e) 70°; f) 80°.

3. A rectangle measures 8 cm by 4 cm. Find the measures of the two acute angles formed at a vertex by a diagonal.

4. Find the value of x:
   a)   b)   c)

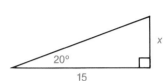
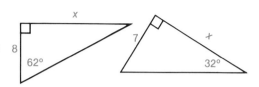

5. In △ABC, find ∠B if AC is: a) 6;   b) 12;   c) 15.

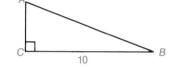

6. A guy wire fastened 50 m from the base of a television tower makes an angle of 55° with the ground. How high up the tower does the guy wire reach?

7. Determine the measures of the acute angles:
   a)   b)   c)

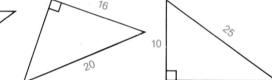

8. Determine the lengths of the sides not given:
   a)   b)   c)

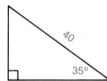
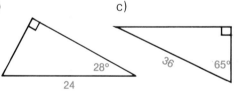

9. In △ABC, ∠C = 90° and AC = 8. Find AB and BC if
   a) ∠A = 28°;   b) ∠A = 54°;   c) ∠B = 48°.

10. Find the measures of the angles in isosceles △ABC.

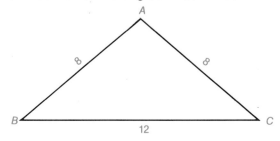

11. Solve each triangle:

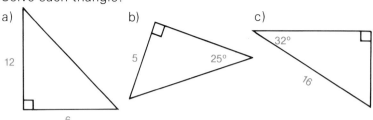

12. From a distance of 60 m at ground level, the angle of elevation of the top of a flagpole is 32°. Determine the height of the flagpole to the nearest tenth of a metre.

13. When the foot of a ladder is 1.8 m from a wall, the angle formed by the ladder and the ground is 72°.
    a) How long is the ladder?
    b) How high up the wall does the ladder reach?

14. A tree casts a shadow 40 m long when the sun's rays are at an angle of 36° to the ground. How tall is the tree?

15. From the top of a building 70 m high, the angle of depression of an automobile on a road is 27°. How far is the automobile from the foot of the building?

16. A broadcasting tower is 255 m high. How far from the base of the tower is a surveyor who observes that the angle of elevation of the top of the tower is 38°?

17. From the data in the diagram, determine the length of the diagonal of the rectangle.

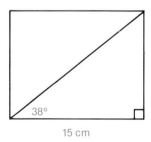

18. Two office towers are 30 m apart. From the 15th floor (40 m up) of the shorter tower, the angle of elevation of the top of the other tower is 70°. Find
    a) the angle of depression of the base of the taller tower from the 15th floor;
    b) the height of the taller tower.

10 Vectors

A Canadian Forces minesweeper leaves Sea Harbour on a practice patrol. It has orders to proceed to Bullion Island, then to Port Karat, and finally to Doubloon to await further orders. The distance and bearing of Bullion Island from Sea Harbour is 40 km and 135°. The distance and bearing of Port Karat from Bullion Island is 60 km and 225°. The distance and bearing of Doubloon from Port Karat is 25 km and 270°. What is the distance and bearing of Doubloon from Sea Harbour? (See *Example 2* in Section 10-4.)

10-1 Vectors as Directed Line Segments

All the quantities studied so far have been **scalar** quantities. This chapter introduces quantities called **vectors**. To understand the difference between scalars and vectors compare the examples in the two columns in the table.

Scalar quantities have magnitude only.	Vectors have magnitude and direction.
Distance is a scalar quantity. Mary lives 100 km from Calgary. Mary lives somewhere on the circle.	Displacement is a vector quantity. Mary lives 100 km northeast of Calgary.
Speed is a scalar quantity. The aircraft is travelling at a speed of 900 km/h.	Velocity is a vector quantity. The velocity of the aircraft is 900 km/h west.
Mass is a scalar quantity. Mr. Dawson has a mass of 100 kg.	Weight is a vector quantity. Mr. Dawson has a weight of 980 N (downward). Weight is understood to be in a downward direction.

N is the SI symbol for newtons, the metric unit of force.

10-1 Vectors as Directed Line Segments

A vector is represented by a directed line segment. The arrowhead indicates the direction of the vector, and the length of the segment its magnitude.

A vector from point A to point B is written $\overrightarrow{AB}$. Points A and B are the tail and the head of the vector, respectively.

Example 1. Use vectors to represent:
 a) a displacement of: i) 20 km due west,
   ii) 15 km to the southeast;
 b) a velocity of: i) 100 km/h due east,
   ii) 60 km/h to the northwest.

Solution. a) Scale: 1 cm : 5 km
  i) ii)

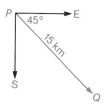

$\overrightarrow{AB}$ represents a displacement of 20 km due west.

$\overrightarrow{PQ}$ represents a displacement of 15 km to the southeast.

b) Scale: 1 cm : 20 km/h
  i) ii)

$\overrightarrow{LM}$ represents a velocity of 100 km/h due east.

$\overrightarrow{CD}$ represents a velocity of 60 km/h to the northwest.

The universal way to give compass directions or course bearings is to consider north as 000°. Then, moving clockwise, all other directions are assigned a three-digit number up to 359°. For example, northeast is 045°; southwest is 225°; and west is 270°.

In navigation, an aircraft or ship at A is said to have a **bearing** of 215° from O; an observer at O has a bearing of 035° from A. The direction in which a ship or aircraft is moving is called its **heading**.

Example 2. Draw a diagram to represent a displacement of 50 km on a heading of 295°.

Solution. Scale: 1 cm : 10 km
$\overrightarrow{OA}$ represents the required vector.

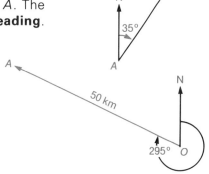

Exercises 10-1

A 1. Which of the following can be described by a vector?
   a) a wind of 30 km/h from the northwest
   b) a distance of 13.5 km
   c) a book containing 448 pages
   d) a current of 5 km/h east
   e) a 65 L tank of fuel
   f) a speed of 90 km/h
   g) a weight on the moon of 200 N
   h) an advance of 15 km due south
   i) a 40 kg bag of fertilizer

2. Use a protractor to find the compass bearing of each vector:

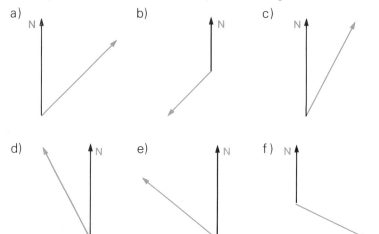

3. Represent each compass bearing by a diagram:
   a) 078° b) 235° c) 150° d) 315° e) 212° f) 012°

4. Represent each displacement by a vector:
   a) 60 m north   b) 35 km southwest c) 400 km on 025°
   d) 250 m on 075°   e) 75 km on 130°   f) 160 km on 240°

"on 025°" means "on a heading of 025°".

5. Draw a vector to represent:
   a) a jet aircraft travelling southeast at 640 km/h;
   b) a ship sailing on a heading of 320° at 30 km/h;
   c) a helicopter flying on a heading of 290° at 200 km/h;
   d) a yacht sailing on a heading of 120° at 90 km/h;
   e) a hovercraft travelling southwest at 100 km/h.

10-2 Equal Vectors

The *Snowbirds* are a precision-flying squadron of the Canadian Armed Forces. They are shown in the photograph flying in formation, that is, flying at exactly the same speed in precisely the same direction. This means that although the planes have different positions, their velocity vectors are equal. If their speeds are 350 km/h, their velocity vectors can be represented as follows (the direction of each vector being parallel to the direction of the aircraft it represents):

Scale: 1 cm : 100 km/h

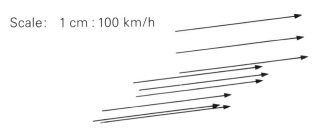

Since a vector is defined by its magnitude and direction, two vectors with the same magnitude and direction are equal.

Example 1. List pairs of vectors that appear to be equal.

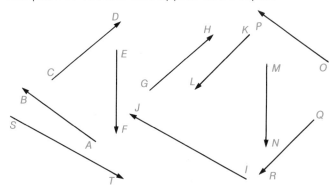

Solution: The vectors that appear to be equal are:
$\overrightarrow{CD}$ and $\overrightarrow{GH}$; $\overrightarrow{EF}$ and $\overrightarrow{MN}$;
$\overrightarrow{AB}$ and $\overrightarrow{OP}$; $\overrightarrow{KL}$ and $\overrightarrow{QR}$.
This can be confirmed by measurement.

Example 2. In the diagram, *ABCD* is a rhombus. List pairs of equal vectors.

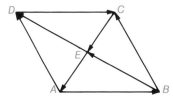

Solution: $\overrightarrow{AB} = \overrightarrow{DC}$ $\overrightarrow{BC} = \overrightarrow{AD}$
$\overrightarrow{BE} = \overrightarrow{ED}$ $\overrightarrow{CE} = \overrightarrow{EA}$

Exercises 10 - 2

A 1. Which of the following pairs of vectors are equal?
   a) a displacement of 10 km southwest
      a displacement of 20 km southwest
   b) a skydiver falling at 150 km/h
      a car speeding west at 150 km/h
   c) a boat sailing on a heading of 045° at 20 km/h
      a cyclist riding northeast at 20 km/h
   d) Alice, in Edmonton, with a weight of 680 N
      Alan, in Halifax, with a weight of 700 N

2. List pairs of vectors that appear to be equal:

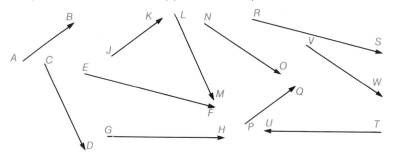

B 3. List pairs of equal vectors:
   a)    b)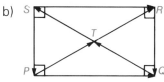
   c) 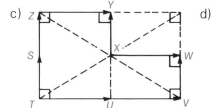   d)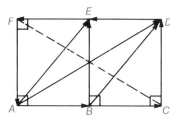

4. How many different pairs of equal vectors are in each diagram?
   a) 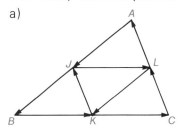   b)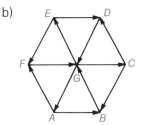

10-3 Vectors as Ordered Pairs

Vectors represented as directed line segments on a grid may also be represented by ordered pairs. Each vector shown in the diagram is represented by the ordered pair [3, 2]. This indicates that starting at the tail, the head may be reached by moving 3 to the right and 2 up. The numbers 3 and 2 are called the *components* of the vector [3, 2].

The magnitude or length of a vector such as $\vec{AB} = [3, 2]$ is represented by the symbol $|\vec{AB}|$ and may be found by using the Pythagorean theorem.

$$|\vec{AB}| = \sqrt{3^2 + 2^2}$$
$$= \sqrt{13}, \text{ or approximately } 3.61$$

Square brackets distinguish vectors from coordinates.

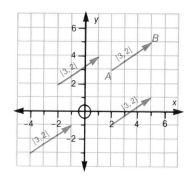

In general, if $\vec{AB} = [a, b]$,
$|\vec{AB}| = \sqrt{a^2 + b^2}$

Example 1. Represent each vector shown as an ordered pair and find its length.

Solution.
$\vec{AB} = [4, 1]$ $|\vec{AB}| = \sqrt{4^2 + 1^2}$, or $\sqrt{17}$
$\vec{CD} = [-3, 3]$ $|\vec{CD}| = \sqrt{(-3)^2 + 3^2}$, or $3\sqrt{2}$
$\vec{EF} = [-4, -3]$ $|\vec{EF}| = \sqrt{(-4)^2 + (-3)^2}$, or 5
$\vec{GH} = [0, -5]$ $|\vec{GH}| = \sqrt{0^2 + (-5)^2}$, or 5

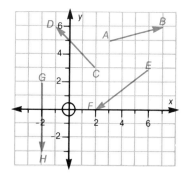

Example 2. Given the points $P(-2, 3)$ and $Q(4, 1)$, express the vector $\vec{PQ}$ as an ordered pair.

Solution. Graph the points P and Q.
Q is 6 units to the right of P and 2 down.
$\vec{PQ} = [6, -2]$

In *Example 2*, the components of $\vec{PQ}$ may also be found by subtracting the coordinates of P from those of Q.

$$\vec{PQ} = [4 - (-2), 1 - 3]$$
$$= [6, -2]$$

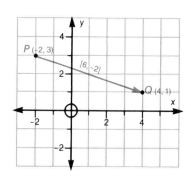

If $A(x_1, y_1)$ and $B(x_2, y_2)$ are any two points, the components of the vector $\vec{AB}$ are found by subtracting the coordinates of A from those of B.
$\vec{AB} = [x_2 - x_1, y_2 - y_1]$

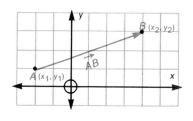

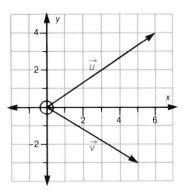

A vector may be represented by a single letter, such as $\vec{v}$. Its length is then represented by $|\vec{v}|$.

Example 3. Using (0, 0) as a common tail, graph the vectors $\vec{u} = [6, 4]$ and $\vec{v} = [5, -3]$. Find the length of each vector.

Solution. Vectors $\vec{u}$ and $\vec{v}$ are shown on the grid.
$$|\vec{u}| = \sqrt{6^2 + 4^2} \qquad |\vec{v}| = \sqrt{5^2 + (-3)^2}$$
$$= \sqrt{52}, \text{ or } 2\sqrt{13} \qquad = \sqrt{34}$$

Exercises 10 - 3

A 1. Represent each vector shown as an ordered pair and find its length.

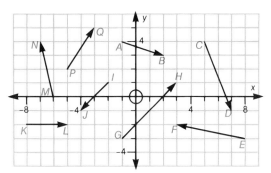

2. In the diagram, ABCD is a parallelogram. Represent the following vectors as ordered pairs:

   a) $\vec{AB}$ b) $\vec{DC}$ c) $\vec{AD}$
   d) $\vec{BC}$ e) $\vec{AC}$ f) $\vec{DB}$

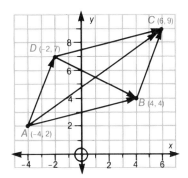

3. In the diagram, PQRS is a rhombus. Represent the following vectors as ordered pairs and find their lengths:

   a) $\vec{PQ}$ b) $\vec{QR}$ c) $\vec{SR}$
   d) $\vec{PS}$ e) $\vec{QO}$ f) $\vec{OS}$
   g) $\vec{RO}$ h) $\vec{OP}$ i) $\vec{RP}$

4. The coordinates of the head and tail of a vector, $\vec{PQ}$, are given. Represent $\vec{PQ}$ as an ordered pair:

   a) P(3, 4), Q(4, 0)   b) P(0, −3), Q(6, 0)
   c) P(−3, 1), Q(0, 5)   d) P(3, −1), Q(0, 2)
   e) P(−4, 3), Q(−4, −1)   f) P(11, 1), Q(6, −3)

10-3 Vectors as Ordered Pairs

B 5. Given the points $A(-5, 2)$, $B(-2, 6)$, $C(3, 6)$, $D(6, 2)$, $E(3, -2)$, $F(-2, -2)$, represent the following vectors as ordered pairs and illustrate graphically:

a) $\vec{AC}$ b) $\vec{DB}$ c) $\vec{EA}$
d) $\vec{CF}$ e) $\vec{BE}$ f) $\vec{FD}$

6. Vector $\vec{AB} = [3, 5]$. If A is the point $(2, -1)$, find the coordinates of B.

7. Vector $\vec{PQ} = [-6, 2]$. Find the coordinates of Q if P is the point
a) $(8, 5)$; b) $(3, 7)$; c) $(-2, -1)$; d) $(-4, 3)$.

8. A is the point $(5, -3)$. Find the coordinates of B if $\vec{AB}$ is
a) $[8, -5]$; b) $[-2, 4]$; c) $[-11, 7]$; d) $[3, -12]$.

9. Given the points $A(4, 8)$, $B(-6, 2)$, $C(8, -4)$, $D(1, -1)$, $E(6, 2)$, $F(-1, 5)$, write two vectors that are equal to
a) $\vec{AF}$; b) $\vec{BD}$; c) $\vec{CE}$.

10. The vector $\vec{AB} = [5, 2]$ represents one side of a square $ABCD$. Write vectors to represent the other three sides of the square.

11. Rectangle $PQRS$ is twice as long as it is wide. If $\vec{PQ} = [3, -2]$ is the width, write vectors for the other three sides.

C 12. a) If $\vec{u} = \vec{v}$, is it always true that $|\vec{u}| = |\vec{v}|$?
b) If $|\vec{u}| = |\vec{v}|$, is it always true that $\vec{u} = \vec{v}$?

13. Use a scale diagram to determine the angle of inclination and the length of: a) $\vec{u} = [5, 2]$; b) $\vec{v} = [3, 7]$.

14. Use a scale diagram to express as an ordered pair:

a) a vector with length 6 and inclination 50°;
b) a vector with length 10 and inclination 25°.

15. The diagram shows right triangle ABC with squares drawn on the three sides. Points P, Q, and R are the centres of the three squares. S is the outer vertex of the rectangle formed by the sides of the two smaller squares.

a) Copy the diagram and show that $\vec{PQ} = \vec{RS}$ and $\vec{PR} = \vec{QS}$.

b) Investigate the results obtained with other right triangles.

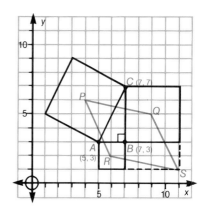

THE MATHEMATICAL MIND

The Four Squares Problem

1
More than 2000 years ago, the Greeks discovered that they could express any given natural number as the sum of four or fewer perfect squares.

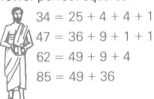

$34 = 25 + 4 + 4 + 1$
$47 = 36 + 9 + 1 + 1$
$62 = 49 + 9 + 4$
$85 = 49 + 36$

2
Although they could not find a single exception to this rule, they were unable to prove that *all* natural numbers can be expressed in this way.

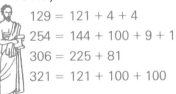

$129 = 121 + 4 + 4$
$254 = 144 + 100 + 9 + 1$
$306 = 225 + 81$
$321 = 121 + 100 + 100$

3
For a long time this was an unsolved problem in mathematics. One of the greatest mathematicians of the 18th century, Leonard Euler, tried unsuccessfully over a period of 40 years to solve this problem.

4
Finally, in 1770, Euler's successor, Joseph Lagrange, succeeded in proving that every natural number, no matter how great, can be expressed as the sum of four or fewer perfect squares.

$$n = a^2 + b^2 + c^2 + d^2$$

Joseph Lagrange 1736-1813

1. Express as the sum of four or fewer perfect squares:
   a) 29   b) 56
   c) 87   d) 115
   e) 130   f) 168
   g) 204   h) 435

2. How many natural numbers less than 100 can be written as the sum of
   a) three or fewer perfect squares?
   b) two or fewer perfect squares?

3. How many natural numbers less than 100 can be expressed as the sum of two squares in two different ways?

Mathematics Around Us

Designing Carton Sizes

In warehouses, goods are usually loaded onto wooden skids, or pallets, for ease of moving and storing. The standard-size pallet measures 120 cm by 100 cm.

For the most efficient use of space, cartons should be designed so that they can be stacked in layers on a pallet without gaps and without overhanging the edge. Therefore, only cartons having certain lengths and widths can be used. And all cartons on a layer should have the same length, width, and height.

Questions

1. If five cartons fit on one layer, find
   a) the area of the pallet occupied by each carton;
   b) the length and width of each carton;
   c) how to stack the cartons.

2. If the minimum length or width of a carton is 30 cm, find all possible carton sizes that can be stacked on a pallet.

10-4 Addition of Vectors: The Triangle Law

Since vectors represent displacements, they can be added by combining them head to tail. The result of combining $\overrightarrow{AB}$ and $\overrightarrow{BC}$ in this way is $\overrightarrow{AC}$. This is called the **triangle law** of addition.

$$\overrightarrow{AB} + \overrightarrow{BC} = \overrightarrow{AC}$$

Example 1. Express as a single vector:
a) $\overrightarrow{AE} + \overrightarrow{EF}$
b) $\overrightarrow{AG} + \overrightarrow{GC}$
c) $\overrightarrow{DA} + \overrightarrow{AG}$

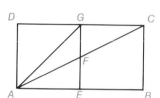

Solution.
a) $\overrightarrow{AE} + \overrightarrow{EF} = \overrightarrow{AF}$
b) $\overrightarrow{AG} + \overrightarrow{GC} = \overrightarrow{AC}$
c) $\overrightarrow{DA} + \overrightarrow{AG} = \overrightarrow{DG}$

The triangle law applies even when more than two vectors are added because each pair of vectors drawn head-to-tail can be considered as resulting in a single equivalent vector.

Example 2. A Canadian Forces minesweeper leaves Sea Harbour on a practice patrol. It has orders to proceed to Bullion Island, then to Port Karat, and finally to Doubloon to await further orders. The distance and bearing of Bullion Island from Sea Harbour is 40 km and 135°. The distance and bearing of Port Karat from Bullion Island is 60 km and 225°. The distance and bearing of Doubloon from Port Karat is 25 km and 270°. What is the distance and bearing of Doubloon from Sea Harbour?

Solution. The distances and bearings given are represented by the vectors shown on the diagram. The distance and bearing of Doubloon from Sea Harbour are represented by the colored vector and, by measurement, are found to be about 82 km and 209°.

10 - 4 Addition of Vectors: The Triangle Law   **389**

Example 3. If $\vec{u} = [5, 2]$ and $\vec{v} = [1, 3]$, represent the following as ordered pairs and illustrate on a grid:

a) $\vec{u} + \vec{v}$ b) $\vec{v} + \vec{u}$

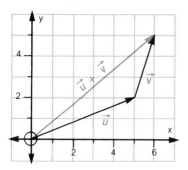

Solution. a) From the diagram, which shows $\vec{u}$ and $\vec{v}$ joined head to tail on a grid, with $\vec{u}$ starting from (0, 0),
$\vec{u} + \vec{v} = [6, 5]$.

b) From the diagram, which shows $\vec{u}$ and $\vec{v}$ joined head to tail on a grid, with $\vec{v}$ starting from (0, 0),
$\vec{v} + \vec{u} = [6, 5]$.

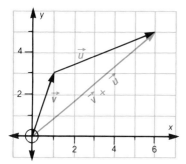

Example 3 suggests that two vectors may be added in either order. That is, $\vec{u} + \vec{v} = \vec{v} + \vec{u}$. The example also shows that when vectors are represented as ordered pairs, they may be added by adding their components.

> For any vectors $[a, b]$ and $[c, d]$,
> $[a, b] + [c, d] = [a + c, b + d]$.

Exercises 10 - 4

A 1. Express as a single vector:

a) $\overrightarrow{AB} + \overrightarrow{BE}$ b) $\overrightarrow{AB} + \overrightarrow{BC}$ c) $\overrightarrow{AE} + \overrightarrow{ED}$

d) $\overrightarrow{DE} + \overrightarrow{EC}$ e) $\overrightarrow{AE} + \overrightarrow{EC}$ f) $\overrightarrow{AB} + \overrightarrow{BC} + \overrightarrow{CD}$

2. Express as a single vector:

a) $\overrightarrow{PT} + \overrightarrow{TQ}$ b) $\overrightarrow{QR} + \overrightarrow{RU}$ c) $\overrightarrow{RV} + \overrightarrow{VS}$

d) $\overrightarrow{PV} + \overrightarrow{VS}$ e) $\overrightarrow{PQ} + \overrightarrow{QR}$ f) $\overrightarrow{RW} + \overrightarrow{WS}$

g) $\overrightarrow{UQ} + \overrightarrow{QW} + \overrightarrow{WV}$ h) $\overrightarrow{SW} + \overrightarrow{WQ} + \overrightarrow{QP}$

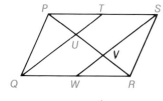

3. Write as the sum of two or more vectors:

a) $\overrightarrow{BG}$ b) $\overrightarrow{BE}$ c) $\overrightarrow{BC}$

d) $\overrightarrow{AF}$ e) $\overrightarrow{FC}$ f) $\overrightarrow{GE}$

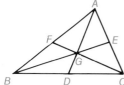

4. Express as a single vector:

a) $[5, 1] + [2, 4]$ b) $[1, 5] + [3, -2]$

c) $[-2, 4] + [6, -1]$ d) $[3, -6] + [-2, -1]$

e) $[-9, -2] + [-5, -6]$ f) $[4, -7] + [-9, -11]$

g) $[3, 5] + [2, -1] + [4, 6]$ h) $[-8, 5] + [-3, -2] + [1, 0]$

B 5. Express as a single vector and illustrate on a grid:
   a) $[0, 7] + [5, -3]$
   b) $[3, 2] + [6, 4]$
   c) $[-5, -2] + [4, -6]$
   d) $[2, -1] + [6, -3]$
   e) $[5, -1] + [-2, 4] + [3, -2]$
   f) $[4, 1] + [-7, 3] + [3, -4]$

6. *ABCD* is a quadrilateral. Show that $\overrightarrow{AB} + \overrightarrow{BC} = \overrightarrow{AD} + \overrightarrow{DC}$.

7. Chris and Jerry sail due west for 10 km, then turn to sail on a heading of 035° for 7 km. What is their distance and bearing now from the starting point?

8. A riding trail starts at point *A* and leads in a north-westerly direction for 1200 m to point *B*, where it alters direction to a heading of 020°. Point *C* is reached after 650 m on this new heading. At *C*, the trail swings due east for a further 1450 m to point *D*. What is the distance and bearing of *A* from *D*?

9. A reconnaissance aircraft flies from its base on a heading of 210° for 40 min at 480 km/h. It then flies east for half an hour at 500 km/h and then turns on a heading of 020° for 1 h at 450 km/h. At this point, what is the distance and bearing from its base?

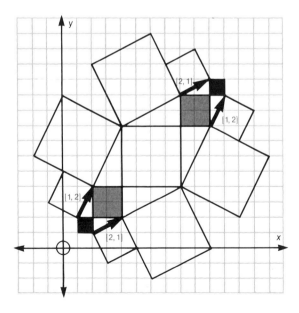

C 10. In the design on the cover of this book, the hypotenuse of each small right triangle may be represented by one of the vectors $[1, 2]$ or $[2, 1]$.
   a) Make a similar design in which the hypotenuse of each small right triangle is represented by
      i) one of the vectors $[1, 3]$ or $[3, 1]$;
      ii) the vector $[1, 1]$.
   b) Find the side lengths of all squares for each design in (a).

11. Given that $\vec{u} = [-3, 6]$, $\vec{v} = [9, -4]$, and $\vec{w} = [-5, -7]$, in how many different ways can $\vec{u}$, $\vec{v}$, and $\vec{w}$ be drawn head-to-tail starting at (0, 0)? Illustrate your answer with a diagram.

10-5 Addition of Vectors: The Parallelogram Law

In the previous section, vectors arranged head to tail were added using the triangle law. If two vectors are arranged tail-to-tail, they may be added by the following method.

In the diagram, $\vec{OA}$ and $\vec{OB}$ are tail-to-tail. Point C is determined by drawing $AC \parallel OB$ and $BC \parallel OA$. Then OACB is a parallelogram, and
$$\vec{OA} + \vec{OB} = \vec{OC}.$$
This is called the **parallelogram law** of addition.

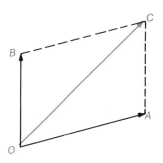

Example 1. In the diagram, ACDH and ABFG are parallelograms. Express as a single vector:
 a) $\vec{AB} + \vec{AH}$ b) $\vec{BA} + \vec{BF}$ c) $\vec{DH} + \vec{DC}$

Solution. a) $\vec{AB} + \vec{AH} = \vec{AE}$
 b) $\vec{BA} + \vec{BF} = \vec{BG}$
 c) $\vec{DH} + \vec{DC} = \vec{DA}$

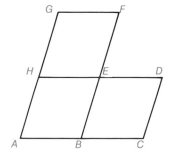

Example 2. If O represents the origin, and $\vec{OA} = [6, 4]$ and $\vec{OB} = [-2, 5]$, represent $\vec{OA} + \vec{OB}$ as an ordered pair and draw $\vec{OA} + \vec{OB}$ on a grid.

Solution. $\vec{OA} + \vec{OB} = [6, 4] + [-2, 5]$
$= [6 - 2, 4 + 5]$
$= [4, 9]$

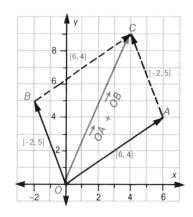

In *Example 2*, the vector [4, 9] is shown with tail O(0, 0) and head C(4, 9). The point C completes the parallelogram, OACB, determined by $\vec{OA}$ and $\vec{OB}$.

Occasionally, it is necessary to replace a vector with an equivalent vector to perform the addition. Either the triangle law or the parallelogram law may be used.

Example 3. Copy the vectors and draw $\vec{u} + \vec{v}$.

Solution. Using the triangle law   Using the parallelogram law

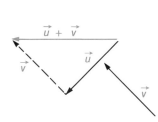
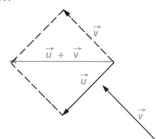

Exercises 10 - 5

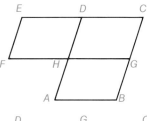

A 1. In the diagram, *ABCD* and *CEFG* are parallelograms. Express as a single vector:
   a) $\overrightarrow{HG} + \overrightarrow{HD}$
   b) $\overrightarrow{HG} + \overrightarrow{HA}$
   c) $\overrightarrow{FG} + \overrightarrow{FE}$
   d) $\overrightarrow{CD} + \overrightarrow{CG}$

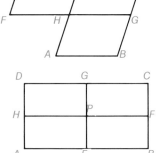

2. In the diagram, *ABCD* is a rectangle and *E, F, G, H* are the midpoints of its sides. Express as a single vector:
   a) $\overrightarrow{AE} + \overrightarrow{AH}$
   b) $\overrightarrow{HF} + \overrightarrow{HD}$
   c) $\overrightarrow{BC} + \overrightarrow{BE}$
   d) $\overrightarrow{BA} + \overrightarrow{BC}$

3. Express as a single vector. Show the addition on a grid to illustrate the parallelogram law.
   a) $[5, 0] + [2, 4]$
   b) $[4, -1] + [3, 5]$
   c) $[4, 1] + [0, -3]$
   d) $[-3, 2] + [1, 2]$
   e) $[-1, 3] + [-5, 1]$
   f) $[-2, -3] + [4, 2]$

B 4. Copy each pair of vectors and show $\vec{u} + \vec{v}$:

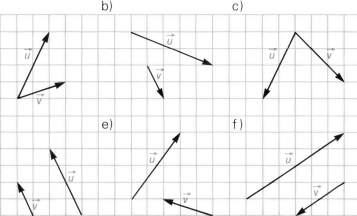

5. In the diagram, rectangle ABCD is divided into eight congruent squares. Express as a single vector:
   a) $\overrightarrow{AE} + \overrightarrow{NL}$
   b) $\overrightarrow{EG} + \overrightarrow{AD}$
   c) $\overrightarrow{KC} + \overrightarrow{FP}$
   d) $\overrightarrow{NH} + \overrightarrow{EL}$

C 6. In the diagram, $\triangle ABC$ is equilateral and D, E, F are the midpoints of its sides. Express as a single vector:
   a) $\overrightarrow{AF} + \overrightarrow{DB}$
   b) $\overrightarrow{FA} + \overrightarrow{EB}$
   c) $\overrightarrow{DA} + \overrightarrow{EC}$
   d) $\overrightarrow{AF} + \overrightarrow{ED}$

7. A helicopter is flying north in still air at 150 km/h. A 50 km/h wind springs up. If uncorrected, what would the helicopter's speed and heading be if the wind direction is:
   a) 000°?
   b) 180°?
   c) 090°?
   d) 030°?
   e) 120°?
   f) 225°?

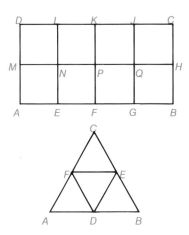

10 - 6 Scalar Products of Vectors

If k is any real number and $\vec{v} = [a, b]$ is any vector, then the vector $k\vec{v}$ is defined by

$$k\vec{v} = [ka, kb].$$

$k\vec{v}$ is called a **scalar product** because it is the product of a scalar and a vector.

The following example shows the geometric relationship between vectors which are scalar multiples of one another.

Example 1. If $\vec{v} = [3, 6]$, find $2\vec{v}$, $\frac{1}{3}\vec{v}$, and $-1.5\vec{v}$, and illustrate on a grid.

Solution. $2\vec{v} = 2[3, 6]$, or $[6, 12]$
$\frac{1}{3}\vec{v} = \frac{1}{3}[3, 6]$, or $[1, 2]$
$-1.5\vec{v} = -1.5[3, 6]$, or $[-4.5, -9]$

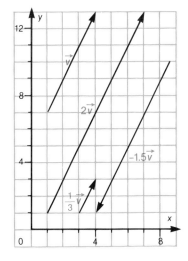

Example 1 shows that:
- if $k > 0$, $k\vec{v}$ has the same direction as $\vec{v}$ and is k times as long as $\vec{v}$;
- if $k < 0$, $k\vec{v}$ is opposite in direction to $\vec{v}$ and is $|k|$ times as long as $\vec{v}$.

If $k = -1$, $k\vec{v}$ becomes $-\vec{v}$. The vectors $\vec{v}$ and $-\vec{v}$ have the same length but opposite direction. That is, the opposite of a vector is formed by multiplying it by -1.

Example 2. If $\vec{u} = [5, -3]$, find the opposite of $\vec{u}$ and illustrate on a grid.

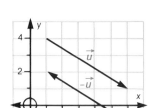

Solution. The opposite of $\vec{u} = [5, -3]$ is $-\vec{u} = -[5, -3]$, or $[-5, 3]$.

Scalar products of vectors can be added.

Example 3. If $\vec{u} = [2, 4]$ and $\vec{v} = [6, 2]$, represent the following vectors as ordered pairs and illustrate on a grid.

a) $\vec{u} + 2\vec{v}$
b) $2\vec{u} + \frac{1}{2}\vec{v}$

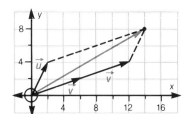

Solution. a) $\vec{u} + 2\vec{v} = [2, 4] + 2[6, 2]$
$= [2, 4] + [12, 4]$
$= [14, 8]$

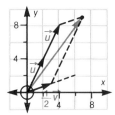

b) $2\vec{u} + \frac{1}{2}\vec{v} = 2[2, 4] + \frac{1}{2}[6, 2]$
$= [4, 8] + [3, 1]$
$= [7, 9]$

Example 4. Express in terms of $\overrightarrow{OA}$ and $\overrightarrow{OB}$:
a) $\overrightarrow{OC}$
b) $\overrightarrow{OD}$
c) $\overrightarrow{OE}$
d) $\overrightarrow{OF}$

Solution. a) $\overrightarrow{OC} = \overrightarrow{OH} + \overrightarrow{OB}$, or $\frac{1}{2}\overrightarrow{OA} + \overrightarrow{OB}$

b) $\overrightarrow{OD} = \overrightarrow{OH} + \overrightarrow{OJ}$, or $\frac{1}{2}\overrightarrow{OA} + \frac{1}{2}\overrightarrow{OB}$

c) $\overrightarrow{OE} = \overrightarrow{OA} + \overrightarrow{OJ}$, or $\overrightarrow{OA} + \frac{1}{2}\overrightarrow{OB}$

d) $\overrightarrow{OF} = \overrightarrow{OG} + \overrightarrow{OJ}$, or $\frac{3}{2}\overrightarrow{OA} + \frac{1}{2}\overrightarrow{OB}$

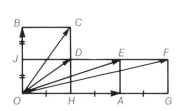

Exercises 10 - 6

A 1. If $\vec{u} = [4, -6]$, find:
a) $2\vec{u}$;
b) $5\vec{u}$;
c) $-3\vec{u}$;
d) $\frac{1}{2}\vec{u}$;
e) $-\frac{3}{2}\vec{u}$;
f) $-\vec{u}$;

10-6 *Scalar Products of Vectors* **395**

2. If $\vec{u} = [5, 2]$ and $\vec{v} = [2, 4]$, illustrate the following on a grid:
   a) $\vec{u}$ b) $\vec{v}$ c) $3\vec{u}$ d) $2\vec{v}$
   e) $-2\vec{u}$ f) $-\vec{v}$ g) $3\vec{u} + 2\vec{v}$ h) $-2\vec{u} + \vec{v}$

3. If $\vec{a} = [1, 3]$ and $\vec{b} = [4, 2]$, represent these vectors as ordered pairs:
   a) $2\vec{a} + \vec{b}$ b) $\vec{a} + 3\vec{b}$ c) $3\vec{a} + \vec{b}$
   d) $2\vec{a} + \frac{1}{2}\vec{b}$ e) $-5\vec{a} + 2\vec{b}$ f) $3\vec{a} + 2\vec{b}$

4. If $\vec{u} = [4, -6]$, find each scalar product and compare its length with that of $\vec{u}$:
   a) $2\vec{u}$ b) $-3\vec{u}$ c) $\frac{1}{2}\vec{u}$
   d) $-4\vec{u}$ e) $\frac{5}{2}\vec{u}$ f) $7\vec{u}$

5. Express in terms of the given vector, $\vec{u}$:
   a) $\vec{AB}$ b) $\vec{AC}$ c) $\vec{AD}$ d) $\vec{BC}$
   e) $\vec{BD}$ f) $\vec{BA}$ g) $\vec{CA}$ h) $\vec{DA}$

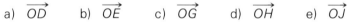

6. From the diagram below (left), express in terms of $\vec{OA}$ and $\vec{OB}$:
   a) $\vec{OD}$ b) $\vec{OE}$ c) $\vec{OG}$ d) $\vec{OH}$ e) $\vec{OJ}$

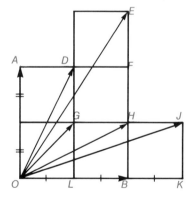
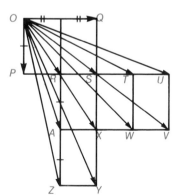

B 7. For the diagram above (right), express in terms of $\vec{OP}$ and $\vec{OQ}$:
   a) $\vec{OR}$ b) $\vec{OS}$ c) $\vec{OT}$ d) $\vec{OU}$ e) $\vec{OV}$
   f) $\vec{OW}$ g) $\vec{OX}$ h) $\vec{OY}$ i) $\vec{OZ}$ j) $\vec{OA}$

8. If $\vec{u} = [5, 2]$, $\vec{v} = [-3, 1]$ and $\vec{w} = [2, -4]$, represent these vectors as ordered pairs:
   a) $3\vec{u} + \vec{v}$
   b) $2\vec{u} + 3\vec{v}$
   c) $3\vec{u} + 2\vec{w}$
   d) $\vec{u} + \vec{v} + \vec{w}$
   e) $-\vec{u} + 2\vec{v} + 3\vec{w}$
   f) $2\vec{u} + \vec{v} + 2\vec{w}$
   g) $4\vec{u} + 2\vec{v} + 5\vec{w}$
   h) $3\vec{u} + \vec{v} + \frac{1}{2}\vec{w}$
   i) $-2\vec{u} + 3\vec{v} + 3\vec{w}$

C 9. $\overrightarrow{AB} = [4, 2]$ represents one side of rectangle $ABCD$. If the length of the rectangle is double the width, find vectors to represent the other three sides of the rectangle.

10. If $\vec{u} = [5, 2]$ and $\vec{v} = [1, -3]$, find two numbers, a and b, such that:
    a) $a\vec{u} + b\vec{v} = [2, 11]$;
    b) $a\vec{u} + b\vec{v} = [6, 16]$;
    c) $a\vec{u} + b\vec{v} = [-17, 0]$;
    d) $\frac{1}{2}a\vec{u} + b\vec{v} = [7.5, 3]$.

10-7 Vector Proofs in Geometry

Since vectors have both magnitude and direction, a statement that two vectors are equal gives two facts about the corresponding line segments:

Vector statement	Equivalent geometric statements
$\overrightarrow{AB} = \overrightarrow{CD}$	$AB = CD$ $AB \parallel CD$
$\overrightarrow{PQ} = 2\overrightarrow{RS}$	$PQ = 2RS$ $PQ \parallel RS$

10-7 Vector Proofs in Geometry

The conclusions in the table suggest that vectors can be useful for solving problems in geometry involving parallel line segments.

Example 1. In the quadrilateral $ABCD$, $AB = DC$ and $AB \parallel DC$. Prove that $BC = AD$ and $BC \parallel AD$.

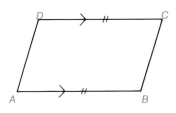

Solution. *Analysis:* In terms of vectors, it is given that $\overrightarrow{AB} = \overrightarrow{DC}$ and it must be proved that $\overrightarrow{BC} = \overrightarrow{AD}$. If AC is drawn, the vector $\overrightarrow{AC}$ can be expressed in terms of these vectors in two different ways: i) by using $\triangle ABC$, ii) by using $\triangle ADC$.

Proof: Join AC.

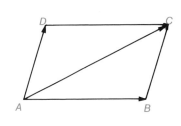

Using $\triangle ABC$: $\overrightarrow{AC} = \overrightarrow{AB} + \overrightarrow{BC}$
Using $\triangle ADC$: $\overrightarrow{AC} = \overrightarrow{AD} + \overrightarrow{DC}$
Since the above expressions must be equal,
$\overrightarrow{AB} + \overrightarrow{BC} = \overrightarrow{AD} + \overrightarrow{DC}$.
But $\overrightarrow{AB} = \overrightarrow{DC}$
Therefore, $\overrightarrow{BC} = \overrightarrow{AD}$
That is, $BC = AD$ and $BC \parallel AD$

In the solution to *Example 1*, the triangle law was used twice. This is a useful strategy when solving problems in geometry with vectors. It is used in the next example.

Example 2. Prove that the line segment joining the midpoints of two sides of a triangle is parallel to the third side and half its length.

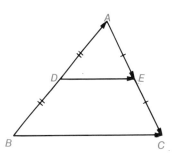

Solution. *Analysis:* In $\triangle ABC$, D and E are the midpoints of AB and AC respectively, and it must be proved that $DE = \frac{1}{2}BC$ and $DE \parallel BC$. In vector terms, it must be proved that $\overrightarrow{DE} = \frac{1}{2}\overrightarrow{BC}$. These two vectors can be expressed in terms of other vectors using $\triangle DAE$ and $\triangle BAC$.

Proof: Using $\triangle DAE$: $\overrightarrow{DE} = \overrightarrow{DA} + \overrightarrow{AE}$
Using $\triangle BAC$: $\overrightarrow{BC} = \overrightarrow{BA} + \overrightarrow{AC}$
$= 2\overrightarrow{DA} + 2\overrightarrow{AE}$ *(Since D and E are midpoints)*
$= 2(\overrightarrow{DA} + \overrightarrow{AE})$
$= 2\overrightarrow{DE}$
That is, $\overrightarrow{DE} = \frac{1}{2}\overrightarrow{BC}$
Therefore, $DE = \frac{1}{2}BC$ and $DE \parallel BC$

Exercises 10 - 7

A 1. In parallelogram *ABCD* (below, left), *E* and *F* are the midpoints of *AB* and *DC* respectively. Prove that *ED* = *BF* and *ED* ∥ *BF*.

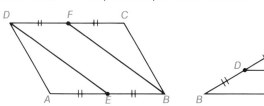

2. In △*ABC* (above, right), *D* and *E* are trisection points of *AB* and *AC* as shown. Prove that $DE = \frac{2}{3}BC$ and *DE* ∥ *BC*.

3. In parallelogram *ABCD* (below, left), *E* and *F* are trisection points of *AB* and *DC* as shown. Prove that *ED* = *BF* and *ED* ∥ *BF*.

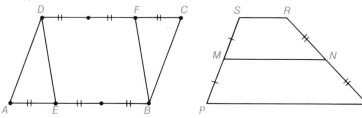

B 4. *PQRS* is a trapezoid (above, right) in which *PQ* ∥ *SR*, and *PQ* = 3*SR*. If *M* and *N* are the midpoints of *SP* and *RQ* respectively, prove that *MN* ∥ *SR* and *MN* = 2*SR*.

5. Prove that if the diagonals of a quadrilateral bisect each other, it is a parallelogram.

6. Prove that the midpoints of the sides of any quadrilateral are the vertices of a parallelogram.

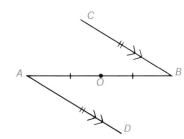

7. In the diagram, *CB* = *AD*, *CB* ∥ *AD*, and *O* is the midpoint of *AB*. Prove that *O* is the midpoint of the line segment joining *C* and *D*.

8. In parallelogram *ABCD*, *M* is the midpoint of the diagonal *BD*. Prove that *M* is also the midpoint of diagonal *AC*.

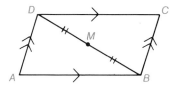

C 9. In △OAB, M is the midpoint of AB. Prove that $\overrightarrow{OM} = \frac{1}{2}\overrightarrow{OA} + \frac{1}{2}\overrightarrow{OB}$.

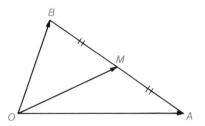

Review Exercises

1. Which of the following can be represented by a vector?
   a) a velocity of 120 km/h southeast
   b) a 70 L tank of propane
   c) a move of 600 m west
   d) a 100 kg sack of sugar
   e) a wind of 25 km/h from the southwest

2. Represent by a vector:
   a) 80 m east   b) 120 m on 225°   c) 60 m on 310°

3. Represent each vector shown on the grid as an ordered pair, and find its length.

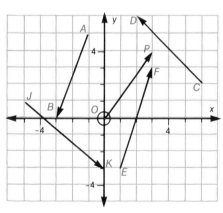

4. Find the length of each vector:
   a) [3, −2]   b) [−6, 8]
   c) [7, 0]   d) [0, −6]
   e) [−12, 5]   f) [−8, −15]

5. The coordinates of the head and tail of $\overrightarrow{PQ}$ are given. Represent $\overrightarrow{PQ}$ as an ordered pair:
   a) P(5, 7), Q(6, 3)   b) P(−2, −5), Q(3, 6)
   c) P(4, −6), Q(−1, −2)   d) P(0, −7), Q(3, −1)

6. Given the points A(−3, 2), B(0, 6), C(5, 6), D(8, 2), E(5, −2), F(0, −2), represent the following vectors as ordered pairs and draw them on the same grid:
   a) $\overrightarrow{AC}$ b) $\overrightarrow{DB}$ c) $\overrightarrow{EA}$ d) $\overrightarrow{CF}$ e) $\overrightarrow{BE}$ f) $\overrightarrow{FD}$

7. $\overrightarrow{AB} = [4, 6]$. If A is the point (1, −2), find the coordinates of B.

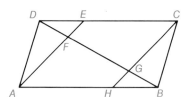

8. Express as a single vector:
   a) $\overrightarrow{DE} + \overrightarrow{EA}$
   b) $\overrightarrow{AB} + \overrightarrow{BF}$
   c) $\overrightarrow{BG} + \overrightarrow{GC}$
   d) $\overrightarrow{DG} + \overrightarrow{GC}$
   e) $\overrightarrow{FA} + \overrightarrow{AH} + \overrightarrow{HG}$
   f) $\overrightarrow{CH} + \overrightarrow{HA} + \overrightarrow{AD}$

9. Express as a single vector:
   a) $[4, 2] + [5, 1]$
   b) $[3, -4] + [2, -1]$
   c) $[-3, 6] + [2, -5]$
   d) $[-3, -5] + [-1, -4]$
   e) $[2, 6] + [1, -2] + [5, 7]$
   f) $[-3, 6] + [-5, 2] + [7, -1]$

10. Solve for $\vec{v}$: $[5, 2] + \vec{v} = [7, 8]$

11. If $\vec{u} = [6, -4]$, find:
    a) $2\vec{u}$
    b) $\frac{1}{2}\vec{u}$
    c) $-\frac{3}{2}\vec{u}$
    d) $1.5\vec{u}$

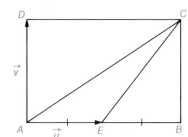

12. In rectangle $ABCD$, E is the midpoint of AB, $\overrightarrow{AE} = \vec{u}$, and $\overrightarrow{AD} = \vec{v}$. Express the following in terms of $\vec{u}$ and $\vec{v}$:
    a) $\overrightarrow{EB}$
    b) $\overrightarrow{BC}$
    c) $\overrightarrow{EC}$
    d) $\overrightarrow{AC}$

13. If $\vec{u} = [2, 4]$ and $\vec{v} = [-3, 3]$, represent as ordered pairs:
    a) $2\vec{u} + \vec{v}$
    b) $3\vec{u} + 2\vec{v}$
    c) $-\frac{1}{2}\vec{u} + \frac{2}{3}\vec{v}$

14. From the diagram, express the following in terms of $\overrightarrow{OA}$ and $\overrightarrow{OB}$:
    a) $\overrightarrow{OC}$
    b) $\overrightarrow{OE}$
    c) $\overrightarrow{OJ}$
    d) $\overrightarrow{OG}$
    e) $\overrightarrow{OH}$

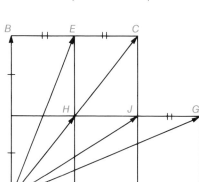

Table of Square Roots

Table of Trigonometric Ratios

Glossary

Selected Answers

Index

Table of Square Roots

n	$\sqrt{n}$	n	$\sqrt{n}$	n	$\sqrt{n}$	n	$\sqrt{n}$
1.0	1.000	5.5	2.345	10	3.162	55	7.416
1.1	1.049	5.6	2.366	11	3.317	56	7.483
1.2	1.095	5.7	2.387	12	3.464	57	7.550
1.3	1.140	5.8	2.408	13	3.606	58	7.616
1.4	1.183	5.9	2.429	14	3.742	59	7.681
1.5	1.225	6.0	2.449	15	3.873	60	7.746
1.6	1.265	6.1	2.470	16	4.000	61	7.810
1.7	1.304	6.2	2.490	17	4.123	62	7.874
1.8	1.342	6.3	2.510	18	4.243	63	7.937
1.9	1.378	6.4	2.530	19	4.359	64	8.000
2.0	1.414	6.5	2.550	20	4.472	65	8.062
2.1	1.449	6.6	2.569	21	4.583	66	8.124
2.2	1.483	6.7	2.588	22	4.690	67	8.185
2.3	1.517	6.8	2.608	23	4.796	68	8.246
2.4	1.549	6.9	2.627	24	4.899	69	8.307
2.5	1.581	7.0	2.646	25	5.000	70	8.367
2.6	1.612	7.1	2.665	26	5.099	71	8.426
2.7	1.643	7.2	2.683	27	5.196	72	8.485
2.8	1.673	7.3	2.702	28	5.292	73	8.544
2.9	1.703	7.4	2.720	29	5.385	74	8.602
3.0	1.732	7.5	2.739	30	5.477	75	8.660
3.1	1.761	7.6	2.757	31	5.568	76	8.718
3.2	1.789	7.7	2.775	32	5.657	77	8.775
3.3	1.817	7.8	2.793	33	5.745	78	8.832
3.4	1.844	7.9	2.811	34	5.831	79	8.888
3.5	1.871	8.0	2.828	35	5.916	80	8.944
3.6	1.897	8.1	2.846	36	6.000	81	9.000
3.7	1.924	8.2	2.864	37	6.083	82	9.055
3.8	1.949	8.3	2.881	38	6.164	83	9.110
3.9	1.975	8.4	2.898	39	6.245	84	9.165
4.0	2.000	8.5	2.915	40	6.325	85	9.220
4.1	2.025	8.6	2.933	41	6.403	86	9.274
4.2	2.049	8.7	2.950	42	6.481	87	9.327
4.3	2.074	8.8	2.966	43	6.557	88	9.381
4.4	2.098	8.9	2.983	44	6.633	89	9.434
4.5	2.121	9.0	3.000	45	6.708	90	9.487
4.6	2.145	9.1	3.017	46	6.782	91	9.539
4.7	2.168	9.2	3.033	47	6.856	92	9.592
4.8	2.191	9.3	3.050	48	6.928	93	9.644
4.9	2.214	9.4	3.066	49	7.000	94	9.695
5.0	2.236	9.5	3.082	50	7.071	95	9.747
5.1	2.258	9.6	3.098	51	7.141	96	9.798
5.2	2.280	9.7	3.114	52	7.211	97	9.849
5.3	2.302	9.8	3.130	53	7.280	98	9.899
5.4	2.324	9.9	3.146	54	7.348	99	9.950
		10.0	3.162			100	10.000

Table of Trigonometric Ratios

θ	sin θ	cos θ	tan θ	θ	sin θ	cos θ	tan θ
0	0.000	1.000	0.000	45	0.707	0.707	1.000
1	0.017	1.000	0.017	46	0.719	0.695	1.036
2	0.035	0.999	0.035	47	0.731	0.682	1.072
3	0.052	0.999	0.052	48	0.743	0.669	1.111
4	0.070	0.998	0.070	49	0.755	0.656	1.150
5	0.087	0.996	0.087	50	0.766	0.643	1.192
6	0.105	0.995	0.105	51	0.777	0.629	1.235
7	0.122	0.993	0.123	52	0.788	0.616	1.280
8	0.139	0.990	0.141	53	0.799	0.602	1.327
9	0.156	0.988	0.158	54	0.809	0.588	1.376
10	0.174	0.985	0.176	55	0.819	0.574	1.428
11	0.191	0.982	0.194	56	0.829	0.559	1.483
12	0.208	0.978	0.213	57	0.839	0.545	1.540
13	0.225	0.974	0.231	58	0.848	0.530	1.600
14	0.242	0.970	0.249	59	0.857	0.515	1.664
15	0.259	0.966	0.268	60	0.866	0.500	1.732
16	0.276	0.961	0.287	61	0.875	0.485	1.804
17	0.292	0.956	0.306	62	0.883	0.469	1.881
18	0.309	0.951	0.325	63	0.891	0.454	1.963
19	0.326	0.946	0.344	64	0.899	0.438	2.050
20	0.342	0.940	0.364	65	0.906	0.423	2.145
21	0.358	0.934	0.384	66	0.914	0.407	2.246
22	0.375	0.927	0.404	67	0.921	0.391	2.356
23	0.391	0.921	0.424	68	0.927	0.375	2.475
24	0.407	0.914	0.445	69	0.934	0.358	2.605
25	0.423	0.906	0.466	70	0.940	0.342	2.747
26	0.438	0.899	0.488	71	0.946	0.326	2.904
27	0.454	0.891	0.510	72	0.951	0.309	3.078
28	0.469	0.883	0.532	73	0.956	0.292	3.271
29	0.485	0.875	0.554	74	0.961	0.276	3.487
30	0.500	0.866	0.577	75	0.966	0.259	3.732
31	0.515	0.857	0.601	76	0.970	0.242	4.011
32	0.530	0.848	0.625	77	0.974	0.225	4.331
33	0.545	0.839	0.649	78	0.978	0.208	4.705
34	0.559	0.829	0.675	79	0.982	0.191	5.145
35	0.574	0.819	0.700	80	0.985	0.174	5.671
36	0.588	0.809	0.727	81	0.988	0.156	6.314
37	0.602	0.799	0.754	82	0.990	0.139	7.115
38	0.616	0.788	0.781	83	0.993	0.122	8.144
39	0.629	0.777	0.810	84	0.995	0.105	9.514
40	0.643	0.766	0.839	85	0.996	0.087	11.430
41	0.656	0.755	0.869	86	0.998	0.070	14.301
42	0.669	0.743	0.900	87	0.999	0.052	19.081
44	0.682	0.731	0.933	88	0.999	0.035	28.636
44	0.695	0.719	0.966	89	1.000	0.017	57.290
45	0.707	0.707	1.000	90	1.000	0.000	

Glossary

Absolute value: the absolute value of x is written $|x|$. It is always positive or zero for all real values of x.

Acute angle: an angle which measures less than 90°.

Alternate angles: $\angle ABC$ and $\angle BCD$ are alternate angles.

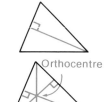

Altitude: a line segment from a vertex perpendicular to the opposite side of a triangle; the length of such a segment (also called the **height**).

The point of intersection of the lines containing the three altitudes of a triangle is called the **orthocentre**.

Angle bisector: a line which divides an angle into two equal angles.

The point of intersection of the bisectors of the three angles of a triangle is called the **incentre**. This point is the centre of the **incircle**, which touches the three sides of the triangle.

Axiom: a statement accepted as true without proof.

Bar graph: a graph in which data is represented by bars.

Broken-line graph: a graph in which plotted points are joined by line segments. Only the end points of the segments represent actual data.

Coefficient: the numerical part of a term, such as 3 in $3x$.

Common factor: a number or expression which is a factor of two or more numbers or expressions. Example: $2x$ is a common factor of $6x^2$ and $10x$.

Composite number: a natural number greater than 1 that is not prime.

Congruent: figures having the same size and shape.

Continuous-line graph: a graph that displays the value of one variable, such as speed, corresponding to the value of another variable, such as stopping distance, for all values over a given interval. All points on a continuous-line graph represent actual data.

Converse: a statement formed by interchanging the "if" and "then" clauses of a given statement.

Coordinates: an ordered pair of numbers that represents a point on a plane.

Corresponding angles: $\angle ABC$ and $\angle DEF$ are corresponding angles.

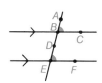

Cosine: in right-triangle trigonometry, the ratio of the side adjacent to an acute angle to the hypotenuse.

Deduction: a succession of statements in which the last statement follows from the others by logical reasoning.

Dilatation: a transformation that enlarges or reduces all dimensions of a figure by a factor k.

Distributive law: a law relating multiplication to addition and subtraction. Examples: $a(b + c) = ab + ac,$ and $a(b - c) = ab - ac.$

Exponent: see **Power**.

Factor: any number or expression that divides a given number or expression. Example: a is a factor of $3a^2b$.

Frequency: the number of times an event occurs.

Glide reflection: a transformation which is the result of a translation and a reflection in a line parallel to the direction of the translation.

Greatest common factor (g.c.f.): the greatest number that is a common factor of two or more numbers.

Histogram: a graph that records the number of times a variable has a value in a particular interval.

Image: the figure obtained by a transformation of a given figure.

Inequality: a mathematical sentence that uses the sign $<$ or $>$ to relate two expressions.

Integers: the set of numbers:
$\{\ldots, -3, -2, -1, 0, 1, 2, 3, \ldots\}$

Invariant property: a property of a figure that remains unchanged under a transformation.

Inverse operations: addition and subtraction are inverse operations. Multiplication and division are inverse operations.

Irrational number: see **rational number**.

Isometry: any transformation that preserves length.

Least common multiple (l.c.m.): the least number that is a common multiple of two or more numbers.

Like terms: terms that have exactly the same variable. Example: $3x$ and $7x$.

Linear programming: a technique for solving problems in which particular solutions of systems of inequalities are found.

Linear relation: a relation in which the terms containing the variables are all of the first degree.
Example: $\{(x, y) \mid 2x + 3y = 12\}$

Mapping notation: a method of writing a relation. Example $x \rightarrow 2x - 1$.

Mapping rule: a method of specifying a transformation in the coordinate plane. For example, the translation: $(x, y) \rightarrow (x + 3, y - 2)$

Mean: the sum of the numbers in a set divided by the number of numbers.

Mean deviation: a measure of dispersion that indicates how the numbers in a set differ from the mean. It is computed by finding the mean of the absolute values of the differences from the mean.

Measure of central tendency: any one of the mean, median, or mode.

Measure of dispersion: a number such as the **range** or **mean deviation** that serves as a measure of the spread of the numbers in a set of data.

Median: the middle number when a set of numbers is arranged in order.
A line segment joining a vertex of a triangle to the mid point of the opposite side.
The point of intersection of the three medians is called the **centroid**.

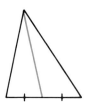

Mode: the most frequently occurring number, or numbers, in a set of numbers.

Monte Carlo methods: the simulations of experiments to determine the probability of outcomes.

Natural numbers: the set of numbers:
$\{1, 2, 3, 4, \ldots\}$

Obtuse angle: an angle which measures greater than 90° but less than 180°.

Ordered pair: two numbers written in a specific order. Example: $(2, -5)$

Parallel lines: two lines in the same plane that do not intersect.

Perpendicular bisector: a line which passes through the midpoint of a line segment and is perpendicular to the segment.

The point of intersection of the perpendicular bisectors of the three sides of a triangle is called the **circumcentre**. This point is the centre of the **circumcircle**, which passes through the three vertices of the triangle.

Polygon: a closed plane figure formed by line segments that do not cross each other. The segments are called **sides**. The points where two sides meet are the **vertices**. They are named by the number of sides they have.
A **regular polygon** has all sides equal and all angles equal.

Polyhedron: a solid bounded by polygons. The polygons are the **faces**. The faces meet at **edges**. The points where three or more edges meet are **vertices**.

Polynomial: the sum or difference of two or more terms in which the variable occurs to positive integral powers only. If there is only one variable, the highest power determines the **degree**. Example: the polynomial $5x^3 - 7x^2 + 4x - 1$ has degree 3.

Population: the set of all things being considered.

Power: an expression of the form a^n. a is called the **base** and n is called the **exponent**.

Prime number: a natural number greater than 1 that has only 1 and itself as factors.

Probability: the likelihood of the outcome of an experiment. If an experiment has n equally likely outcomes of which r are favorable to event A, then the probability of event A is $P(A) = \frac{r}{n}$.

Pythagorean theorem: see right triangle.

Quadratic relation: a relation that has at least one second-degree term but no terms of higher degree.

Random numbers: a set of single digit numbers that is generated in such a way that each number has an equal chance of occurring each time.

Random sample: a sample that is chosen in such a way that each item in the population has an equal chance of being selected.

Range: a measure of dispersion — the difference between the greatest and least numbers in a set of numbers.

Rational expression: the indicated quotient of two polynomials.

Rational number: any number that can be written in the form $\frac{m}{n}$, where m and n are integers, and $n \neq 0$. Rational numbers can also be expressed either as terminating or repeating decimals. Examples:

$$\frac{3}{5} = 0.6 \qquad -\frac{4}{3} = -1.3333\ldots$$

Any number that cannot be written in such a form is called an **irrational number**. The decimal form of an irrational number neither terminates nor repeats. Examples:

$$\pi = 3.141\,592\,653\,589\ldots,$$
$$\sqrt{2} = 1.414\,213\,562\,373\ldots$$

Real number: any member of the set of rational or irrational numbers.

Reciprocals: two numbers with a product of 1. Example: $\frac{3}{2}$ and $\frac{2}{3}$.

Reflection: a reflection in a line l is a transformation that maps every point P onto an image point P' such that:
- P and P' are equidistant from line l.
- PP' is perpendicular to line l.

The line l is called the **reflection line**.

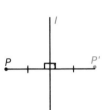

Relation: a set of ordered pairs.

Relative frequency: if outcome, A, occurs r times in n repetitions of an experiment, the relative frequency of A is $\frac{r}{n}$.

Rhombus: a parallelogram with all sides equal.

Right angle: an angle which measures 90°.

Right triangle: a triangle with one right angle. The side opposite the right angle is called the **hypotenuse**. The **Pythagorean theorem** relates the lengths of the sides of a right triangle: $AC^2 = AB^2 + BC^2$.

Rotation: when a shape is turned about a fixed point it is said to have undergone a rotation. A rotation through 180° is called a **half-turn**.

Sample: a representative portion of a population.

Scalar: a quantity that has magnitude only.

Glossary

Scientific notation: a number is written in scientific notation when it is expressed as the product of:
- a number equal to or greater than 1 but less than 10, and
- a power of 10.

Example: $372\,000\,000 = 3.72 \times 10^8$.

Sine: in right-triangle trigonometry, the ratio of the side opposite the acute angle to the hypotenuse.

Slope: the ratio: $\dfrac{\text{rise}}{\text{run}}$.

In a coordinate system,

slope = $\dfrac{\text{difference in } y\text{-coordinates}}{\text{difference in } x\text{-coordinates}}$, or $\dfrac{y_2 - y_1}{x_2 - x_1}$

Slope y-intercept form: a form of the equation of a line, $y = mx + b$, in which the slope, m, and the y-intercept, b, can be read directly.

Sphere: a surface consisting of all points in space at a given distance, the **radius**, from a fixed point, the **centre**.

Square: any expression of the form x^2. The numbers 1, 4, 9, 16, ..., n^2, where n is a natural number, are called **perfect squares**.
A rectangle with four equal sides.

Square root: a square root of a number is a number which, when squared, produces the given number.
The **radical sign**, $\sqrt{}$, denotes the positive square root. Example: $\sqrt{36} = 6$.

Statistics: the branch of mathematics which deals with the collection, organization, and interpretation of data.

Stem-and-leaf diagram: a method of displaying data in which all numbers are shown. The stem is formed by the tens' and higher digits and the leaf is formed by the ones' digits.

Example:
```
6 | 1
7 | 4 8 7
8 | 3 3 5 4 8 2
9 | 9 6
```

Straight angle: an angle which measures 180°.

Tangent: in right-triangle trigonometry, the ratio of the side opposite an acute angle to the side adjacent.

Tessellation: the covering of a plane by the images of a figure under one or more transformations.

Tetrahedron: see polyhedron.

Theorem: a deduction that is used in proving other deductions.

Transformation: whenever the shape, size, appearance, or position of an object is changed, it has undergone a transformation.

Translation: a translation is defined by the mapping $A \to A'$, $B \to B'$, $C \to C'$ in the diagram. The segments AA', BB' and CC' are all equal in length and parallel. The **translation arrow** represents the length and direction of these segments.

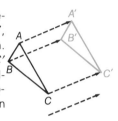

Transversal: a line which intersects two or more lines.

Trapezoid: a quadrilateral with just one pair of parallel sides.

Variable: a letter used to represent a member or members of a set of numbers.

Variation: the relationship between two variables as they take on different values. When two variables vary **directly**, they are related by the equation: $y = kx$. They are in **partial** variation when related by the equation: $y = kx + b$; they vary **inversely** when related by the equation: $xy = k$.

Vector: a quantity that has both magnitude and direction. It is represented by an arrow indicating the direction and with length proportional to the magnitude.

Vertex: see polygon and polyhedron.

Vertically opposite angles: $\angle AOB$ and $\angle COD$ are vertically opposite angles.

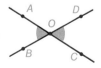

Whole numbers: the set of numbers: $\{0, 1, 2, 3, 4, ...\}$.

Answers

Exercises 1-1

1. b), c), e), f), j)
2. a) 2×3^2 b) 7×2^2 c) $2 \times 3 \times 7$
   d) 7×2^3 e) 2^6 f) 19×2^2
   g) 7×13 h) $2 \times 5 \times 11$ i) 11×13
   j) 11×17
3. a) 1, 2, 3, 4, 6, 12 b) 1, 2, 4, 8, 16 c) 1, 2, 3, 6, 9, 18
   d) 1, 5, 25 e) 1, 2, 4, 8, 16, 32
   f) 1, 2, 3, 4, 6, 9, 12, 18, 36
   g) 1, 2, 3, 4, 5, 6, 10, 12, 15, 20, 30, 60
   h) 1, 2, 4, 5, 10, 20, 25, 50, 100
   i) 1, 2, 3, 4, 6, 8, 9, 12, 16, 18, 24, 36, 48, 72, 144
   j) 1, 2, 4, 5, 8, 10, 20, 25, 40, 50, 100
4. a) 12 b) 15 c) 14 d) 5
   e) 12 f) 13 g) 48 h) 16
   i) 8 j) 18 k) 35 l) 42
5. a) 72 b) 48 c) 140 d) 160
   e) 90 f) 225 g) 420 h) 360
6. No. $65 = 5 \times 13$, $51 = 3 \times 17$, $52 = 2^2 \times 13$
7. The perfect squares.
8. 2, 71, 73
9. a) $7 + 8$; $4 + 5 + 6$; $1 + 2 + 3 + 4 + 5$
   b) $9 + 10 + 11$; $6 + 7 + 8 + 9$; $4 + 5 + 6 + 7 + 8$.
   c) $52 + 53$; $34 + 35 + 36$; $19 + 20 + 21 + 22 + 23$;
      $15 + 16 + 17 + 18 + 19 + 20$;
      $12 + 13 + 14 + 15 + 16 + 17 + 18$;
      $6 + 7 + 8 + 9 + 10 + 11 + 12 + 13 + 14 + 15$;
      $1 + 2 + \ldots + 14$.
10. a) $6^2 + 1$, $10^2 + 1$, $14^2 + 1$
    b) If n is odd, n^2 is odd and $n^2 + 1$ is even and not prime.
11. a) $2^5 - 1$, $2^7 - 1$, $2^{11} - 1$
    b) If n is even, then $n = 2k$ where k is a natural number. $2^{n-1} = 2^{2k} - 1$, or $(2^k + 1)(2^k - 1)$
       That is, $2^n - 1$ is the product of two consecutive odd numbers.
12. 2, 17, 19
13. 80 s
14. 224
15. a) 61 b) 301
16. a) $A2, B3$ b) $A4, B5$ c) $A3, B4$
    d) $A16, B25$ e) $A7, B8$ f) $A4, B5$
17. a) 15 min b) 12 min
18. 72
19. $729 = 9^3 = 27^2$, $4096 = 16^3 = 64^2$
20.

a, b	g.c.f.	l.c.m.	g.c.f. × l.c.m.	ab
9, 12	3	36	108	108
18, 45	9	90	810	810
20, 35	5	140	700	700
40, 48	8	240	1920	1920

(g.c.f.) × (l.c.m.) = ab

21. a) i) 1 ii) 1 iii) 1
    b) The g.c.f. of consecutive numbers is 1.

Exercises 1-2

1. a) -5 b) 45 c) -26 d) 9
   e) 9 f) 28
2. a) -20 b) 40 c) -4 d) -9
   e) 120 f) -20
3. a) 26 b) -32 c) -15 d) 36
   e) -76 f) -43
4. a) -34 b) -4 c) -3.5 d) 87
   e) 5 f) -4
5. a) 21 b) 18 c) -24 d) 85
6. a) 16 b) -33 c) -10 d) 24
   e) -21
7. a) 29 b) -11 c) 107 d) 5
   e) -15 f) 89
8. a) -3 b) 0 c) 8 d) 13
   e) -56 f) 0
9. a) 17 b) -24 c) 21 d) 16
   e) 38 f) -15
10. a) 0 b) -34 c) 65 d) 33
    e) -20 f) -76
11. a) -36 b) -23 c) 30 d) 21
    e) -22 f) -4 g) -53 h) -118
    i) 4 j) 2
12. a) 35 units^2 b) 12 units^2 c) 38 units^2 d) 49.5 units^2
13. a) 29 units^2 b) 50 units^2

Answers **409**

14. 45 ha   **15.** 22 km²
16. a) 14 units²   **b)** 41 units²
17. a) 6, −8   **b)** 5, −4   **c)** None

MAU

• 2 080 000 km²

MM

1. Answers will vary. Some examples are:
  a) 11 + 5   **b)** 11 + 13   **c)** 61 + 3
  d) 41 + 31   **e)** 59 + 29   **f)** 11 + 89
  g) 113 + 31   **h)** 97 + 103
2. No. Example: 11
3. a) Yes   **b)** No
4. a) 7   **b)** 7   **c)** 5   **d)** 0

Exercises 1-3

1. a) $-\dfrac{7}{12}$ **b)** $\dfrac{1}{6}$ **c)** $\dfrac{18}{55}$
  d) $-\dfrac{22}{3}$ **e)** $-\dfrac{125}{8}$ **f)** $\dfrac{39}{8}$
  g) $-\dfrac{2}{3}$ **h)** $\dfrac{39}{8}$ **i)** $-\dfrac{33}{14}$
2. a) -1 **b)** $-\dfrac{5}{6}$ **c)** $-\dfrac{10}{9}$
  d) $-\dfrac{5}{8}$ **e)** $-\dfrac{79}{63}$ **f)** $\dfrac{17}{12}$
3. a) $\dfrac{97}{12}$ **b)** $\dfrac{31}{30}$ **c)** $-\dfrac{307}{18}$
  d) $-\dfrac{37}{12}$ **e)** $-\dfrac{149}{12}$ **f)** $\dfrac{1}{12}$
4. a) $-\dfrac{1}{6}$ **b)** $-\dfrac{1}{3}$ **c)** $\dfrac{23}{12}$
  d) $-\dfrac{65}{72}$ **e)** $\dfrac{13}{8}$ **f)** $-\dfrac{13}{2}$
5. a) -0.625 **b)** $-0.\overline{4}$
  c) $3.\overline{142\,857}$ **d)** $-3.91\overline{6}$
  e) 5.4 **f)** $-1.\overline{54}$
  g) $-0.\overline{538\,461}$ **h)** $-2.8\overline{3}$
6. a) $\dfrac{137}{100}$ **b)** $\dfrac{5}{11}$ **c)** $-\dfrac{61}{9}$
  d) $\dfrac{23}{8}$ **e)** $-\dfrac{517}{999}$ **f)** $-\dfrac{961}{80}$
  g) $\dfrac{4091}{990}$ **h)** $\dfrac{662}{165}$ **i)** $\dfrac{10}{9}$
  j) $-\dfrac{15\,697}{4995}$ **k)** $\dfrac{518}{999}$ **l)** $-\dfrac{2081}{6660}$
  m) $-\dfrac{1}{1}$ **n)** $\dfrac{1}{2}$ **o)** $\dfrac{18}{7}$

7. a) $\dfrac{1}{12}$ **b)** $\dfrac{15}{16}$ **c)** $\dfrac{9}{5}$
  d) 0 **e)** $-\dfrac{9}{5}$ **f)** $\dfrac{4}{5}$
  g) $-\dfrac{47}{12}$ **h)** $\dfrac{13}{60}$ **i)** $-\dfrac{91}{24}$
8. a) $-\dfrac{1}{8}$ **b)** -1 **c)** $-\dfrac{28}{27}$
9. $3980, $3486   **10.** 38.72 MW
11. a) 20%   **b)** 16%
12. a) i) 40 cm   **ii)** 26.7 cm   **iii)** 25 cm
  b) i) 30 cm   **ii)** 26.7 cm   **iii)** infinite
  c) The image of a distant object is at the principal focus.
13. a) $0.\overline{5}$ **b)** $0.\overline{9}$ **c)** $0.\overline{2}$
  d) $0.08\overline{3}$ **e)** $0.\overline{2}$ **f)** $0.\overline{148}$

Exercises 1-4

1. b), c), d), e), h), i)   **2. a), d), e), g), l)**
3. a), b), c), e), f), h), i), j), k)   **4.** Answers will vary.
5. a) Rational   **b)** Rational
  c) Integer   **d)** Natural
  e) Irrational   **f)** Rational
  g) Irrational   **h)** Natural
  i) Irrational   **j)** Rational
  k) Irrational   **l)** Natural
6. a) $1 - 7\sqrt{2}$. No.   **b)** 6. Yes.
  c) 5. Yes.   **d)** $16 - 2\sqrt{2}$. No.
7. a) $-4 - 2\sqrt{3}$. No.   **b)** $16 + 8\sqrt{3}$. No.
  c) 0. Yes.   **d)** $23 + 12\sqrt{3}$. No.
8. No. When representing irrational numbers as decimals, rational approximations are used.
9. a) $\dfrac{1}{9}, \dfrac{1}{3}; \sqrt{\dfrac{1}{9}} = \dfrac{1}{3}$
  b) $\sqrt{0.444\,444\,4} = 0.666\,666\,6$

CP

1.

2^{29}	536870912	2^{-2}	.25
2^{30}	1.07374182 E + 09	2^{-7}	7.8125 E − 03
2^{64}	1.84467441 E + 19	10^{-3}	1 E − 03
10^{19}	1 E + 19	10^{-14}	1 E − 14

2. 60
3. a) Yes. 1 light-year ≐ 9.4608×10^{12} km
  b) Yes. Distance ≐ 4.0776×10^{13} km
4. 90 years

410 Answers

Exercises 1-5

1. a) 8.9 b) 10.7 c) 12.0 d) 1.7 e) 4.8 f) 9.2
2. 30
3. a) $2\sqrt{2}$ cm b) $\sqrt{13}$ cm c) $\sqrt{10}$ cm d) $\sqrt{29}$ cm
4. 27.5 m
5. About 87 m
6.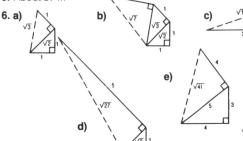
7. 24 km
8. a) $2\sqrt{3}$ cm b) $4\sqrt{3}$ cm²
9. $4\sqrt{2}$ cm
10. $\sqrt{39}$
11. 40 m
12. 246 m
13. a) $\sqrt{38}$ b) $d = \sqrt{a^2 + b^2 + c^2}$
14. a) $\sqrt{2}, \sqrt{3}$
    b) Edge 12; Face diagonal 12; Body diagonal 4
15. a) 4 b)

1	$\sqrt{2}$	$\frac{1}{2}\sqrt{2}$	$\frac{1}{9}\sqrt{6}$
12	6	24	24

16. 187 m

Exercises 1-6

1. a) $2\sqrt{14}$ b) $\sqrt{154}$ c) -12 d) $6\sqrt{10}$ e) -18 f) $70\sqrt{6}$
2. a) $-108\sqrt{6}$ b) $15\sqrt{15}$ c) $-\frac{2}{3}\sqrt{42}$ d) 12 e) -21 f) 26.4
3. a) $\sqrt{2} \times \sqrt{12}$ b) $\sqrt{3} \times \sqrt{6}$ c) $\sqrt{3} \times \sqrt{15}$ d) $\sqrt{2} \times \sqrt{14}$ e) $\sqrt{3} \times \sqrt{24}$ f) $\sqrt{10} \times \sqrt{6}$ g) $\sqrt{3} \times \sqrt{13}$ h) $\sqrt{5} \times \sqrt{13}$ i) $\sqrt{8} \times \sqrt{12}$ j) $\sqrt{8} \times \sqrt{15}$ k) $\sqrt{6} \times \sqrt{21}$ l) $\sqrt{7} \times \sqrt{15}$
4. a) $4\sqrt{2}$ b) $5\sqrt{2}$ c) $3\sqrt{3}$ d) $4\sqrt{6}$ e) $2\sqrt{2}$ f) $5\sqrt{3}$ g) $6\sqrt{3}$ h) $4\sqrt{5}$
5. a) $7\sqrt{3}$ b) $3\sqrt{6}$ c) $2\sqrt{19}$ d) $6\sqrt{5}$ e) $6\sqrt{5}$ f) $15\sqrt{2}$ g) $18\sqrt{10}$ h) $77\sqrt{2}$
6. a) $12\sqrt{3}$ b) $105\sqrt{2}$ c) $-192\sqrt{3}$ d) $40\sqrt{15}$ e) $84\sqrt{2}$ f) $165\sqrt{2}$
7. a) $12\sqrt{3}$ b) 120 c) 60 d) $280\sqrt{6}$ e) $60\sqrt{21}$ f) 1008
8. a) $2\sqrt{10}, 3\sqrt{5}, 4\sqrt{3}, 5\sqrt{2}, 2\sqrt{13}, 3\sqrt{6}$
   b) $-2\sqrt{21}, -4\sqrt{5}, -5\sqrt{3}, -6\sqrt{2}, -2\sqrt{17}, -3\sqrt{7}$
   c) $7\sqrt{0.05}, 2\sqrt{0.8}, 6\sqrt{0.1}, 3\sqrt{0.7}, 5\sqrt{0.3}, 4\sqrt{0.5}$
9. a) $108\sqrt{2}$ b) 1200 c) 1080 d) $4320\sqrt{2}$ e) 518 400 f) 30
10. $AC = \sqrt{2^2 + 2^2}$, or $\sqrt{8}$
    $AB = \sqrt{1^2 + 1^2}$, or $\sqrt{2}$
    But $AC = 2AB$
    Therefore, $\sqrt{8} = 2\sqrt{2}$
11. 3
12. a) $4, \sqrt{5}, 2\sqrt{5}$
    b) i) 2:1 ii) $2:\sqrt{5}$ iii) $2:\sqrt{5}$ iv) $1:\sqrt{5}$ v) $1:\sqrt{5}$

Exercises 1-7

1. a) $2\sqrt{7}$ b) $16\sqrt{6}$ c) $-6\sqrt{13}$ d) $-25\sqrt{19}$ e) $33\sqrt{3}$ f) $5\sqrt{15}$
2. a) $-4\sqrt{5}$ b) $3\sqrt{10}$ c) $18\sqrt{2}$ d) $13\sqrt{6} - 6\sqrt{2}$ e) $3\sqrt{5} - 6\sqrt{10}$ f) $8\sqrt{5} - 11\sqrt{2}$
3. a) $5\sqrt{10}$ b) $6\sqrt{2}$ c) $-3\sqrt{3}$ d) $-\sqrt{5}$ e) $2\sqrt{2}$ f) $-2\sqrt{6}$
4. a) $7\sqrt{6}$ b) $3\sqrt{7}$ c) $2\sqrt{5}$ d) $9\sqrt{3}$ e) $11\sqrt{3}$ f) $-4\sqrt{2}$
5. a) $10\sqrt{3}$ b) $13\sqrt{3}$ c) $24\sqrt{2}$ d) $-8\sqrt{2}$ e) $13\sqrt{6}$ f) $6\sqrt{5}$ g) $15\sqrt{2}$ h) $9\sqrt{7}$
6. 0.97 km
7. a) $-33\sqrt{3}$ b) $7\sqrt{7} + \sqrt{11}$ c) $\sqrt{3} - 5\sqrt{5}$ d) $47\sqrt{2} - \sqrt{7}$ e) $-6\sqrt{3} - 9\sqrt{2}$ f) $41\sqrt{6} + 36\sqrt{7}$ g) $-2\sqrt{5} - 41\sqrt{3}$
8. a) $2\sqrt{2}$ b) $2\sqrt{5}$ c) $31\sqrt{2}$ d) $\sqrt{5} - 4\sqrt{3}$ e) $-\sqrt{2} - 4\sqrt{6}$

Answers

9. a) $\sqrt{20}$ b) No.
10. a) $\sqrt{x+y}$ b) $\sqrt{(\sqrt{x})^2 + (\sqrt{y})^2} = \sqrt{x+y}$
11. Yes. $\sqrt{x^2} = x, x \geq 0$; $|x| = x, x \geq 0$
    Therefore, $\sqrt{x^2} = |x|$

Exercises 1-8

1. a) $\sqrt{10} - \sqrt{14}$ b) $\sqrt{33} + \sqrt{6}$
   c) $\sqrt{78} - \sqrt{30}$ d) $2\sqrt{15} + 6\sqrt{21}$
   e) $\sqrt{30} - 5\sqrt{2}$ f) $12\sqrt{22} + 20\sqrt{26}$
2. a) $\sqrt{39} + 13$ b) $9 - 18\sqrt{2}$
   c) -48 d) $6\sqrt{10} + 8\sqrt{15}$
   e) $36\sqrt{3} - 72\sqrt{2}$ f) $36 - 40\sqrt{3}$
3. a) $10\sqrt{21} - 12 + 18\sqrt{2}$ b) $8\sqrt{21} + 12\sqrt{42} - 28$
   c) 80 d) $24\sqrt{15} - 216\sqrt{2} - 216$
4. a) $35 + 140\sqrt{6}$
   b) $48\sqrt{3} - 96\sqrt{2} - 96$
   c) $20\sqrt{6} + 12 - 8\sqrt{15} + 18\sqrt{2}$
   d) $8\sqrt{10} - 24\sqrt{6} + 128$
5. a) $\sqrt{35} - \sqrt{15} + \sqrt{14} - \sqrt{6}$
   b) $9\sqrt{2} - 12$
   c) $14 + 7\sqrt{2}$
   d) $52 - 38\sqrt{2}$
6. a) $12\sqrt{15} + 18\sqrt{6}$
   b) $18\sqrt{2} - 12\sqrt{30} - 10\sqrt{6} + 20\sqrt{10}$
   c) $8 + 3\sqrt{6}$
   d) $44\sqrt{3} + 44\sqrt{7} - 18\sqrt{42} - 126\sqrt{2}$
7. a) $38 + 12\sqrt{10}$ b) -43
   c) $67 - 42\sqrt{2}$ d) 80
   e) 33 f) $24 - 10\sqrt{3}$
   g) $197 - 70\sqrt{6}$ h) -2
   i) $220 - 120\sqrt{2}$ j) 162
8. a) $16 + 4\sqrt{15}, 16 + 2\sqrt{39}, 16 + 2\sqrt{15}, 16 + 2\sqrt{55}$
   b) $\sqrt{6} + \sqrt{10}, \sqrt{5} + \sqrt{11}, \sqrt{13} + \sqrt{3}, \sqrt{15} + \sqrt{1}$
9. $\sqrt{20} - \sqrt{6}, \sqrt{18} - \sqrt{8}, \sqrt{15} - \sqrt{11}, \sqrt{14} - \sqrt{12}$

Exercises 1-9

1. a) $2\sqrt{2}$ b) $\sqrt{7}$ c) $2\sqrt{3}$ d) $3\sqrt{5}$
   e) $3\sqrt{3}$ f) $\frac{5}{2}\sqrt{2}$ g) 4 h) 1
2. a) $6\sqrt{2}$ b) 2 c) 6 d) 14
3. a) 2 b) 2 c) $2\sqrt{2}$

d) $\frac{3}{2}\sqrt{2}$ e) $\frac{5}{3}\sqrt{3}$ f) $\frac{4\sqrt{14}}{15}$
g) $\sqrt{15}$ h) 2 i) $\frac{2\sqrt{30}}{9}$
j) $\sqrt{5}$ k) $\frac{\sqrt{15}}{3}$ l) $\sqrt{10}$

4. a) $\frac{\sqrt{6}}{6}$ b) $\frac{3}{5}\sqrt{5}$ c) $\frac{12\sqrt{17}}{17}$
   d) $2\sqrt{5}$ e) $\sqrt{3}$ f) $\frac{3\sqrt{7}}{14}$
   g) $\frac{2\sqrt{2}}{5}$ h) $2\sqrt{2}$ i) $\frac{\sqrt{15}}{4}$
   j) $\frac{8\sqrt{6}}{15}$ k) $\frac{20\sqrt{3}}{9}$ l) $\frac{4}{5}\sqrt{2}$

5. a) $\frac{3\sqrt{5} + 5\sqrt{3}}{15}$ b) $\frac{3\sqrt{2} - \sqrt{6}}{6}$
   c) $\frac{2\sqrt{3} + \sqrt{6}}{6}$ d) $\frac{10\sqrt{7} - 21\sqrt{5}}{35}$
   e) $\frac{20\sqrt{3} + 3\sqrt{10}}{15}$ f) $\frac{6\sqrt{7} - 21\sqrt{6}}{14}$

6. a) $\frac{3\sqrt{3} + 2\sqrt{2}}{6}$ b) $\frac{15\sqrt{2} - 4\sqrt{6}}{12}$
   c) $\frac{21\sqrt{5} - 20\sqrt{3}}{30}$ d) $-\frac{3\sqrt{6}}{4}$
   e) $\frac{2\sqrt{6} - 2\sqrt{3}}{3}$ f) $\frac{17}{6}$

Review Exercises

1. a) 2×13 b) $2 \times 2 \times 11$
   c) $2 \times 3 \times 5 \times 7$ d) $2 \times 7 \times 13$
   e) $3 \times 11 \times 13$
2. a) 3 b) 13 c) 5 d) 7
3. a) 240 b) 210 c) 600 d) 504
4. 210 s
5. a) 22 b) -17 c) -9 d) -24
6. a) -18 b) -96
7. a) 30 b) 15
8. a) $\frac{1}{8}$ b) $-\frac{23}{12}$ c) $-\frac{1}{24}$
9. a) $-\frac{35}{12}$ b) $\frac{19}{144}$
10. a) $0.\overline{2}$ b) $-0.\overline{54}$ c) $3.\overline{6}$ d) $-5.\overline{16}$
11. a) $\frac{323}{100}$ b) $\frac{43}{99}$ c) $\frac{14}{11}$ d) $\frac{1541}{495}$
12. a), d)
13. a) $3\sqrt{3} - 6$, irrational   b) -4, rational
14. a) 8.6 b) 5.2 c) 17.9
15. 15.8 km

16. a) $7\sqrt{2}$ **b)** $4\sqrt{7}$ **c)** $6\sqrt{5}$ **d)** $-18\sqrt{10}$
17. a) $24\sqrt{2}$ **b)** 72 **c)** -100
d) $-150\sqrt{2}$ **e)** $30\sqrt{3}$ **f)** 144
18. a) $5\sqrt{5}$ **b)** $3\sqrt{2}$ **c)** $8\sqrt{2}$
d) $3\sqrt{3}$ **e)** $6\sqrt{5}$
19. a) $4 - \sqrt{6}$ **b)** $\sqrt{6} - 27$ **c)** $2\sqrt{10} - \sqrt{15}$
20. a) $\sqrt{6}$
b) -19
21. a) $2\sqrt{3}$ **b)** 12 **c)** 2 **d)** $\dfrac{1}{2}$
22. a) $\sqrt{6}$ **b)** $\dfrac{\sqrt{3}}{3}$ **c)** $\dfrac{\sqrt{2}}{2}$ **d)** 2

Exercises 2-1

1.

Term	Variable	Coefficient	Exponent
$-ab^3$	a, b	-1	1, 3
$4x^2y$	x, y	4	2, 1
$-5m^3n^2$	m, n	-5	3, 2
$7x^4y^2z$	x, y, z	7	4, 2, 1
$\frac{1}{3}p^6q^9$	p, q	$\frac{1}{3}$	6, 9
$6x$	x	6	1
8	—	8	—

2. a) -6 **b)** 12 **c)** -18 **d)** 72
e) $-\dfrac{8}{9}$ **f)** $-\dfrac{4}{9}$ **g)** 18 **h)** 9
3. a) $3w$ **b)** $3s^2$ **c)** x^3
d) $-4y$ **e)** $-3a^2b$ **f)** $\dfrac{-3p^3}{q}$
4. a) 16 **b)** 64 **c)** 2
d) 1 **e)** 24 **f)** 64
5. a) $\dfrac{4}{5}x$ **b)** $\dfrac{3x}{2y}$ **c)** m^3n^2
d) $\left(\dfrac{p}{7q}\right)^3$
6. a) i) $A = lw$ **ii)** $P = 2l + 2w$
b) 180 m^2
7. a) $C = 2\pi r$ **b)** 157 cm
8. a) $A = \pi r^2$ **b)** 706.5 cm^2
9. 2.4 m^3
10. $S = 2\pi rl$
11. a) 500 kg **b)** 250 kg
12. a) 0.8 m^3 **b)** 8.4 m^3
13. a) 184 kPa **b)** 1500 kPa
14. $\dfrac{x^2}{y}$

Exercises 2-2

1. a) 16 **b)** $\dfrac{1}{25}$ **c)** $\dfrac{1}{3}$
d) 4 **e)** $\dfrac{3}{2}$ **f)** $\dfrac{16}{9}$
g) 2 **h)** 1 **i)** $\dfrac{4}{9}$
2. a) 1 **b)** $\dfrac{1}{9}$ **c)** $-\dfrac{1}{8}$
d) $-\dfrac{3}{2}$ **e)** $\dfrac{25}{9}$ **f)** 1
g) 10 000 **h)** 8 **i)** 10
3. a) $(-7)^3$ **b)** b^{-5} **c)** x^{-4} **d)** $(1.5)^{-2}$
4. a) x^2y^3 **b)** x^3y^2 **c)** x^3z^{-2} **d)** a^4b^{-4}
5. a) $8\dfrac{1}{8}$ **b)** $\dfrac{3}{4}$ **c)** $\dfrac{9}{100}$ **d)** $\dfrac{1}{25}$
e) 8 **f)** $-\dfrac{27}{100}$ **g)** $\dfrac{1}{2}$ **h)** $\dfrac{9}{8}$
i) $\dfrac{1}{200}$
6. a) $8; -8$ **b)** $-8; 8$ **c)** $-8; 8$
d) $\dfrac{1}{8}; -\dfrac{1}{8}$ **e)** $8; -8$ **f)** $-\dfrac{1}{8}; \dfrac{1}{8}$
7. a) $\dfrac{3}{2}$ **b)** 0 **c)** $-\dfrac{3}{2}$
d) $\dfrac{99}{10}$ **e)** $-\dfrac{3}{2}$ **f)** $-\dfrac{99}{10}$
8. a) $\dfrac{3}{2}$ **b)** $\dfrac{99}{25}$ **c)** $\dfrac{25}{2}$
d) 400 **e)** $-\dfrac{1}{5}$ **f)** 576
9. a) $-\dfrac{1}{16}$ **b)** $\dfrac{1}{4}$ **c)** -2
d) $-\dfrac{1}{8}$ **e)** $\dfrac{3}{4}$ **f)** -1
10. a) $-\dfrac{3}{16}$ **b)** $\dfrac{1}{36}$ **c)** 1728
d) 0 **e)** $-\dfrac{4}{9}$ **f)** $\dfrac{9}{4}$
11. a) 4^3 **b)** 3^{-2} **c)** $(-2)^{-5}$
d) 10^{-3} **e)** $(-5)^{-2}$ **f)** $(-2)^{-3}$
g) 5^0 **h)** $\left(\dfrac{1}{2}\right)^{-1}$ **i)** $\left(\dfrac{4}{3}\right)^{-1}$
12. a) 10^2 **b)** $(-10)^3$ **c)** $\left(\dfrac{3}{4}\right)^3$
d) $(0.4)^2$ **e)** $(-2)^5$ **f)** $(-0.1)^3$
g) $\left(\dfrac{1}{7}\right)^3$ **h)** 2^6 **i)** $\left(\dfrac{4}{3}\right)^3$
13. a) 3^{-2} **b)** 6^{-2} **c)** 10^{-4}
d) $\left(\dfrac{3}{2}\right)^{-3}$ **e)** 5^{-2} **f)** $(-2)^{-3}$
g) 5^{-2} **h)** $\left(\dfrac{1}{4}\right)^{-3}$ **i)** $\left(\dfrac{3}{2}\right)^{-3}$
14. a) $(1.5)^{-1}$ **b)** $\left(\dfrac{1}{2}\right)^2$ **c)** 2^{-3}

Answers **413**

d) $\left(\frac{1}{2}\right)^{-5}$ e) $(-5)^{-3}$ f) $\left(\frac{2}{7}\right)^{-3}$

15. a) 0   b) -1 c) -3
    d) -4 e) -4 f) $\frac{2}{3}$
    g) $\frac{1}{2}$ h) 5   i) $\frac{1}{25}$

Exercises 2-3

1. a) x^7 b) a^7 c) b^8
   d) m^5 e) c^6 f) y^9
2. a) x^2 b) y^4 c) n
   d) a^3 e) x^3 f) y^2
3. a) x^9 b) y^6 c) n^{12}
   d) a^4 e) c^{10} f) a^6b^6
   g) x^2y^6 h) x^8y^4 i) m^3n^6
4. a) x^{-3} b) c^{-1} c) y^{-5}
   d) a^{-4} e) b^{-3} f) x^{-6}
5. a) x b) m^{-4} c) a^4
   d) 1   e) x f) b^3
6. a) x^{-6} b) y^5 c) m^{10}
   d) 1   e) c f) x^2
7. a) x^{-6} b) y^2 c) m^{-6}
   d) c^{-9} e) a^{-4} f) xy^{-2}
   g) x^4y^{-6} h) ab^2 i) a^4b^{-4}
8. a) b^{-1} b) 1   c) x^2y^{-1}
   d) xy^{-6} e) c^7d f) m^4n^{-4}
9. a) -32 b) -4 c) -64
   d) $-\frac{1}{32}$ e) -16 f) 64

10. a) Exponents should be added.
    b) Base is unchanged.
    c) Exponents should be subtracted.
    d) Exponents should be multiplied.
    e) Coefficients should be squared.
    f) Bases must be the same.

11. a) $5^{-8} = 0.000\,002\,56$ $5^1 = 5$
    $5^{-7} = 0.000\,012\,8$ $5^2 = 25$
    $5^{-6} = 0.000\,064$ $5^3 = 125$
    $5^{-5} = 0.000\,32$ $5^4 = 625$
    $5^{-4} = 0.001\,6$ $5^5 = 3125$
    $5^{-3} = 0.008$ $5^6 = 15\,625$
    $5^{-2} = 0.04$ $5^7 = 78\,125$
    $5^{-1} = 0.2$ $5^8 = 390\,625$
    $5^0 = 1$ $5^9 = 1\,953\,125$
    $5^{10} = 9\,765\,625$

    b) i) 78 125   ii) 390 625   iii) 125
       iv) 390 625   v) 0.000 012 8   vi) 25

12. a) $27x^{-15}$ b) $-6a^{-8}$ c) $2x^2$
    d) q^2 e) $-\frac{27}{8}x^{-5}y^3$ f) $a^{11}b^{-5}c^6$

MM

1. $3^2 + 4^2 = 5^2$; $5^2 + 12^2 = 13^2$; $8^2 + 15^2 = 17^2$
2. a) $2^2 + 3^2 + 6^2 = 7^2$; $3^2 + 4^2 + 12^2 = 13^2$
   b) $3^3 + 4^3 + 5^3 = 6^3$; $1^3 + 6^3 + 8^3 = 9^3$
   c) $9^2 + 2^2 = 7^2 + 6^2$; $8^2 + 1^2 = 7^2 + 4^2$
   d) $12^3 + 1^3 = 10^3 + 9^3$
3. $x^6 + y^6 = z^6$ can be written: $(x^2)^3 + (y^2)^3 = (z^2)^3$

Exercises 2-4

1. a) $2x - 3y$ b) $-4a + 15b$
   c) $-5c + 7d$ d) $-7x - 7y$
   e) $30m - 12n$ f) $-5b$
2. a) $-9r - 4t + 16s$ b) $-5x - 27y + 14z$
   c) $-6a + 3b + 11c$ d) $24m + 12n - 9p$
   e) $2h - 21k - 6l$ f) $-18x - 2y - 6z$
3. a) $4a^2 - 2a - 4$ b) $-5m^2 - 7m - 7$
   c) $-5s^2 - 14s + 5$ d) $6x^2 - 9x + 5$
   e) $5p^2 - 4p + 13$ f) $6g^2 - 11g - 9$
4. a) $xy + 3xz$ b) $10ab - 6bc + 4ac$
   c) $-5pq + 8pr + 3qr$ d) $6mn + 5mp - 11np$
   e) $-5de + 3df - 15ef$ f) $39xy - 29xz$
5. a) $12ab + 2ac$ b) $2x^2 - 13xy$
   c) $-4y^2 - xy$ d) $58xy$
   e) $-7m^2 - 5mn$ f) $-18c^2 - 14cd$
6. a) $-4x^2y^2 + 7x^2y$
   b) $y^3z^3 - y^2z^3 - 3yz^3$
   c) $-2m^2n^2 - 2mn - 7m + 2n$
   d) $3p^2q^3 - 7p^2q^2$
   e) $a^3b^2 - a^3b - ab^3$
   f) $a^2 - 3c^2 - 2ab + 2$

Exercises 2-5

1. a) $35x^6$ b) $12a^2$ c) $-54m^5$
   d) $10n^3$ e) $56y^9$ f) $72p^9$

2. a) $18xy$ **b)** $32p^2q$ **c)** $35m^3n$
  d) $14a^2b$ **e)** $-45r^2s^3$ **f)** $-44c^2d$
3. a) $-8ab^3$ **b)** $-12m^5n^7$ **c)** $21a^3b^5$
  d) $24p^3q^7$ **e)** $24x^6y^3$ **f)** $-6a^4b^4$
4. a) $4x^3$ **b)** $9m^4$ **c)** $-9a^6$
  d) $4d^2$ **e)** $-4n^2$ **f)** $8b^4$
5. a) $\frac{9}{2}r^4s^2$ **b)** $-3x^4y^3$
  c) $\frac{5}{2}p^6q^4$ **d)** $2a^2b$
  e) $-8m^6$ **f)** $6x^5y^6$
6. a) $54x^{11}$ **b)** $-12a^{10}b$
  c) $45s^{10}t^5$ **d)** $-100p^7q^3$
  e) $-24m^5n^9$ **f)** $675x^{10}y^{14}$
7. a) $6x^6y^5$ **b)** $-6q^2$
  c) $\frac{7}{2}a^2b^3$ **d)** $2x^5$
  e) $-16m^8n^8$ **f)** $a^{12}b^{13}$
8. a) $A = \pi\frac{d^2}{4}$ **b)** $A = \pi d^2$
  c) $V = \pi\frac{d^3}{6}$
9. $\pi:4$ **10.** $2:\pi$
11. $4:3$ **12.** $2:1$
13. $\frac{3}{2}xy$; 126 units **14.** $9k^2$
15. a) $\frac{7}{9}x^2$ **b)** $\frac{7}{4}y^2$
16. a) $-6a^3b^3$ **b)** $-16a^5b^7$
  c) $-80a^8b^7$ **d)** $100a^6b^6$
  e) $4a^4b^5$
17. $\frac{1}{3}$

Exercises 2-6

1. a) $10a + 4$ **b)** $10x - 10y$
  c) $-6m^2 - 3$ **d)** $3p + 7$
  e) $11x^2 - 4x$ **f)** $-5k + 11l$
2. a) $11m^2 - 3m + 2$ **b)** $4a^3 + 4a^2 + 3a$
  c) $-3x^2 - 20x + 17y$ **d)** $9t^3 - 18t^2 - 15t - 1$
  e) $8a^3 - 18$ **f)** $-5x^2 - 8x + 5$
3. a) $10x^2 - 6x^2y$; third **b)** $p^2q^2 - 3p + 2q$; fourth
  c) $3m^2 - 3m - 2n^2 + 2n$; second
  d) 3; zero **e)** $-3xy^2 + 2xy$; third
4. a) $10x^2 - 8x + 16$ **b)** $5z^3 - 10z^2 + 2z - 7$
  c) $8x^2y + xy - xy^2$ **d)** $4m^2n^2 - 6mn + 7mn^3$
5. a) $-3x^2 - y^2 - 2$; second **b)** $3m^2$; second
  c) $-5x^2y^2 + 7x^2y - xy - 6xy^2$; fourth
  d) $2a^2b^2 - 2a^2b - 2b^2a - 2b^2$; fourth

Exercises 2-7

1. a) $20x^2 + 40$ **b)** $14a - 35$
  c) $16k - 24k^2$ **d)** $24b^2 - 36b + 108$
  e) $45m^2 - 63m + 27$ **f)** $24p^2 - 15p + 21$
2. a) $3x + 19$ **b)** $24 - 5a$
  c) $10 - 2y$ **d)** $28m - 33$
  e) $-13p^2 - 12p$ **f)** $-8x - 21y$
3. a) $5x - 6$ **b)** $-x - 3$
  c) $42b - 36c$ **d)** $16m - 37n + 14$
  e) $23x^2 - 69x$ **f)** $-14c + 12d - 57e$
4. a) $60x^2 - 48x$ **b)** $6a - 21a^2$
  c) $12p^2 - 6pq$ **d)** $135n^3 - 90n^2$
  e) $21m^4n + 42m^3$ **f)** $56x^2y - 40x^3$
5. a) $9x^3 + 13x^2y$ **b)** $-2a^4 - 27a^3b$
  c) $4p^3 + 51p^2q$ **d)** $8m^2 - 17mn + 13m$
  e) $8x^3 - 32x^2y - 26x^2$ **f)** $-34a^4 + 66a^3b - 80a^3$
6. a) $2a^3b^2 - 2a^2b^3 - 2ab^3$ **b)** $3x^3y^3 + 3x^3y^2 - 3x^2y^2$
  c) $5m^2n^3 - 5m^3n^3 - 15m^3n^2$ **d)** $5x^2 - y^2 - 7xy + 2y$
  e) $2b^3 - 2b^2c - 2c^2 + 5bc$
  f) $7x^3 - 7xy^2 - 2xy - 2x^2y - 2y^3$
7. a) $8a^2b - 7a^2 + 15a$; third **b)** $21mn^2 - 33m^2n$; third
  c) $2xy^2 - 15x^2y$; third
  d) $7s^3 - 87s^2 + 34s$; third
  e) $5x^3 - 14x^2y - 53xy^2$; third
  f) $55xy - 15y^2 + 2x^2$; second
8. a) $99a^2 + 27a$ **b)** $10x - 6x^2$
  c) $40x^2 + 30x$ **d)** $-20a^3 - 8a^2 + 12a$
  e) $3.6n^2 - 8.4n$ **f)** $10x^3 - 30x^2 + 15x$
9. a) $29p - 18q + 29$ **b)** $-29k - 4l - 75$
  c) $3x^2 + 5xy - 9x - 2y^2 - 14y$ **d)** $5a^2 - 100a$
  e) $45x^2 - 8x^2y - 2x^3$ **f)** $30m^2n - m^2 - 94mn$
10. a) $a + 5b$
  b) $10a + 4b$
  c) $a^2 + b^2$
  d) $a^3 - a^2b - b^2a - b^3$
11. a) $A = 2\pi r(r + h)$
  b) i) 911 cm² **ii)** $8\pi r^2$ **iii)** $8\pi r^2 - 10\pi r$ **iv)** $3\pi r^2 + 14\pi r$

Exercises 2-8

1. a) $8x^2 + 14x + 3$
   b) $30m^2 - 7m - 2$
   c) $8c^2 - 22c + 15$
   d) $9x^2 - 49$
   e) $16t^2 - 12t - 40$
   f) $24x^2 - 62x + 14$

2. a) $g^2 + 2gh + h^2$
   b) $x^2 + 2xy + y^2$
   c) $m^2 + 2mn + n^2$
   d) $p^2 - 2pq + q^2$
   e) $k^2 - 2kl + l^2$
   f) $s^2 - 2st + t^2$

3. a) $9x^2 + 6xy + y^2$
   b) $x^2 + 10xy + 25y^2$
   c) $9a^2 + 12a + 4$
   d) $25x^2 - 90x + 81$
   e) $16m^2 - 56m + 49$
   f) $16y^2 + 88y + 121$

4. a) $6x^2 + 5xy + y^2$
   b) $15a^2 + 2ab - b^2$
   c) $6x^2 + 5xy - 4y^2$
   d) $x^2 + 7xy + 12y^2$
   e) $35m^2 - 19mn + 2n^2$
   f) $18x^2 - 48xy + 14y^2$

5. a) $9a^2 + 30ab + 25b^2$
   b) $4m^2 - 28mn + 49n^2$
   c) $36s^2 - 24st + 4t^2$
   d) $16p^2 - 24pq + 9q^2$
   e) $64x^2 - 48xy + 9y^2$
   f) $25y^4 + 70y^2z^2 + 49z^2$

6. a) $x^2 - y^2$
   b) $4m^2 - 25n^2$
   c) $81a^2 - 16b^2$
   d) $36x^2 - 4y^2$
   e) $49p^2 - 4q^2$
   f) $144m^2 - 81$

7. a) $3x^2 - 8x - 1$
   b) $21x^2 - 18x$
   c) $12m^2 - 26m + 36$
   d) $6cd$
   e) $-18a^2 - 51ab + 2b^2$
   f) $9x^2 - 26xy + 7y^2$

8. a) $7x^2 - 11x - 1$
   b) $-3s^2 + 13st - 11t^2$
   c) $3x^2 + 4xy - 5y^2$
   d) $5m^2 + 117mn - 131n^2$
   e) $-40xy$
   f) $-7x^4 - 19x^2 - 19$

9. a) $6x^3 + 7x^2 - 15x + 4$
   b) $4m^3 - 19m^2 + 33m - 36$
   c) $6a^3 + a^2 - 58a + 45$
   d) $15p^3 - 8p^2 - 6p + 4$
   e) $4x^3 - 24x^2 + 27x + 20$
   f) $24y^3 + 41y^2 - 9y - 28$

10. a) $6x^3 + 2x^2 - 128x - 160$
    b) $40p^3 - 218p^2 + 309p - 126$
    c) $24x^3 - 62x^2 - 7x + 30$
    d) $18x^3 - 3x^2 - 88x - 80$
    e) $50a^3 - 235a^2 + 228a - 63$
    f) $125m^3 - 150m^2 + 60m - 8$

11. a) $10x + 2y, 4x^2 + 2xy - 4y^2$
    b) $14x + 6y, 10x^2 + 14xy + 2y^2$
    c) $16x + 18, 12x^2 + 38x + 20$

12. a) $3m^2 - 9m - 30$
    b) $6x^2 - 4xy - 10y^2$
    c) $30p^2 - 145p + 140$
    d) $8x^3 - 68x^2y + 84xy^2$
    e) $18a^3 - 39a^2b - 15ab^2$
    f) $12x^3 + 48x^2y + 48xy^2$

13. a) $30x^2 + 108x + 70$
    b) $6m^2 - 2m - 48$
    c) $52x^2 + 107xy - 60y^2$
    d) $-3x^2y^2 - 5xy + 28$
    e) $-12a^2 + 15ab - 20b^2$
    f) $7x^2y^2 + 14xy + 91$

14. a) $8x^2 + 57x - 69$
    b) $-4s^2 - 39s - 11$
    c) $8x^3 - 6x^2 - 26x + 40$
    d) $-12m^2 + 18mn + 47m - 28n - 20$
    e) $-12y^3 + 91y^2 - 39y - 70$
    f) $42x^3 - 70x^2 + 98x - 54$

15. a) $m^2 + 11m + 26$
    b) $3m^2 + 16m + 28$
    c) $m^3 + 5m^2 + 6m - 5$

16. a) $4a^2 + 4ab + 4b^2$
    b) $-5a^2 + 12ab - 3b^2$
    c) $8a^2 + 6ab + 2b^2$

Exercises 2-9

1. a) $7x$
   b) y^4
   c) $4a^2b$
   d) $8m^2n^2$
   e) $3x^2y^2$
   f) $18s^2t$

2. a) $3x(x + 2)$
   b) $4y^2(2y - 1)$
   c) $5p^2(p - 3)$
   d) $8mn(3m + 2n)$
   e) $6a^2b^2(2 + 3a)$
   f) $-7x^2y^2(4y + 5x)$

3. a) $w(3w - 7w^2 + 4)$
   b) $-2x(x^4 - 2x + 3 - x^2)$
   c) $4(2x^2 - 3x^4 + 4)$
   d) $5ab(b + 2 - 3a)$
   e) $3xy(17x + 13y - 24)$
   f) $3m^2n^2(3m^2 - 2mn + 4n^2)$

4. a) $5(y - 2)$
   b) $8(m + 3)$
   c) $6(1 + 2x^2)$
   d) $5a(7 + 2a)$
   e) $7b^2(7 - b)$
   f) $7z^2(5 - 2z^4)$

5. a) $3(x^2 + 4x - 2)$
   b) $x(3x + 5x^2 + 1)$
   c) $a(a^2 + 9a - 3)$
   d) $3x(x + 2x^2 - 4)$
   e) $8y(2y - 4 + 3y^2)$
   f) $8xy(x - 4y + 2xy)$

6. a) $3(2b^7 - b + 4)$
   b) $y(5y^2 + 6y + 3)$
   c) $16x(1 + 2x + 3x^3)$
   d) $12y^2(y^2 - 1 + 2y)$
   e) $a(9a^2 + 7a + 18)$
   f) $5z(2z^2 - 3z + 6)$

7. a) $5x(5y + 3x)$
   b) $7mn(2m - 3n)$
   c) $3a^2b^2(3b - 4)$
   d) $4xy(x - 4y)$
   e) $6pq(2p + 3q)$
   f) $3m^2n^2(9m - 5n)$

8. a) $5a^2b^2(2a + 3b^2 - 1)$
   b) $4mn(3n - 2 - 5m)$
   c) $5x^2y(4 - 3y + 5xy)$
   d) $7ab^2(a^2b + 2a - 3)$
   e) $8x^2y^2(x^2y^2 - 2xy + 4)$
   f) $6abc(3a - 1 + 5c - 4bc)$

9. a) $(a + b)(3x + 7)$
   b) $(2x - y)(m - 5)$
   c) $(x + 4)(x^2 + y^2)$
   d) $(a + 3b)(5x - 9y)$
   e) $(x - 3)(10y + 7)$
   f) $(w + x)(7w - 10)$

10. a) $(x - 7)(3x^2 + 2x + 5)$
    b) $(a - b)(2m + 3n - 7)$

c) $(x^2 + y)(5a^2 - 7a + 8)$ d) $(b - a)(6a - 4b - 3)$
e) $(2x + y)(4m^2 - 3m - 7)$ f) $(3a - 2b)(2x^2 + 5x + 9)$
11. a) $\frac{\pi}{4}(x^2 - y^2)$ b) $\pi(x^2 + 2xy)$
12. $x^2\left(1 + \frac{3}{4}\pi\right)$
13. a) $\pi r^2\left(h + \frac{2}{3}r\right)$ b) $2\pi r(h + r)$
14. a) $(x + 3)(x + y)$ b) $(x + 1)(x^2 + 1)$
c) $(a + 2b)(5m + 1)$ d) $(x - 2y)(3x + 5)$
e) $(5m - 3)(m + 2n)$ f) $(a - 3b)(2a - 3)$

CP

1. a) 17   b) 1   c) 116
2. a), e)
3. a) 9   b) 16

MAU

1. 1.2×10^6 m³   2. a) 9×10^7 m³   b) About 56 days

Exercises 2-10

1. a) $(x + 6)(x + 4)$ b) $(m + 3)(m + 2)$
c) $(a + 8)(a + 2)$ d) $(m + 4)(m + 4)$
e) $(y + 7)(y + 6)$ f) $(d + 9)(d + 2)$
2. a) $(x - 4)(x - 3)$ b) $(c - 9)(c - 8)$
c) $(a - 6)(a - 1)$ d) $(b - 8)(b - 4)$
e) $(s - 10)(s - 2)$ f) $(x - 7)(x - 4)$
3. a) $(y + 5)(y - 1)$ b) $(b + 20)(b - 1)$
c) $(p + 18)(p - 3)$ d) $(x + 14)(x - 2)$
e) $(a + 6)(a - 4)$ f) $(k + 6)(k - 3)$
4. a) $(x - 4)(x + 2)$ b) $(n - 8)(n + 3)$
c) $(a - 5)(a + 4)$ d) $(d - 9)(d + 5)$
e) $(m - 15)(m + 6)$ f) $(y - 8)(y + 6)$
5. a) $(x + 12y)(x + 2y)$ b) $(x + 6y)(x + 3y)$
c) $(p + 13q)(p + 2q)$ d) $(a + 9b)(a + 8b)$
e) $(k + 6l)(k + 5l)$ f) $(v + 9w)(v + 4w)$
6. a) $(m - 10n)(m - 5n)$ b) $(a - 6b)(a - 5b)$
c) $(c - 4d)(c - 5d)$ d) $(t - 12v)(t - v)$
e) $(r - 9s)(r - s)$ f) $(x - 10y)(x - 6y)$
7. a) $(a + 10b)(a - 2b)$ b) $(x + 12y)(x - 8y)$
c) $(b + 9c)(b - 2c)$ d) $(s + 25t)(s - 4t)$
e) $(x + 41y)(x - y)$ f) $(y + 5z)(y - 4z)$
8. a) $(m + 8n)(m - 9n)$ b) $(x + 2y)(x - 8y)$

c) $(p + 9q)(p - 10q)$ d) $(c + 4d)(c - 20d)$
e) $(m + 5n)(m - 12n)$ f) $(x - 15y)(x + 3y)$
9. a) $(a + 6)(a + 3)$ b) $(m + 4)(m - 9)$
c) $(xy - 9)(xy - 6)$ d) $(pq - 17)(pq - 3)$
e) $(cd - 12)(cd + 3)$ f) $(xy + 3)(xy - 24)$
g) $(ab + 16)(ab + 3)$ h) $(mn + 2)(mn - 30)$
10. a) $\pm 9, \pm 11, \pm 19$ b) $\pm 1, \pm 4, \pm 11$
c) $\pm 12, \pm 13, \pm 20, \pm 37$ d) $\pm 4, \pm 7, \pm 11, \pm 17, \pm 28, \pm 59$
e) $\pm 16, \pm 40$ f) $\pm 14, \pm 50$
11. Answers will vary. Examples are:
a) $4, 6, -14, -24, -36$ b) $3, 4, -5, -12, -21$
c) $3, 8, 15, 24, 35$ d) $13, 28, 45, 64, 85$
e) $4, 10, 18, 28, 40$ f) $6, 10, 12, -8, -18$
12. a) $4(x + 4)(x + 3)$ b) $9(x + 5)(x + 1)$
c) $2(x + 3)(x + 1)$ d) $5(x - 4)(x - 2)$
e) $3(5x - y)(x - y)$ f) $3(4m + n)(5m + n)$
13. a) $5(a + 4)(a - 1)$ b) $2(a + 2)(a - 6)$
c) $3(p + 9q)(p - 4q)$ d) $4(y + 2z)(y - 6z)$
e) $2(m + 8n)(m - 7n)$ f) $7(x + 2y)(x - 14y)$
14. a) $3m(n + 4)(n + 2)$ b) $5a(m - 7)(m - 1)$
c) $15a(a + 3b)(a + 3b)$ d) $4y(x - 3)(x - 3)$
e) $3q^2(p + 4)(p + 8)$ f) $7d(c - 2d)(c - 3d)$
15. a) $4y(x + 2)(x - 7)$ b) $-6(p + 6q)(p - 5q)$
c) $7xy(xy - 4)(xy - 5)$ d) $3q(x + 10y)(x - 2y)$
e) $4a(a + 3b)(a - 4b)$ f) $5xy^2(x + 6y)(x - 4y)$
16. a) $2(x + 1)(2x + 3)$ b) $3(a + 1)(3a + 4)$
c) $7(p + 1)(7p + 6)$ d) $(5m + 6)(5m - 3)$
e) $2(2y - z)(4y + 7z)$ f) $3(x + y)(3x - 7y)$
17. a) $(x + y + 10)(x + y - 1)$ b) $(p - 2q - 8)(p - 2q - 3)$
c) $3(y + 1)(3y - 13)$ d) $(x + 1)(x + 3)(x^2 + 4x + 5)$
e) $(2m - n - 5p)(2m - n + 4p)$
f) $12(x + 3y + 2)(x - y + 2)$
18. a) $4a^2$ b) $2(a - 3b)(a - 2b)$
c) $(7a - 5b)(3b - a)$ d) $(a^2 - b^2 - 2)(a^2 - b^2 + 1)$

Exercises 2-11

1. a) $3, 2$ b) $6, 3$ c) $-5, -3$
d) $-5, 3$ e) $-3, 10$ f) $-6, 4$
2. a) $(2x + 3)(x + 4)$ b) $(2k + 1)(k + 1)$
c) $(2x + 1)(x + 2)$ d) $(3x + 1)(x + 4)$
e) $(3s + 1)(s + 1)$ f) $(3t + 7)(t + 4)$

Answers 417

3. a) $(2x + 3)(3x + 1)$ **b)** $(4x + 3)(2x + 1)$
c) $(2x + 3)(3x + 4)$ **d)** $(2x + 3)(x + 5)$
e) $(4x + 7)(x + 3)$ **f)** $(3x + 2)(2x + 7)$
4. a) $(2m - 3)(m - 4)$ **b)** $(2x - 5)(x - 3)$
c) $(3x - 2)(x - 4)$ **d)** $(3y - 1)(y - 7)$
e) $(3x - 2)(2x - 5)$ **f)** $(4t - 3)(t - 5)$
5. a) $(4m - 1)(2m - 3)$ **b)** $(5x - 2)(2x - 5)$
c) $(7x - 3)(2x - 1)$ **d)** $(3m - 4)(2m - 3)$
e) $(3a - 4)(a - 4)$ **f)** $(8x - 3)(4x - 1)$
6. a) $(3x - 2)(x + 3)$ **b)** $(5y - 1)(y + 4)$
c) $(3x - 4)(2x + 5)$ **d)** $(3k + 4)(2k - 1)$
e) $(3x - 2)(2x + 7)$ **f)** $(5x - 2)(3x + 7)$
7. a) $(2m - 7)(m + 3)$ **b)** $(3x + 2)(4x - 5)$
c) $(5x + 3)(2x - 5)$ **d)** $(3x + 5)(2x - 7)$
e) $(3x + 4)(2x - 3)$ **f)** $(8s + 1)(3s - 2)$
8. a) $3x^2(3y - 4)(y - 4)$ **b)** $2n(3m - 4)(m - 3)$
c) $5y^3(4x + 3)(x - 5)$ **d)** $xy(4x - 3y)(x + 6y)$
e) $10m^2n^2(2m - 7)(m + 3)$ **f)** $-3(3x + 1)(2x - 3)$
9. a) 25, 12 **b)** 8, 9 **c)** 12, −10
d) −22, 10 **e)** −22, −12 **f)** −22, 21
10. a) $(3k - 4)(3k - 4)$ **b)** $(3t + s)(5t - 2s)$
c) $(4x + y)(3x - 2y)$ **d)** $(x + 6y)(4x - 3y)$
e) $(4m + n)(5m - 3n)$ **f)** $(3x + 4y)(7x - y)$
11. a) $(7x - 6)(3x + 5)$ **b)** $(9x - 2)(8x + 3)$
c) $(5x + 4)(3x - 8)$ **d)** $(8x + 3y)(6x - 5y)$
e) $(2m - 5)(m + 7)$ **f)** $(5y + 2z)(8y - 3z)$
g) $(7xy - 1)(5xy - 3)$ **h)** $(5mn + 3)(2 - 9mn)$
12. a) $3y(3x - 5)(2x + 9)$ **b)** $2xy^2(5x - 6)(2x - 7)$
c) $3mn(4m + 3n)(3m - 4n)$ **d)** $2a^2b^2(3cd - 5)(5cd + 7)$
e) $2x(8x - 3)(3x + 5)$ **f)** $10x^2(5y + 6)(7 - 2y)$

Exercises 2-12

1. a) $(x + 7)(x - 7)$ **b)** $(2b + 11)(2b - 11)$
c) $(3m + 8)(3m - 8)$ **d)** $(5y + 12)(5y - 12)$
e) $(7x + 6)(7x - 6)$ **f)** $(4y + 9)(4y - 9)$
2. a) $(10m + 7)(10m - 7)$ **b)** $(8b + 1)(8b - 1)$
c) $(11a + 20)(11a - 20)$ **d)** $(5p + 9)(5p - 9)$
e) $(12m + 7)(12m - 7)$ **f)** $(6x + 11)(6x - 11)$
3. a) $(2s + 3t)(2s - 3t)$ **b)** $(4x + 7y)(4x - 7y)$
c) $(9a + 8b)(9a - 8b)$ **d)** $(p + 6q)(p - 6q)$
e) $(12y + 9z)(12y - 9z)$ **f)** $(5m + 13)(5m - 13)$

4. a) $(4m + 9n)(4m - 9n)$ **b)** $(8x + 15y)(8x - 15y)$
c) $(7a + 11b)(7a - 11b)$ **d)** $(14x + 5z)(14x - 5z)$
e) $(20a + 18b)(20a - 18b)$ **f)** $(16p + 25q)(16p - 25q)$
5. a) $8(m + 3)(m - 3)$ **b)** $6(x + 5)(x - 5)$
c) $5(2x + y)(2x - y)$ **d)** $3(2a + 5)(2a - 5)$
e) $2(3p + 7)(3p - 7)$ **f)** $5(4s + 9)(4s - 9)$
6. a) $3x(2x + 3)(2x - 3)$ **b)** $2m(4m + 7)(4m - 7)$
c) $7b(3a + 2)(3a - 2)$ **d)** $5y^2(5x + 6)(5x - 6)$
e) $2m(4m + 7)(4m - 7)$ **f)** $2y(8x + 5y)(8x - 5y)$
7. a) $(x - y - z)(x - y + z)$ **b)** $(2a + b + 9)(2a + b - 9)$
c) $4(c + 3)(c - 8)$ **d)** $(x - y - z)(x + y + z)$
e) $(6a - b)(12a + b)$ **f)** $(4x - 2y + 5z)(4x - 2y - 5z)$
8. a) $-5(2x + 9)$ **b)** $4(m + 1)(4m - 3)$
c) $24a$ **d)** $96yz$
e) $-(11p - 5)(5p + 9)$ **f)** $-15(x + 1)(3x + 1)$
9. a) $-4x^2y^2$ **b)** $(x + 1)(x - 4)(x^2 + 3x + 4)$
c) $(x - 6)(x + 2)(x^2 - 4x + 12)$
d) $(x + 6)(x - 1)(x + 3)(x + 2)$
e) $(a - 12)(a + 2)(a - 6)(a - 4)$
f) $(x + 20)(x - 3)(x - 5)(x - 12)$
10. About 48.5 m².
11. a) $V = \frac{4}{3}h(s^2 - h^2)$
b) 2 585 173 m³
12. The two pieces can be rearranged to form a rectangle $(x + y)$ by $(x - y)$

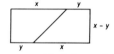

13. a) $x^2 - 2x^2y + x^2y^2$ **b)** $2x^2y - x^2y^2$
14. a) $8(d + 2e)(d - 2e)$ **b)** $\left(5m + \frac{1}{2}n\right)\left(5m - \frac{1}{2}n\right)$
c) $2y^2(3x + 5y)(3x - 5y)$ **d)** not possible
e) not possible
f) $\left(p + \frac{1}{3}q\right)\left(p - \frac{1}{3}q\right)$
g) $5(x + 2)(x - 2)(x^2 + 4)$
h) $\left(\frac{x}{4} + \frac{y}{7}\right)\left(\frac{x}{4} - \frac{y}{7}\right)$
i) $7(m^2 + n^2)(m + n)(m - n)$
15. 3, 7

Exercises 2-13

1. a) 369 **b)** −1296 **c)** 864
d) 459 **e)** 729 **f)** 54

2. a) −12 **b)** −1000 **c)** −168
   d) −1975 **e)** 288 **f)** 330
3. a) −655 **b)** 9120 **c)** −8730
   d) −1760 **e)** 24 300 **f)** −1940
4. a) 600 **b)** 600 **c)** 651
   d) 10 000 **e)** 2240 **f)** 1540
5. a) 10 000 **b)** 1600 **c)** 5625
   d) 900 **e)** −784 **f)** 0
6. a) 1599 **b)** 3591 **c)** 249 996
   d) 896 **e)** 8084 **f)** 22 331
7. a) 332 **b)** 453 **c)** −693
   d) 132 **e)** 7
8. a) i) 1280 **ii)** 1836 **iii)** 3344 **iv)** 5460
   b) $2x$ by $(x + 8)$ by $(x − 3)$ **c)** $10x^2 + 30x − 48$
9. a) 24, 39, 56, 75, 96, 119, 144, 171, 200, 231
   b) 7 **c) i)** 3 **ii)** 9 **iii)** 12
10. a) 285 **b)** 1240 **c)** 4900 **d)** 14 910
11. a) $h = \frac{l}{2}\sqrt{x^2 − 1}$

b)

Metal	h
Steel	4.33 cm
Brass	5.59 cm
Aluminum	6.12 cm

d) $\frac{x − 3}{x + 3}$ **e)** $\frac{r + 3}{r − 1}$ **f)** $\frac{x + 4}{x − 2}$
7. a) $\frac{b}{2c}$ **b)** $\frac{3ac}{4b}$ **c)** $−\frac{6x}{y}$
   d) $\frac{3ac^2}{4b}$ **e)** $\frac{x}{3z}$ **f)** $\frac{−5c^2}{9b^2}$
   g) $\frac{9n}{m^3}$ **h)** $\frac{5a^2c}{8b^5}$
8. a) $\frac{5}{3}$ **b)** −1 **c)** $−\frac{2}{3}$
   d) $−\frac{3}{2}$ **e)** $−\frac{5y}{2}$ **f)** $−\frac{3}{2}$
9. a) $\frac{x + 3y}{2x}$ **b)** $\frac{m − .4n}{3m}$ **c)** $\frac{3a + 4b}{2a}$
   d) $\frac{a − 3b}{a}$ **e)** $\frac{3x + 2y}{x}$ **f)** $\frac{m − n}{3m}$
10. a) $\frac{c − 8d}{c + 4d}$ **b)** $\frac{x − 5y}{x + 5y}$ **c)** $\frac{a + 4b}{a − 6b}$
   d) $\frac{2u − 5v}{2u + 5v}$ **e)** $\frac{m + 6n}{m + 8n}$ **f)** $\frac{x − 6y}{x − 9y}$
11. a) $\frac{x}{x − 1}$ **b)** $\frac{5(x − 2)}{x + 12}$ **c)** $\frac{x − 4}{x + 5}$
   d) $−\frac{a + 4}{2(a − 7)}$ **e)** $\frac{x − 5}{2x}$ **f)** $\frac{x − 17}{3 − x}$
12. a) $2x + 1$ **b)** 2 **c)** $4x$
   d) 1 **e)** −1 **f)** Not possible
13. a) $\frac{2a − b}{3a + 2b}$ **b)** $\frac{a}{b}$ **c)** $\frac{a + 2b}{2a + b}$
14. a) $\frac{1}{2x − 3}$ **b)** $\frac{1}{x + 3}$ **c)** $\frac{1}{y − 4}$
   d) $\frac{2x + y}{3x − y}$

Exercises 2-14

1. a) $8x$ **b)** $−\frac{9x}{y}$ **c)** $11a^2$
   d) $−4m^2$ **e)** $\frac{x}{3}$ **f)** $\frac{4}{7}$
   g) $\frac{6a^2}{b}$ **h)** $−\frac{12m}{n}$
2. a) $3m + 2$ **b)** $a^2 + b$ **c)** $3x − 2$
   d) $4 − 2n$ **e)** $\frac{3x + 4}{y}$ **f)** $3b − 4$
3. a) $\frac{m}{m − 2}$ **b)** $\frac{3}{2x − 5}$ **c)** $\frac{b}{a + 2b}$
   d) $\frac{3y}{2x − 5y}$ **e)** $\frac{2s}{4s + 3t}$ **f)** $\frac{2mn}{3m − 5}$
4. a) 2 **b)** $\frac{1}{3}$ **c)** $\frac{2}{3}$
   d) $−\frac{1}{2}$ **e)** $−\frac{3}{5}$ **f)** $\frac{1}{2}$
5. a) $\frac{2(x + 3)}{5}$ **b)** $\frac{m}{2}$ **c)** $\frac{5x + 7}{3}$
   d) $4x$ **e)** $−\frac{3x}{7}$ **f)** −3
6. a) $\frac{a + 5}{a + 3}$ **b)** $\frac{x − 5}{x + 1}$ **c)** $\frac{x − 4}{x + 8}$

Exercises 2-15

1. a) $\frac{5a}{12}$ **b)** $\frac{m}{2}$ **c)** $−\frac{x^2}{6}$
   d) $\frac{2c}{9}$ **e)** $\frac{−4t^2}{5}$ **f)** $6r$
2. a) $\frac{5a}{6b}$ **b)** $\frac{x}{7}$ **c)** $−\frac{y}{x}$
   d) $\frac{3a}{2b}$ **e)** $\frac{14}{5}$ **f)** $\frac{9a^3}{10c^2}$
3. a) $\frac{2t}{7s}$ **b)** $\frac{9x^2}{2}$ **c)** $\frac{3f^2}{7e^2}$
   d) $−x$ **e)** $\frac{2n}{3}$ **f)** $\frac{7x}{6y}$
4. a) $\frac{2x}{3}$ **b)** $2b$ **c)** $−\frac{n}{6}$
   d) $\frac{2x}{5}$ **e)** $\frac{20n^5}{3m^3}$ **f)** 1
5. a) $\frac{a^5b^2}{3}$ **b)** $−\frac{5}{4xy}$ **c)** $6x^2y$
   d) $\frac{8n}{9m}$ **e)** $\frac{2}{5}$ **f)** $\frac{16m^2}{15}$
6. a) $\frac{7}{2}$ **b)** $\frac{1}{12}$ **c)** $x − 3$

Answers **419**

d) $8m$
e) $\dfrac{1}{12}$
f) $-\dfrac{2(a+1)}{a}$
6. a) $\dfrac{9}{2x}$
b) $\dfrac{47}{5x}$
c) $\dfrac{29}{30x}$

7. a) $\dfrac{2b}{5}$
b) $\dfrac{xy^3}{2}$
c) $\dfrac{3y}{x+3}$
d) $\dfrac{13}{4a}$
e) $\dfrac{41}{24m}$
f) $-\dfrac{11}{18k}$

d) $\dfrac{3x+4}{12}$
e) $\dfrac{2m}{15n}$
f) $\dfrac{4(a+3b)}{3}$
7. a) $\dfrac{3a}{5}$
b) $\dfrac{5m}{24}$
c) $\dfrac{11x}{18}$

8. a) $\dfrac{9}{4}$
b) $\dfrac{x+11}{x-2}$
c) $\dfrac{5}{12}$
d) $-\dfrac{29c}{36}$
e) $\dfrac{7e}{12}$
f) $\dfrac{13m}{24}$

d) 1
e) $\dfrac{3x(x-2)}{4y(x+2)}$
f) $\dfrac{x-2}{x-1}$
8. a) $\dfrac{13}{12a}$
b) $-\dfrac{7}{12x}$
c) $\dfrac{13}{24m}$

9. a) -1
b) 1
c) $\dfrac{x+2}{x-1}$
d) $-\dfrac{5}{3x}$
e) $\dfrac{7}{12y}$
f) $-\dfrac{5}{24y}$

d) -1
e) $\dfrac{a-3}{a-1}$
9. a) $\dfrac{2x-2}{x}$
b) $\dfrac{m-9}{m}$
c) $\dfrac{11a+9}{3a}$

f) $\dfrac{(x+3)^2}{(x-3)^2}$
d) $\dfrac{4x+26}{5x^2}$
e) $\dfrac{-5m-7}{2m}$

10. a) $\dfrac{2xy(x+4y)}{3(x+5y)}$
b) $\dfrac{1}{2}$
f) $\dfrac{-8x-1}{4x}$

c) $\dfrac{x+2y}{x-y}$
d) $\dfrac{4(m-2n)}{3m(m+3n)}$
10. a) $\dfrac{k-43}{20}$
b) $\dfrac{5c-14}{6}$
c) $\dfrac{x+10}{12}$

11. a) $\dfrac{(x-1)(x-6)}{2x^2}$
d) $\dfrac{5m-9}{12}$
e) $\dfrac{25-14a}{24}$
f) $\dfrac{16x-35}{18}$

b) $\dfrac{x+y}{x-y}$
c) 1
11. a) $\dfrac{8x-11}{6x}$
b) $\dfrac{-6n-5}{24n}$
c) $\dfrac{9a-29}{18a}$

12. a) $\dfrac{2a-b}{2b-a}$
b) $\dfrac{4b-3a}{2(4a-3b)}$
d) $\dfrac{11x+5}{12x}$
e) $\dfrac{11m-13}{8m}$
f) $\dfrac{17-8a}{30a^2}$

12. a) $\dfrac{5}{2a}$
b) $\dfrac{4n-3m}{6mn}$
c) $\dfrac{4y+3}{xy}$

d) $\dfrac{2b+a}{2b}$
e) $\dfrac{2b+15a}{3ab}$
f) $\dfrac{9n-10m}{12mn}$

Exercises 2-16

1. a) $\dfrac{-16}{24mn}$
b) $\dfrac{15m}{24mn}$
c) $\dfrac{-20n}{24mn}$
13. a) $\dfrac{3y-2x+xy}{xy}$
b) $\dfrac{2bc-3ac+4ab}{abc}$

d) $\dfrac{13mn}{24mn}$
e) $\dfrac{-9mn}{24mn}$
f) $\dfrac{14m^2n}{24mn}$
c) $\dfrac{6yz+9xz-10xy}{12xyz}$
d) $\dfrac{1-5y-4x}{xy}$

g) $\dfrac{-21m^2}{24mn}$
h) $\dfrac{-22n^2}{24mn}$
e) $\dfrac{8b+3a-10}{12ab}$
f) $\dfrac{9np-16mp+20mn}{24mnp}$

2. a) i) $\dfrac{4x}{6}$
ii) $\dfrac{8x^2}{12x}$
iii) $\dfrac{2x^3}{3x^2}$
14. a) $\dfrac{7x^2+10y^2}{6xy}$
b) $\dfrac{5m^2-4n^2}{3nm}$

b) i) $\dfrac{-9m-3}{-12}$
ii) $\dfrac{6mx+2x}{8x}$
iii) $\dfrac{12mx^2+4x^2}{16x^2}$
c) $\dfrac{9a^2-20b^2}{15ab}$
d) $\dfrac{15x^2-8a^2}{10ax}$

c) i) $\dfrac{3a-21}{3a}$
ii) $\dfrac{a^2-7a}{a^2}$
iii) $\dfrac{5a^4-35a^3}{5a^4}$
e) $\dfrac{54p^2+49q^2}{42pq}$
f) $\dfrac{9x^2-10y^2}{15xy}$

d) i) $\dfrac{15y-9}{6y}$
ii) $\dfrac{10y^3-6y^2}{4y^3}$
iii) $\dfrac{-5y^2+3y}{-2y^2}$
15. a) $x^2+3+\dfrac{2}{x^2}$

e) i) $\dfrac{-8x-20}{-4x^2}$
ii) $\dfrac{40x^2+100x}{20x^3}$
iii) $\dfrac{2x+5}{x^2}$
b) $\dfrac{x^4+3x^2+2}{x^2}$

c) They are equivalent.

3. a) $\dfrac{3x}{6},\dfrac{2x}{6}$
b) $\dfrac{10}{5x},\dfrac{x^2}{5x}$
16. a) $\dfrac{a^4-3a^2+2}{a^2}$
b) $\dfrac{k^4-2k^2-15}{k^2}$

c) $\dfrac{3}{2a},\dfrac{4}{2a}$
d) $\dfrac{5n}{n^2},\dfrac{2}{n^2}$
c) $\dfrac{4a^4-12a^2+9}{a^2}$
d) $\dfrac{y^4-4}{y^2}$

e) $\dfrac{4}{24x},\dfrac{15}{24x}$
f) $\dfrac{4x+4}{20x^2},\dfrac{5x^2-5x}{20x^2}$
17. a) 2
b) 7
c) $\sqrt{10}$

4. a) $-\dfrac{3}{x}$
b) $\dfrac{2}{x}$
c) $\dfrac{7-5x}{4x}$
18. a) $4-\dfrac{a}{3}$
b) $\dfrac{1}{5}-\dfrac{1}{2x}$

d) $\dfrac{2-9m}{3m^2}$
e) 6
f) $\dfrac{x}{y^2}$
c) $\dfrac{x}{y}+1$
d) $7x+1+\dfrac{1}{x}$

5. a) $-\dfrac{2a}{15}$
b) $-\dfrac{2}{15a}$
c) $\dfrac{10a^2-12}{15a}$
19. a) $\dfrac{2}{3}$
b) $\dfrac{5}{2}$
c) $\dfrac{5}{3}$
d) 5

d) $\dfrac{10-12a^2}{15a}$
e) $-\dfrac{2}{15a}$
f) $\dfrac{10-12a}{15a}$

Exercises 2-17

1. a) $\dfrac{4m-1}{m+3}$ b) $\dfrac{-4s+11}{s-5}$
   c) 3   d) $\dfrac{4x+18}{x+6}$
   e) $\dfrac{3m-5}{2m+1}$ f) $\dfrac{-4a-11}{a^2+4}$
2. a) $\dfrac{3a+3}{a(a-3)}$ b) $\dfrac{-4y+30}{y(y-5)}$
   c) $\dfrac{4m-28}{m(m-4)}$ d) $\dfrac{2c^2-5c+5}{c(c-1)}$
   e) $\dfrac{3x^2-6x-12}{x(x+2)}$ f) $\dfrac{8x+6}{x(x+2)}$
3. a) $\dfrac{3-8a}{2a}$ b) $\dfrac{5-2y}{y+1}$
   c) $\dfrac{4n-29}{n-5}$ d) $\dfrac{x^2+4x-2}{x-4}$
   e) $\dfrac{3+16s-2s^2}{s-8}$ f) $\dfrac{-4w^2-10w}{w+3}$
4. a) $\dfrac{2+3x-x^2}{x-1}$ b) $\dfrac{-(x+3)(x-2)}{x-1}$
   c) $\dfrac{x^2-8x+17}{x-3}$ d) $\dfrac{x^2+x-1}{x-2}$
   e) $\dfrac{34-4x-x^2}{x-4}$ f) $\dfrac{5+2x-x^2}{x+2}$
5. a) $\dfrac{5x+19}{(x+5)(x+2)}$ b) $\dfrac{2x+10}{(x-3)(x+1)}$
   c) $\dfrac{x^2-3x-2}{x^2-1}$ d) $\dfrac{3x+2}{x+4}$
   e) $\dfrac{3x^2+17x}{(x-1)(x+3)}$ f) $\dfrac{5x^2+9x}{(x+5)(x-3)}$
6. a) $\dfrac{-2}{x^2-1}$ b) $\dfrac{x^2-5x+7}{(x-2)(x-3)}$
   c) $\dfrac{x^2-25x-4}{(x+7)(x-3)}$ d) $\dfrac{2x^2-3x+10}{(x+2)(x-4)}$
   e) $\dfrac{2x^2-6x-18}{(x-3)(x-5)}$ f) $\dfrac{6x}{x^2-4}$
7. a) $\dfrac{6}{x+2}$ b) $\dfrac{41}{10(x-2)}$
   c) $-\dfrac{17x}{6(x+3)}$ d) $\dfrac{x}{2(2x-3)}$
   e) $\dfrac{8x+5}{3(x-4)}$ f) $\dfrac{5x+6}{6(x+4)}$
8. a) $\dfrac{2x^2-3x+3}{(x-6)(x-3)}$ b) $\dfrac{-3x^2-8x-7}{(x-5)(x+3)}$
   c) $\dfrac{2x^2+4x-3}{x^2-4}$ d) $\dfrac{4x^2-10x-9}{x^2-9}$
   e) $\dfrac{3x^2+4x}{(x+2)(x-1)}$ f) $\dfrac{13x-2x^2}{(x-3)(x-4)}$
9. a) $\dfrac{2x-4-3x^2}{x(x-2)}$ b) $\dfrac{27}{10(x+3)}$
   c) $\dfrac{19y}{6(y+9)}$ d) $\dfrac{3a+19}{3(a-7)(a+1)}$
   e) $\dfrac{4a-2}{a^2-1}$ f) $\dfrac{-x^2-x}{(x-2)(x-3)}$
10. a) $\dfrac{m(4-m)}{m^2-1}$ b) $\dfrac{3x^2+5x-14}{x(x+5)(x-1)}$
    c) $\dfrac{49a^2+17a+10}{15(a-2)(a-3)}$ d) $\dfrac{28k-11k^2}{4(k-3)(k-4)}$
    e) $\dfrac{13x^2+2x-11}{2(x-2)(x+7)}$ f) $\dfrac{18m^2+107m+27}{6(2m-1)(3m+7)}$
11. a) $\dfrac{-x-10}{(x+8)(x+6)}$ b) $\dfrac{4m-7}{(m+5)(m-4)}$
    c) $\dfrac{1}{x+2}$ d) $\dfrac{-5}{m+2}$
    e) $\dfrac{4x^2-37x+21}{(x+7)(x-10)}$ f) $\dfrac{12-x}{(x-2)(x+3)}$
12. a) $2x$ b) $-5y$
    c) $\dfrac{5x-6y}{6}$ d) $\dfrac{2a}{a^2-9b}$
13. a) $\dfrac{y+x}{xy}$ b) $\dfrac{y-x}{xy}$ c) $\dfrac{y+x}{y-x}$
    d) $\dfrac{2xy^2}{y^2-x^2}$ e) $\dfrac{y^2+x^2}{y^2-x^2}$
14. a) $\dfrac{2a+3}{a+2}$ b) $\dfrac{2a+3}{a+1}$ c) $\dfrac{2a+3}{3a+5}$

Exercises 2-18

1. 4.8
2. a) 12 km/h   b) $\dfrac{2xy}{x+y}$
3. $\dfrac{V^2-w^2}{V}$ 4. 6.5 h
5. a) 2 h 52 min   b) $\dfrac{210v-150x}{v(v-x)}$
6. $\dfrac{400x}{x^2-y^2}$
7. a) $R=\dfrac{rs}{r+s}$ b) $r=\dfrac{Rs}{s-R}$ c) $s=\dfrac{Rr}{r-R}$
8. a) 12.5 min   b) $\dfrac{150x}{v+x}$
9. a) $w=\dfrac{4000}{l}$ b) Increase in $w=\dfrac{4000}{l(l-x)}$
   c) i) 8.9 cm   ii) 4.8 cm
10. a) 2985 kg/m³; 3007.5 kg/m³   b) 20°C

Review Exercises

1. a) 1   b) $-\dfrac{1}{8}$ c) $\dfrac{1}{9}$
   d) $9\dfrac{1}{9}$ e) 1   f) $\dfrac{1}{72}$
2. a) $-\dfrac{1}{8}$ b) $-\dfrac{1}{8}$ c) -4
   d) $\dfrac{1}{24}$ e) $-\dfrac{3}{2}$ f) -1
3. a) x^7 b) c^8 c) y^3
   d) a^5 e) x^8 f) a^6b^4
4. a) x^{-1} b) x^5 c) x^{-6}
   d) x e) x^{-6} f) $x^{-1}y^{-2}$
5. a) -32 b) 4   c) 16
6. a) x^2+9xy b) $3a^2b^2-5a^2b-3ab$
7. a) $-40xy^2$ b) $-6a^3b^3$ c) $24x^4y^4$

Answers

8. a) $-7m^3n$ b) $4x^4y^4$ c) $-a^3b^2$
9. a) $-6x^2 + 15xy - 3y^2$ b) $8n^2 + 8mn - 12m^2$
10. a) $11xy - 4x^2$ b) $3a^2 - 12ab$
    c) $-5x^2y - 20xy^2$ d) $5mn^2 - 12m^2n$
11. $12x + 12y$; $7x^2 + 15xy + 7y^2$
12. a) $4m^2(2m - 1)$ b) $4(2y^2 - 3y^4 + 6)$
    c) $7a^2(4 - a)$ d) $3a^2b^2c(2b - 5c)$
    e) $10x^2y(3 - 2y + xy)$ f) $4mn(2n - 3 - 4m)$
13. a) $(m + 4)(m + 4)$ b) $(a - 4)(a - 3)$
    c) $(y - 4)(y + 2)$ d) $(n - 9)(n + 5)$
    e) $(s - 9)(s - 6)$ f) $(k - 15)(k + 6)$
14. a) $(a + 12b)(a + 2b)$ b) $(m + 6n)(m + 3n)$
    c) $(x - 5y)(x + 4y)$ d) $(c + 25d)(c - 4d)$
15. a) $3x(y + 2)(y + 4)$ b) $15m(m + 3n)(m + 3n)$
    c) $7ab(ab - 5)(ab - 4)$ d) $13st(st - 11)(st + 4)$
16. a) $(b + 6)(b - 6)$ b) $(3k + 1)(3k - 1)$
    c) $(6x + 7y)(6x - 7y)$ d) $(2a + 3b)(2a - 3b)$
    e) $(5m + 9n)(5m - 9n)$ f) $(1 + 8s)(1 - 8s)$
17. a) $8(a + 3)(a - 3)$ b) $6(5 + n)(5 - n)$
    c) $7(x + y)(x - y)(x^2 + y^2)$ d) $3m(3m + 2)(3m - 2)$
    e) $\left(\frac{a}{6} + \frac{b}{7}\right)\left(\frac{a}{6} - \frac{b}{7}\right)$ f) $5q^2(5p + 6q)(5p - 6q)$
18. 73.6 m^2
19. a) 3600 b) 3200 c) $-25\,200$ d) $-57\,200$
20. a) 332 b) 2855
21. a) $(4x - 3)(x - 1)$ b) $(4x + 1)(7x - 4)$
22. a) $13ab$ b) $3m + 2$ c) $5y - 2$
    d) $-3(b - 2)$ e) $\frac{2y}{x + 5y}$ f) $\frac{3m}{2}$
23. a) -3 b) $\frac{a + 5}{3a}$ c) $\frac{n - 6}{n - 2}$
    d) $\frac{a - 3}{a + 3}$ e) $\frac{b + 4}{b - 2}$ f) $\frac{m - 3}{m + 2}$
24. a) $6a^2$ b) $\frac{-n}{m}$ c) 1
    d) $\frac{x^4}{y}$ e) $6m^2n$ f) $\frac{5x^2}{8y^2}$
25. a) $8a$ b) $\frac{-4(a + 1)}{a}$ c) $\frac{n}{m}$ d) $\frac{x^2y}{2}$
26. a) $\frac{3b(2a - b)}{2(a^2 - 9)}$ b) $\frac{m - 1}{m - 2}$ c) -1
    d) $\frac{y - 3}{y - 1}$
27. a) $\frac{3}{a}$ b) $\frac{-1}{6x}$ c) $\frac{23}{4m}$
    d) $\frac{5x}{24}$ e) $\frac{5}{12a}$ f) $\frac{-17}{15n}$
28. a) $\frac{2a - 2}{a}$ b) $\frac{x - 9}{x}$ c) $\frac{y - 33}{20}$
    d) $\frac{8a - 11}{6a}$ e) $\frac{x - 5}{24x}$ f) $\frac{1 + 12x}{12x}$
29. a) 3 b) $\frac{2a^2 - 5a + 5}{a(a - 1)}$
    c) $\frac{5 - 2x}{x + 1}$ d) $\frac{m^2 - 3m - 2}{m^2 - 1}$
    e) $\frac{y^2 - 5y + 7}{(y - 2)(y - 3)}$ f) $\frac{a^2 - 25a - 4}{(a + 7)(a - 3)}$
30. a) $\frac{a}{2(2a - 3)}$ b) $\frac{8m + 5}{3(m - 4)}$
    c) $\frac{2k^2 + 4k - 3}{k^2 - 4}$ d) $\frac{-2b^2 + 13b}{(b - 3)(b - 4)}$
    e) $\frac{49x^2 + 17x + 10}{15(x - 2)(x + 3)}$ f) $\frac{18x^2 + 107x + 27}{6(2x - 1)(3x + 7)}$
31. 220 km/h

Cumulative Review (Chapters 1-2)

1. a) $5, 180$ b) $6, 180$ c) $6, 2730$ d) $14, 4620$
2. a) 4 b) 90
3. a) -37 b) -30
4. a) -11 b) $\frac{47}{18}$
5. a) $8\sqrt{5}$ b) $6\sqrt{3}$ c) $-18\sqrt{5}$ d) $12\sqrt{6}$
   e) $60\sqrt{2}$ f) $-36\sqrt{2}$ g) $30\sqrt{3}$ h) 576
6. a) 13.6 b) 12.1 c) 24.3
7. a) $\sqrt{6}$ b) $-5\sqrt{6}$ c) $\sqrt{15} - 2\sqrt{6}$ d) 12
8. a) $\sqrt{3}$ b) $-17\sqrt{2}$
9. a) 16 b) $\sqrt{2}$ c) 2 d) 3
10. a) $\frac{25}{4}$ b) 64 c) $\frac{4}{9}$ d) 5
    e) $\frac{5}{4}$ f) 1
11. a) y^6 b) x^{-3} c) a^{-5} d) x^4
    e) x^2y^5 f) x^2
12. a) $-3x^2 - 11xy + 9y^2$ b) $-19x^2 - 13xy$
13. a) $-42x^3y^3$ b) $12a^3b^4$
    c) $12x^2y^5$ d) $-\frac{a^2b}{2}$
14. a) $9x^2(x - 2)$ b) $7y^3(2y - 1)$
    c) $(m + n)(m + 2)$ d) $(a + 4)(a + 4)$
    e) $(x - 5)(x + 4)$ f) $(y - 7)(y + 3)$
    g) $(m - 9)(m - 8)$ h) $(c - 6d)(c + 4d)$
    i) $(x + 8y)(x - 3y)$ j) $(x - 7y)(x - 13y)$
15. a) $2x^2(y + 4)(y + 2)$ b) $5mn(m + 7)(m - 3)$
    c) $8(x + 3)(x - 3)$ d) $5(6 - m)(6 + m)$
    e) $8(x^2 + y^2)(x - y)(x + y)$

422 Answers

16. a) 6000   **b)** 4800   **c)** 6400
d) -5600

17. a) $14x^2y$ **b)** $\dfrac{x+6}{3x}$ **c)** $\dfrac{m-4}{m-5}$
d) $\dfrac{a-4}{a+4}$ **e)** $\dfrac{x-6}{x-5}$ **f)** $\dfrac{y+6}{y-5}$

18. a) $6xy$ **b)** $\dfrac{-3(x+1)}{x}$ **c)** $\dfrac{8c^2}{3(c+2)}$ **d)** -1

19. a) $\dfrac{19}{10x}$ **b)** $\dfrac{23}{12m}$ **c)** $\dfrac{11}{12a}$ **d)** $\dfrac{29-2x}{15}$

20. a) $\dfrac{x^2-5x-6}{x^2-4}$ **b)** $\dfrac{17a}{12(2a-3)}$
c) $\dfrac{3-4x}{2(x-4)}$ **d)** $\dfrac{13m^2+8m+4}{6(m^2-4)}$

PSS

1.

2. 1, 2, 3, 4, 5, 6   **3.** The back
    1, 2, 7, 8, 9, 0

Exercises 3-1

1. a) 9   **b)** -13 **c)** 11   **d)** -5
     e) 7   **f)** -2 **g)** 7   **h)** 11

2. a) 8   **b)** 7   **c)** 2   **d)** -1
     e) -5 **f)** -6 **g)** 1   **h)** 2

3. a) 3   **b)** 0   **c)** $-\dfrac{1}{2}$ **d)** 8
     e) 12   **f)** $-\dfrac{3}{5}$ **g)** -3 **h)** $-\dfrac{9}{2}$

4. a) 3   **b)** -1 **c)** 2   **d)** $-\dfrac{9}{4}$
     e) 14   **f)** $\dfrac{25}{8}$ **g)** -3 **h)** $\dfrac{5}{7}$

5. a) -3 **b)** 5   **c)** 4   **d)** 26
     e) 2   **f)** -2 **g)** 8   **h)** 5

6. a) 4   **b)** 5   **c)** $-\dfrac{1}{5}$ **d)** 2
     e) -1 **f)** $-\dfrac{3}{5}$ **g)** 3   **h)** -1
     i) 13

7. 7.5 L   **8.** 2.5 cm   **9.** 3 cm   **10.** 100

11. 10 nickels, 20 dimes, 15 quarters   **12.** 7, 9, 35

13. i) a, c, d   **ii)** c

14. a) 12   **b)** 14   **c)** 4   **d)** -1

Exercises 3-2

1. a) i) 11   **ii)** 25   **iii)** -9
     iv) 10   **v)** 6

b) i) 1   **ii)** 5   **iii)** $-\dfrac{5}{2}$
     iv) -4 **v)** $-\dfrac{9}{4}$

2. a) $\{x, x-6\}$ **b)** $\{x, x+6\}$ **c)** $\{x, -x-6\}$
d) $\{x, 2x-8\}$ **e)** $\{x, -3x\}$ **f)** $\{x, 10-5x\}$
g) $\{x, 2x+6\}$ **h)** $\{x, 4x+8\}$ **i)** $\{x, 4x-12\}$

3. a) $\left\{x, \dfrac{4-3x}{2}\right\}$ **b)** $\left\{x, \dfrac{3x+4}{2}\right\}$ **c)** $\left\{x, \dfrac{10-2x}{5}\right\}$
d) $\left\{x, \dfrac{2x}{5}\right\}$ **e)** $\left\{x, \dfrac{10+2x}{5}\right\}$ **f)** $\left\{x, \dfrac{4x-24}{3}\right\}$
g) $\left\{x, \dfrac{9-x}{3}\right\}$ **h)** $\left\{x, \dfrac{3x-30}{5}\right\}$ **i)** $\left\{x, \dfrac{-x}{3}\right\}$

4. a) $\{x, 2x-6\}$ **b)** $\left\{x, \dfrac{3-x}{3}\right\}$ **c)** $\{x, 4-2x\}$
d) $\left\{x, \dfrac{3x-12}{2}\right\}$ **e)** $\left\{x, \dfrac{4-x}{2}\right\}$ **f)** $\left\{x, \dfrac{5x-20}{2}\right\}$
g) $\left\{x, \dfrac{24-4x}{3}\right\}$ **h)** $\left\{x, \dfrac{3}{2}x\right\}$ **i)** $\left\{x, \dfrac{5x-20}{4}\right\}$

5. Answers will vary. Examples are:
  a) $(1, 10), (2, 9), (3, 8), (-2, 13), (-1, 12), (0, 11)$
  b) $(10, 5), (9, 4), (8, 3), (0, 5), (4, -1), (3, -2)$
  c) $(0, 10), (1, 8), (2, 6), (-1, 12), (-2, 14), (-3, 16)$
  d) $(3, 3), (6, 2), (9, 1), (12, 0), (15, -1), (18, -2)$

6. 59, 48, 37, 26, or 15

7. 80, 71, 62, 53, 44, 35, 26, or 17

8. a) i) 5   **ii)** 7   **iii)** -8 **iv)** -11
b) Examples are:
  i) $(0, 3), (2, 0), (-2, 6)$
  ii) $(0, 0), (2, -3), (-2, 3)$
  iii) $(-1, 0), (1, -3), (3, -6)$
  iv) $(1, 2), (-1, 5), (3, -1)$

9. a) $1.30
  b) i) 8   **ii)** 40   **iii)** 90

10. a) i) 12 min, 3 h 20 min   **ii)** 16 h
  b) 21 kL

11. $(5, 1), (3, 4), (1, 7)$

12.

Suzanne		25¢	0	2	4	6
		10¢	17	12	7	2
Ivan	25¢	0	1	2	3	4	5	6
	5¢	34	29	24	19	14	9	4

13. a) $(2, 4)$ **b)** $(5, 1)$ **c)** $(1, 0)$ **d)** $(-1, 2)$

Exercises 3-3

1. a) (3, 1) b) (−2, 2) c) (−1, 3)
   d) (4, −3) e) $\left(-\frac{1}{2}, 2\right)$ f) (2, −5)
   g) $\left(\frac{3}{4}, -2\right)$ h) (−3, −2) i) $\left(\frac{2}{3}, -\frac{1}{3}\right)$
2. b)
3. a) (18, −24) b) (4, −3) c) (2, 5)
   d) (4, −6) e) (3, 4) f) $\left(\frac{7}{2}, -\frac{5}{3}\right)$
   g) (4, 7) h) $\left(\frac{2}{5}, -3\right)$ i) $\left(\frac{1}{6}, \frac{1}{2}\right)$
4. a) 2, 9 b) 19, 7 c) 5, −2
5. a) Infinitely many solutions
   b) Infinitely many solutions
   c) (5, 2) d) (2, 5)
6. a) $400; $8.50 b) $1887.50 c) 188
7. a) $15; $0.12 b) $195
8. a) (12, 15) b) (−6, −3)

MM

1. a) 17.5 b) 16.625
2. The problem means that the 700 loaves are to be divided in the ratio $\frac{2}{3}:\frac{1}{2}:\frac{1}{3}:\frac{1}{4}$.
   Thus, the first person gets $266\frac{2}{3}$ loaves, the second 200, the third $133\frac{1}{3}$, and the fourth 100.
3. 3.1605

Exercises 3-4

1. a) (2, 7) b) (3, −2) c) (−1, −8)
   d) (1, 4) e) (−2, 5) f) (−2, 3)
   g) (3, 5) h) (4, −2) i) (6, −4)
2. c)
3. a) (−1, 3) b) (2, −2) c) (2, 1)
   d) (3, 1) e) (−2, −3) f) (4, −3)
   g) (20, −6) h) (4, −3) i) $\left(\frac{1}{4}, -4\right)$
4. a) $\left(\frac{7}{6}, \frac{1}{12}\right)$ b) (2, −1) c) $\left(\frac{23}{2}, -\frac{15}{2}\right)$
   d) (19, −30) e) (0, 1) f) $\left(-\frac{1}{6}, \frac{7}{12}\right)$
   g) $\left(2, -\frac{7}{3}\right)$ h) $\left(\frac{1}{6}, \frac{1}{12}\right)$ i) (1, 1)
5. a) (5, −1) b) (−1, 1) c) (3, 2) d) (−2, 1)
6. a) $9000; $1200 b) $4200

7. a) $500; $18 b) $770 c) About 5.5 h
8. a) 6, 26
   b) i) 8, 30 ii) 3, 29 iii) 7, 21
9. a) $3x + 2y = 1$ b) $5x − y = −7$
   c) (−1, 2) d) (2, 0)

Exercises 3-5

1. a) (3, 4) b) (−1, 3) c) (2, −3)
   d) (4, 3) e) (−2, 3) f) (2, −4)
   g) (3, −1) h) (3, 7) i) (−1, −3)
2. a) (2, −3) b) (−1, 5) c) (3, −1)
   d) (4, 3) e) (−4, 2) f) (−2, −3)
   g) (3, 4) h) (2, −1) i) (6, −7)
3. a) (4, −2) b) (3, 5) c) $\left(\frac{1}{2}, \frac{1}{3}\right)$
   d) (2, 3) e) (−3, −7) f) (−2, 1)
   g) $\left(\frac{5}{2}, -3\right)$ h) (−6, 3) i) (3, −4)
4. a) $\left(\frac{4}{5}, -\frac{3}{5}\right)$ b) $\left(\frac{23}{11}, \frac{8}{11}\right)$ c) $\left(\frac{3}{14}, \frac{15}{14}\right)$
   d) (2, 1) e) $\left(\frac{25}{11}, \frac{20}{11}\right)$ f) (−3, 5)
   g) (2, −1) h) (4, −6) i) $\left(\frac{35}{16}, -\frac{5}{8}\right)$
5. a) $3850; $0.13/km b) $6450
6. $552.50
7. b) Answers will vary. Examples are: (−4, −6), (11, 4).
   c) Both equations represent the same line.
   d) $\left\{\left(x, \frac{2x − 10}{3}\right) \mid x \in R\right\}$
8. b) No. The equations represent parallel lines.
9. c)
10. a) (4, 3) b) (3, 1) c) (−1, 2) d) (2, 0)
11. a) (3, −4) b) (2, 0) c) (−6, 10) d) (4, −2)
    e) (12, −9) f) (6, 3)

MAU

1.

	Principal of a loan, P	Interest rate, r	Interest for 1 year, I	Amount owing at the end of 1 year, A
a)	$ 200	22%	22% of $200 = $44	$ 244
b)	$ 1000	18%	$ 18	$ 1018
c)	$ 3000	19.5%	$ 58.50	$ 3058.50
d)	$ 3200	16.25%	$ 528	$ 3728
e)	$17 500	21.5%	$3762.50	$21 262.50

Answers **423**

2. a) $I = Prt$ **b)** $A = P(1 + rt)$
3. a) $6.51 **b)** $331.98
4. a) $1.08 **b)** $2.69 **c)** $5.55

5.

Year	Amount Start of Year	Interest	Amount End of Year
1	$800	$160	$960
2	$960	$192	$1152
3	$1152	$230.40	$1382.40
4	$1382.40	$276.48	$1658.88

6. a) 1.2 **b)** $A = 800(1.2)^4$; $A = 800(1.2)^n$
7. a) $304.17 **b)** $847.60 **c)** $1349.47
8. $235.54

Exercises 3-6

1. a) x **b)** $x + 5$ **c)** $x - 4$ **d)** $2x$
2. a) x **b)** y **c)** $x + y$ **d)** $2x + 3y$
3. a) x **b)** y **c)** $5x - 4y$
  d) $\frac{1}{2}x + \frac{1}{3}y$
4. a) $10t + u$ **b)** $10u + t$ **c)** $t + u$
  d) $11t + 11u$
5. a) n **b)** d **c)** $(5n + 10d)$¢ **d)** $n + d$
6. a) a **b)** b **c)** $(2a + 5b)$ **d)** $a + b$
7. a) a **b)** c **c)** $a + c$ **d)** $(4a + 2c)$
8. a) l **b)** w **c)** $2l + 2w$ **d)** lw
9. a) C **b)** $C - 5$ **c)** $C + 2$ **d)** $25 - C$ **e)** C
10. a) D **b)** T **c)** $D + T - 6$ **d)** $D + T + 40$
11. $x + y = 50$ **12.** $x - y = 15$
13. $x + x + 1$ **14.** $x + x + 2 + x + 4$
15. $x + x + 2 + x + 4$ **16.** $2x - 5y = 10$
17. $\frac{1}{4}x - 3 = \frac{1}{2}y$ **18.** $5x + 4y = 40$
19. $x, 10 - x$ **20.** $J = D + 3$
21. $C - 3 = 2(S - 3)$ **22.** $4x \quad 3y = 150$
23. $6a + 4c = 18$ **24.** $5n + 10d = 200$
25. $12w$ **26.** $(4w + 20)$ cm
27. $5(w - 1)$ or $\frac{10}{3}(l + 1)$ cm **28.** $x^2 = 2y$
29. $x^2 + (x + 1)^2$ **30.** $x = 7y + 5$

Exercises 3-7

1. $x + y = 25, x - y = 7$
2. $x + y = 10, x + 3y = 24$
3. $x + y = 7, 3x - y = 15$
4. $x - y = 5, 3x - 4y = 5$
5. $x + y = 10, 2x + 3y = 22$
6. $\frac{x}{3} = \frac{15 - x}{2}$
7. $T = B + 15, T + B = 95$
8. $W + S = 100, 5W = 4S$
9. $A = 2B, A - 4 = 2(B - 4)$
10. $A = L + 6, A + 2 = 2(L + 2)$
11. $d + q = 7, 10d + 25q = 115$
12. $n = d + 3, 5n + 10d = 195$
13. $2x + 3(x + 1) = 38$
14. $5x = 4(x + 1) + 6$
15. $\frac{x}{3} + \frac{x + 2}{4} = 11$
16. $3x = 2(x + 4) + 9$
17. a) $6w = 48$ **b)** $6w - 4 = 32$
18. $x + 0.2 = 0.5(1 + x)$
19. $\frac{x + 1}{y} = \frac{3}{4}, \frac{x}{y + 1} = \frac{2}{3}$
20. $\frac{x + 5}{x + 8} = \frac{3}{4}$

Exercises 3-8

1. 19, 27
2. 18, 38
3. 5, 8
4. 12 cm by 20 cm
5. 17, 59
6. 3, 8
7. 11, 12
8. 54, 56
9. 12, 0
10. 8, 12
11. 18, 30
12. 45 kg, 60 kg
13. 47 kg, 54 kg
14. $4, $1.50
15. $45
16. $1.75, $0.43
17. $80, $60
18. Chris: $450, $340
  Jerry: $375, $425
19. 35
20. $3, $2
21. $\frac{3}{4}$
22. $\frac{7}{19}, \frac{11}{19}$

Exercises 3-9

1. 16 kg at $13, 24 kg at $18
2. a) 200 km **b)** 10 L
3. 58
4. 25 km/h, 5 km/h
5. 83, or 38
6. 71
7. 27
8. 51

9. 24

10. a) $10t + u = 7(t + u)$ **b)** $10u + t = 4u + 6u + t$
　　　　$3t = 6u$ 　　　　　　　　　$= 4u + 3t + t$
　　　　$t = 2u$ 　　　　　　　　　$= 4(u + t)$

11. 80　　**12.** 36, 12
13. 26, 23, 19　　**14.** 7, 9, 35
15. 33, 65　　**16.** $8.75
17. Red: 2 kg; Blue: 1 kg
18. 7 nickels, 10 dimes, 12 quarters
19. 6 nickels, 9 dimes, 12 quarters
20. 10, 12, 14, 16　　**21.** 19, 21, 23, 25
22. 54　　**23.** 14, 25, 36, 47, 58, 69
24. $0.75　　**25.** 12 min
26. 6.6 days
27. a) $9　　**b)** $9
28. a) almost 33 km　　**b)** about 1 h 22 min
29. $10t + u = 4(t + u)$
　　　　$2t = u$
　　　　$t = u - t$
　　　$10t + u = 10(u - t) + u$
　　　　　　$= 10(u - t) + 2t$
　　　　　　$= 10(u - t) + 2(u - t)$
　　　　　　$= 12(u - t)$
30. 18 km　　**31.** 72, 76

Exercises 3-10

1. a) $-3, -5$　**b)** 3, 4　**c)** $5, -4$　**d)** $-8, 3$
　e) $-2, -6$　**f)** $9, -4$　**g)** 6, 4　**h)** $-7, -8$
2. a) $x^2 - 7x - 18 = 0$　**b)** $x^2 - 14x + 40 = 0$
　c) $3x^2 + 16x + 5 = 0$　**d)** $32x^2 - 20x - 7 = 0$
3. a) 4, 5　**b)** 7, 9　**c)** $8, -2$　**d)** $-7, -3$
　e) $7, -2$　**f)** $-5, 3$　**g)** 4　**h)** 6, 2
4. a) $-3, -2$　**b)** 9, 3　**c)** $5, -2$　**d)** 3, 10
　e) $-6, 4$　**f)** 2
5. 1 s and 5 s
6. a) 12 cm × 16 cm　**b)** 14 cm × 14 cm
　c) 8 cm × 20 cm
7. a) 7, 12
　b) Negative terms are not excluded.
　c) 0, 19

8. 17 cm
9. 3, 9, or $-3, -9$
10. 4, 8, and $-6, 18$
11. 76
12. 5 m
13. a) $\frac{1}{5}, -4$　**b)** $\frac{5}{2}, 3$
　c) $\frac{7}{2}, -\frac{5}{3}$　**d)** $\frac{1}{3}, -2$

PSS

1. a) 5　　**b)** 3
2. 12　　**3.** 12
4. a) 30　　**b)** 100
5. a) 27　**b)** 21　**c)** 45　**d)** 39
6. a) 3　**b)** $1, \sqrt{2}, \sqrt{3}$
7. 3　**8. a)** 9　**b)** $1, 2, \sqrt{2}, \sqrt{5}, \sqrt{8}, \sqrt{3}, \sqrt{6}, 3, \sqrt{12}$

Review Exercises

1. a) 1　**b)** -3　**c)** 12　**d)** 8
2. a) $\{(x, 2x - 8) \mid x \in R\}$
　b) $\left\{\left(x, \frac{20 - 3x}{5}\right) \mid x \in R\right\}$
　c) $\left\{\left(x, \frac{4x - 24}{3}\right) \mid x \in R\right\}$
3. a) 6　**b)** 1　**c)** -19　**d)** 34
4. a) $(-3, -5)$　**b)** $\left(\frac{7}{2}, 1\right)$　**c)** $\left(-\frac{14}{5}, \frac{2}{5}\right)$
5. a) $(-2, 3)$　**b)** $(4, 1)$　**c)** $\left(\frac{18}{5}, \frac{3}{5}\right)$
　d) $(-3, 6)$　**e)** $(-4, -3)$　**f)** $\left(\frac{3}{4}, -4\right)$
6. a) $(-1, -1)$　**b)** $\left(\frac{5}{3}, 0\right)$　**c)** $(-2, -1)$
　d) $(3, -2)$　**e)** $(4, 0)$　**f)** $(2, -1)$
7. a) $x + y = 80$　**b)** $y = x - 7$
　c) $3x + 4y = 30$　**d)** $I - 3 = 2(B - 3)$
　e) $5n + 10d = 300$　**f)** $P = 4w + 10$
8. 20, 16　　**9.** 57
10. 24, 39　　**11.** 18
12. 31　　**13.** 38
14. 80 kg at $2, 20 kg at $2.40
15. 35 km/h, 50 km/h
16. 14 km/h, 4 km/h
17. 80 g of sterling, 20 g of pure silver
18. Skilled: $37.50, unskilled: $18.75

426 *Answers*

19. a) 7, −3   **b)** −8, 7   **c)** 6, −4   **d)** 8, −3
20. a) $x^2 − 2x − 35 = 0$ **b)** $x^2 − 5x − 24 = 0$
   c) $3x^2 − 10x − 8 = 0$ **d)** $32x^2 − 20x − 3 = 0$

CP

1. a) (10, 5)   **b)** (−4, 9)
   c) (−0.664 368 121, −1.263 360 3)
   d) (2.313 882 06, −1.318 796 07)
   e) (1.752 697 38, 0.658 674 43)
   f) (−0.949 969 514, −2.694 045 44)
2. $38.75; $0.185

Exercises 4-1

1. a) 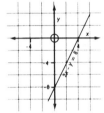   **b)**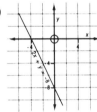

c) 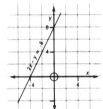   **d)**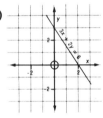

e) 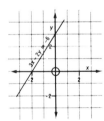   **f)**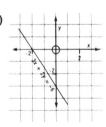

g) 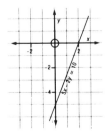   **h)**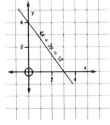

i)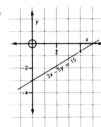

2. a) 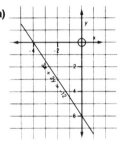   **b)**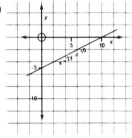

c) 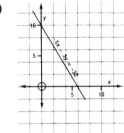   **d)**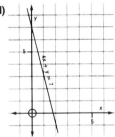

e) 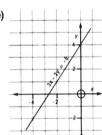   **f)**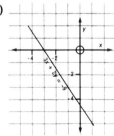

g) 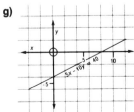   **h)**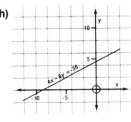

3. a)

b)

c)

d)

e)

f)

g)

h)

4. a)

b)

c)

d)

e)

f)

g)

h)

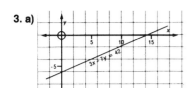

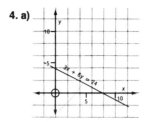

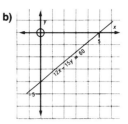

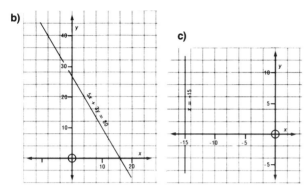

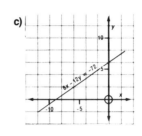

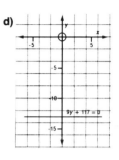

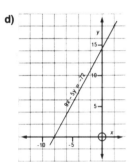

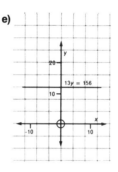

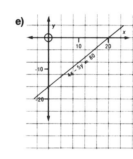

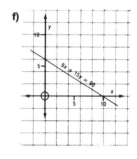

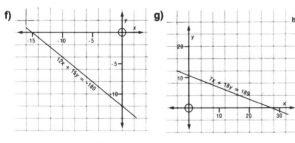

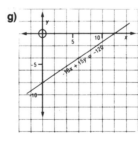

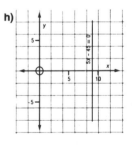

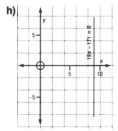

5.

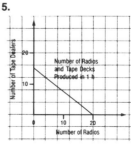

6.

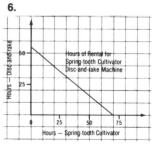

428 Answers

7.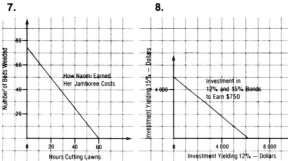

8.

9. a)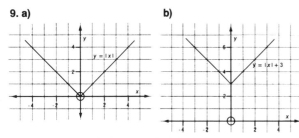

b)

c)

d)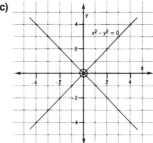

10. a) $x + y = 5$ **b)** $x - y = 3$
c) $x + 2y = 2$ **d)** $2x + y = 4$
e) $3x + 2y = 6$ **f)** $3y - 7x = 10.5$

Exercises 4-2

1. a) (2, 3)   **b)** (2, 4)   **c)** (40, 30)
d) (−3, −5)   **e)** (0, −4)   **f)** (4, 5)
2. a) (30, −20)   **b)** (−2, −6)   **c)** (0, 30)

d) $\left(-\dfrac{32}{13}, -\dfrac{30}{13}\right)$ **e)** (24, −32)   **f)** $\left(-\dfrac{17}{26}, \dfrac{10}{13}\right)$

3. a) $\left(\dfrac{5}{2}, \dfrac{3}{2}\right)$ **b)** $\left(\dfrac{7}{2}, \dfrac{3}{4}\right)$ **c)** $\left(\dfrac{10}{3}, -\dfrac{4}{3}\right)$

d) $\left(\dfrac{3}{4}, \dfrac{1}{2}\right)$ **e)** $\left(\dfrac{240}{7}, -\dfrac{90}{7}\right)$ **f)** $\left(\dfrac{1}{2}, \dfrac{1}{3}\right)$

g) $\left(-\dfrac{54}{23}, \dfrac{79}{23}\right)$

4. a) (3, 2)   **b)** No solution
   c) Infinitely many solutions
   d) Infinitely many solutions
   e) Infinitely many solutions
   f) Infinitely many solutions
   g) No solution

5. a), c), d)

6. a, d

7. a) When they are equations for the same line.
   b) When they are equations for parallel lines.
   c) When they are equations for intersecting lines.

8. a) **i)** $4x + y = 11$ **ii)** $5x + 3y = 19$
   iii) $3y - 2x = 5$ **iv)** $8y - 3x = 18$

   b) 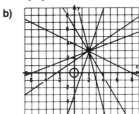   **c)** The lines are concurrent. They all pass through (2, 3).

MM

The following are partial answers:

1. a)

x	−2	0	2	4	6
y	9	6	3	0	−3

b)

x	−4	−1	2	5	8
y	−2	−1	0	1	2

c)

x	−1	0	1	2	3
y	21	17	13	9	5

d)

x	−7	0	7	14	21
y	−7	−3	1	5	9

e)

x	−8	−3	2	7	12
y	5	3	1	−1	−3

f) No solution

2.

q	2	4	6
d	12	7	2

3.

q	1	2	3	4	5	6
n	29	24	19	14	9	4

4.

s	0	3	6	9	12	15
f	23	19	15	11	7	3

5.

t	1	4	7	10	13	16
s	25	20	15	10	5	0

6. 84

Exercises 4-3

1. a) i) $x \geq -3$ ii) $x \leq -3$
  b) i) $y < 4$ ii) $y > 4$
  c) i) $x + y > 4$ ii) $x + y < 4$
  d) i) $4x - y \leq -2$ ii) $4x - y \geq -2$
  e) i) $3x + 5y \leq -15$ ii) $3x + 5y \geq -15$
  f) i) $2x - y < 24$ ii) $2x - y > 24$

2. a) 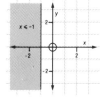 **b)**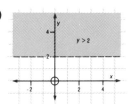

c) 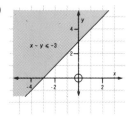 **d)**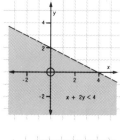

e) 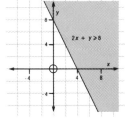 **f)**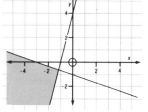

g) 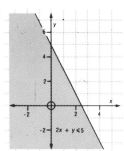 **h)**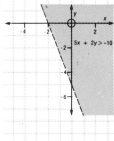

i)

3. a) $x \geq 0, y \geq 0$
  b) $x \geq -4, y \geq -2$
  c) $x + y \leq 5, x \geq 0$
  d) $x - y \leq -1, x + y \leq 5$
  e) $5x + 3y \geq 15, x - 2y \leq -4$
  f) $2x - 5y \geq 100, 2x + 5y \leq 100$

4. a) 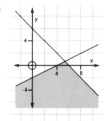 **b)**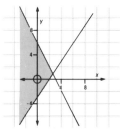

c) **d)**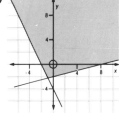

430 Answers

e)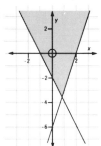

f)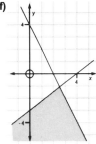

c)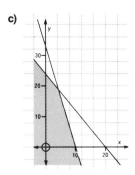

d)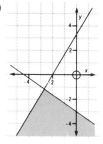

g)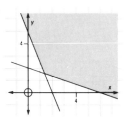

h)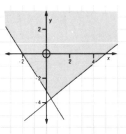

e)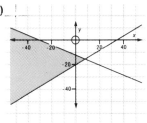

f)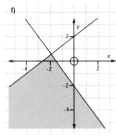

i)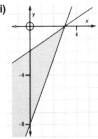

g)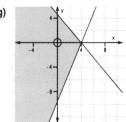

h)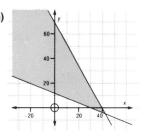

5. a)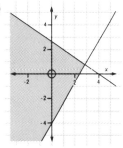

b)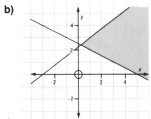

6. a) $y \geq -1, y \leq 2$
   b) $x \leq 3, x \geq 2$
   c) $x - y \geq -3, x - y \leq 2$
   d) $3x + 2y \leq 6, 3x + 2y \geq -6$
   e) $x - y \geq -2, x + 2y \leq 4, x - y \leq 4$
   f) $3x - y \geq -3, 3x + 2y \geq 6, x - 2y \leq 2$

7. a)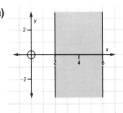

b)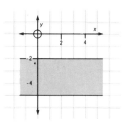

Answers **431**

c)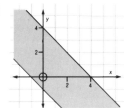

d)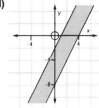

e)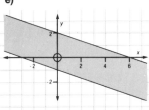

f)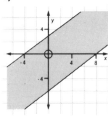

g)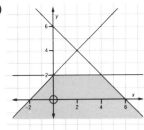

h)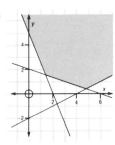

i)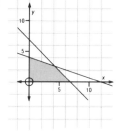

8.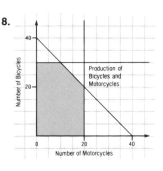

9.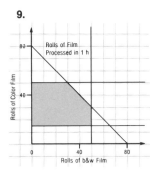

10.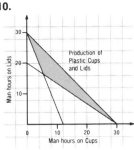

11.

Exercises 4-4

1. **a)** 34, −14   **b)** 56, −7
2. **a)** 40, 5   **b)** 51, 5   **c)** 77, 9   **d)** 24.5, 7
3. **a)** 9, −14   **b)** 49, 3   **c)** 26, 1   **d)** 28, −1
4. **a)** 20, 12   **b)** 5, 2   **c)** 23, 19 **d)** 33, 24
5. **a)** 29, 11   **b)** 9, 0   **c)** 15, 3   **d)** 16, −8
6. **a)** 27, 11   **b)** 24, 8   **c)** 19, 2
7. **a)** 41, 0   **b)** 32, 7   **c)** 43, −3 **d)** 35, 8

PSS

1. 103   2. $\frac{1}{16}$ 3. 171 cm

MAU

1. 380 351 m²   2. 1 521 404 m³

Exercises 4-5

1. **a)** 30 footballs, 0 soccer balls; profit: $600
   b) 0 footballs, 15 soccer balls; profit: $300
2. **a)** 20 ha wheat, 20 ha barley
   b) 26.7 ha wheat, 13.3 ha barley
3. **a)** All 80 ha wheat
   b) 20 ha corn, 60 ha wheat
   c) 40 ha corn, 0 ha wheat
4. **a)** 20 of each
   b) 10 motorcycles, 30 bicycles
5. **a)** 30 b&w, 50 color
   b) 50 b&w, 30 color
6. 30 TV sets, 40 stereos
7. 40 steers, 30 heifers
8. Brand A: Brand B = 1:2
9. **a)** 30 *Cheetahs*, 42 *Gazelles*
   b) 50 *Cheetahs*, 30 *Gazelles*

Review Exercises

1. a) 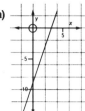 **b)**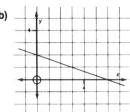

c)

2. a) 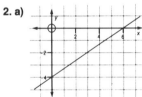 **b)**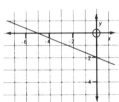

3. 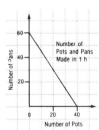 **4.**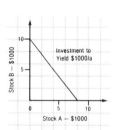

7. a) 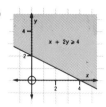 **b)**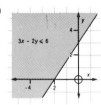

c)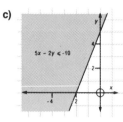

8. a)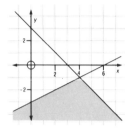

b) 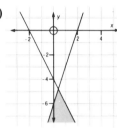 **c)**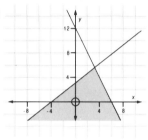

d) 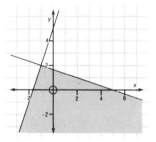 **e)**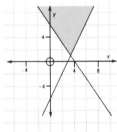

f)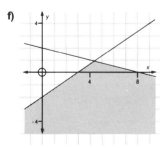

5. a) (2, 5) **b)** (−1, 4) **c)** (3, −2) **d)** (4, 3)
 e) Infinitely many solutions **f)** No solution
 g) (−1, 2) **h)** (−1, 2) **i)** (−1, 2)

6. a) i) $y \geq -2$ **ii)** $y \leq -2$
 b) i) $x \leq 1$ **ii)** $x \geq 1$
 c) i) $x - y \leq -2$ **ii)** $x - y \geq -2$

9. a) $y \geq x$ **b)** $x + 2y \leq 4$ **c)** $x + y \geq -2$
  $y \geq -x$ $x \geq 0$ $x + y \leq 2$
    $y \geq 0$ $x - y \leq 2$

10.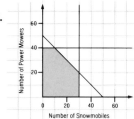

11. a) 3, 0 **b)** 24, 19 **c)** 25, 18 **d)** 18, 17
12. 12 of each; $324
13. a) Sheila 3 h, Ken 2 h; $20
  b) Sheila 0 h, Ken 8 h; $24, or 3 h, and 2 h; $24
  c) Sheila 6 h, Ken 0 h; $18

Exercises 5-1

1. a) 5 **b)** 7 **c)** 11 **d)** 12
  e) 8 **f)** 14
2. $AB = 7$, $CD = \sqrt{40}$, $EF = \sqrt{90}$, $GH = \sqrt{73}$, $JK = \sqrt{85}$
3. a) $2\sqrt{5}$ **b)** $2\sqrt{13}$ **c)** $5\sqrt{2}$
  d) $\sqrt{137}$ **e)** $\sqrt{29}$ **f)** $\sqrt{28.25}$
4. a) $\triangle AHT$: 5, 12, 13; **b)** $\triangle AHT$: 30
  $\triangle BEG$: 5, 3, $\sqrt{34}$; $\triangle BEG$: 7.5
  $\triangle IDP$: 3, 4, 5; $\triangle IDP$: 6
  $\triangle LJS$: $2\sqrt{5}$, $2\sqrt{5}$, $2\sqrt{10}$; $\triangle LJS$: 10
  $\triangle RKM$: $\sqrt{10}$, $2\sqrt{10}$, $5\sqrt{2}$; $\triangle RKM$: 10
5. a) $7\sqrt{2}$ **b)** $4\sqrt{5}$ **c)** $\sqrt{89}$ **d)** $\sqrt{85}$
6. a) Isosceles **b)** Scalene **c)** Isosceles
7. a) i) 9, 8 **ii)** 34 **iii)** $\sqrt{145}$ **iv)** 72
  b) i) $\sqrt{10}$, $3\sqrt{10}$ **ii)** $8\sqrt{10}$ **iii)** 10 **iv)** 30
  c) i) $4\sqrt{13}$, $2\sqrt{13}$ **ii)** $12\sqrt{13}$ **iii)** $2\sqrt{65}$ **iv)** 104
8. a) i) $\sqrt{34}$, $\sqrt{34}$, $2\sqrt{34}$
    ii) $4\sqrt{5}$, $2\sqrt{5}$, $6\sqrt{5}$
    iii) $5\sqrt{5}$, $\sqrt{5}$, $6\sqrt{5}$
  b) They are collinear.
9. The coastguard cutter

10. Diagonals: 525.6 m, 410.0 m; Perimeter: 1364.6 m; Area: 105 312.5 m²
11. P, R, S
12. (2, 0) and (8, 0)
13. a) (0, 3) **b)** (0, 2) **c)** (0, 1)
14. 75.5

Exercises 5-2

1. a) (3, 4) **b)** (−2, 2) **c)** (4, 5)
  d) $\left(1, \dfrac{3}{2}\right)$ **e)** $\left(-\dfrac{3}{2}, \dfrac{3}{2}\right)$ **f)** $\left(-2, \dfrac{5}{2}\right)$
  g) $\left(2, -\dfrac{1}{2}\right)$ **h)** (−2, −2)
2. a) (−1, 0), (2, −4), (5, −8)
  b) $\left(\dfrac{3}{2}, -4\right)$, (−2, −1), $\left(-\dfrac{11}{2}, 2\right)$
  c) $\left(\dfrac{1}{2}, -\dfrac{7}{2}\right)$, (3, −1), $\left(\dfrac{11}{2}, \dfrac{3}{2}\right)$
  d) $\left(-\dfrac{5}{2}, \dfrac{-3}{2}\right)$, (0, 0), $\left(\dfrac{5}{2}, \dfrac{3}{2}\right)$
3. a) (1, 4)
  b) Half length of hypotenuse: $\dfrac{1}{2}\sqrt{8^2 + 6^2} = 5$
    Length of median with endpoint M: 5
    Therefore, M, is equidistant from all three vertices.
    (It is the *circumcentre* of the triangle.)
4. $\left(5, \dfrac{1}{2}\right)$ The diagonals bisect each other.
5. a) (6, −6) **b)** (2, −9) **c)** (−6, −2) **d)** (0, 0)
6. a) (14, −4) **b)** (10, 10) **c)** (−8, 9) **d)** (0, 0)
7. a) $M(-1, 2)$, $N(5, 1)$
  b) $BC = 2MN$
8. From P: $3\sqrt{5}$ From Q: $6\sqrt{2}$ From R: 9
9. a) $E(9, 10)$, $F(-3, -2)$
  b) $CG = GH = HA = 2\sqrt{10}$
  c) i) 1:2 **ii)** 1:3
10. a) i) 6 **ii)** 8
  b) C is the intersection of the diagonals.
11. (2, 2)

Exercises 5-3

1. a) $-\dfrac{11}{8}$ **b)** 3 **c)** $-\dfrac{5}{2}$
  d) Undefined **e)** $\dfrac{2}{3}$ **f)** 0
2. $\dfrac{1}{3}$ **3.** $\dfrac{5}{3}$ **4.** $\dfrac{4}{125}$

434 Answers

5.

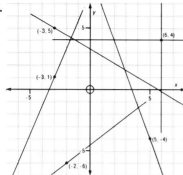

6. a) $B \frac{5}{8}$, $C \frac{3}{8}$, $D \frac{1}{8}$, $E\ 0$, $F -\frac{1}{4}$, $G -\frac{5}{8}$

b) $Q -\frac{7}{5}$, $R -\frac{7}{2}$, S Undefined, $T \frac{7}{2}$, $U \frac{7}{4}$, $V\ 1$

7. a) i) 0   **ii)** Undefined

b) i) Undefined   **ii)** 0

8. a) $-1, \frac{9}{2}, \frac{4}{7}$ **b)** $\frac{1}{3}, \frac{5}{2}, -\frac{7}{5}$

c) $-\frac{5}{4}$, 0, undefined   **d)** $-5, 2, 1$

9. a) Yes   **b)** No   **c)** Yes

10. 2.4

11. a) 34.4 m   **b)** $\frac{12}{5}$

12. a) 6   **b)** $-\frac{15}{2}$

c) -6 **d)** -2

13. Slopes are equal for all triangles.

14. The opposite sides of the figure formed by joining the midpoints of adjacent sides have equal slopes. Therefore, the midpoints of the sides are the vertices of a parallelogram.

15. a) $(-2, 0)$ **b)** $(1, 0)$

c) $(2, 0)$ **d)** $(-8, 0)$

e) $(7, 0)$

16. $(-2, 0), (8, 0)$

17. a) $(4, 5)$, or $(6, -5)$, or $(-12, -3)$

b) $(3, 1)$, or $(-11, 7)$, or $(13, -11)$

18. a) i) $\frac{4}{5}, \frac{7}{9}, \frac{11}{14}$

ii) $-\frac{1}{2}, -\frac{1}{2}, -\frac{1}{2}$

iii) $-\frac{1}{4}, -\frac{2}{7}, -\frac{1}{5}$

iv) $-\frac{2}{5}, -\frac{4}{9}, -\frac{3}{7}$

b) In (ii), A, B, C are collinear

19. 0.24 approximately

20. a) $(0, 0), (10, 0), (4, 6)$

b) $(5, 0), (5, 10), (-11, 6)$

21. a) Yes

b) i) Yes   **ii)** Yes

Exercises 5-4

1. a) $-\frac{3}{2}$ **b)** $-\frac{8}{5}$ **c)** $\frac{4}{3}$ **d)** 2

e) $-\frac{1}{3}$ **f)** $\frac{1}{2}$ **g)** -5 **h)** -1

2. a), d), f)

3. a), b)

4. a), c)

5. They are parallel.

6. a) -8 **b)** 4   **c)** $-\frac{3}{2}$

d) 3   **e)** -10 **f)** $-\frac{7}{3}$

7. a) $(-2, 0)$ **b)** $(4, 2)$ **c)** $(5, 1)$

8. a) $(4, 5), (5, 1)$ **b)** $(7, 10), (10, 3)$

c) $(1, 10), (10, 9)$ **d)** $(b, a+b)(a+b, a)$

9. $(3, 0)$, or $(7, 0)$, or $(7 \pm 4\sqrt{6}, 0)$

10. a) A rhombus   **b)** $(1, 1); \frac{1}{2}, -2$

c) The diagonals of a rhombus bisect each other at right angles.

11. a) $(5, -2), (7, 3)$, or $(-3, 7), (-5, 2)$, or $\left(-\frac{3}{2}, \frac{7}{2}\right), \left(\frac{7}{2}, \frac{3}{2}\right)$

b) $(4, 8), (7, 4)$, or $(-1, -2), (-4, 2)$, or $\left(\frac{7}{2}, \frac{9}{2}\right), \left(-\frac{1}{2}, \frac{3}{2}\right)$

12. a)

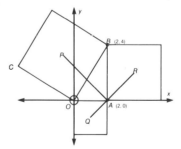

b) Coordinates of C: $(-4, 2)$

Coordinates of midpoint, P, of BC: $(-1, 3)$

Coordinates of Q and R can be found similarly.

$AP = QR = 3\sqrt{2}$

Slope of PA: -1 Slope of QR: 1

c) i) $BQ = PR = \sqrt{26}$ **ii)** $PQ = OR = 2\sqrt{5}$
Slope of BQ: 5   Slope of PQ: -2
Slope of PR: $-\frac{1}{5}$ Slope of OR: $\frac{1}{2}$

d) This property is true for all triangles, not just right triangles.

Exercises 5-5

1. a) 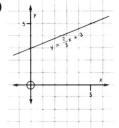 **b)**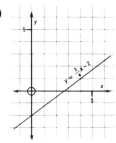

c) 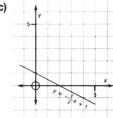 **d)**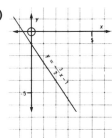

e) 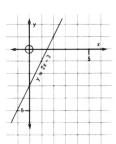 **f)**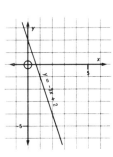

5. a)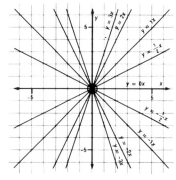

b) It rotates about (0, 0)

6. a)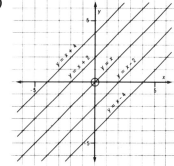

b) It moves parallel to itself.

7. a) 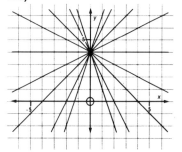 **b)**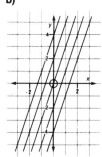

c) i) It rotates about $(0, b)$.
ii) It moves parallel to itself.

2. a) $y = \frac{1}{2}x + 1$ **b)** $y = \frac{3}{2}x - 2$ **c)** $y = -2x + 1$

3. a) $y = -x + 2$ **b)** $y = -\frac{3}{2}x - 3$ **c)** $y = \frac{2}{3}x$

4. a) 3, 5   **b)** $-2, 3$ **c)** $\frac{2}{5}, -4$

d) $-\frac{1}{2}, 6$ **e)** $-4, -7$ **f)** $\frac{3}{8}, -\frac{5}{2}$

g) $\frac{4}{3}, -2$ **h)** $\frac{9}{5}, 1$ **i)** $-\frac{2}{3}, 0$

8. a) $y = 2x + 3$ **b)** $y = -x + 4$
c) $y = \frac{2}{3}x - 1$ **d)** $y = -\frac{4}{5}x + 8$

9. a) $y = 2x + 5$ **b)** $y = \frac{2}{3}x - 2$

10. a) $\frac{3}{4}, -3$ **b)** $\frac{5}{2}, 5$ **c)** $-2, 3$

d) $-\frac{3}{5}, -4$ **e)** $-\frac{1}{2}, \frac{5}{2}$ **f)** $\frac{4}{7}, \frac{15}{7}$

11. a) 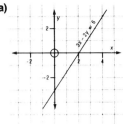 **b)**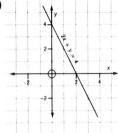

c) 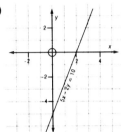 **d)**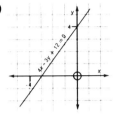

e) 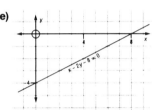 **f)**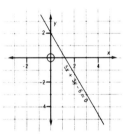

g)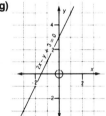

12. i) b and g, a and h, c and i
   ii) b and d, g and d, c and f, i and f

13. a) Yes **b)** No **c)** Yes

14. Slope: $-\dfrac{A}{B}$, y-intercept: $-\dfrac{C}{B}$

15. a) $y = \dfrac{1}{2}x + 1$ **b)** $y = \dfrac{2}{3}x + 2$

   c) $y = -2x + 3$ **d)** $y = -\dfrac{1}{3}x - 2$

16. $y = 2x + 8$. or $y = 2x - 32$

MAU

1. a)

Year	1915	'26	'31	'43	'54	'68
b) Increase ¢	1	−1	1	1	1	1
c) % increase	50	−33.3	50	33.3	25	20

	'71	'72	'76	'77	'78	'79	'82
	1	1	2	2	2	3	13
	16.7	14.3	25	20	16.7	21.4	76.5

2. a) 1926   **b)** $33\dfrac{1}{3}\%$

3.

Stamp	% Increase
2¢ Map	250
3¢ King George V	122
2¢ Confederation	114
3¢ King George VI	67
4¢ Alexander Graham Bell	67
5¢ Queen Elizabeth II	40
6¢ Sir Isaac Brock	167
7¢ Pierre Laporte	100
8¢ Indians	20
10¢ Chicora	75

Exercises 5-6

1. a) −5 **b)** 7 **c)** −11 **d)** 5
   e) −3 **f)** −6

2. a) 4 **b)** −5 **c)** $\dfrac{19}{3}$ **d)** 4
   e) $\dfrac{4}{3}$ **f)** 0

3. a) $y = 3x - 1$ **b)** $y = 7x + 30$
   c) $y = -4x + 16$ **d)** $y = 2x + 3$
   e) $y = -\dfrac{3}{5}x - 3$ **f)** $y = \dfrac{2}{3}x - \dfrac{16}{3}$
   g) $y = -\dfrac{7}{2}x - \dfrac{7}{2}$ **h)** $y = \dfrac{8}{3}x + \dfrac{34}{3}$
   i) $y = \dfrac{3}{4}$

4. a) $y = 3x + 11$ **b)** $y = 2x + 8$
   c) $y = x + 5$ **d)** $y = \dfrac{1}{2}x + \dfrac{7}{2}$
   e) $y = 3x - 7$ **f)** $y = -2x - 4$
   g) $y = -x - 1$ **h)** $y = -\dfrac{1}{2}x + \dfrac{1}{2}$

5. a) $y = 2x + 5$ **b)** $y = 2x + 3$
   c) $y = 2x + 1$ **d)** $y = 2x - 1$
   e) $y = 2x - 3$ **f)** $y = 2x - 5$
   g) $y = 2x - 7$ **h)** $y = 2x - 9$

Answers

6. a) $y = 2x - 3$ **b)** $y = -3x - 7$
  c) $2y = 7x - 39$ **d)** $y = 2x - 1$
  e) $3y = 2x - 11$ **f)** $3y = 8x + 20$
  g) $y = -3x + 16$ **h)** $y = -2x - 14$
7. a) $y = 3x - 5$ **b)** $2x + 5y = 18$
  c) $x + y = 2$ **d)** $3y = 2x + 11$
  e) $4y = 3x - 12$ **f)** $y + 2x = 5$
  g) $y = 2x + 4$ **h)** $3y + x = 6$
8. a) $y = x + 7$ **b)** $x + y = 3$
  c) $5y + 6x = 13$ **d)** $y + 3x = -1$
  e) $3y = x + 17$
9. $y = \frac{1}{5}x + 5$; $y = 3x - 9$; $y = -\frac{5}{3}x + 5$
10. a) $3y = x - 6$ **b)** $3y = x + 24$
  c) $3y = x + 9$ **d)** $13y = -9x + 59$
11. a) $y = 3x - 11$ **b)** $y = -\frac{3}{4}x + 4$
  c) $3y = 4x - 13$ **d)** $y = -2x + 9$
  e) $5y = -3x + 17$ **f)** $3y = 8x - 29$
12. a) $y = \frac{2}{3}x + 4$; $y = -\frac{3}{2}x + 4$;
  $y = -\frac{3}{2}x - 9$; $3y = 2x - 14$
  b) $y = 5x + 4$; $5y = -x - 6$
13. $2y + 5x = 20$
14. b) AB: $y = 2x$; CD: $2y = x$;
  EF: $y = -3x$; GH: $3y = -x$.
  All pass through $(0, 0)$.
  c) AI: $2y = x + 12$; CJ: $y = 2x - 12$;
  KL: $3y + x = 48$; MN: $y + 3x = 48$.
  $(12, 12)$
15. $3x + 2y = 8$; $x + 6y = 8$
16. a) $10y = 3x + 19$
  b) $y = 4x - 27$
  c) $2y + 3x = 13$
  d) $y + 3x = 19$
17. a) $y = 3x + 14$; $y = \frac{1}{2}x - 1$; $4y + 3x = 26$
  b) $3y + x = 2$; $y + 2x = -1$; $3y = 4x + 7$
  c) $(-1, 1)$; $AP = AQ = AR = \sqrt{50}$
18. a) QS: $y + 2x = 6$; PR: $2y = x + 7$
  b) No
  c) Yes
  d) i) No **ii)** No

Exercises 5-7

1. a)

x	1	2	3	4	5	
y	2	4	8	6	10	$y = 2x$

b)

x	1	2	3	4	5	
y	5	10	15	20	25	$y = 5x$

c)

x	1	2	3	4	5	
y	-3	-6	-9	-12	-15	$y = -3x$

d)

x	1	2	3	4	5	
y	$\frac{1}{5}$	$\frac{2}{5}$	$\frac{3}{5}$	$\frac{4}{5}$	1	$y = \frac{1}{5}x$

2. a) $y = 4x$ **b)** 48 **c)** 7.5 **d)** 4
3. a) $y = \frac{2}{3}x$ **b)** 20 **c)** 22.5 **d)** $\frac{2}{3}$
4. a) $h = \frac{4}{5}s$ **b)** 20 m **c)** 11.25 m
5. a) $d = 8t$ **b)** 32 cm, 8 cm **c)** 1.9 min
6. a) y is doubled. **b)** y is halved.
7. a)

x	3	6	9	12	15	
y	4	8	12	16	20	$y = \frac{4}{3}x$

b)

x	10	18	8	22	6	
y	15	27	12	33	9	$y = \frac{3}{2}x$

c)

x	7	14	28	42	70	
y	5	10	20	30	50	$y = \frac{5}{7}x$

d)

x	5	25	85	95	105	
y	2	10	34	38	42	$y = \frac{2}{5}x$

8. a) $y = \frac{5}{6}x$ **b)** $10, \frac{50}{3}, 30$
  c) 6, 15, 30 **d)** $\frac{5}{6}$
9. a) $m = 1.4d$
  b) $28
  c) 28.6 km
10. a) 0.9 m
  b) 5 m
11. a) 1500
  b) 200
  c) 31.25
12. $y = x\sqrt{2}$
13. a) $r = \frac{x}{2}\sqrt{5}$ **b)** $8:5\pi$

Exercises 5-8

1. b) 25 **c)** 12

2. a)
x	0	2	4	6	8	10
y	2	4	6	8	10	12

b) 7
c) 10

3. a)
n	0	100	200	500	1000
T	200	225	250	325	450

b) $200; $25/100 copies
c) $825
d) 1200

4. a)
d	0	1	2	3	4	5	6	7	8
n	72	63	54	45	36	27	18	9	0

c) 21.6 L   **d)** 467 km

5. a)
n	0	500	1000	1500
C	2	4.80	7.60	10.40

c) $C = 2 + 0.56n$
d) About 840

6. a) 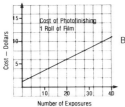   **b)** $C = \frac{1}{4}n + 1$

7. a)
x	1	2	3	4	5
y	5	8	11	14	17

b)
x	1	2	3	4	5
y	8	13	18	23	28

c)
x	2	4	6	8	10
y	-2	2	6	10	14

d)
x	1	2	3	4	5
y	4	2	0	-2	-4

8. b) $t = 20 - \frac{13}{2000}a$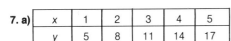
c) 10.25°C
d) About 3000 m

9. a) 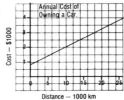   **b)** $800
c) $C = \frac{3}{25}d + 800$
d) $2600

10. $P = 2\ell + 10$

11. a) i) $300; $267; $250   **ii)** $300; $367; $400
b)    **c)** $6000

12. a)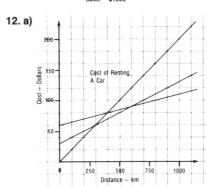

b) Rent from A for $d > 600$;
from B for $300 < d < 600$;
from C for $d < 300$.
c) $c = 60 + 0.5d$; $c = 30 + 0.1d$; $c = 0.2d$

13. c) i) $y = 180 - 2x$
y decreases as x increases.
ii) $y = x + 90$
y increases as x increases.
iii) $y = x - 75$
y increases as x increases.

Exercises 5-9

1. a)
x	2	3	4	6	8	xy = 24
y	12	8	6	4	3	

b)
x	10	12	15	20	30	xy = 60
y	6	5	4	3	2	

Answers **439**

c)

x	4	8	12	16	24	
y	12	6	4	3	2	$xy = 48$

d)

x	44	33	22	12	11	
y	3	4	6	11	12	$xy = 132$

2. a) $xy = 36$
  b) i) 3 **ii)** 18
3. a) $xy = 84$
  b) i) 14 **ii)** 21
4. a) i) $96 **ii)** $32
  b) i) 3 **ii)** 5
5. a) 8 min **b)** 20 L/min

6. a)

x	6	8.47	9	36	48	
y	24	17	16	4	3	$xy = 144$

b)

x	6	7.5	9	10	15	
y	30	24	20	18	12	$xy = 180$

c)

x	$2\frac{1}{2}$	$3\frac{1}{3}$	24	30	32	
y	64	48	$6\frac{2}{3}$	$5\frac{1}{3}$	5	$xy = 160$

d)

x	15	10	9	7.5	2.5	
y	15	22.5	25	30	90	$xy = 225$

7. a) i) 4 h **ii)** 2.4 h **b)** 24 km/h
  c)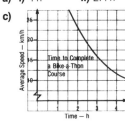

8. a) i) 30 r/min **ii)** 24 r/min **b)** 16
  c)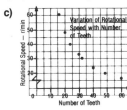

9. 5 **10.** 75 r/min
11. a) y is decreased by 20%. **b)** y is increased by 25%.

12. a) b)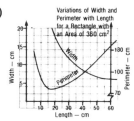

  c) i) Yes **ii)** No
  $$P = 2\left(\ell + \frac{360}{\ell}\right)$$
  When ℓ is large, $\frac{360}{\ell}$ is small and P varies directly as ℓ.

13. a) 12 **b)** 24 **c)** 36

PSS

1. 1002 **2.** 301 **3.** $\frac{37}{64}$
4. 3893 **5.** 16

Exercises 5-10

1. a) Parabola **b)** Hyperbola

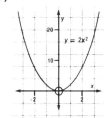 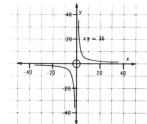

c) Circle **d)** Ellipse

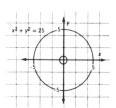 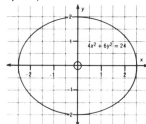

e) Hyperbola

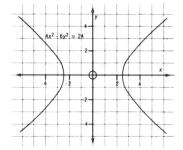

2. a) Parabola

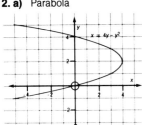

b) Hyperbola

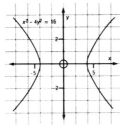

5. a) $x^2 + y^2 = 65$

b)

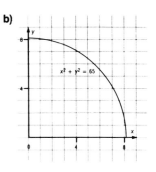

c) Circle

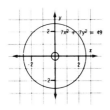

e) Ellipse

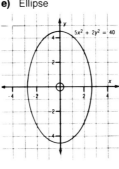

6. a) i) $25 - x$
ii) $x(25 - x)$

b)

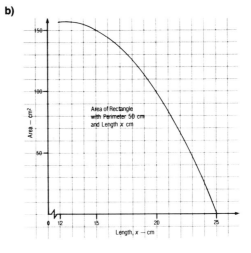

d) Hyperbola

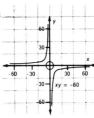

3. a) Yes

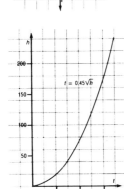

b) 6.3 s
c) 36 m

7. a) i) $\dfrac{36}{x}$
ii) $2x + \dfrac{72}{x}$

b)

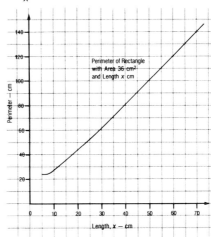

4. a)

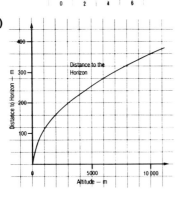

b) 312 km
c) About 6950 m

Answers **441**

8. a) $h^2 = r^2 - 16$

b)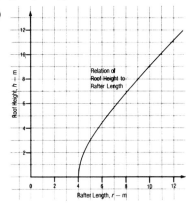

9.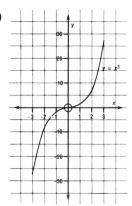

10. a)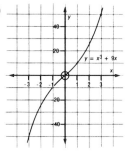

b) 32 L
c) 26.7 kPa

Exercises 5-11

1. a) Quadratic **b)** Linear **c)** Other **d)** Quadratic
e) Quadratic **f)** Quadratic **g)** Linear **h)** Quadratic

2. a)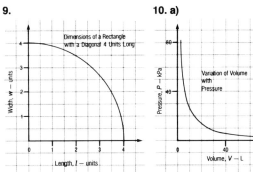

c)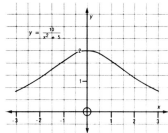

3. a)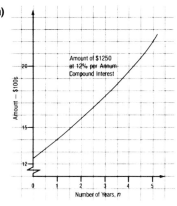

b) $1966.90
c) About 6 years

4. a) i) 1.5 m² **ii)** 2.3 m²
b) i) 30 kg **ii)** 75 kg
5. a) i) 22 m **ii)** 30 m **iii)** 15 m
b) i) 4 s, 42 s, 49 s, 86 s, 93 s
ii) 6 s, 11 s, 17 s, 28 s, 35 s, 39 s, 51 s, 56 s, 62 s, 73 s, 79 s, 83 s, 96 s
iii) 23 s, 68 s
c) 45 s

6. a)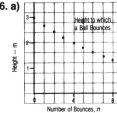

b) 2.2 m, 1.04 m
c) Sixth

8. a) $y = \dfrac{x}{x - 1}$

7. a)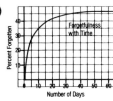

b) 4
c) 54%

b)

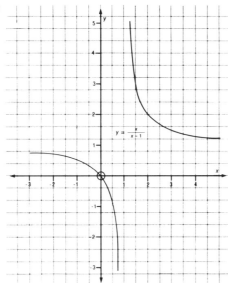

c) $(0, 0)$, $\left(-1, \dfrac{1}{2}\right)$, $\left(\dfrac{1}{2}, -1\right)$, $(2, 2)$, $\left(3, \dfrac{3}{2}\right)$

9. a) $V = x(50 - 2x)^2$

b)

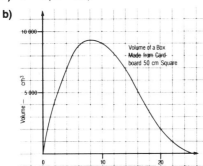

c) $8\dfrac{1}{3}$

10. a)

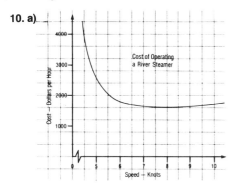

b) 8 knots

Review Exercises

1. a) 15 **b)** 14 **c)** $10\sqrt{2}$
  d) $5\sqrt{2}$ **e)** $4\sqrt{5}$ **f)** $8\sqrt{2}$

2. a) 6, 8 **b)** 28 **c)** 10
  d) 48

3. a) $(3, 2)$ **b)** $(-3, -1)$ **c)** $(-4, -3)$
  d) $(9, -5)$ **e)** $\left(\dfrac{3}{2}, 6\right)$ **f)** $\left(\dfrac{7}{2}, \dfrac{5}{2}\right)$

4. a) $KL: (-1, 4)$, $LM: (1, -2)$, $MK: (5, 1)$
  b) 10

5. a) $\dfrac{5}{2}$ **b)** 0 **c)** $-\dfrac{13}{7}$
  d) $\dfrac{2}{3}$ **e)** $-\dfrac{5}{7}$ **f)** -3

6. a) $\dfrac{3}{8}$, 4 **b)** $-\dfrac{9}{5}$ **c)** $-\dfrac{6}{13}$

7.

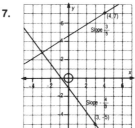

8. a) $\dfrac{5}{3}$ **b)** $-\dfrac{1}{4}$ **c)** $\dfrac{4}{7}$ **d)** $-\dfrac{10}{3}$

9. Slope of PQ: $\dfrac{3}{5}$; slope of PR: $-\dfrac{5}{3}$.
  Therefore, $PQ \perp PR$, and $\angle PQR = 90°$.

10. a) $4, -3$ **b)** $-\dfrac{5}{3}, 7$ **c)** $-\dfrac{9}{4}, -3$
  d) $-\dfrac{2}{5}, 3$ **e)** $\dfrac{3}{4}, -4$ **f)** $\dfrac{5}{3}, -6$

11. a) $y = -\dfrac{1}{2}x - 4$ **b)** $y = \dfrac{4}{3}x - 6$

12. a) $y = x + 2$ **b)** $9y + 11x = 19$
  c) $y = 4x + 2$ **d)** $7y + 4x = -11$

13. a) $4y + 3x = 35$
  b) $5y - 2x = 15$

14. a)

x	3	5	9	12	4	20
y	12	20	36	48	16	80

$y = 4x$

b)

x	-3	3	6	12	15
y	1	-1	-2	-4	-5

$y = -\dfrac{1}{3}x$

15. a) $3x = 4y$ **b)** 22.5 **c)** 80

16. a)

x	4	6	8	10	14
y	7	11	15	19	27

b)

x	−3	0	3	6	9
y	−4	5	14	23	32

17. a)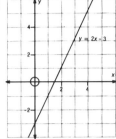

b) 77   **c)** 20

18. a)

x	4	6	12	18	24
y	18	12	6	4	3

b)

x	72	54	36	24	18
y	3	4	6	9	12

19. a) 10 min   **b)** 12 L/min

Cumulative Review (Chapters 3-5)

1. a) −1   **b)** 2

2. a) $\{(x, 3x + 5) \mid x \in R\}$ **b)** $\left\{\left(x, -\frac{4}{3}x + 8\right) \mid \varepsilon R\right\}$

c) $\left\{\left(x, -\frac{5}{2}x + 10\right) \mid x \in R\right\}$

3. a) (2, 3)   **b)** (−1, 4)   **c)** (−1, −3)

4. a) $\left(\frac{1}{2}, -1\right)$ **b)** (4, −5)   **c)** (3, 5)

5. a) (4, 3)   **b)** (2, −1)   **c)** $\left(-\frac{1}{2}, \frac{2}{3}\right)$

6. 18, 21

7. 89, 91, 93

8. adult: 40, student: 80

9. 37

10. a) 6, −5   **b)** −7, 4

11. a)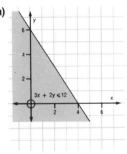

b)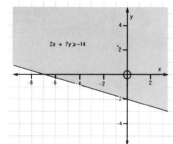

12. a) (−1, 2)   **b)** (−1, 2)   **c)** (−1, 2)

13. a)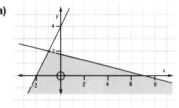

b) 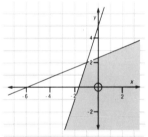   **c)**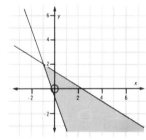

14. a) 10   **b)** 18   **c)** $12\sqrt{2}$ **d)** $7\sqrt{2}$

15. a) (7, −4)   **b)** (2, 2)   **c)** $\left(\frac{3}{2}, \frac{5}{2}\right)$ **d)** $\left(\frac{11}{2}, \frac{1}{2}\right)$

16. a) −1   **b)** $\frac{1}{2}$ **c)** $-\frac{8}{9}$ **d)** $\frac{3}{5}$

17. a) $\frac{11}{2}$ **b)** (−2, −1)   **c)** $\sqrt{61}$

18. a)

y varies directly as x.					
x	5	10	20	40	80
y	−2	−4	−8	−16	−32

$y = -\frac{2}{5}x$

b)

y varies inversely as x.				
x	2	3	4	6	12
y	54	36	27	18	9

$xy = 108$

19. a) −2, 7   **b)** $\frac{3}{4}, -3$ **c)** $-\frac{5}{2}, -3$ **d)** $-2, \frac{5}{2}$

20. a) $y = -\frac{2}{3}x + 4$ **b)** $y = \frac{6}{5}x - 3$

444 *Answers*

21. a) $7y = 2x + 48$ **b)** $9y + 16x = 10$
22. a) $5y + 2x = 8$ **b)** $3y + 5x = 1$
23. a) 16 min   **b)** 4 L/min

Exercises 6-1

1. a) (8, 0)   **b)** (−5, 2)   **c)** (10, 2)
d) (3, 6)   **e)** (5, 2)   **f)** (−15, −1)
g) (11, 3)   **h)** (11, 11)

2. a) b)

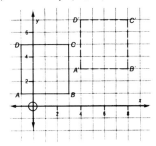

c) A translation 5 to the right and 2 up.

3. a) b)

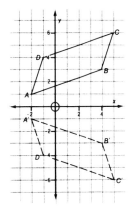

c) A reflection in the x-axis.

4. a)

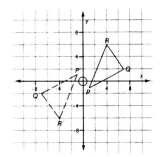

b) A rotation about (0, 0).

5. a) (0, −6)   **b)** (2, 8)   **c)** (0, 6)
d) (0, 3)   **e)** (0, 7)   **f)** (−4, 2)

6. a) (3, −1)   **b)** (3, 2)   **c)** (4, −3)
d) (1, 7)   **e)** (6, −1)   **f)** (2, 3)

7. 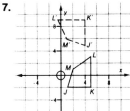   **8.**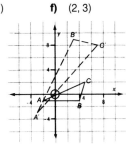

9. a) (−6, 11)   **b)** (−2, 4)   **c)** (−4, 2)
d) (8, −4)   **e)** (−8, −4)   **f)** (9, 8)

10. a) Translation   **b)** Translation

   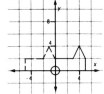

c) Reflection   **d)** Rotation

   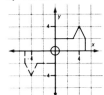

e) Rotation   **f)** Dilatation

   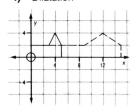

Exercises 6-2

1.

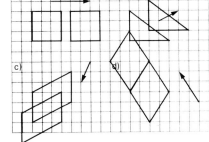

Answers

2.

Mapping rule	Point	Image
$(x, y) \to (x - 2, y)$	$(5, -1)$	$(3, -1)$
$(x, y) \to (x, y + 3)$	$(-2, 1)$	$(-2, 4)$
$(x, y) \to (x + 4, y)$	$(3, 2)$	$(7, 2)$
$(x, y) \to (x - 3, y - 6)$	$(1, 3)$	$(-2, -3)$
$(x, y) \to (x + 4, y - 1)$	$(0, 0)$	$(4, -1)$
$(x, y) \to (x + 5, y + 1)$	$(4, -3)$	$(9, -2)$
$(x, y) \to (x + 5, y - 2)$	$(-2, 1)$	$(3, -1)$
$(x, y) \to (x, y - 2)$	$(-3, 2)$	$(-3, 0)$
$(x, y) \to (x - 3, y + 3)$	$(4, -2)$	$(1, 1)$

3. a) $(-2, 5), (1, 6), (0, -1)$ **b)** $(2, -2), (3, 2), (-1, -4)$

c) 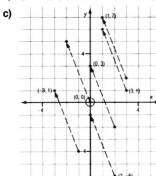   **d)** $\sqrt{29}, -\dfrac{5}{2}$

4. b, d, f

5.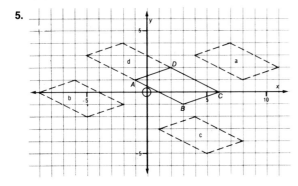

6.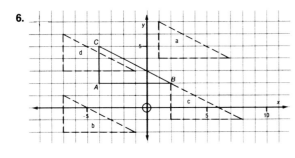

7. a)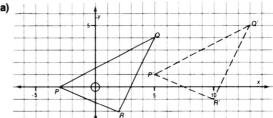

b) i) Equal   **ii)** Same   **iii)** Equal

8. a)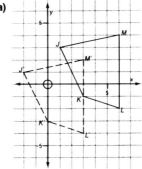

b) i) Equal   **ii)** Same   **iii)** Equal

9. a)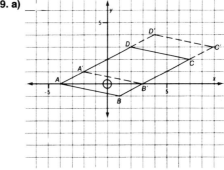

b) i) Same   **ii)** Equal

10. c) Under a translation, length and slope are invariant.

11.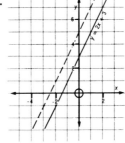

12.

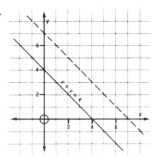

13. $2y = 3x - 23$

Exercises 6-3

1.

Object	2	3	4	5	6	7	8
After a $\frac{1}{4}$-turn	↻2	↻3	↻4	↻5	6	7	↻8
After a $\frac{1}{2}$-turn	ᄅ	Ɛ	ᔭ	ގ	9	ㄥ	8
After a $\frac{3}{4}$-turn	↺2	↺3	4	↺5	6	7	↺8
After a full turn	2	3	4	5	6	7	8

2. a, c, d, f

3. a) $(-1, 4), (-5, 0), (-2, -3)$ **b)** $(6, -2), (0, -3), (-4, 1)$

c)

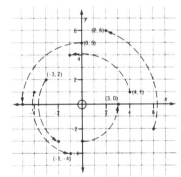

4. a) b) c)

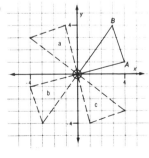

5. a) b) c)

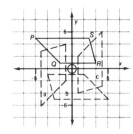

6. a)

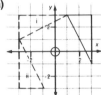

b)

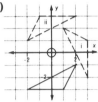

c)

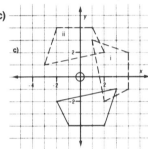

7. a)

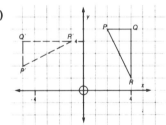

b) i) Equal   **ii)** Negative reciprocals   **iii)** Equal

8. a)

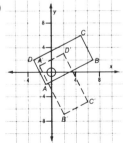

b) i) Equal   **ii)** Negative reciprocals   **iii)** Equal

Answers **447**

9. A 90° rotation about (3, 0).

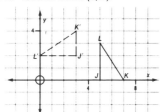

16. a) 53°
b) About 1:32 a.m.

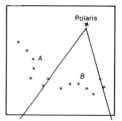

10. a)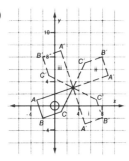

b) i) $\frac{1}{4}$-turn about (3, 3)
ii) $\frac{1}{2}$-turn about (3, 3)
iii) $\frac{3}{4}$-turn about (3, 3)

17. a)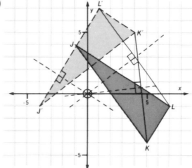

11. a)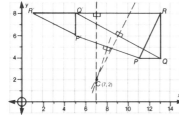

b) Perpendicular bisectors of segments joining matching points pass through the turn centre.

b) (0, 0), the turn centre

c) All are right angles.

12. 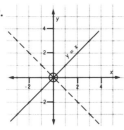 **13.**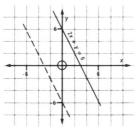

18. a) 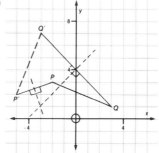 **b)** (−3, 1)

14. $2y = 3x + 6$

15. a)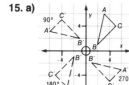
b) 90°: $(x, y) \rightarrow (-y, x)$
180°: $(x, y) \rightarrow (-x, -y)$
270°: $(x, y) \rightarrow (y, -x)$

19. a) 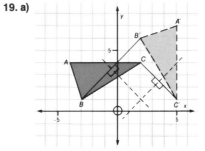 **b)** (2, 1)

448 Answers

20. a) 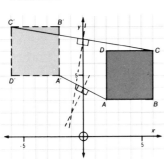   **b)** $(-1, 2)$

4. a) $(-1, 3), (-4, -2), (1, 5)$

b) $(-2, -3), (1, 2), (-3, 0)$

c)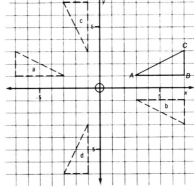

Exercises 6-4

1.

2. a) 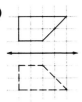   **b)**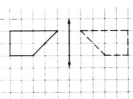

c) 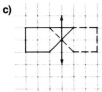   **d)**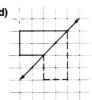

5.

e)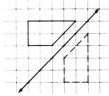

6.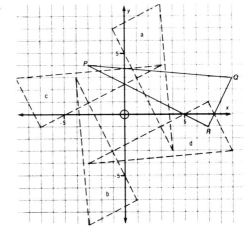

f)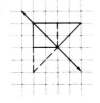

3. *a, b, d*

Answers **449**

7. a)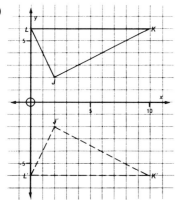

b) i) Equal **ii)** Not invariant **iii)** Equal

8. a)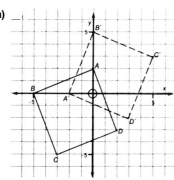

b) i) Equal **ii)** Not invariant **iii)** Equal

9.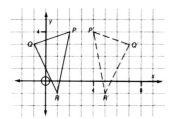

10. 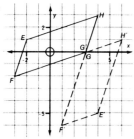 **11.**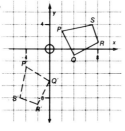

12.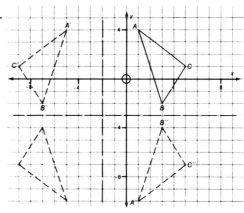

13. b) i) Slope of reflection line: -2
Slope of PP' and QQ': $\frac{1}{2}$
Therefore, reflection line is perpendicular to PP' and QQ'.

ii) Midpoint of PP': $(7, 3)$
Midpoint of QQ': $(5, 7)$
Both lie on the reflection line.

c) A reflection line is the perpendicular bisector of segments joining matching points.

14. b) $x = 4$

c) Lines and their reflection images intersect on the reflection line.

15. a) 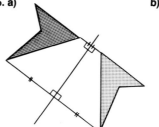 **b)**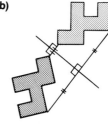

16. 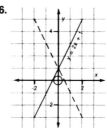 **17.**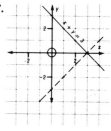

450 Answers

18. $3x + 2y = 6$

19. a)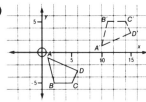

b) Line segments joining matching points are not parallel. Therefore, the transformation is not a reflection or a translation. It is not a rotation because the orientation is reversed.

c) Length, angle measure, and area are invariant.

20. a) 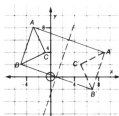   b) $y = 3x - 3$ c) $(3, 4)$

21. a)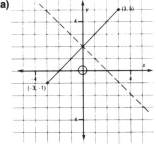

b) (−3, 1), (3, 0), (4, −2), (0, −1), (−4, 6), (5, 3)
c) $x + y = 2$

Exercises 6-5

1. a) b) 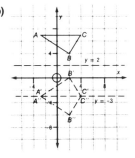   c) A translation

2. a) b) 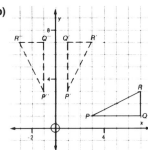   c) $\frac{1}{4}$-turn about (0, 0)

3. a) b) 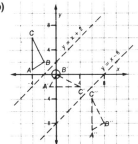   c) A translation

4. a) b) 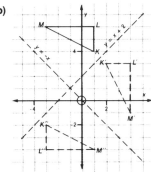   c) $\frac{1}{2}$-turn about (−1, 1)

5. a) b)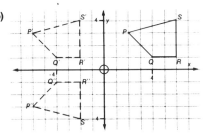

c) A glide reflection

6. a: rotation + translation

b: rotation + translation
c: rotation + translation
d: rotation + reflection
e: rotation + translation + reflection
f: rotation + translation
g: rotation + translation
h: rotation + translation

By virtue of the symmetry, all could be glide reflections.

7. a) **b)**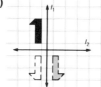

c) **d)**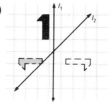

8. a) Glide reflection
b) Reflection
c) Rotation
d) Rotation
e) Reflection
f) Glide reflection

9. a) 3 rotations
b) 3 reflections
c) 3 reflections
d) Translations
e) Translations
f) Rotations

10. a) i) **iii)**

b) i) **iii)**

c) i) **iii)**

a) ii) A translation 8 units right
iii) A translation 8 units left
b) ii) $\frac{1}{2}$-turn **iii)** $\frac{1}{2}$-turn
c) ii) $\frac{3}{4}$-turn **iii)** $\frac{1}{4}$-turn

11. i) a) **b)**

c)

ii) A translation to the right twice the distance between l_1 and l_2.
iii) No

12. i) 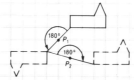 **a)** A translation

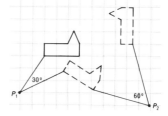 **b)** A rotation

13. a)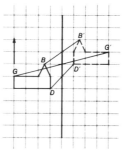

c) Their midpoints lie on the reflection line.
d) No.

452 Answers

14. a)

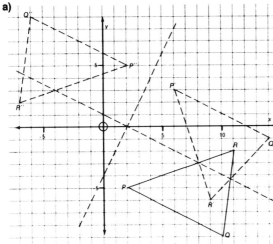

b) $y = 2x - 4$
c) $(2, 0)$
d) $\frac{1}{2}$-turn about $(2, 0)$

15. a)

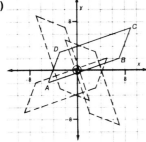

b)

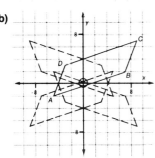

c) Lengths, angles, and areas are preserved by both rotations and reflections. Orientation is preserved by rotations and by two successive reflections. The image under a $\frac{1}{2}$-turn is the same as the image under successive reflections in the x- and y-axes. This suggests that these transformations are equivalent.

Exercises 6-6

1. a) Reflection   **b)** Glide reflection   **c)** Rotation
d) Rotation   **e)** Translation   **f)** Reflection

2. a) Reflection   **b)** Rotation   **c)** Reflection
d) Reflection   **e)** Rotation   **f)** Reflection
g) Rotation   **h)** Reflection   **i)** Rotation
j) Reflection

3. a) i) Rotation   **b)** i) Reflection
   ii) Rotation   ii) Reflection
   iii) Rotation   iii) Rotation

4. i) c, d, f, i
ii)
a) Reflection   **b)** Glide reflection
c) Rotation   **d)** Translation
e) Glide reflection   **f)** Rotation
g) Reflection   **h)** Glide reflection
i) Rotation   **j)** Reflection

5. a) Reflection   **b)** Translation   **c)** Rotation
d) Rotation + Reflection   **e)** Translation

6. a)   **b)**   **c)**

7. 1. Translation   4. Translation
      2. Rotation   5. Rotation
      3. Translation

a) 1. Rotation   4. Rotation + reflection
      2. Rotation   5. Rotation + reflection
      3. Translation

 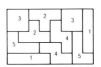

b) Each arrangement creates a new set of isometries, and the symmetries of the shapes make alternatives possible.

8. a) i) Rotation + reflection
        ii) Rotation
        iii) Rotation + reflection
b) i) 18
      ii) 13

Answers **453**

9. i)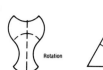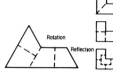

ii) b) Reflection
Rotation

c) Rotation   Rotation
Reflection   Reflection
Reflection   Translation
Translation

Exercises 6-7

1. a)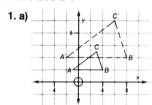

b) i) Same ii) Twice as long iii) Four times as great

c) 2; (0, 0)

2. a)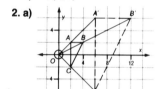

c) i) $6\sqrt{2}, 2\sqrt{2}, 6\sqrt{5}, 2\sqrt{5}, 6\sqrt{2}, 2\sqrt{2}$

ii) 3, 3, 3

d) i) 6, 2, $6\sqrt{5}, 2\sqrt{5}$, 12, 4

ii) 3, 3, 3

e) 3

3. a)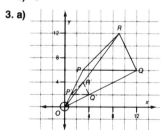

b) Area of image is $\frac{1}{9}$ that of original triangle.

c) i) $\sqrt{5}, 3\sqrt{5}, 2\sqrt{5}, 6\sqrt{5}, 5, 15$

ii) $\frac{1}{3}, \frac{1}{3}, \frac{1}{3}$

d) i) Same   ii) $\frac{1}{3}$ as long

e) $\frac{1}{3}$

4. a)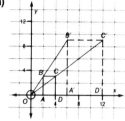

b) Area of image is 9 times that of original triangle.

c) i) $3\sqrt{13}, \sqrt{13}, 15, 5$

ii) 3, 3

d) i) 3 times as long
ii) Equal
iii) 9 times as great

5.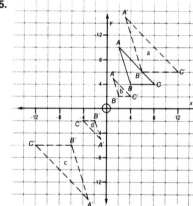

6. a) 1.5   b) 2.5   c) $\frac{5}{3}$

d) $\frac{2}{5}$ e) $\frac{3}{5}$ f) $\frac{2}{3}$

7. a)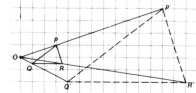

b)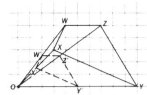

8. a)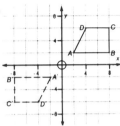

b) $\frac{1}{2}$-turn about (0, 0)
c) -1

9. a)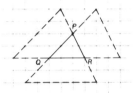

b) 4:1

10. a)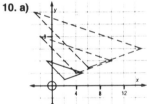

b)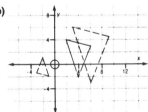

c)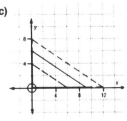

d)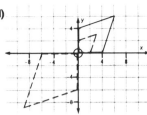

11. a) i) 2, 4, 8   ii) 4, 16, 64

b) Lines through the corresponding vertices of the trapezoids intersect in one point, and the sides of the trapezoids preserve slope and angle measure.

13.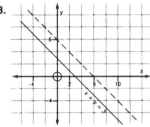

14. $y = 2x + 6$

MAU

1. Choose different positions for B and C
2. Interchange marker, M, and the pencil.
3. The image is inverted.
4. The pantograph preserves ratios of lengths.

Exercises 6-8

1.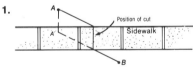

2. a) (8, 0)   b) $5\sqrt{13}$

3. a)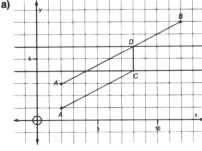

Answers **455**

b) 8   **c)** $5\sqrt{5} + 2$

4.

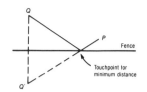

5. a)

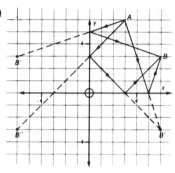

b) i) $3\sqrt{10}$ **ii)** $3\sqrt{10}$ **iii)** $9\sqrt{2}$

6. a) B

b) About 110 m, taking side of grid square to measure 4.5 mm.

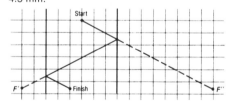

7.

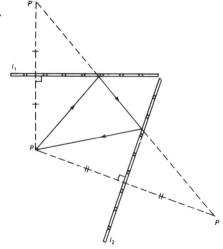

8. a) D

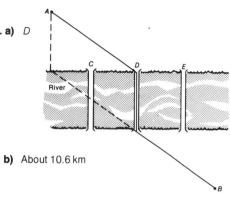

b) About 10.6 km

CP

1. The minimum distance has the least y-coordinate.

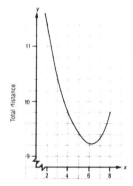

2. a) 13   **b)** About 16.55
c) About 14.14   **d)** 9

Review Exercises

1. a) $(8, 3)$ **b)** $(-6, -4)$ **c)** $(4, 6)$ **d)** $(9, 1)$
2. a) $(6, 11)$ **b)** $\left(\dfrac{3}{2}, 13\right)$ **c)** $(6, 7)$ **d)** $(-5, 2)$
3.

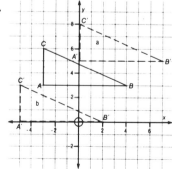

4.

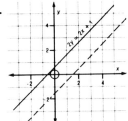

5. a)

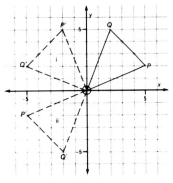

  i) $\frac{1}{4}$-turn about (0, 0)   **ii)** $\frac{1}{2}$-turn about (0, 0)

6.

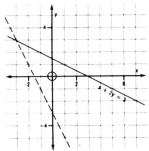

7. a)

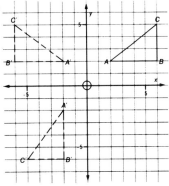

  i) A reflection in the y-axis   **ii)** A reflection in the line $x + y = 0$

8.

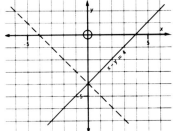

9.

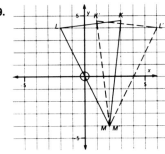

10. a) b)

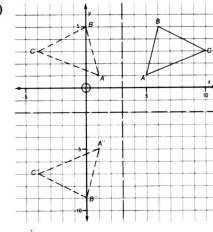

  c) $\frac{1}{2}$-turn about (3, −2)

11.

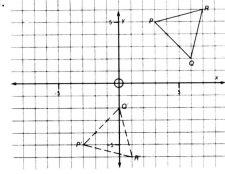

Answers **457**

12. a) Reflection
    b) Glide reflection
    c) Translation

13. i) a, c
    ii) a) Rotation
        b) Rotation + reflection
        c) Translation
        d) Reflection

14.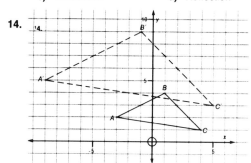

15.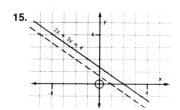

16. 15

h) $AB \perp CD$
i) $x = -5$
j) 2 is the only even prime number.

2. a) $\angle DEC$ and $\angle BEA$ b) $\angle TQR$ and $\angle SQP$
      $\angle DEA$ and $\angle BEC$ $\angle TQP$ and $\angle SQR$

3. a) $x, 115°; y, 115°$ b) $x, 50°; y, 70°$
   c) $x, 75°; y, 75°$

4. a) Yes b) No c) Yes d) No
   e) No f) Yes g) Yes h) No
   i) No j) Yes

5. $\angle EBD = 90°$ ... vertically opposite $\angle ABC$
   $\angle ABD = 90°$ ... $\angle DBC = 180°$
   $\angle EBC = 90°$ ... vertically opposite $\angle ABD$

6. $\angle ABC + \angle ABD = 180°$
   $\angle PQR + \angle PQS = 180°$
   Therefore, $\angle ABD = \angle PQS$ ... $\angle ABC = \angle PQR$

7. $\angle ADC = \angle BDC = 90°$ ... $\angle ADB = 180°$
   $\angle ADE = 90° - \angle EDC$
   $\angle BDF = 90° - \angle FDC$
   But $\angle FDC = \angle EDC$ ... given
   Therefore, $\angle ADE = \angle BDF$

8. a) 42° b) 14° c) 18°
   d) 68° e) 32° f) 25°

9. $\angle ABC = \dfrac{180}{n+1}$, $\angle ABD = \dfrac{180n}{n+1}$

10. Eddie, Bob, Alec, Dan, Carl

11. Use 3 L pail twice to fill the 5 L pail. 1 L will remain in the 3 L pail.

12. a) If equals are decreased by equals and multiplied by equals, the results are equal.
    b) If equals are divided by equals, the results are equal. If $a \geq c$ and $b \leq c$, then $b \leq a$.
    c) Angles are invariant under a rotation.
    d) The product of two quantities is the sum of the products of their parts.
    e) If $a = b$ and $c = d$, then $a + c = b + d$. If equals are divided by equals, the results are equal.

Exercises 7-2

1. a) SAS b) SAS c) SAS
      $\angle A = \angle D$ $\angle P = \angle S$ $\angle J = \angle M$
      $\angle C = \angle F$ $\angle Q = \angle T$ $\angle K = \angle N$

MM

1. 12 and 1, 10 and 9
2. $50 = 5^2 + 5^2$
       $= 7^2 + 1^2$
3. $1^2 + 3^2 + 10^2 = 110$
   $5^2 + 6^2 + 7^2 = 110$
   $2^2 + 5^2 + 9^2 = 110$
   $2^2 + 4^2 + 9^2 = 101$
   $4^2 + 6^2 + 7^2 = 101$

Exercises 7-1

1. a) Lynn is a teenager.
   b) Manuel stays fit.
   c) Sharon has long hair.
   d) The number is 7.
   e) A square has four equal sides.
   f) A square has both pairs of opposite sides parallel.
   g) Some rectangles have four equal sides.

458 *Answers*

∠B = ∠E   ∠R = ∠U   ∠L = ∠O
AC = DF   PQ = ST   JK = MN
AB = DE   QR = TU   JL = MO
BC = EF   RP = US   KL = NO

2. a) △PRS ≅ △PRQ   **b)** △ABD ≅ △CDB
∠S = ∠Q   ∠A = ∠C
∠SPR = ∠QPR   ∠ABD = ∠CDB
∠SRP = ∠QRP   ∠ADB = ∠CBD
PS = PQ   AB = CD
RS = RQ   AD = CB

c) ∠B = ∠ACB   **d)** △PTQ ≅ △PTR
∠D = ∠CAD   △STQ ≅ △STR
AB = AC = CD   △QSP ≅ △RSP

3. ∠AED = ∠BEC   ...vertically opposite
AE = BC and DE = CE   ...given
Therefore, △AED ≅ △BEC   ...SAS
So that AD = BC and ∠D = ∠C

4. ∠ABC = ∠ACB   ...Isosceles Triangle theorem
∠ABE + ∠ABC = 180°   ...∠EBC = 180°
∠BCD + ∠ACB = 180°   ...∠ACD = 180°   Therefore,
∠ABE = ∠BCD

5. a)   OA = OB   ...radii of circle
Therefore, ∠A = ∠B   ...Isosceles Triangle theorem

b) Yes, except for the end points of a diameter.

6. ∠A = ∠B } Isosceles Triangle
and ∠B = ∠C } ...theorem
That is, ∠A = ∠B = ∠C
An equilateral triangle is equiangular.

7. △PCA ≅ △PCB   ...SAS
Therefore, PA = PB

8. PB = PC and PC = PA   ...result of Exercise 7
Therefore, P is equidistant from all three vertices.

9. ∠QPR = ∠QPS + ∠RPS
∠QPS = ∠Q } ...Isosceles Triangle theorem
∠RPS = ∠R }
Therefore, ∠QPR = ∠Q + ∠R

10. a) ∠ABC = ∠ACB } ...Isosceles Triangle theorem
∠DCB = ∠DBC }
∠ABC + ∠DBC = ∠ACB + ∠DCB
That is, ∠ABD = ∠ACD

b) Yes

11. a) △APC ≅ △APB   ...SAS
Therefore, PC = PB
That is, △PBC is isosceles.

b) Yes

12. ∠ABC = ∠ACB   ...Isosceles Triangle theorem
∠ABE = 180° − ∠ABC, and ∠ACD = 180° − ∠ACB
Therefore, ∠ABE = ∠ACD
and △ABE ≅ △ACD   ...SAS
so that AE = AD

13. AC + CD = AE + EB
That is, AD = AB
Then △DAE ≅ △BAC
and BC = DE

14. $AE = \frac{1}{2}AB$, and $AD = \frac{1}{2}AC$
Since AB = AC, AE = AD,
and △ABD ≅ △ACE   ...SAS
Therefore, BD = CE

15. In △ADC and △ACB,
AC = AC
∠A = ∠A
CD = CB
But △ADC ≇ △ACB
So that SSA is not a congruence axiom.

Exercises 7-3

1. a) SSS   **b)** SSS   **c)** SAS
AC = FD   GH = KL   PQ = ST
CB = DF   HJ = LM   PR = SV
BA = EF   JG = MK   RQ = VT
∠A = ∠F   ∠G = ∠K   ∠P = ∠S
∠C = ∠D   ∠H = ∠L   ∠Q = ∠T
∠B = ∠E   ∠J = ∠M   ∠R = ∠V

2. a) △PQS ≅ △PRS   **b)** △CDE ≅ △ABE
PQ = PR   CD = AB
QS = RS   DE = BE
∠Q = ∠R   EC = EA
∠PSQ = ∠PSR   ∠C = ∠A
∠QPS = ∠RPS   ∠D = ∠B
                             ∠DEC = ∠BEA

Answers **459**

c) △JMK ≅ △JML
△JNK ≅ △JNL
△MNK ≅ △MNL
JK = JL
MK = ML
NK = NL
∠JNK = ∠JNL
∠JMK = ∠JML
∠JKN = ∠JLN
∠KJN = ∠LJN
∠JKM = ∠JLM
JN ⊥ KL

3. △ABC ≅ △CDA ...SSS
Therefore, ∠B = ∠D

4. PS ⊥ QR ...Perpendicular Bisector theorem
△PQS ≅ △PRS ...SSS
Therefore, ∠QPS = ∠RPS

5. △BDA ≅ △BDC ...SSS
Therefore, ∠ABD = ∠CBD

6. △OAB ≅ △OCD ...SSS
Therefore, ∠AOB = ∠COD

7. △WOX ≅ △ZOY ...SAS
Therefore, XW = YZ

8. a) AD ⊥ BC ...Perpendicular Bisector theorem
△ABD ≅ △ACD ...SSS
Therefore, ∠BAD = ∠CAD
b) Yes

9. △DAB ≅ △DCB ...SSS
Therefore, ∠A = ∠C,
∠ADB = ∠CDB,
and ∠ABD = ∠CBD

10. Since BC = BD and AC = AD, AB ⊥ CD ...
Perpendicular Bisector theorem
△CAB ≅ △DAB ...SSS
Therefore,
∠CAB = ∠DAB
and ∠CBA = ∠DBA

11. △AOD ≅ △COD ...SAS
△AOD ≅ △AOB ...SAS
△COD ≅ △COB ...SAS
Therefore, AD = DC,

AB = AD, DC = BC
ABCD is a rhombus.

12. Corresponding angles of △ABC and △DEC are equal, but △ABC ≆ △DEC. Therefore, AAA is not a congruence axiom.

13. △QPS ≅ △QRS ...SSS
Therefore, ∠PQT = ∠RQT
△PQT ≅ △RQT ...SAS
Therefore, PT = RT

Exercises 7-4

1. a) ASA   b) ASA   c) SAS

2. a) △ABC ≅ △CDA   b) △PQR ≅ △TSR
   c) △JLM ≅ △JKM   d) △AEB ≅ △CED

3. ∠ACB = ∠DCE ...vertically opposite
△ACB ≅ △DCE ...ASA
Therefore, ∠A = ∠D

4. △ABD ≅ △CBD ...ASA
Therefore, AB = CB and AD = CD.

5. a) ii), iii), ASA   b) i), ii), SAS
   c) i), iii), SSS   d) ii), iii), SAS
   e) i), ii), ASA   f) ii), iv), SAS

6. a) △AEB ≅ △ADC   ...SAS
      △BED ≅ △CDE   ...SSS
   b) △RQT ≅ △PST   ...ASA
      △PQT ≅ △RQT   ...ASA
      △PST ≅ △PQT   ...SAS
   c) △JKM ≅ △LKM   ...SSS
      △JKN ≅ △LKN   ...SAS
      △JMN ≅ △LMN   ...SAS

7. △KLM ≅ △NML   ...ASA
Therefore, KL = NM, and KM = NL.

8. △ABD ≅ △ACE ...ASA
Therefore, BD = CE

9. In △PQR, ∠Q = ∠R ...Isosceles Triangle theorem
Therefore, ∠SQR = ∠TRQ
△QRT ≅ △RQS... ASA
Therefore, QS = RT

10. △CAD ≅ △CBD   ...SAS
Therefore, CA = CB

460 Answers

That is, △ABC is isosceles.

11. ∠WOX = ∠ZOY ... vertically opposite
Therefore, △WOX ≅ △YOZ ... SAS
and, ∠WXY = ∠WZY

12. QS = RS ... Perpendicular Bisector theorem
△PQS ≅ △PRS ... SSS
Therefore, ∠QPS = ∠RPS

13. ∠POQ = ∠ROS ... vertically opposite
△POQ ≅ △ROS ... SAS
Therefore, PQ = RS

14. Let E be the midpoint of BC.
Then, △OEB ≅ △OEC ... SAS
and OE ⊥ BC
Therefore, AE = DE ... Perpendicular Bisector
Since AB = AE − BE   theorem
and CD = DE − CE
AB = CD

15. △PQR ≅ △SRQ ... SAS
Therefore, PR = SQ
△PSR ≅ △SPQ ... SSS
Therefore, ∠P = ∠S

16.

Let O be the centre of the circle.
∠LOG = ∠MOH ... vertically opposite
△LOG ≅ △MOH ... SAS
Therefore, LG = MH
∠LOJ = ∠MOK ... vertically opposite
△LOJ ≅ △MOK ... SAS
Therefore, JL = KM
∠JOG = ∠KOH ... vertically opposite
△JOG ≅ △KOH ... SAS
Therefore, JG = KH
and △GJL ≅ △HKM ... SSS

17.

OX = OW ... Perpendicular Bisector theorem
OY = OZ ... Perpendicular Bisector theorem
XY = WZ ... given
Therefore △OXY ≅ △OWZ

18.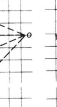

∠ABC = ∠ACB ... Isosceles Triangle theorem
Since ∠CBD = 180° − ∠ABC
and ∠ECA = 180° − ∠ACB
∠CBD = ∠ECA
△CBD ≅ △ECA ... SAS
Therefore, DC = AE

MM

1. An axiom is a self-evident property. A theorem is a property established on other proven properties.

2. No

3. Answers will vary.

Exercises 7-5

1. Alternate angles: ∠AGF and ∠GFD; ∠CFG and ∠FGB
Corresponding angles: ∠AGH and ∠CFG; ∠AGF and ∠CFE
∠HGB and ∠GFD; ∠BGF and ∠DFE

2. a) x = 38°, y = 142°   b) x = 55°, y = 125°
   c) x = 120°, y = 60°

3. a) x = 130°   b) x = 58°   c) x = 56°
   y = 50°   y = 128°   y = 35°
   z = 130°   z = 128°   z = 89°
   d) w = 120°   e) w = 65°   f) w = 125°
   x = 70°   x = 115°   x = 55°
   y = 70°   y = 65°   y = 125°
   z = 50°   z = 115°   z = 55°

4.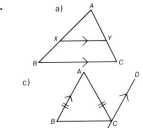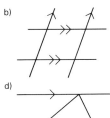

   a)   b)
   c)   d)

5. ∠CYX = ∠BXY ... Parallel Lines theorem
Therefore, XY ⊥ AB

6.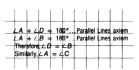

∠A + ∠D = 180° ... Parallel Lines axiom
∠A + ∠B = 180° ... Parallel Lines axiom
Therefore, ∠D = ∠B
Similarly, ∠A = ∠C

7. ∠AEC = ∠BED ... vertically opposite
∠A = ∠B   ... Parallel Lines theorem
AE = BE   ... given
Therefore, △AEC ≅ △BED ... ASA
and, EC = ED

8. ∠B = ∠C   ... Isosceles Triangle theorem
∠B = ∠DAE   ... Parallel Lines theorem
∠C = ∠CAE   ... Parallel Lines theorem
Therefore, ∠DAE = ∠CAE
That is, AE bisects ∠DAC.

Answers **461**

9. a) 50°   **b)** 70°   **c)** 90°
    130°   110°   90°
    130°   110°   90°

10. Draw PR.
    ∠SPR = ∠QRP   ...Parallel Lines theorem
    ∠SRP = ∠QPR   ...Parallel Lines theorem
    PR = RP
    Therefore, △SPR ≅ △QRP   ...ASA
    and SR = QP

11. In △DOC and △BOA
    AB = CD   ...Parallelogram's Sides theorem
    ∠D = ∠B and ∠C = ∠A   ...Parallel Lines theorem
    Therefore, △DOC ≅ △BOA   ...ASA
    and DO = BO and AO = CO
    That is, the diagonals bisect each other.

12. AP = QC   ...Parallelogram's Sides theorem
    ∠RAP = ∠RCQ ⎫ Parallel Lines
    ∠RPA = ∠RQC ⎭ theorem
    Therefore, △RAP ≅ △RCQ
    and AR = CR and RP = RQ
    That is, PQ and AC bisect each other.

13. ∠B = ∠AFE
    ∠BFD = ∠A   ...Parallel Lines theorem
    ∠BDF = ∠C = ∠FEA

Exercises 7-6

1. a) 77°   **b)** 110°   **c)** 117°
   d) 114° **e)** 25°   **f)** 33°
   g) 62°   **h)** 85°   **i)** 120°

3.   ∠A = ∠B   ...Isosceles Triangle theorem
    ∠A + ∠B = ∠ACD   ...Exterior Angle theorem
    Therefore, ∠ACD = 2∠A

4. a) $x = 35°, y = 125°$
   b) $x = 45°, y = 105°$
   c) $x = 97\frac{1}{2}°, y = 147\frac{1}{2}°$
   d) $x = 120°, y = 100°$
   e) $x = 135°$
   f) $x = 37\frac{1}{2}°, y = 145°$

5. a) 105°   **b)** 90°   **c)** 105°

6. ∠EAB = 180° − ∠DAB
    ∠FCB = 180° − ∠DCB
    ∠EAB + ∠FCB = 360° − (∠DAB + ∠DCB)
                = ∠B + ∠D

7. a)

Polygon	Number of Sides	Number of Triangles	Sum of Angles
Triangle	3	1	180°
Quadrilateral	4	2	360°
Pentagon	5	3	540°
Hexagon	6	4	720°
Octagon	8	6	1080°
Decagon	10	8	1440°

b) $S = (n − 2)180$

8. a)

Measure of Each Angle	Polygon	Number of Sides	Number of Triangles	Sum of Angles
60°	Triangle	3	1	180°
90°	Quadrilateral	4	2	360°
108°	Pentagon	5	3	540°
120°	Hexagon	6	4	720°
135°	Octagon	8	6	1080°
144°	Decagon	10	8	1440°

b) $A = \dfrac{(n − 2)180}{n}$

9. $x = 36°, y = 36°, z = 72°$

10. $(a + b) + (c + d) + (e + f) = 3 \times 180°$, or 540°
    But $a + c + e = 180°$
    Therefore, $b + d + f = 540° − 180°$, or 360°

11. ii)

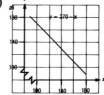 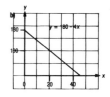

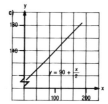

462 Answers

iii) a) $y = 270 - x$ b) $y = 180 - 4x$ c) $y = 90 + \frac{x}{2}$
As x increases,   As x increases,   As x increases,
y decreases. y decreases. y increases.

12. a) i) 360°   ii) 360°   iii) 360°
   b) The sum of the exterior angles of any polygon is 360°.
13. a) 360°   b) 360°
14. $\angle XBC + \angle YCB = 180°$ ...Parallel Lines axiom
   Therefore, $\angle EBC + \angle ECB = 90°$
   and $\angle BEC = 90°$ ...Angle Sum theorem
15. $\angle C = \angle F$ ...Angle Sum theorem
   Therefore, $\triangle ABC \cong \triangle DEF$ ...ASA
16. $\angle A = \angle OPA$, and $\angle B = \angle OPB$...Isosceles Triangle theorem
   Therefore, $\angle A + \angle B = \angle APB$
   But $\angle A + \angle B + \angle APB = 180°$
   $2\angle APB = 180°$
   $\angle APB = 90°$

Exercises 7-7

1. a) AAS   b) AAS   c) ASA   d) SSS
2. a) $CR = YP$ or, another pair of angles
   b) Any pair of corresponding sides equal
   c) $\angle D = \angle Q$, or $SM = PX$
   d) Any pair of corresponding sides equal
3. BC and EF are not corresponding sides.
4. a) $\triangle QPR \cong \triangle STR$ b) $\triangle ABC \cong \triangle EDC$
   c) $\triangle MJN \cong \triangle LKN$ d) $\triangle XUZ \cong \triangle XVY$
5. $\triangle ABC \cong \triangle DEC$ ...AAS
   Therefore, $BC = EC$
6. $\left.\begin{array}{l}\angle P = \angle S \\ \angle Q = \angle R\end{array}\right\}$ ...Parallel Lines theorem
   Therefore, $\triangle PQT \cong \triangle SRT$ ...ASA
   So that $PT = ST$ and $QT = RT$
   That is, PS and QR bisect each other.
7. $\angle ABC = \angle ACB$ ...Isosceles Triangle theorem
   $\triangle CBD \cong \triangle BCE$ ...AAS
   Therefore, $BD = CE$

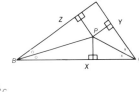

8. $\left.\begin{array}{l}PX = PY \\ PX = PZ\end{array}\right\}$ ...Angle Bisector theorem
   Therefore, $PY = PZ$ and P is equidistant from all three sides.
   So that $PT = ST$ and $QT = RT$
9. Given: $\triangle ABC \cong \triangle DEF$
   Therefore, $\angle B = \angle E$
   and, $\triangle CGB \cong \triangle FHE$ ...AAS
   Therefore, $CG = FH$

10. $\triangle AEC \cong \triangle BCD$ ...AAS
    Therefore, $AC = BC$
    That is, $\triangle ABC$ is isosceles

PSS Exercises

1. If $PQ = PR$, then $\angle Q = \angle R$.
   But $\angle Q \neq \angle R$. Therefore, $PQ \neq PR$.
2. $\angle A + \angle B + \angle C = 180°$
   a) If $\angle A = \angle B = 90°$, then $\angle C = 0°$
      Therefore, either $\angle A$ or $\angle B \neq 90°$
   b) If $\angle A > 90°$ and $\angle B > 90°$,
      then $\angle C < 0°$.
      Therefore, a triangle cannot have two obtuse angles.
3. If $PQ \perp l$, and $PR \perp l$,
   then $\angle Q = \angle R = 90°$
   But in $\triangle PQR$, $\angle P + Q + R = 180°$
   So that $\angle P = 180° - 90° - 90°$, or $0°$
   Therefore, Q and R must coincide,
   and there is only one perpendicular
   from P onto line l.

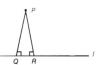

4. Assume that n is odd.
   Then, $n = 2k + 1$ where k is an integer and
   $n^2 = 4k^2 + 4k + 1$
   The first two terms are even, therefore n^2 is odd. But n^2 is even. Therefore, n must be even.
5. If both m and n are even, mn is even.
   If m is odd and n is even, mn is even.
   Therefore, if mn is odd, both m and n must be odd.
6. If $AB = AC$, then $\triangle AMB \cong \triangle AMC$ ...SSS
   and $\angle AMB = \angle AMC = 90°$

But ∠AMC = 60°

Therefore, AB ≠ AC

7. If PQ = PR,

then △PQS ≅ △PRS ...HS Congruence theorem

and QS = RS.

But QS ≠ RS.

Therefore, PQ ≠ PR.

8. If EH and FG bisect each other at O, △EGO ≅ △FHO ... SAS

and EG = FH. This means that DE = DF.

But DE ≠ DF.

Therefore, EH and FG do not bisect each other.

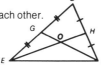

Exercises 7-8

1. a) If you live in Canada, you live in Vancouver.
  b) If water is frozen, it is ice.
  c) If $x^3 = 8$, then $x = 2$.
  d) If $|x| > 5$, then $x^2 > 25$.
  e) If two triangles have equal angles, they are congruent.
  f) If a right triangle has a 60° angle, then it also has a 30° angle.
  g) Every student who takes mathematics is in this room.
  h) All that swim are fish.
  i) Lines that intersect at right angles are perpendicular.
  j) Every parallelogram is a rhombus.

2. a) b, c, d, f, i
  b) Water is ice if, and only if, it is frozen.
  c) $x^3 = 8$ if, and only if, $x = 2$.
  d) $x^2 > 25$ if, and only if, $|x| > 5$.
  f) A right triangle has a 30° angle if, and only if, it has a 60° angle.
  i) Lines are perpendicular if, and only if, they are at right angles.

3.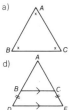

4. a) If an animal is a dog, then it is four-legged.
  b) If persons are students, then they like hamburgers.
  c) If a prime number is greater than 2, it is odd.
  d) If a number is a perfect square, then it has an odd number of factors.
  e) If a triangle is equilateral, then it is isosceles.
  f) If a triangle is right, then it has two acute angles.

5. a) If an animal is four-legged, it is a dog. F
  If persons like hamburgers they are students. F
  If a number is odd, it is a prime number greater than 2. F
  If a number has an odd number of factors, it is a perfect square. T
  If a triangle is isosceles, it is equilateral. F
  If a triangle has two acute angles, it is a right triangle. F

  b) A number is a perfect square if, and only if, it has an odd number of factors.

6. a) If you can drive a car, then you have insurance.
  If you have insurance, then you can drive a car. F
  b) If shadows are seen, the sun is shining.
  If the sun is shining, shadows are seen. T
  c) If $x + 5$ is even, then x is odd.
  If x is odd, then $x + 5$ is even. T
  d) If a line is perpendicular to $y = 2x + 3$, its slope is $-\frac{1}{2}$.
  If a line's slope is $-\frac{1}{2}$, it is perpendicular to $y = 2x + 3$. T
  e) If a quadrilateral has four right angles, it is a rectangle.
  If a quadrilateral is a rectangle, it has four right angles. T
  f) If two rectangles have equal areas, they have equal lengths.
  If two rectangles have equal lengths, they have equal areas. F

7. ∠ABE + ∠ABC = 180°,

and ∠ACD + ∠ACB = 180°

Since ∠ABE = ∠ACD ...given

∠ABC = ∠ACB

Therefore, AB = AC ...Converse of Isosceles Triangle theorem

8. ∠ACD = ∠A + ∠ABC ⎫
∠EBC = ∠A + ∠ACB ⎬ ... Exterior Angle theorem

Therefore, ∠A + ∠ABC = ∠A + ∠ACB

That is, ∠ABC = ∠ACB

and, $AB = AC$ ...Converse of Isosceles Triangle theorem

9. $\triangle CAB \cong \triangle CED$ ...SAS

Therefore, $\angle A = \angle E$

and, $AB \parallel DE$ ...Converse of Parallel Lines theorem

10. In $\triangle ABD$, $\angle ABD = \angle ADB$ ...Isosceles Triangle theorem

In $\triangle CDB$, $\angle CDB = \angle ADC - \angle ADB$

and, $\angle CBD = \angle ABC - \angle ABD$

Therefore, $\angle CDB = \angle CBD$

and, $BC = DC$ ...Converse of Isosceles Triangle theorem

11. $\angle CBD = \angle CDB$ ...Isosceles Triangle theorem

Since $\angle ABD = \angle CBD$, $\angle ABD = \angle CDB$

and $\angle AB \parallel DC$ ...Converse of Parallel Lines theorem

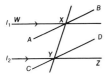

12. $\angle WXY = \angle ZYX$ ...Parallel Lines theorem

Therefore, $\angle AXY = \angle DYX$

and, $AB \parallel CD$ ...Converse of Parallel Lines theorem

13.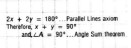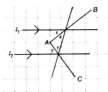

14. a)

b)

15. a)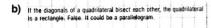

b) If the diagonals of a quadrilateral bisect each other, the quadrilateral is a rectangle. False. It could be a parallelogram.

16.
$\angle B = \angle C$ ...Isosceles Triangle theorem
$\angle EAC = \angle B + \angle C$ ...Exterior Angle theorem
$2\angle EAD = 2\angle B$
$\angle EAD = \angle B$
Therefore, $AD \parallel BC$ ...Converse of Parallel Lines theorem

17. a) $2x = 180° - \angle A$
$2y = 180° - \angle A$ ...Angle Sum theorem

Therefore, $x = y$

and $BC \parallel DE$ ...Converse of Parallel Lines theorem

b) $2x = 180° - z$
$2y = 180° - z$ ...Angle Sum theorem

Therefore, $x = y$

and $BC \parallel DE$ ...Converse of Parallel Lines theorem

18. Answers will vary.
  a) Isosceles Triangle theorem
  b) If you live in Canada, you live in Singapore.
  c) All right angles are equal.
  d) If you live in Canada, you live in Edmonton.

MAU

1. a) 145
   b) Answers will vary.
   c) Age 15: 550 m
2. a) Age 15: 180 m
   b) Answers will vary.
   c) 440 years
3. 48.5 m/year
4. Rates of flow and recession are constant.

Exercises 7-9

1. a) 4.70   b) 26.66   c) 26.83
2. c), d), f), h), i)

3. 164.5 cm
4. 11 m
5. a) Yes b) No c) No
6. a) $4\sqrt{2}$ cm b) 62.5%
7. a) $\sqrt{10}$ cm b) 63.7%
8. a) $(s^2 - t^2)^2 + 4s^2t^2 = s^4 + t^4 - 2s^2t^2 + 4s^2t^2$
$= s^4 + t^4 + 2s^2t^2$
$= (s^2 + t^2)^2$

   b) Answers will vary.
   3, 4, 5;   5, 12, 13;   6, 8, 10;
   8, 15, 17;   24, 45, 51

9. a) $AB = \sqrt{12^2 + 16^2}$, or 20   b) $AB = \sqrt{12^2 + 9^2}$, or 15
   $AE = \sqrt{12^2 + (16-7)^2}$, or 15 $AE = \sqrt{12^2 + (9+7)^2}$, or 20
   $BE = \sqrt{20^2 + 15^2}$, or 25 $BE = \sqrt{20^2 + 15^2}$, or 25
   $= 16 + 9$ $= 18 + 7$
   Therefore, $\angle BAE = 90°$ Therefore, $\angle BAE = 90°$

10. a) $PQ = \sqrt{13^2 - 12^2}$, or 5
    $5^2 = 3^2 + 4^2$
    $\angle A = 90°$

    b) $PS = \sqrt{17^2 - (21-6)^2}$, or 8
    $PA^2 = 6^2 + 8^2$
    $= 26^2 - 24^2$
    Therefore, $\angle A = 90°$

    c) $PQ^2 = 7^2 + 4^2$, or 65
    $= 8^2 + 1^2$
    Therefore, $\angle A = 90°$

11. $\triangle PSR \cong \triangle PQR$ ...HS Congruence theorem
    Therefore, $RS = RQ$

12. a) $AD^2 = 13^2 - 5^2$
    $= 15^2 - 9^2$
    Therefore, $\angle BAD = 90°$
    and $AB \parallel CD$ ...Converse of Parallel Lines theorem

    b) $AD^2 = 10^2 - 6^2$
    $= 17^2 - 15^2$
    Therefore, $\angle BAD = 90°$
    and $AB \parallel CD$ ...Converse of Parallel Lines theorem

13. a) $c^2 = 4\left(\frac{1}{2}ab\right) + (a-b)^2$
    $= 2ab + a^2 + b^2 - 2ab$
    $= a^2 + b^2$

    b) $(a+b)^2 = 4\left(\frac{1}{2}ab\right) + c^2$
    $a^2 + b^2 + 2ab = 2ab + c^2$
    $a^2 + b^2 = c^2$

14. $\triangle BAF \cong \triangle BAE$ ...SAS
    $\triangle BAE = \triangle BCA + \triangle ACE$
    $\triangle BCA \cong \triangle BCD$ ...SAS
    Therefore, $\triangle BAF = \triangle BCD + \triangle ACE$

15. a) $c^2 = a^2 + b^2$ ...right triangle
    $c^2 > a^2 + b^2$ ...obtuse triangle
    $c^2 < a^2 + b^2$ ...acute triangle

    b) i) Right   ii) Acute   iii) Obtuse

Exercises 7-10

1. Under the translation mapping DC onto AB, $D \to A$ and $C \to B$. Length is invariant. Therefore, $AD = BC$ and $AD \parallel BC$.

2. Under the translation mapping A onto D, $B \to E$ and $C \to F$. Therefore, $\triangle ABC \cong \triangle DEF$.

3. Translate QR so that $Q \to P$. Then, sum of the angles about P equals the sum of the external angles, or 360°.

4. Under a reflection about the perpendicular bisector, one endpoint of a line segment maps onto the other. Therefore, any point on this perpendicular bisector of the segment is equidistant from its endpoints.

5. Under a reflection about PR, $\triangle PSR \to \triangle PQR$. Angle measure is invariant. Therefore, $\angle SPR = \angle QPR$.

6. Since $PT = QT$ and $RT = ST$, a line through T perpendicular to PQ must also be perpendicular to SR. Reflected in this line, $P \to Q$ and $S \to R$. Therefore, $\triangle PQS \cong \triangle QPR$.

7. a) Under a reflection about AC, $E \to F$ and $CE \to CF$. Length is invariant. Therefore, $CE = CF$.

    b) Under a reflection about AC, $E \to F$ and $BE \to DF$. Length is invariant. Therefore, $BE = DF$.

8. Under a rotation, angle measure is invariant. Therefore, under $\frac{1}{2}$-turn about O, $\angle ZOX \to \angle YOW$. That is, vertically opposite angles are equal.

9. Under $\frac{1}{2}$-turn about O, the midpoint of AB, $YZ \to XW$. Angle measure is invariant. Therefore, $\angle BAY = \angle ABX$ and $\angle BAZ = \angle ABW$.

10. Under $\frac{1}{2}$-turn about O, $\triangle ABC \to \triangle CDA$

Therefore, $AB \parallel CD$ and $AB = CD$
Also, $BC \parallel DA$ and $BC = DA$

11. Under a 120° rotation about the centroid of $\triangle ABC$,
    $CDB \to AEC$ and $A \to B$.
    Therefore, $AD = BE$. $\triangle CAD \to \triangle ABE$.
    Therefore, $\angle AFB = 120°$ and $\angle BFD = 60°$.

12. Under a 60° rotation about C, $\triangle ADC \to \triangle BEC$. Therefore,
    $BE = AD$ and $\angle AFB = 60°$.

13. $PS = \frac{1}{2}PQ$, $PT = \frac{1}{2}PR$.
    Therefore, $ST = \frac{1}{2}QR$.
    $ST \parallel QR$ since slope is invariant under a dilatation.

14. With scale factor 2 and centre B, $\triangle BFD \to \triangle BAC$.
    With scale factor $\frac{1}{2}$ and centre A, $\triangle BAC \to \triangle FAE$.
    Therefore, $\triangle BFD \cong \triangle FAE$.
    Similarly, $\triangle BFD \cong \triangle DEC$.
    Since slope is invariant under a dilatation, $FD \parallel EC$ and $FE \parallel DC$.
    Therefore, $\triangle DEC \cong \triangle EFD$.

Review Exercises

1. a) Jeffrey is poor.
   b) $\angle P = \angle R$, and $\angle Q = \angle S$.
   c) $\triangle KLM$ is isosceles

2. a) Yes   b) No   c) No

3. a) $\angle PQR = \angle PRQ$ ...Isosceles Triangle theorem
      Therefore, $\angle PQS = \angle PRT$
      $\triangle PQS \cong \triangle PRT$ ...SAS
      Therefore, $PS = PT$
   b) $\angle K = \angle NMK$
      $\angle L = \angle NML$ ...Isosceles Triangle theorem
      But $\angle NMK + \angle NML = \angle KML$
      Therefore, $\angle KML = \angle K + \angle L$

4. a) $\triangle LMJ \cong \triangle JKL$ ...SSS
      Therefore, $\angle M = \angle K$
   b) $\triangle ABD \cong \triangle ACD$ ...SSS
      Therefore, $\angle BAD = \angle CAD$
      That is, AD bisects $\angle A$.

5. $\triangle BEC \cong \triangle BDA$ ...ASA
   Therefore, $EB = DB$

6. $\angle P = \angle Q$ ...Isosceles Triangle theorem

   $\triangle TPS = \triangle UQS$ ...ASA
   Therefore, $ST = SU$

7. a) $w = 125°$, $x = 65°$, $y = 65°$, $z = 60°$
   b) $w = 100°$, $x = 80°$, $y = 100°$, $z = 80°$

8. $\angle P = \angle S$ and $\angle Q = \angle R$ ...Parallel Lines theorem
   $PT = ST$ ...given
   Therefore, $\triangle PTQ \cong \triangle STR$ ...AAS
   and $RT = QT$

9. a) $x = 60°$, $y = 110°$
   b) $x = 90°$, $y = 120°$
   c) $x = 45°$, $y = 150°$

10. $\triangle PQT \cong \triangle SQR$ ...AAS
    Therefore, $SQ = PQ$

11. a) If angles are right angles, then they are equal.
    b) If triangles are congruent, then their corresponding angles are equal.
    c) If lines are perpendicular to each other, then they form right angles.
    d) If triangles are congruent, then their corresponding sides are equal.

12. a) If angles are equal, then they are right angles. False
       If their corresponding angles are equal, then triangles are congruent. False
       If intersecting lines form right angles, then they are perpendicular. True
       If their corresponding sides are equal, triangles are congruent. True
    b) Intersecting lines form right angles if, and only if, they are perpendicular to each other.
       All corresponding sides of triangles are equal if, and only if, the triangles are congruent.

13. $\angle PQS = \angle RQS$ ...given
    $\angle PQS = \angle RSQ$ ...alternate angles
    Therefore, $\angle RSQ = \angle RQS$
    and $SR = QR$ ...Converse of Isosceles Triangle theorem

14. a), c)

15. $\triangle ABC \cong \triangle ABD$ ...HS theorem
    Therefore, $\angle CAE = \angle DAE$
    $\triangle CAE \cong \triangle DAE$ ...SAS
    Therefore, $CE = DE$

Exercises 8-1

1. a) i) 1 ii) 3 iii) 2
   b) 7 c) 67 d) 25
2. a) i) 1 ii) 0 iii) 2
   b) 6 c) 3 d) 8
3. a) i) 2 ii) 0 iii) 3 iv) 1
   b) i) 5 ii) 9 c) 1809 kg
4. a) 235 s b) i) 9 ii) 13 iii) 23
   c) 19 d) 5
5. a) 14 | 9 8 9
      15 | 0 0 5 9 5 6 3 8 7
      16 | 2 8 1 8 4 8 3 0
      17 | 6 1 0 2 5 2 3

   b) 14 | 9 8 9
      15 | 0 0 3
      15 | 5 9 5 6 8 7
      16 | 2 1 4 3 0
      16 | 8 8 8
      17 | 1 0 2 2 3
      17 | 6 5

6. a) 34 b) 14 c) Answers will vary.
7. a) 30 b) 6.7%
8. a) i) 3 ii) 11 iii) 50
   b) i) 4 and 35 ii) 67 c) 120
9. a) i) −25°C ii) −33°C b) 40 km/h
10. a) i) 22.5 km ii) 28 km
    b) i) 18 m ii) 70 m
11. a) i) 152, 161 ii) 140, 150 iii) 130, 140
    b) i) 39, 51 ii) 52, 64

Exercises 8-2

1.
	Mean	Median	Mode
a)	5.6	6	6
b)	31	32	25
c)	5.3	5.3	6.3
d)	100.5	99	—
e)	65.7	65	58

2. 179, 178, 182
3. $245.24, $200, $200
4. a) 52, 53.5, 55 b) 188, 188.5, 192
5. a) 0.276 b) 0.283 c) No
6. a) $700 000 b) 770 c) $909
7. a) $52.93 b) $25 c) $25
8. a) i) 322.5 ii) 319 b) i) 296.9 ii) 308
9. a) 0.394 b) 0.390 c) 5 d) 0.406

10. Each is multiplied by 3.

Exercises 8-3

1. a) 24°C, 24°C, 23°C
   b) 180.3 cm, 180.5 cm, 180.5 cm
2. a) $224, $225, $175
   b) 2.74 years, 2.5 years, 3.5 years
3. a) $21, $15, $15 b) 16, 15, 15
4. a) $48.50, $50, $30 b) 5.1, 5, 1
5. a) 9.7, 10, 10 b) $1.63, $1.50, $1.75
6. 21
7. Mean: no change
   Median: may change
   Mode: may change
8. Answers will vary. Examples are:
   a) 10, 11, 11, 13, 13, 13, 13
   b) 58, 58, 58, 65, 65, 65, 65
   c) 0.5, 5, 5, 10, 11, 12, 12.5
   d) 9, 10, 11, 12, 15, 15, 33

Exercises 8-4

1. a) 8, 2.4 b) 10, 2.4 c) 15, 4
   d) 8, 2.7 e) 15, 3.14 f) 6, 1.56
2. a) 16, 4.4 b) 32, 7.6 c) 32, 10
   d) 16, 5.6 e) 7, 2.4 f) 20, 7.2
3. 11, 2.47
4. a) 40, 8.16 b) 36, 8.4 c) 24, 5.2
   d) 4, 0.94 e) 8, 1.71
5. 26, 5.48
6.
	Mean	Median	Mode	Range	Mean Deviation
a)	16.3	16	15	16	3.8
b)	6	7	7	10	2.1
c)	9	8	8, 13	10	2.9
d)	$320	$300	$300	$645	$46
e)	16	15	14	9	2.15

MAU

1. a) Buffaloes remain within the park; none die and none are born; tagged and untagged buffaloes are thoroughly mixed.
   b) 1275

468 Answers

2. 1250
3. 3800

Exercises 8-5

1. 0.49
2. 0.163, 0.170, 0.159, 0.162, 0.177, 0.169. It was a fair die because each outcome was close to $\frac{1}{6}$.
3. 0.026
4. d) It should approximate 0.5 more and more closely.
5. 0.38
6. Answers will vary.
7. e) Answers will vary, but should approximate:
  i) 200, **ii)** 400, **iii)** 40.
8. c) Answers will vary, but should approximate:
  i) 125, **ii)** 42, **iii)** 83.
9. 621
10. O — 889, A — 700, B — 166, AB — 95
11. a) A **b)** A
12. Red — 391, Blue — 234, Yellow — 625
13. Answers will vary.

Exercises 8-6

1. a) $\frac{1}{4}$ **b)** $\frac{1}{2}$ **c)** $\frac{1}{13}$ **d)** $\frac{1}{26}$ **e)** $\frac{3}{26}$
2. a) $\frac{2}{11}$ **b)** $\frac{1}{11}$ **c)** $\frac{4}{11}$ **d)** $\frac{3}{11}$ **e)** $\frac{6}{11}$
3. a) $\frac{1}{120}$ **b)** $\frac{1}{200}$ **c)** $\frac{1}{315}$
4. a) $\frac{1}{100\,000}$ **b)** $\frac{1}{10\,000}$ **c)** $\frac{3}{2500}$
  d) 0 **e)** 1
5. a) $\frac{1}{6}$ **b)** $\frac{1}{2}$ **c)** $\frac{1}{3}$ **d)** $\frac{1}{2}$
6. a) $\frac{1}{8}$ **b)** $\frac{1}{2}$ **c)** $\frac{1}{2}$ **d)** $\frac{1}{4}$
  e) $\frac{3}{8}$ **f)** $\frac{7}{8}$ **g)** 0 **h)** 1
7. a) $\frac{1}{8}$ **b)** $\frac{1}{2}$ **c)** $\frac{1}{12}$ **d)** $\frac{1}{24}$
8. a) 8 **b)** 83 **c)** 833
9. a) 6 **b)** 6 **c)** 2
10. No. The outcomes are not equally likely.
11. a) $\frac{3}{10}$ **b)** $\frac{1}{4}$ **c)** $\frac{9}{20}$
  d) $\frac{11}{20}$ **e)** $\frac{3}{4}$ **f)** 0
12. B

13. $\frac{2}{5}$
14. a) 500 **b)** 77 **c)** 250 **d)** 19
15. a) $\frac{6}{25}$ **b)** $\frac{12}{25}$ **c)** 0 **d)** $\frac{2}{25}$
16. a) i) $\frac{1}{25}$ **ii)** $\frac{51}{100}$ **iii)** $\frac{3}{20}$
  b) 11 000 **c)** 350
17. a) 0.85 **b)** 0.56 **c)** 1

Exercises 8-7

1. a) $\frac{1}{8}$ **b)** $\frac{1}{4}$

2. $\frac{1}{36}$ **3.** $\frac{1}{16}$
4. a) $\frac{1}{8}$ **b)** $\frac{1}{64}$
5. a) $\frac{3}{10}$
  b) i) $\left(\frac{3}{10}\right)^4$, or 0.0081 **ii)** $\left(\frac{7}{10}\right)^4$, or 0.24
6. a) 0.64
  b) 0.04
7. $\left(\frac{1}{5}\right)^4$, or $\frac{1}{625}$
8. a) $\frac{25}{64}$ **b)** $\frac{5}{14}$
9. a) $\frac{9}{25}; \frac{1}{3}$ **b)** $\frac{16}{81}; \frac{1}{6}$ **c)** $\frac{9}{25}; \frac{33}{95}$
10. a) $\frac{1}{3}$ **b)** $\frac{1}{6}$ **c)** $\frac{1}{18}$
11. a) $\frac{1}{16}; \frac{1}{17}$ **b)** $\frac{1}{169}; \frac{1}{221}$ **c)** $\frac{9}{169}; \frac{11}{221}$
12. a) 0.0002
  b) 0.013
  c) 0.174 or, if Y is a vowel, 0.191
13. a) $\frac{6}{55}$ **b)** $\frac{21}{55}$ **c)** $\frac{1}{55}$
14. a) $\frac{1}{4}$ **b)** $\frac{3}{16}$ **c)** $\frac{9}{16}$
15. a) $\frac{1}{144}$ **b)** $\frac{1}{12}$ Assuming months have the same number of days.
16. a) $\left(\frac{1}{6}\right)^5$ **b)** $\left(\frac{5}{6}\right)^5$ **c)** $\left(\frac{2}{3}\right)^5$ **d)** $\left(\frac{1}{6}\right)^4$
17. a) $\frac{16}{221}$ **b)** $\frac{376}{5525}$
18. a) $\frac{13}{68}$ **b)** $\frac{247}{1700}$

Answers

Exercises 8-8
Answers will vary.

Exercises 8-9
1. 6 **2.** 0.375 **3.** 0.48 **4.** $\frac{5}{12}$
5. 0.210 **6.** 0.246 **7.** 0.383 **8.** 0.185
9. a) 0.692 **b)** 0.838 **10. a)** 0.008 **b)** 0.240
11. 0.691 **12. a)** 0.3125 **b)** 0.656

Review Exercises
1. a) i) 2 **ii)** 3 **iii)** 0
  b) 10 **c)** 67 **d)** 30
2. a) 175 **b)** 85 **c)** 45
3. a) 5.8, 6, 6 **b)** 42.55, 42, 45 **c)** 4.31, 4.05, 3.8
4. a) 51.17, 53.5, 42 **b)** 10.03, 10, 10

	Range	Mean Deviation
5. a)	5	1.24
b)	14	3.69
c)	1.6	0.51
6. a)	24	6.138
b)	4	0.743

7. 0.027
8. a) $\frac{1}{4}$ **b)** $\frac{1}{2}$ **c)** $\frac{1}{13}$ **d)** $\frac{1}{26}$ **e)** $\frac{3}{26}$
9. a) $\frac{1}{5}$ **b)** $\frac{3}{10}$ **c)** $\frac{1}{2}$ **d)** $\frac{7}{10}$
  e) $\frac{4}{5}$ **f)** 0
10. a) $\frac{1}{6}$ **b)** $\frac{1}{3}$ **c)** $\frac{1}{2}$ **d)** $\frac{1}{3}$
11. a) $\frac{49}{144}$ **b)** $\frac{7}{22}$
12. a) $\frac{1}{16}, \frac{1}{17}$ **b)** $\frac{1}{169}, \frac{1}{221}$ **c)** $\frac{9}{676}, \frac{5}{442}$
13. a) $\frac{3}{14}$ **b)** $\frac{3}{14}$ **c)** $\frac{1}{28}$
14. $\frac{2}{2235}$
15. a) $\frac{1}{4}$ **b)** $\frac{3}{16}$ **c)** $\frac{3}{8}$

Cumulative Review (Chapters 6–8)
1. a) (7, 6) **b)** (−7, −5)

2. a) b)

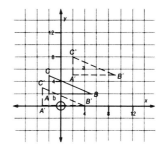

3.

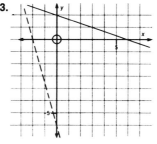

4. a) i) A reflection in the y-axis.
  ii) A reflection in $x + y = 0$.
b)

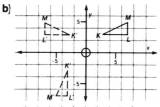

5.

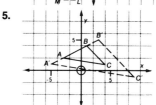

6. a) No **b)** Yes
7. a) $\triangle YXZ \cong WXZ$ …SAS
  Therefore, $\angle YXZ = \angle WXZ$
b) $\triangle MKN \cong \triangle MLN$ …SSS
  Therefore, $\angle MNK = \angle MNL = 90°$
   …since $\angle KNL = 180°$
  That is, $MN \perp KL$
8. a) $\angle CAB = \angle CBA$ …Isosceles Triangle theorem
  $\angle EAB = \angle DBA$ …given
  $AB = AB$
  Therefore $\triangle DAB \cong \triangle EBA$ …ASA
  and $BD = AE$

b) $\triangle ABC \cong \triangle EBD$ ...AAS
Therefore, $CB = DB$

9. a) $w = 70°, x = 60°$
$y = 70°, z = 60°$
b) $w = 105°, x = 75°$
$y = 105°, z = 75°$

10. a) $x = 75°, y = 130°$ b) $x = 37.5°, y = 145°$

11. a) If a quadrilateral is a rectangle, then it has four right angles.
b) If triangles are congruent, then they are equal in area.
c) If an angle is obtuse, then its measure is between 90° and 180°.

12. a) If a quadrilateral has four right angles, then it is a rectangle.
If triangles are equal in area, then they are congruent.
If the measure of an angle is between 90° and 180°, then the angle is obtuse.
b) A quadrilateral is a rectangle if, and only if, it has four right angles.
An angle is obtuse if, and only if, its measure is between 90° and 180°.

13. a), b)

14. a) 30   b) 14   c) 72.27, 73, 66
    d) 43, 9.72

15. 0.027

16. a) $\frac{1}{4}$ b) $\frac{1}{2}$ c) $\frac{1}{13}$ d) $\frac{1}{26}$ e) $\frac{3}{26}$

17. a) $\frac{14}{25}$ b) $\frac{6}{25}$ c) $\frac{1}{5}$
    d) $\frac{11}{25}$ e) $\frac{4}{5}$ f) 0

18. a) $\frac{4}{25}$ b) $\frac{1}{7}$

MAU

1. 776 m²
2. a) 28 m²   b) 47 m²   c) 78.5 m²
3. 3104
4. a) 113   b) 188   c) 314
5. a) 0.036   b) 0.061   c) 0.101

Exercises 9-1

1. a) 34°   b) 45°   c) 59°
   d) 83°   e) 23°   f) 50°
2. a) 37°   b) 22°   c) 31°
   d) 53°   e) 68°   f) 59°
3. a) 0.84   b) 2.1   c) 3.7
   d) 0.36   e) 0.58   f) 1.43
4. a) 39°   b) 35.5°   c) 59°
   d) 49.6°   e) 20°   f) 15°
5. a) 33.7°   b) 32°   c) 24°
   d) 67.4°   e) 16°   f) 73.7°
6. a) 0.466   b) 2.605   c) 0.087
   d) 1.000   e) 1.732   f) 57.290
7. 27°, 63°
8. 39°, 39°, 102°
9. 55 mm
10. a) 45°   b) 33.7°   c) 36.9°
11. a) 153°   b) 124°   c) 112°
12. 67.4°
13. a) 79°, 76°, 72°, 64°, 45°, 27°, 18°, 14°, 11°
    b) i) 6   ii) 12   iii) 29   iv) 58
    c) i) 6   ii) 12   iii) 29   iv) 58

Exercises 9-2

1. a) 0.533, 1.875   b) 0.750, 1.333   c) 0.952, 1.050
2. a) 28°, 62°   b) 37°, 53°   c) 44°, 46°
3. a) 6.46   b) 15.51   c) 37.32
   d) 4.08   e) 3.54   f) 1.69
4. a) i) 3.90   ii) 8.72   iii) 16.52
   b) i) 14°   ii) 26.6°   iii) 36.9°
5. a) i) 1.82   ii) 2.89   iii) 4.20
   b) i) 58°   ii) 67.4°   iii) 72.6°
6. a) 69.3 m   b) 80.0 m   c) 40 m
7. a) About 6.4 km   b) 1°
8. a) 1176 m   b) 63.4°
9. a) About 179 m   b) 57°
10. About 107 m
11. 38°
12. $\tan \theta, \sqrt{1 + \tan^2 \theta}$

Exercises 9-3

1.

		sin	cos	tan
a)	i) ∠A	0.385	0.923	0.417
	ii) ∠C	0.923	0.385	2.400
b)	i) ∠A	0.800	0.600	1.333
	ii) ∠C	0.600	0.800	0.750
c)	i) ∠A	0.862	0.508	1.697
	ii) ∠C	0.508	0.862	0.589

2. a) 23°, 67°   b) 53°, 37°   c) 59°, 31°
3. a) 24°, 66°   b) 25°, 65°   c) 42°, 48°
   d) 48°, 42°   e) 53°, 37°   f) 56°, 34°
4. a) 5.3, 8.5   b) 5.1, 14.1   c) 2.5, 7.6
   d) 3.5, 11.5   e) 13.4, 14.9   f) 15.1, 9.8
5. a) 5.26   b) 23.85   c) 28.31
   d) 9.01   e) 17.93   f) 31.73
6. a) i) 21.4, 12.9   ii) 24.5, 5.2
   b) i) 24°, 66°   ii) 47°, 43°
7. a) i) 28.3, 24.0   ii) 22.4, 16.7
   b) i) 58°, 32°   ii) 59°, 31°
8. a) 9.56 m   b) 2.92 m
9. a) 14.27 m   b) 53°
10. 0.812, 0.718
11. 0.733, 1.078
12. 0.818, 0.576
13. 97°, 41.5°, 41.5°, 39°, 70.5°, 70.5°
14. 7.3 m
15. 17.3 cm
16. a) In a right triangle, the hypotenuse is always the longest side.
    b) The side opposite θ is longer than the side adjacent to θ.
    c) i) Both are 0.5.   ii) Both are 0.985. The sine of an angle is the same as the cosine of its complement.
17. a) $AB = \cos\theta$; $BC = \sin\theta$.
    b) By the Pythagorean theorem,
    $AB^2 + BC^2 = AC^2$
    $\cos^2\theta + \sin^2\theta = 1$

Exercises 9-4

1. a) 9.4, 32°, 58°   b) 10.4, 60°, 30°   c) 14.7, 52°, 38°
   d) 21.0, 51°, 39°   e) 26.1, 42°, 48°   f) 20.6, 39°, 51°
2. a) 40°, 10.9, 8.3   b) 25°, 35.3, 14.9   c) 55°, 14.7, 10.3
   d) 65°, 34.2, 14.4   e) 35°, 14.3, 20.5   f) 68°, 20.4, 8.2
3. a) 63°, 8.9, 4.5   b) 55°, 18.3, 10.5   c) 40°, 15.3, 12.9
   d) 28°, 213, 188   e) 34°, 56°, 48.3   f) 43°, 47°, 32.8
4. 26.0 m
5. 432 m
6. 73 m, assuming the string is straight.
7. 10.2 cm
8. 5.1 cm
9. 24 km
10. a) 74°   b) 74°
11. 2.9 m, 7.7 m
12. 58.8 cm
13. 119°

MAU

Banff Sulphur Mountain Gondola Lift, 27.8°; Lake Louise Gondola, 8.6°; Jasper Tramway, 17.2°

Exercises

1. a) $0.\overline{291\,139\,240\,506\,3}$
   b) $2.2\overline{26\,190\,47}$
   c) $1.\overline{652\,173\,913\,043\,478\,260\,869\,5}$
   d) $0.\overline{560\,105\,680\,317\,040\,951\,122\,853\,368}$
2. a) 17ths have one group of 16 repeating digits:
   $0.\overline{058\,823\,529\,411\,764\,7}$
   b) 41sts have 8 groups of 5 repeating digits:
   $0.\overline{024\,39}$ for numerators: 1, 10, 16, 18, 37
   $0.\overline{048\,78}$ for numerators: 2, 20, 32, 33, 36
   $0.\overline{073\,17}$ for numerators: 3, 7, 13, 29, 30
   $0.\overline{097\,56}$ for numerators: 4, 23, 25, 31, 40
   $0.\overline{121\,95}$ for numerators: 5, 8, 9, 21, 39
   $0.\overline{146\,34}$ for numerators: 6, 14, 17, 19, 26
   $0.\overline{268\,29}$ for numerators: 11, 12, 28, 34, 38
   $0.\overline{365\,85}$ for numerators: 15, 22, 24, 27, 35

Exercises 9-5

1. 55.9 m, 63.4°
2. 32.8 m
3. 90 m
4. 8.57 m
5. 5.3 m
6. 1.42 m

7. 23 m, 21.7 m, 18 m, 13 m, 8 m, 4.3 m, 3 m

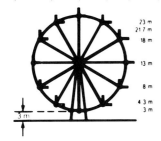

8. 23.5 cm
9. 8.5 cm
10. a) 4.0 m   **b)** 3.5 m
11. 841 m
12. 7.4°
13. 84.3 m
14. 21 m
15. 130.1 m
16. 3.7 m

Review Exercises

1. a) 21.8°   **b)** 20.6°   **c)** 56.3°   **d)** 51.3°
e) 63.4°   **f)** 69.7°
2. a) 0.213   **b)** 0.510   **c)** 0.577   **d)** 1.428
e) 2.747   **f)** 5.671
3. 26.6°, 63.4°
4. a) 5.46   **b)** 15.05   **c)** 11.20
5. a) 31°   **b)** 50.2°   **c)** 56.3°
6. 71.4 m
7. a) 14.5°, 75.5°   **b)** 36.9°, 53.1°   **c)** 23.6°, 66.4°
8. a) 22.9, 32.8   **b)** 11.3, 21.2   **c)** 32.6, 15.2
9. a) 9.1, 4.25   **b)** 13.6, 11.0   **c)** 10.8, 7.2
10. 97.2°, 41.4°, 41.4°
11. a) 13.4, 26.6°, 63.4° **b)** 10.7, 11.8, 65°   **c)** 13.6, 8.5, 58°
12. 37.5 m
13. a) 5.8 m   **b)** 5.5 m
14. 29 m
15. 137.4 m
16. 326.4 m
17. 19 cm
18. a) 53°   **b)** 122.4 m

Exercises 10-1

1. a), d), g), h)
2. a) 045°   **b)** 225°   **c)** 028°
d) 332°   **e)** 309°   **f)** 115°

3. a)    **b)**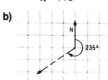

c) 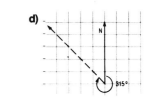   **d)**

e) 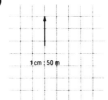   **f)** (figure)

4. a) (figure)   **b)** (figure)

c) (figure)   **d)** (figure)

e) 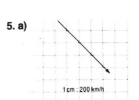   **f)**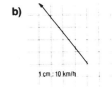

5. a) (figure)   **b)** (figure)

c) (figure)   **d)**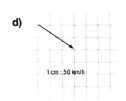

e)

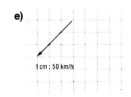

1 cm : 50 km/h

Exercises 10-2

1. c)

2. $\vec{AB}$, $\vec{JK}$, and $\vec{PQ}$;
$\vec{CD}$ and $\vec{LM}$; $\vec{EF}$ and $\vec{RS}$;
$\vec{VW}$ and $\vec{NO}$.

3. a) $\vec{AD}$ and $\vec{BC}$; $\vec{PQ}$ and $\vec{SR}$; $\vec{QT}$ and $\vec{TS}$;
  b) $\vec{AB}$ and $\vec{DC}$; $\vec{PT}$ and $\vec{TR}$; $\vec{SP}$ and $\vec{RQ}$.
  c) $\vec{ZY}$, $\vec{TU}$, $\vec{UV}$, and $\vec{XW}$;
  $\vec{TS}$, $\vec{SZ}$, $\vec{XY}$, and $\vec{VW}$.
  d) $\vec{AB}$ and $\vec{BC}$; $\vec{DE}$ and $\vec{EF}$; $\vec{BE}$ and $\vec{CD}$; $\vec{AE}$ and $\vec{BD}$.

4. a) $\vec{AJ}$, $\vec{JB}$, and $\vec{LK}$;
  $\vec{CL}$, $\vec{LA}$, and $\vec{KJ}$;
  $\vec{BK}$, $\vec{KC}$, and $\vec{JL}$.
  b) $\vec{ED}$, $\vec{FG}$, $\vec{GC}$, and $\vec{AB}$;
  $\vec{EF}$, $\vec{DG}$, $\vec{GH}$, and $\vec{CB}$;
  $\vec{AF}$, $\vec{BG}$, $\vec{GE}$, and $\vec{CD}$.

Exercises 10-3

1. $\vec{AB} = [3, -1]$, $\sqrt{10}$; $\vec{CD} = [2, -5]$, $\sqrt{29}$;
$\vec{EF} = [-5, 1]$, $\sqrt{26}$; $\vec{GH} = [4, 4]$, $4\sqrt{2}$;
$\vec{IJ} = [-2, -2]$, $2\sqrt{2}$; $\vec{KL} = [3, 0]$, 3;
$\vec{MN} = [-1, 4]$, $\sqrt{17}$; $\vec{PQ} = [2, 3]$, $\sqrt{13}$.

2. a) [8, 2]   **b)** [8, 2]   **c)** [2, 5]
  d) [2, 5]   **e)** [10, 7]   **f)** [6, −3]

3. a) [−1, −7], $5\sqrt{2}$ **b)** [5, −5], $5\sqrt{2}$ **c)** [−1, −7], $5\sqrt{2}$
  d) [5, −5], $5\sqrt{2}$ **e)** [3, 1], $\sqrt{10}$ **f)** [3, 1], $\sqrt{10}$
  g) [−2, 6], $2\sqrt{10}$ **h)** [−2, 6], $\sqrt{10}$ **i)** [−4, 12], $4\sqrt{10}$

4. a) [1, −4]   **b)** [6, 3]   **c)** [3, 4]
  d) [−3, 3]   **e)** [0, −4]   **f)** [−5, −4]

5. a) [8, 4]   **b)** [−8, 4]   **c)** [−8, 4]
  d) [−5, −8]   **e)** [5, −8]   **f)** [8, 4]

6. (5, 4)

7. a) (2, 7)   **b)** (−3, 9)   **c)** (−8, 1)   **d)** (−10, 5)
8. a) (13, −8)   **b)** (3, 1)   **c)** (−6, 4)   **d)** (8, −15)
9. a) $\vec{FB}$, $\vec{ED}$ **b)** $\vec{DC}$, $\vec{FE}$ **c)** $\vec{EA}$, $\vec{DF}$
10. $\vec{BC} = \vec{AD} = [-2, 5]$
  $\vec{DC} = [5, 2]$
11. $\vec{SR} = [3, -2]$; $\vec{QR} = \vec{PS} = [4, 6]$
12. a) Yes   **b)** No
13. a) 22°, $\sqrt{29}$ **b)** 67°, $\sqrt{58}$
14. a) [3.9, 4.6]   **b)** [9.1, 4.2]
15. a) $\vec{PQ} = \vec{RS} = [5, -1]$
  $\vec{PR} = \vec{QS} = [2, -4]$
  b) PQRS is always a parallelogram.

MM

1. a) 25 + 4   **b)** 36 + 16 + 4
  c) 81 + 4 + 1 + 1   **d)** 81 + 25 + 9
  e) 121 + 9   **f)** 100 + 64 + 4
  g) 100 + 100 + 4   **h)** 400 + 25 + 9 + 1
2. a) 56
  b) 43
3. 3 (50, 65, 85)

MAU

1. a) 0.24 m²
  b) 60 cm × 40 cm   **c)**
2. 60 cm × 50 cm
  60 cm × 40 cm
  50 cm × 40 cm
  50 cm × 30 cm
  40 cm × 30 cm

474 Answers

Exercises 10-4

1. a) $\vec{AE}$ b) $\vec{AC}$ c) $\vec{AD}$
   d) $\vec{DC}$ e) $\vec{AC}$ f) $\vec{AD}$

2. a) $\vec{PQ}$ b) $\vec{QU}$ c) $\vec{RS}$
   d) $\vec{PS}$ e) $\vec{PR}$ f) $\vec{RS}$
   g) $\vec{UV}$ h) $\vec{SP}$

3. Answers may vary.
   a) $\vec{BD} + \vec{DG}$ b) $\vec{BC} + \vec{CE}$
   c) $\vec{BG} + \vec{GC}$ d) $\vec{AG} + \vec{GF}$
   e) $\vec{FA} + \vec{AC}$ f) $\vec{GC} + \vec{CE}$

4. a) [7, 5] b) [4, 3]
   c) [4, 3] d) [1, −7]
   e) [−14, −8] f) [−5, −18]
   g) [9, 10] h) [−10, 3]

5. a) [5, 4] b) [9, 6]
   c) [−1, −8] d) [8, −4]
   e) [6, 1] f) [0, 0]

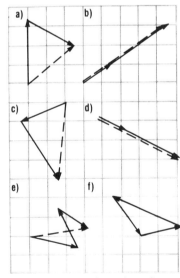

6. $\vec{AB} + \vec{BC} = \vec{AC}$, and $\vec{AD} + \vec{DC} = \vec{AC}$
   Therefore, $\vec{AB} + \vec{BC} = \vec{AD} + \vec{DC}$

7. 8.25 km, 315°
8. 1670 m, 210°
9. 285 km, 059°
10. b) i) 1, 3, 8, $\sqrt{10}$, $\sqrt{73}$
    ii) 1, $\sqrt{2}$

11. 6

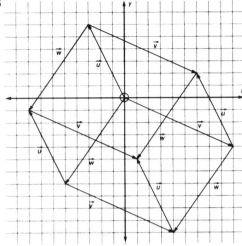

Exercises 10-5

1. a) $\vec{HC}$ b) $\vec{HB}$ c) $\vec{FC}$ d) $\vec{CH}$
2. a) $\vec{AP}$ b) $\vec{HC}$ c) $\vec{BG}$ d) $\vec{BD}$
3. a) [7, 4] b) [7, 4] c) [4, −2] d) [−2, 4]
   e) [−6, 4] f) [2, −1]

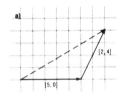
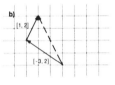
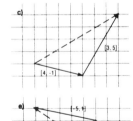
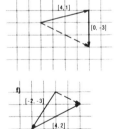

4. a)  b)

Answers **475**

e)   f)

5. a) $\overrightarrow{AN}$ b) $\overrightarrow{EJ}$ c) $\overrightarrow{FH}$ d) $\overrightarrow{EC}$
6. a) $\overrightarrow{AE}$ b) $\overrightarrow{CD}$ c) $\overrightarrow{BF}$ d) —
7. a) 200 km/h, 000°   b) 100 km/h, 180°
   c) 160 km/h, 018°   d) 193 km/h, 007°
   e) 133 km/h, 018°   f) 120 km/h, 343°

7. a) $\overrightarrow{OP} + \frac{1}{2}\overrightarrow{OQ}$ b) $\overrightarrow{OP} + \overrightarrow{OQ}$
   c) $\overrightarrow{OP} + \frac{3}{2}\overrightarrow{OQ}$ d) $\overrightarrow{OP} + 2\overrightarrow{OQ}$
   e) $2\overrightarrow{OP} + 2\overrightarrow{OQ}$ f) $2\overrightarrow{OP} + \frac{3}{2}\overrightarrow{OQ}$
   g) $2\overrightarrow{OP} + \overrightarrow{OQ}$ h) $3\overrightarrow{OP} + \overrightarrow{OQ}$
   i) $3\overrightarrow{OP} + \frac{1}{2}\overrightarrow{OQ}$ j) $2\overrightarrow{OP} + \frac{1}{2}\overrightarrow{OQ}$

8. a) [12, 7]   b) [1, 7]   c) [19, −2]
   d) [4, −1]   e) [−5, −12]   f) [11, −3]
   g) [24, −10]   h) [13, 5]   i) [−13, −13]

9. If AB is the width:
   $\overrightarrow{BC} = \overrightarrow{AD} = [-4, 8]$, $\overrightarrow{DC} = [4, 2]$
   If AB is the length:
   $\overrightarrow{BC} = \overrightarrow{AD} = [-1, 2]$, $\overrightarrow{DC} = [4, 2]$
10. a) 1, −3   b) 2, −4   c) −3, −2   d) 3, 0

Exercises 10-6

1. a) [8, −12]   b) [20, −30]   c) [−12, 18]
   d) [2, −3]   e) [−6, 9]   f) [−4, 6]

2.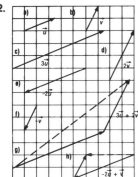

3. a) [6, 8]   b) [13, 9]   c) [7, 11]
   d) [4, 7]   e) [3, −11]   f) [8, 13]
4. a) [8, −12], Twice   b) [−12, 18], Three times
   c) [2, −3], Half   d) [−16, 24], Four times
   e) [10, −15], $2\frac{1}{2}$ times   f) [28, −42], Seven times
5. a) $\vec{u}$ b) $2\vec{u}$ c) $3\vec{u}$ d) $\vec{u}$
   e) $2\vec{u}$ f) $-\vec{u}$ g) $-2\vec{u}$ h) $-3\vec{u}$
6. a) $\overrightarrow{OA} + \frac{1}{2}\overrightarrow{OB}$ b) $\frac{3}{2}\overrightarrow{OA} + \overrightarrow{OB}$ c) $\frac{1}{2}\overrightarrow{OA} + \frac{1}{2}\overrightarrow{OB}$
   d) $\frac{1}{2}\overrightarrow{OA} + \overrightarrow{OB}$ e) $\frac{1}{2}\overrightarrow{OA} + \frac{3}{2}\overrightarrow{OB}$

Exercises 10-7

1. $\overrightarrow{DA} + \overrightarrow{AE} = \overrightarrow{DE}$, and $\overrightarrow{FC} + \overrightarrow{CB} = \overrightarrow{FB}$
   But $\overrightarrow{DA} = \overrightarrow{CB}$, and $\overrightarrow{AE} + \overrightarrow{FC}$
   Therefore, $\overrightarrow{DE} = \overrightarrow{FB}$
   That is, $DE = FB$, and $DE \parallel FB$.

2. $\overrightarrow{AB} + \overrightarrow{AC} = \overrightarrow{BC}$, and $\overrightarrow{AD} + \overrightarrow{AE} = \overrightarrow{DE}$
   But $\overrightarrow{AD} = \frac{2}{3}\overrightarrow{AB}$, and $\overrightarrow{AE} = \frac{2}{3}\overrightarrow{AC}$
   Therefore, $\overrightarrow{DE} = \frac{2}{3}\overrightarrow{BC}$
   That is, $DE = \frac{2}{3}BC$, and $DE \parallel BC$.

3. The answer is the same as that for Exercise 1.

4. $\overrightarrow{PS} + \overrightarrow{SR} + \overrightarrow{RQ} = \overrightarrow{PQ}$
   or $2\overrightarrow{MS} + \overrightarrow{SR} + 2\overrightarrow{RN} = \overrightarrow{PQ}$
   But, $\overrightarrow{MS} + \overrightarrow{SR} + \overrightarrow{RN} = \overrightarrow{MN}$
   Therefore, $2\overrightarrow{MN} - \overrightarrow{SR} = \overrightarrow{PQ}$
            $2\overrightarrow{MN} = 3\overrightarrow{SR} + \overrightarrow{SR}$
            $\overrightarrow{MN} = 2\overrightarrow{SR}$
   That is, $\overrightarrow{MN} \parallel \overrightarrow{SR}$, and $\overrightarrow{MN} = 2\overrightarrow{SR}$

5. $\overrightarrow{AE} + \overrightarrow{ED} = \overrightarrow{AD}$, and $\overrightarrow{BE} + \overrightarrow{EC} = \overrightarrow{BC}$
   But, $\overrightarrow{AE} = \overrightarrow{EC}$,

476 *Answers*

and $\vec{ED} = \vec{BE}$

Therefore, $\vec{AD} = \vec{BC}$

That is, $AD = BC$,

and $AD \parallel BC$.

$ABCD$ is a parallelogram.

6. Opposite sides of the inscribed figure are each parallel to the same diagonal of the quadrilateral. (See *Example 2*.) Therefore, the midpoints of the sides of any quadrilateral are the vertices of a parallelogram.

7. $\vec{DA} + \vec{AO} = \vec{DO}$, and $\vec{OB} + \vec{BC} = \vec{OC}$

   But, $\vec{AO} = \vec{OB}$, and $\vec{DA} = \vec{BC}$

   Therefore, $\vec{DO} = \vec{OC}$

   That is, $DO = OC$, and O is the midpoint of CD.

8. $\vec{AD} + \vec{DM} = \vec{AM}$, and $\vec{MB} + \vec{BC} = \vec{MC}$

   But, $\vec{DM} = \vec{MB}$, and $\vec{AD} = \vec{BC}$

   Therefore, $\vec{AM} = \vec{MC}$

   That is, $AM = MC$, and M is the midpoint of AC.

9. Complete the parallelogram $OAPB$. M is the midpoint of OP.

   $\vec{OP} = \vec{OA} + \vec{OB}$

   $2\vec{OM} = \vec{OA} + \vec{OB}$

   $\vec{OM} = \frac{1}{2}\vec{OA} + \frac{1}{2}\vec{OB}$

e) [5, −8]   f) [8, 4]

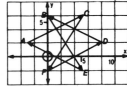

7. (5, 4)

8. a) $\vec{DA}$ b) $\vec{AF}$ c) $\vec{BC}$
   d) $\vec{DC}$ e) $\vec{FG}$ f) $\vec{CD}$

9. a) [9, 3]   b) [5, −5]   c) [−1, 1]
   d) [−4, −9]   e) [8, 11]   f) [−1, 7]

10. [2, 6]

11. a) [12, −8]   b) [3, −2]   c) [−9, 6]   d) [9, −6]

12. a) $\vec{u}$ b) $\vec{v}$ c) $\vec{u} + \vec{v}$ d) $2\vec{u} + \vec{v}$

13. a) [1, 11]   b) [0, 18]   c) [−3, 0]

14. a) $\vec{OA} + \vec{OB}$ b) $\frac{1}{2}\vec{OA} + \vec{OB}$ c) $\vec{OA} + \frac{1}{2}\vec{OB}$
    d) $\frac{3}{2}\vec{OA} + \frac{1}{2}\vec{OB}$ e) $\frac{1}{2}\vec{OA} + \frac{1}{2}\vec{OB}$

Review Exercises

1. a), c), e)

2.

3. $\vec{AB} = [−2, −5]$

   $\vec{CD} = [−4, 4]$

   $\vec{EF} = [2, 6]$

   $\vec{JK} = [5, −4]$

   $\vec{OP} = [3, 4]$

4. a) $\sqrt{13}$ b) 10   c) 7
   d) 6   e) 13   f) 17

5. a) [1, −4]   b) [5, 11]   c) [−5, 4]   d) [3, 6]

6. a) [8, 4]   b) [−8, 4]   c) [−8, 4]   d) [−5, −8]

AAS, 292–294
Absolute value, 6–9, 178
Alternate angles, 284, 299
Angle(s),
 alternate, 284, 299
 bisector, 292
 complementary, 271
 corresponding, 273–281, 284, 299
 exterior, 288
 of inclination, 356–359
 opposite, 288, 297
 sum in a polygon, 290
 sum in a triangle, 287
 supplementary, 271
 vertically opposite, 269
Angle bisector, 292
Area,
 of a triangle on a grid, 7, 10
ASA, 279
Axiom, 268, 282–283

Base, of a power, 41
Bearing, 379–380
Bermuda Triangle, 10
Binomial(s), 59–62, 68–69
 difference of squares, 75–77
 product of, 59–60, 72
Bisector,
 angle, 292
 perpendicular, 276, 298–299
Broken-line graph, 316–320

Cartesian coordinate system, 178
Centroid, 182
Coefficient, 38, 50, 52, 62, 120, 150
Composite number, 2
Computer Power, 19, 66, 150, 263, 348–349, 371
Cone, 38–39, 52
Congruence axioms,
 AAS, 292
 ASA, 279
 HS, 305
 SAS, 273
 SSS, 276
Congruent triangles, 234, 238, 249–250, 273–281, 292–294, 305–308
Constant of proportionality, 200, 208

Continuous-line graph, 316–320
Converse, of a theorem, 297–301, 304, 308
Coordinate(s), 152–153, 156, 165, 181, 183
Corresponding angles,
 of congruent triangles, 273–281
 of parallel lines, 284–299
Cosine, 363
Cube, 147
Cylinder, 40, 52

Data, 316–334
Decagon, 290
Decimal(s), 12–16
 form, 12–13
 patterns in, 13
 repeating, 13, 16, 371
 terminating, 13, 16, 371
Deduction, 268
De Fermat, 48–49
Degree, of polynomial, 55
Denominator, 33, 81, 85, 92–94
 common, 88, 92–94, 108–109
 lowest common, 88, 92–94
 rationalizing the, 33
Deviation, 327–329
Dilatation(s), 254–258, 309–312
 centre, 255
Diophantine equation(s), 158–159
Diophantus, 158–159
Distance, formula, 178-179
 minimum, 263
Distributive law, 27, 30, 55–57, 62, 108

Ellipse, 215
Equation(s), 108–126, 132–145, 150–159, 190–193, 196–199
 containing fractions, 108–109
 in two variables, 112–126, 132–142, 150
 linear, 116–126, 150, 152–155, 190–193
 of a line, 152–153, 155–157, 190–193, 196–199
 quadratic, 143–145
 solving problems with, 109–110, 134–142, 152–155
 writing, 132–142

Equations, pairs of, 116–126, 132–142, 155–157
 solution by addition or subtraction, 123–126
 solution by comparison, 116–118
 solution by graphing, 155–157
 solution by substitution, 120–122
Equilateral triangle, 147
Euclid, 282
Euclidean algorithm, 66
Euclid's *Elements*, 282-283
Euler, 48–49, 386
Event, 335, 339
Exponent(s), 38, 41–47
 in formulas, 52
 laws of, 45–47, 52

Factor, 2–3, 62–63, 66, 68–69, 72
 common, 2–3, 62–63, 69, 72, 75
 greatest common (g.c.f.), 2, 62, 66
 prime, 2
Factoring, 62–65, 68–80, 82, 85, 93–94, 143–145
 by decomposition, 72–73
 difference of squares, 75–77
 trinomials, 68–73
 trinomials that are perfect squares, 68–71
Formula(s), 38–40, 52, 77, 97–99, 113
 exponents in, 52
 substituting in, 52, 97–99
Frequency, 326, 328–329

Geometry, 268–312
Glide reflection, 244
Goldbach, 11
Graph(s),
 broken-line, 316–320
 continuous-line, 316–320
 histogram, 316–320
Gwennap Pit, 354

Half-plane, 160
Heading, 379–380
Hemisphere, 54, 65
Hexagon, 290
Histogram, 316–320
HS, 305
Hyperbola, 215

Hypotenuse, 20, 178, 181, 304–305, 363

Image, 226
Inequality(ies), 160–164, 170–174
   graphing, 160–164, 170–174
   solving, 160–164, 170–174
Integer(s), 6–9, 17, 68–69, 72
Interest, 127–128
   compound, 128
   simple, 127
Intersect(ion), 155–156
Invariant, 226
Irrational numbers, 16–18, 20, 24
Isometry, 249–253
Isosceles triangle, 179, 273, 279

Kummer, 48–49

Lagrange, 386
Lamé, 48–49
Legendre, 48–49
Lindemann, 48–49
Line(s), 152–153, 155–157
   equation of, 152–153, 155–157
   parallel, 156, 184, 243, 284–287, 299
   perpendicular, 187–188, 233
Linear programming, 170–174
Linear relation(s), 152–157

Mapping, 226
   notation, 226
   rule, 226
Mathematics Around Us, 10, 67, 127–128, 169, 194–195, 259, 302–303, 330–331, 354, 370, 387
Mathematical Mind, The, 11, 48–49, 119, 158–159, 266, 282–283, 386
Mean, 320–326
Mean deviation, 327–329
Measures of central tendency, 320–326
Measures of dispersion, 327–329
Median,
   of a set of numbers, 320–326
   of a triangle, 182
Midpoint, 181–182
Mode, 320–326
Monomials, 50–54, 56–58, 62–65, 68–69, 88–89
   addition and subtraction, 50–51
   division and multiplication, 52–54
   product of a polynomial and, 56–58

Monte Carlo methods, 342–349
Multiple, least common (l.c.m.), 3

Nested form, 78–80
Newton (unit of force), 378
Non-linear relation(s), 217–220
Numbers,
   irrational, 16–18, 20, 24
   natural, 2, 17
   negative, 6–9
   positive, 6–9
   rational, 12–15, 17, 81, 88
   real, 17
Numerator, 33, 81, 85

Octagon, 290
Operations,
   inverse, 32
   with powers, 41–47
Ordered pair, 112–113, 116, 123, 152–153, 165, 171–172, 200, 383–384
Orientation, 239, 244
Outcome, 335, 339

Pantograph, 259
Parabola, 214
Parallel lines, 156, 184, 243, 284–287, 299
Parallelogram Law, 391–393
Pentagon, 290
Percent, 127–128
Perpendicular bisector, 276, 298–299
Perpendicular lines, 187–188, 233
Plane, 178
Polygon,
   regular, 290
   sum of the angles, 290
Polynomials, 55–65, 68–80
   addition of, 55–56
   division by monomials, 56–58
   evaluating, 78–80
   multiplication by monomials, 56–58
   nested form, 78–80
   simplifying, 55–56
   subtraction of, 55–56
Population, 332
Power, 41
Predicting, 332–334
Prime number, 2, 11
Probability, 335–349

Problem-Solving Strategy, 106, 146–147, 168, 212–213, 295–296
Proof(s),
   by deduction, 269
   by transformations, 309–312
   by vectors, 396–399
   indirect, 295–296
Pythagoras, 20
Pythagorean theorem, 20–23, 75–76, 178, 304–308, 361, 367, 383
Pythagorean triple, 307

Quadratic relation(s), 208–211, 214–216

Radical(s), 24–34
   addition and subtraction, 27–30
   combined operations, 30–32
   division, 32–34
   entire, 24–25
   like and unlike, 27–28
   mixed, 24–25
   multiplication, 24–27
   sign, 24
   simplifying, 24–25, 30–34
Ramanujan, 266
Random numbers, 342, 344–349
Random sample, 332
Range, 327–329
Rational expression(s), 81–99
   addition, 88–96
   applications, 97–99
   division, 84–87
   equivalent, 81–84, 88
   multiplication, 84–87
   subtraction, 88–96
Rational numbers, 12–15, 81, 88
   $\frac{m}{n}$ in decimal form, 13
   set of, 12
Real numbers, 17
Reciprocal, 41, 46, 85, 187–188
Rectangular prism, 52
Reflection(s), 226, 238–242, 309–312
Reflection line, 239
Relation(s), 152–157, 160–167, 200–211
   linear, 152–157, 160–167, 200
   non-linear, 217–220
   quadratic, 208–211, 214–216
Relative frequency, 332–334, 348
Rhind papyrus, 119

Index

Right triangle, 20–23, 188
    solving, 367–369
    trigonometric ratios in, 360–366
Rotation(s), 226, 233–237, 309–312

Sample, 330–334
Sampling, 330–334
SAS, 273
Scalar,
    product, 393–396
    quantity, 378
Scale factor, 254
Scientific notation, 19
Set, 12, 112–113, 152–153, 160, 200
Set-builder notation, 113, 152–153, 160
Side(s),
    adjacent, 360, 363
    of a polygon, 290
    opposite, 285, 360, 363
Simulation, 342
Sine, 363–366
Slope(s), 183–193, 196–199, 229–230, 233–234, 238, 255, 356–359
    of parallel lines, 184, 187
    of perpendicular lines, 187–189
    y-intercept form, 190–193, 196
Solution set, 112–113, 155–156, 160
Sphere, 52
SSS, 276
Statistics, 316–334

Stem-and-leaf diagram, 316–320
Surface area, 63

Tangent, 356, 360–362
Terms, 38
    collecting like, 50–51, 55
    like, 50
Tessellation, 244
Tetrahedron, 340
Theorem, 269
Transformation(s), 226–262, 309–312
    successive, 243–248
Translation(s), 226, 229–232, 309–312
Translation arrow, 243
Transversal, 284
Tree diagram, 339
Triangle(s),
    angle sum in a, 287
    congruent, 234, 238, 273–281, 292–294, 305
    equilateral, 147
    exterior angles of a, 288
    isosceles, 179, 273, 279
    solving, 367-369
Triangle Law, 388–390
Trigonometry, 356–376
Trinomials, 59–62, 68–74
    factoring, 68–74
    perfect squares, 59, 68–69

Turn centre, 233

Variable, 38, 50, 62, 78, 108–109, 112–113, 116, 120, 123–124, 129–130, 132–133, 170–171
    isolating the, 108
Variation, 200–211
    direct, 200–203
    inverse, 208–211
    partial, 204–208
Vector(s), 378–400
    as directed line segment, 378–380
    addition of, 388–393
    as ordered pairs, 383–384, 389
    components, 383
    equal, 381–382
    magnitude of, 383
    proofs in geometry, 396–399
    scalar products of, 393–396
Vertically opposite angles, 269
Volume,
    of a cone, 38–39
    of a cylinder, 40
    of a rectangular prism, 52
    of a sphere, 52
    of a square-based pyramid, 77

Wagstaff, 48–49

y-intercept, 190–193, 196